P9-DFI-973

Applied Anatomy and Biomechanics in Sport

SECOND EDITION

Timothy R. Ackland
PhD, FASMF

Bruce C. Elliott
PhD, FACHPER, FAAKPE, FISBS

John Bloomfield
AM, PhD, FACHPER, FASMF, FAAESS, FI Biol, FAI Biol

The University of Western Australia

Editors

Human Kinetics

Library of Congress Cataloging-in-Publication Data

Ackland, Timothy R., 1958-
Applied anatomy and biomechanics in sport / Timothy R. Ackland, Bruce C. Elliott, John Bloomfield. -- 2nd ed.
p. cm.
Includes bibliographical references and index.
ISBN-13: 978-0-7360-6338-8 (hard cover)
ISBN-10: 0-7360-6338-2 (hard cover)
1. Human anatomy. 2. Human mechanics. 3. Sports--Physiological aspects. I. Elliott, Bruce, 1945- II. Bloomfield, J. (John), 1932- III. Title.
QP303.A25 2009
612--dc22

2008022743

ISBN-10: 0-7360-6338-2 (print) ISBN-10: 0-7360-8771-0 (Adobe PDF)
ISBN-13: 978-0-7360-6338-8 (print) ISBN-13: 978-0-7360-8771-1 (Adobe PDF)

Permission notices for material reprinted in this book from other sources can be found on page x.

The first edition of *Applied Anatomy and Biomechanics in Sport* was published in 1994 by Blackwell Publishing.

The Web addresses cited in this text were current as of May 2008, unless otherwise noted.

Acquisitions Editor: Loarn D. Robertson, PhD; **Developmental Editor:** Kathleen Bernard; **Assistant Editor:** Jillian Evans; **Copyeditor:** Joyce Sexton; **Proofreader:** Joanna Hatzopoulos Portman; **Indexer:** Betty Frizzell; **Permission Manager:** Dalene Reeder; **Graphic Designer:** Nancy Rasmus; **Graphic Artist:** Yvonne Griffith; **Cover Designer:** Keith Blomberg; **Photographer (cover):** Getty Images; **Photographer (interior):** see page x; **Photo Office Assistant:** Jason Allen; **Art Manager:** Kelly Hendren; **Associate Art Manager:** Alan L. Wilborn; **Printer:** Thomson-Shore, Inc.

Printed in the United States of America. 10 9 8

The paper in this book is certified under a sustainable forestry program.

Human Kinetics
Web site: www.HumanKinetics.com

United States: Human Kinetics
P.O. Box 5076
Champaign, IL 61825-5076
800-747-4457
e-mail: info@hkusa.com

Canada: Human Kinetics
475 Devonshire Road, Unit 100
Windsor, ON N8Y 2L5
800-465-7301 (in Canada only)
e-mail: info@hkcanada.com

Europe: Human Kinetics
107 Bradford Road
Stanningley
Leeds LS28 6AT, United Kingdom
+44 (0)113 255 5665
e-mail: hk@hkeurope.com

Australia: Human Kinetics
57A Price Avenue
Lower Mitcham, South Australia 5062
08 8372 0999
e-mail: info@hkaustralia.com

New Zealand: Human Kinetics
P.O. Box 80
Mitcham Shopping Centre, South Australia 5062
0800 222 062
e-mail: info@hknewzealand.com

contents

contributors

Timothy R. Ackland, PhD, FASMF
The University of Western Australia

Jacqueline A. Alderson, PhD
The University of Western Australia

Anthony J. Blazevich, PhD
Edith Cowan University

John Bloomfield, AM, PhD, FACHPER, FASMF, FAAESS, FI Biol, FAI Biol
The University of Western Australia

J.E. Lindsay Carter, PhD
San Diego State University

Jodie Cochrane, PhD
Australian Institute of Sport

John Cronin, PhD
Auckland University of Technology

Bruce C. Elliott, PhD, FACHPER, FAAKPE, FISBIS
The University of Western Australia

Damian Farrow, PhD
Australian Institute of Sport

Jason P. Gulbin, PhD
Australian Institute of Sport

Peter Hamer, PhD
University of Notre Dame Australia

Patria Hume, PhD
Auckland University of Technology

Deborah A. Kerr, PhD
Curtin University of Technology

Duane Knudson, PhD
California State University at Chico

William J. Kraemer, PhD
University of Connecticut

David Lloyd, PhD
The University of Western Australia

Andrew Lyttle, PhD
W.A. Institute of Sport

Michael McGuigan, PhD
Edith Cowan University

Robert U. Newton, PhD
Edith Cowan University

Timothy S. Olds, PhD
University of South Australia

Nicholas Ratamess, PhD
The College of New Jersey

Duncan Reid, MHSc (Hons)
Auckland University of Technology

J. Hans de Ridder, PhD
North-West University

Arthur D. Stewart, PhD
The Robert Gordon University

Grant R. Tomkinson, PhD
University of South Australia

preface

The application of sport science to coaching has become the single most important factor behind the rapid advances made in international sport performances during the past 20 years, yet few publications exist to document these advances. The first edition of this book helped to fill the gap, and now this revised edition has been comprehensively updated, incorporating the latest principles and practices to assist high-level coaches and sport scientists prepare their athletes for competition.

Applied Anatomy and Biomechanics in Sport examines coaching from a unique perspective, focusing on the individual rather than using the en masse approach of the past. It in no way resembles the majority of the sport books currently in print; instead it is designed to advise coaches how to appraise the body structure of their athletes so that their strengths can be fully utilized and their weaknesses improved through the use of specially designed programs. To this end, the concept of modifying the individual's body or technique (or both) is referred to throughout, in simple language that coaches, teachers and sport scientists at all levels will find easy to understand.

During their careers, the authors of this book have consulted a large number of highly trained and internationally successful coaches from a wide range of sports in order to glean much of the information that appears in this work. These coaches continually stress that an athlete can reach only a general level of proficiency with *group* coaching and that many of the most successful coaches understand the need to tailor sport technique to suit each individual if an optimal level of performance is to be attained. These coaches also expressed dissatisfaction with the generalist instructional books, citing them as no longer relevant for the modern coach.

This book comprises 18 chapters within four sections:

- An introduction outlines the **fundamental concepts** related to assessment and modification of athletes' capacities, as well as talent identification and profiling.
- An **applied anatomy** section provides a theoretical framework for valid and reliable assessment and evidence-based modification of an athlete's body structure and physical capacity. Individual chapters cover the following specific topics: body size, somatotype, composition, proportionality, posture, strength, power, speed, flexibility, balance and agility.
- An **applied biomechanics** section explains how sport technique may be analysed and the resulting information used to improve athletic performance. Most importantly, this section explains the interrelationships between the athlete's structure, physical capacity and his or her unique technique—such that modification of one aspect may affect another, or certain deficiencies may be accommodated with a variation in technique. Commonly employed analysis tools are explained, and practical information on the use of modern video analysis systems is provided to help readers become proficient users of this technology.
- A **practical example** is analysed in-depth to provide a specific case for assessment of physical capacity and biomechanics of an individual athlete, the intervention program and technique modification employed, and the resultant influence on the athlete's performance.

In the past, the majority of books in the coaching field have dealt with only one sport. However, single sports no longer exist in isolation, as they share many common features with other activities. This book systematically deals with nine sport groups: racquet sports; aquatic sports; gymnastics and power sports; track, field and cycling; mobile field sports; set field sports; court sports; contact field sports; and the martial arts. Within these groups, specific examples are presented from over 30 international sports so as to promote discussion of their various characteristics and allow *cross-fertilization* of ideas to occur.

Special features of this book include information from recent research on the development of training methods that will increase explosive power; the utilization of elastic energy in the development of power and speed; and the modification of strength and body composition to improve performance. Moreover, information available in this publication on proportionality and posture as applied to sport performance has rarely appeared in other books or recent journals in the coaching or sport science area.

The majority of this text emphasizes application, as it concentrates on the various ways in which both the human body and the individual's technique can be modified to achieve optimal skilled performance, with reference to either the average competitor or the elite athlete.

The contents at all times take into account the growth, development and gender of the individual while also making the reader aware of the mechanical overstresses that cause various injuries and impair performance. In order to achieve this, the book has a preventive medical theme throughout.

Finally, this book comprehensively covers the latest assessment techniques currently used by exercise and sport science specialists to evaluate human physiques, physical capacities and sport techniques. The specific laboratory-based techniques and field tests conform to the standards of such international bodies as the International Society for the Advancement of Kinanthropometry (ISAK) and the National Strength and Conditioning Association (NSCA). Wherever possible, normative data are provided for comparison with the readers' athletes. These aspects of the publication will be especially useful for current practitioners who wish to evaluate their athletes objectively, as well as for students who are pursuing their professional training.

photo credits

Photos on pages 1, 7, 8, 13, 20, 27, 277, and 325, and figures 14.1, 17.2-17.7, and 17.10-17.14: © Human Kinetics

Figure 3.11: Reprinted, by permission, from J. Wilmore, D. Costill, and L. Kenney, 2008, *Physiology of sport and exercise,* 4th ed. (Champaign, IL: Human Kinetics), 321.

Figures 3.13, 3.15, 3.17, and 3.18: Photo courtesy of Bill Ross, rosscraft@shaw.ca.

Figure 4.1: *(a)* Photo courtesy of B.H. Heath, *(b)* Photo courtesy of J. Borms, reprinted with permission of Cambridge University Press; *(c)* Photo courtesy of D. Roberts, reprinted with permission of Cambridge University Press.

Figures 4.12, 6.7-6.15, and 7.10-7.14: Reprinted from J. Tanner, 1964. *The physique of the Olympic athlete* (London: George Allen and Unwin). By permission of J. Tanner.

Figure 4.13: From J.E.L. Carter and B. Heath, 1990, *Somatotyping: development and applications* (New York: Cambridge University Press). Adapted with the permission of Cambridge University Press.

Figure 6.6: Reproduced courtesy of the West Australian.

Figures 8.11-8.14: Photo courtesy of Michael McGuigan.

Figure 10.7: Photo courtesy of J. Cronin and A. Blazevich.

Figure 11.2: Photo courtesy of New Zealand Academy of Sport.

Figures 11.3 and 11.5: Photo courtesy of Patria Hume.

Figure 11.4: Photo courtesy of Tim Ackland.

Figure 11.6: Photo courtesy of Duncan Reid.

Figure 12.4: Reprinted, by permission, from J. Bloomfield, P.A. Fricker, and K.D. Fitch, 1992, *Textbook of science and medicine in sport* (Melbourne, Australia: Blackwell Scientific Publications), 391.

Photo for exercise 1 in chapter 13: Photo courtesy of Cybex.

Photo for exercise 26 in chapter 13: Photo courtesy of Nautilus, Inc.

Figures 13.2, 13.3, and 13.5: Reproduced courtesy of the West Australian.

All other photos in chapter 13: Photo courtesy of Tim Ackland.

Figures 14.4, 14.5, 15.3: Image courtesy of SiliconCOACH, Dunedin, New Zealand.

Figure 14.8: Photograph courtesy of Tom Rovis-Hermann from the Western Australian newspaper.

Figure 15.1: Photo courtesy of Jacqueline Alderson.

Figures 15.4 and 15.6: Image courtesy of Vicon.

Figure 16.2: Photo courtesy of B. Elliott and D. Farrow.

Figure 17.1: © Icon Sports Media

Figure 17.8: Ron Chapple/Taxi/Getty Images

Figure 17.9: Ryan McVay/Allsport Concepts/Getty Images

Figure 18.1: Photo courtesy of Athletics Australia.

part I

Fundamental Concepts

one

The Assessment and Modification Model

Timothy R. Ackland, PhD; John Bloomfield, PhD; and Bruce C. Elliott, PhD

While "raw athletic talent" is the most important factor in the attainment of high levels of sporting performance, coaching based on sound sport science principles is also an essential ingredient if this talent is to be fully developed. Sport science, therefore, focuses first on the identification of sporting talent and then on its development, so that an athlete's optimal performance can be achieved. It is generally accepted by coaches and sport scientists that high-level performance is dependent upon an identifiable set of basic factors, each one carrying a relative importance for the given activity (Pollock, Jackson, and Pate 1980). Athletes therefore will reach their full potential only if the following factors are combined:

- The physical characteristics of the athlete that are important in a particular sport must be present (physical capacities).
- Appropriate techniques for the sport need to be developed (biomechanics).
- A level of fitness that is specific to the particular activity must be attained (physiological capacity).
- The psychological factors that enable the performer to compete successfully need to be developed and maintained (psychological makeup).
- A work ethic that includes an appropriate attitude toward training must be present.
- The opportunity to compete with athletes of a similar or superior level must be available, particularly during the development process.

In order to maximize an athlete's inherent abilities, such factors as a coach's rich set of experiences, as well as specialized knowledge of current world trends in a particular sport, are important in the formulation of a structured development plan. Past experience provides coaches with a personal knowledge of biological or behavioural factors (or both), as in the following examples:

- The appropriate grip to be used in a forehand drive in tennis on a clay court
- The need for anaerobic training to compete successfully in squash or American football
- The need for specific psychological skills in shooting sports
- The need to foster group cohesion in interactive team sports
- The progressive sequence of drills that should be used while gymnasts are learning a new and complex skill

Current world trends in performance are also of vital importance, and coaches need to be aware of the techniques of international-level performers so that they can constantly improve the skills of their charges. Examples of these trends are the following:

- Using the "track start" to provide greater stability on the starting block in swimming starts
- The use of upper arm internal rotation in the generation of power in the tennis serve
- The use of the "short grip" in squash
- The role of pronation in the badminton smash
- Deemphasis on an athlete's physical bulk in favour of good balance and glide speed across the circle in the throwing field events of discus and shot put

Coaches must also be aware of the role of athletic flair in performance. One of the most puzzling aspects of coaching international-level athletes is that it is often difficult to determine whether their natural abilities have made them champions or whether it is their training that has differentiated them from other talented competitors.

The coaches who offer the best guidance to their athletes are those who can integrate these various factors with a thorough understanding of sport science, which generally includes the subdisciplines of sport physiology, sport psychology, motor control and learning and biomechanics.

> The high-level coach has to determine which aspects of performance to foster and which to modify in order for athletes to reach the highest levels of achievement.

The Assessment and Modification Model

During the past 30 years there has been a rapid development in sport performance standards. There are many reasons for this, including such factors as better living conditions, a much larger number of people playing sport worldwide, and more participants from those racial types whose physiques are highly suitable for certain sports or events within those sports. However, the contribution of sport science and medicine in the majority of sports has also been integral to these improvements, because the knowledge base in this field is now extensive. Furthermore, coaching standards have risen dramatically during this time, and much of the information that successful coaches now have to pass on to their athletes has its basis in sport science and sports medicine.

One area of sport science that has received increasing attention is the modification of an individual's physique, technique or both to improve performance. In the past, coaches have tended to teach standard techniques, especially those that have been used by world or Olympic champions, whether these suited their own athletes or not. Bloomfield and Blanksby, as far back as 1973, stated that "it is sometimes disturbing to be able to identify the coach of a certain group of swimmers by the similarity of their stroke technique regardless of the obvious differences in their body type". The authors further suggested that performances would be improved considerably if coaches were able to objectively assess the physical strengths and weaknesses of their swimmers, as this would enable athletes to develop a stroke better suited to their specific physical capacities.

In the modern era, high-level coaches adopt these concepts to provide training interventions that are tailored specifically to the individual athlete. Whether the focus is on talent identification, talent development or athlete preparation for competition, all three aspects of a high-performance sport program follow the general process shown in figure 1.1. Valid and reliable assessments are followed by a system of athlete profiling that helps to expose athletes' current strengths and weaknesses. Careful consideration is then given to some coordinated intervention strategy that may involve physical capacity modification or technical and tactical intervention (or both). Then, after a reassessment of the athlete to establish the effect of the intervention, a coach will consider the next step in promoting the athlete's advancement.

Sport Science Theoretical Base

Coaches and sport scientists must have a thorough understanding of the critical factors for success in their particular sport prior to implementing the model shown in figure 1.1. Though a relatively new field of science, exercise and sport science has accumulated a sound body of knowledge over the past 30 years, with results of applied research communicated via academic and professionally oriented books and journals. Without such knowledge, those responsible for athlete development cannot determine the important technical performance elements of a skill or the critical physical capacities that must be assessed and monitored.

Observation and Assessment

The next stage requires observation and assessment of the individual athlete on a range of parameters that relate specifically to performance success in a given sport.

Assessment of Body Morphology and Physical Capacity

Coaches with a good knowledge of sport science, compared to those who are untrained, are usually more aware of the various physical characteristics of their athletes and are often able to appraise them more quickly. Many are able to assess strength and power by simply observing the individual's size, shape, proportionality, posture and musculature. Likewise, flexibility levels can be assessed through examination of technique, while speed and agility can be appraised reasonably well through observation of game performance. This subjective appraisal is of great value, and experienced coaches may need only this level of information in order to select or develop their athletes.

However, coaches will generally need an objective measure for appraising the development of their athletes even if they can "see" or be aware of a certain capacity,

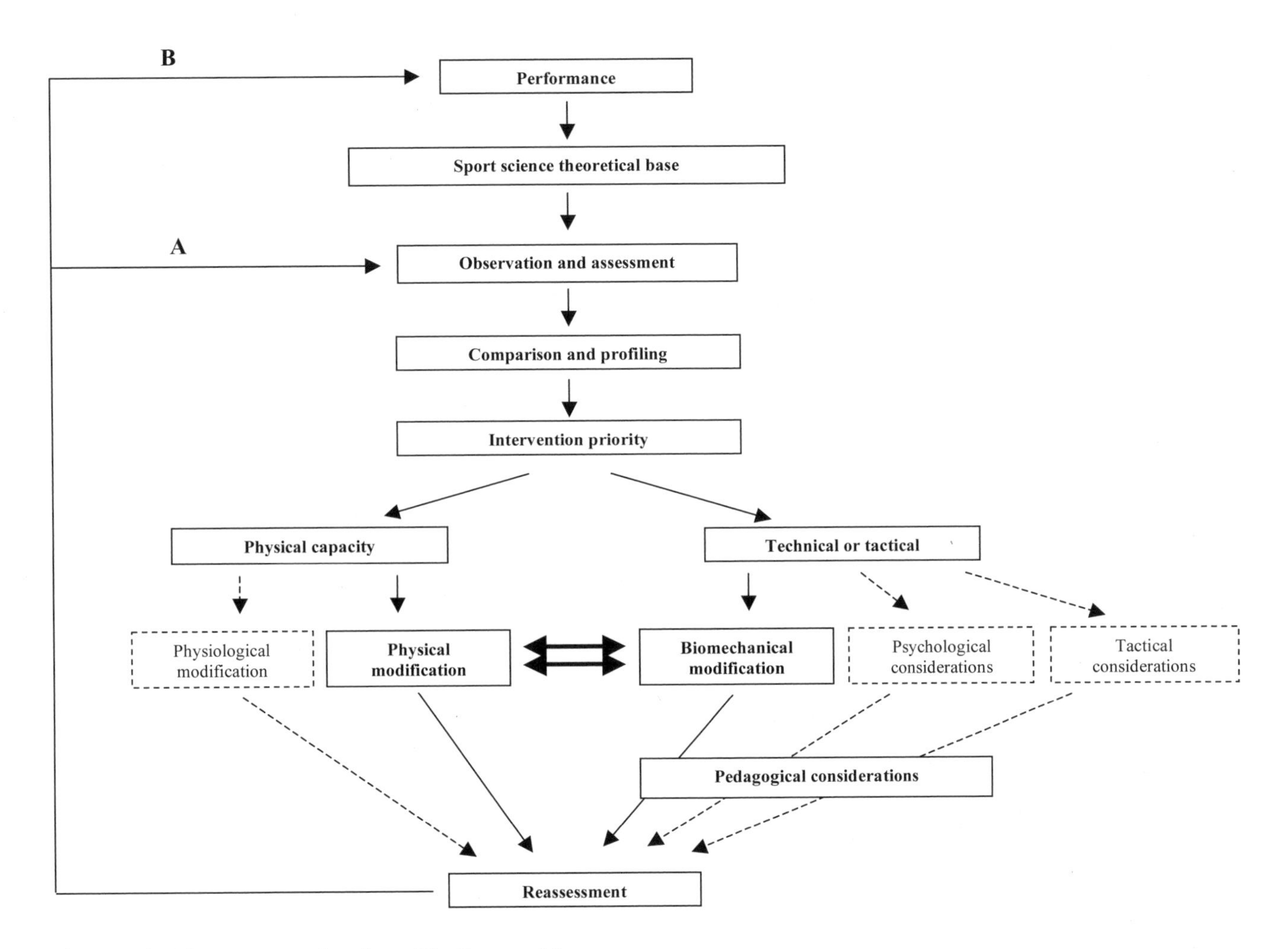

Figure 1.1 The assessment and modification model.

and this may require the assistance of a sport scientist. This is often possible in locations where members of university departments are available to assist coaches with such problems, or in institutes of sport or intensive training centres that have sport scientists on staff.

At the international level, it is useful for the coach to collect appropriate data for each athlete on such physical capacities as shape, body composition, proportionality, posture, strength and power, flexibility, speed and agility. Each chapter in part II of this text covers the various tests and measurements that are needed to do this thoroughly using the most recent internationally approved testing protocols. On the other hand, chapter 2 demonstrates how to apply these data in a practical way.

Observation of Techniques in a Tactical Environment

Coaches who understand the mechanical base of a skill, who are able to analyse movement and who are also able to communicate provide the best opportunity for athlete development. Coaches need this understanding of biomechanics to be able to assist athlete development from its earliest stages to the Olympic or World Championship level. Part III of the book provides the tools needed to use a biomechanical approach to this development.

First coaches must learn to approach analysis in a more holistic manner as discussed in chapter 14. That chapter, when read in conjunction with chapter 17, provides the reader with an appreciation of how the mechanical principles of performance may be categorized and "models of performance" may be formulated. In many ways, the key to successful analysis is a comprehensive understanding of the skill to be taught and the use of a system to assess performance prior to intervention.

While video linked to appropriate software and a computer may be used with young athletes, more sophisticated objective analysis tools and the ways in which they may be used for athlete development are discussed in chapters 15 and 16. While coaches may not actually use all the protocols presented in this section, an understanding of how objective data can be integrated into coaching will hopefully remove the fear factor related to biomechanics.

> Setting priorities for intervention not only permits a coordinated approach to training, but also allows for consideration of the interaction between physical and technical factors.

Comparison and Profiling

Some elite athletes at the international level have almost the perfect physique for their sport, combined with all the other characteristics that contribute to optimal performance. These specific physical characteristics, especially in the technique-oriented sports, are often the very reason that the athlete selected that particular sport in the first place. However, athletes at all levels of competition generally have some weaknesses that need to be eradicated or limited in their effect. Therefore, some form of assessment, technique comparison or profiling will be necessary in order for the coach to prioritise and implement the appropriate remedial action to strengthen these deficiencies.

Profiling allows the coach to compare the athlete's assessment data with his or her previous scores or with norms from other elite performers. In the biomechanics field, profiling relates to the comparison of technique with an athlete's past performances, or perhaps to some performance model that the athlete is endeavouring to replicate. A formalized system of profiling allows the coach to monitor the athlete's progress at various stages in the training regimen. A variety of data can be used for this profiling exercise (depending upon the critical factors associated with success for a particular sport), including medical history and physical capacities as well as technique, skill and strategy or decision-making information. These concepts are developed and several examples provided in chapter 2.

Intervention Priority

Before launching into wholesale modifications, some consideration of intervention priorities, as well as the athlete's circumstances, is essential. When one is considering an athlete's performance, the critical facets to be addressed for each individual (those that are most influential on that person's current abilities) can be revealed only after careful consideration of

- the physical, physiological, tactical and technical requirements of the sport;
- the age and maturity of the athlete;
- the level of competition; and
- the athlete's aspirations.

Clearly, the various critical factors for successful performance are not immune to other changes, and modifications made to such physical capacities as strength, power and flexibility, for example, could have an immediate effect on an athlete's technique. Similarly, improvements in speed and agility might influence strategic and tactical opportunities.

Modification

In this model we have recognized that coaches may intervene to modify a variety of factors. These have been grouped under the headings *physical capacities,* including physiological and structural and functional (physical) parameters, and *technical-tactical* aspects, which incorporate biomechanical and psychological elements as well as strategies and tactical considerations. While we acknowledge the importance of all these facets, the primary focus of this book is on the two interrelated factors of physical and biomechanical interventions.

Modification of the Physical Body

Athletes who do not have the physique advantages of other competitors can undergo an intervention program to modify their structure and physical function in order to improve performance. The following are human physical capacities that can be modified to a greater or lesser extent:

- *Strength, power, body bulk* and *flexibility* are the most easily modified capacities, and very good results can be obtained in many sports if these can be improved. Chapters 8, 9 and 11 cover these capacities in detail.
- *Body type, body composition, speed* and *agility* are more difficult than the first group to change; however, intensive intervention programs have been developed during the past 20 years that have been shown to improve these capacities. Chapters 4, 5, 10 and part of chapter 12 explain ways in which these characteristics can be modified.
- *Posture* and *dynamic balance* are not easy to modify, and a long-term intervention program over several years may be needed to accomplish such modification. Chapter 7 and part of chapter 12 cover these capacities in detail.
- *Proportionality* or body lever lengths are capacities that can be changed only marginally during childhood and early adolescence, because bone lengths are finite once the adult has reached full maturity. However, it is important to mention at this point that with intelligent

coaching, these capacities can be modified within the context of the athlete's technique. Chapter 6 covers proportionality in detail.

> In some situations, individuals are not able to modify their physical capacities enough to give them an optimal performance; if this is the case, technique modification should be considered.

Before we give examples of physical modification, it is important to state that the capacities just mentioned are in some cases closely related. For example, body type and body composition have many factors in common, while strength, power, speed and agility are closely related. Flexibility and posture are partially related, while proportionality has some relationship to body type, strength, power, speed, balance and agility.

The following examples are cited to illustrate the fact that various physical characteristics can be modified with intensive intervention programs in order for athletes to improve their performance:

- A *butterfly swimmer* with all the other physical capacities required to swim at the top level may lack the shoulder flexibility that is needed for an effective technique. By embarking on a long-term, intensive stretching program, particularly for the shoulder joint, such a swimmer can greatly improve flexibility and technique. Chapters 11 and 13 outline such an approach.
- It is desirable for a *male gymnast* to have long arms in order to perform well on the pommel horse. Unfortunately, long arms present strength problems when a gymnast is performing on the rings. In order to perform optimally in this event, it is necessary to develop very high levels of strength in the upper trunk and arms. Chapters 8 and 13 address these issues.

Development of upper body strength is essential for high performance gymnastics on the rings.

- *Basketball, volleyball* and *netball* players all need to develop their jumping skill in order to be effective in their sports. Modern power training methods can improve their performances, especially if combined with flexibility training, by increasing the amount of energy recovered for the concentric movement from the stretch-shorten cycle, in which elastic energy is stored and then released during various ballistic skills. Chapters 9, 10 and 13 cover such issues.
- *Cricket fast bowlers* may wish to increase the speed of their delivery. To do this and protect themselves from injury, they should improve their strength and power as well as their flexibility levels, especially in the trunk and shoulder joints. The combination of these programs will give them the body bulk, power and elastic energy needed to increase the velocity of their ball delivery. Chapters 9, 11 and 13 address these issues.
- A *female lightweight rower* may find that as she progresses through adolescence, her body fat level rises, thus affecting her power to weight ratio. She will, at this point, need professional advice from a sport scientist and nutritionist to enable her to reduce her fat mass yet maximize strength and power and still achieve the weight limit for competition. Chapters 5, 8, 9 and 13 assist in outlining such a program.

Technique Modification

Many athletes will be able to advance toward their goal by simply undergoing a body modification program. However, there are situations in which individuals are not able to modify their physical capacities enough to give them an optimal performance; and if this is the case, other approaches need to be adopted. Technique modification should then be considered, but before we discuss this, it is important to point out one of the problems that may occur if the coach is not aware of the *mechanical principles* associated with a given skill. It is necessary to observe these principles at all times in order to obtain an optimal result. However, because humans demonstrate many anatomical differences, the way in which these principles can be applied varies considerably.

When seeking to modify the biomechanical aspects of an athlete's performance, the coach must develop a *set of preferences* that can be applied in a particular sport; and these must be chosen to suit an individual or particular circumstance. These preferences are the ways in which athletes will perform the skill within the framework of the mechanical principles we have mentioned. Many such preferences are available to the athlete in every sport, and their selection often relates to the physique of the player, the environmental conditions (e.g., wind, heat, cold, humidity), the playing surface, the opposition and myriad other variables that need to be taken into account in every sport. The following two examples illustrate these technique preferences.

- Tennis players of differing stature must also play on a number of court surfaces with varying bounce heights. In response to these variables, different players may choose different grips so as to "square" the racket to the ball at a given impact location. The taller player may select an "eastern grip", with the hand more behind the handle, whereas the shorter player may select a "western grip", with the hand more "behind and under" the handle. The shorter player, by holding the racket in this manner, is able to contact the ball at the same height as the taller opponent.
- Right-handed golfers who wish to "move the ball" from right to left during flight (a draw) because of a dogleg or windy conditions can adopt a "closed position" at address. This means that the left foot will be farther forward than the right one and the feet will be aligned with the body up to 15° to the right of the target. If right-handed golfers wish to hit a straight ball, the feet and the body should be square to the target line; but when a left to right shot (a fade) is needed, they should set up with the left foot farther behind the right foot and with the body facing about 15° to the left of the target.

These preferences can be determined only when coaches have a good knowledge of the physique of their athletes as well as of their movement and psychological capacities. This approach is the very essence of scientific coaching, whereby the coach decides what suits the individual best, according to his or her physical strengths and weaknesses, rather than teaching a general technique that is mechanically efficient for other champions but may not suit that athlete particularly well.

The following examples illustrate technique modification undertaken when the body's physical capacities cannot be modified further:

- If a *butterfly swimmer's* flexibility level has not improved to the point where the style is economical and mechanically effective, despite an intensive stretching program for the shoulder joints, then the coach should teach that competitor the side-breathing technique.
- In combative sports such as *judo* and *wrestling,* individuals with long lower limbs compared to their trunk length should flex them at the knees a little more than shorter athletes in order to lower their centre of gravity, thus increasing dynamic stability.

Shorter athletes have a mechanical advantage in combative sports like wrestling, though taller athletes may crouch to lower their centre of gravity.

- The *breaststroke swimmer* with a poor kicking posture (i.e., slightly inverted feet or a conventional foot posture) should concentrate more on the synchronization of the body and the arm pull in order to overcome this postural deficiency in the stroke.
- *Contact sport* players with long legs tend to stride out during on-field running, and this lowers their level of dynamic balance. In order to increase it, they should flex a little more at the knees and consciously take short, very fast steps, especially when near or contacting opponents. By adopting such a technique, they keep their base of support more under their body, thereby improving their dynamic balance.

Pedagogical Considerations

It is often easy to obtain objective measures on selected aspects of performance from biomechanical tools (e.g., video—chapters 14 and 15) or from sport science or coaching publications. However, every coach or teacher knows that modification of performance, particularly at the elite end of the spectrum, is a difficult task. Intervention, whereby feedback appropriate to each athlete assists in the development of the skill, is an art. A biomechanical approach to intervention strategies is included in chapter 16.

Figure 1.1 shows how technical-tactical aspects of performance should be linked in the athlete development pathway. In this way athlete flair can also be encouraged as different players will approach a given situation differently while still adhering to key mechanical requirements of performance as detailed in chapters 14 and 17.

Reassessment

Modification of physical capacities or technique (or both) does not provide a one-time solution to affect the athlete's competitive performance. Instead, this is a process that often takes years of careful management involving many cycles of assessment, profiling, modification and reassessment to develop and refine the athlete. Thus, the process of reassessment following some intervention is crucial so that the coach can understand the influence and impact of past modification strategies on performance (feedback pathway B in figure 1.1).

It is important to make the point here that following an intervention, the coach does not always seek a direct effect on the athlete's performance. That is, many interventions serve as building blocks or precursors to improved performance at a later time. For this reason, one of the feedback pathways (pathway A in figure 1.1) bypasses the *performance* criterion altogether. The following are two examples of situations in which such a strategy is appropriate:

- A *female rower* who needs to make the lightweight category for competition embarks on a regimen of aerobic exercise and dietary modification to minimize body fat and maximize lean mass within the competition standards. Body composition needs to be monitored over several months to ensure that the intervention is effective, but its effect on performance is not at issue during this preparation phase.
- The *tennis player* lacking a powerful serve after technique modification may need to follow a strength or power training program designed to improve shoulder internal rotation speed. It is relatively easy to place such a player on a specific program and check if service speed improves. However, this book stresses the importance of a holistic approach to performance, so after the technique modification program it is necessary to measure speed of internal rotation. Knowing this, a coach can then slowly build the desired "more powerful serve".

Summary

The aim of this chapter has been to assist coaches to apply the assessment and modification concept. Coaches do this by either modifying the body to suit a mechanically sound technique or altering the technique to suit the individual or both. The combination of the two approaches is frequently the best solution to the problem. Throughout the history of sport we have seen various "technique fads" sweep through the sporting world. New jumping, hitting, throwing, running and swimming techniques have evolved at a rapid rate, especially during the past 30 years. These techniques have obviously suited the athletes for whom they were developed, and other individuals with similar physiques have adopted them successfully. However, those competitors with different physical capacities have often copied these techniques blindly and have been disappointed with their poor performances.

It is the responsibility of coaches to equip themselves with enough scientific knowledge to enable them to tailor a technique to suit each individual athlete. The remainder of this book is therefore dedicated to this end.

two

Talent Identification and Profiling

Jason P. Gulbin, PhD; and Timothy R. Ackland, PhD

During the past 50 years, coaches have identified talent and informally profiled athletes; but it was not until the early 1970s that Eastern European countries, especially Russia, East Germany, Hungary and Czechoslovakia, and later China began the systematic programs that were to help them win a large number of medals internationally in the 1970s and the 1980s. Both Alabin, Nischt, and Jefimov (1980) and Hahn (1990) suggested that efficient talent identification procedures play a very important role in modern sport and were a major factor in Eastern Europe's domination of many Olympic sports during the 1970s and 1980s.

Similar programs emerged in Western Europe, North America and some Commonwealth countries in the 1980s, but it is extremely difficult for countries that do not have sport institutes or centres of excellence for sport to compete with those that do. Where such sport systems are in operation, sport scientists and sports medicine specialists have been able to form sophisticated testing teams for talent identification and profiling, gathering systematic information from the athletes and integrating it into scientifically based training programs for the benefit of both individuals and teams.

These programs have underpinned the recent successful sporting performances of small countries like Australia, which has a population of approximately 20 million. The development of the Australian sport system in general, and talent identification programs specifically, began after poor rankings at the 1976 Montreal Olympic Games, in which Australia did not win a single gold medal. With steady progress through the 1980s and 1990s, the Australian team achieved fourth place overall at Sydney in 2000 with 58 medals and finished first on a *medals per capita* basis. These successful results were mirrored recently in Athens in 2004.

With a relatively small population, Australia cannot rely on a *natural selection* or *self-selection* approach to producing elite athletes (Ackland 2001). Competition for junior athletes is high between the major sporting codes such as soccer, other football codes, basketball and baseball. In small countries, this often leaves the "minor" sports with a minute talent pool from which to draw. By way of comparison, China had a population of 1.3 billion in 2001, with 200 million registered basketball players—10 times the entire population of Australia! Therefore, if small nations wish to compete successfully on the international stage in a variety of sports, some system of talent identification and development is essential.

Talent Identification

The terminology used in the field of talent identification can be confusing, mainly due to the different terms preferentially adopted by various countries and their sporting systems (Hohmann 2004; Williams and Reilly 2000; Ziemainz and Gulbin 2002). For the purposes of this chapter, *talent identification* is a general term that describes the scientific process used to identify talent. Specifically, *talent detection* refers to the process used to identify talent from outside the sport, and *talent selection* is the approach that allows identification of talent from within the sport. Common to both approaches is the critical *talent development* phase in which the identified talent has the opportunity to achieve full potential in an environment that ideally includes, but is not limited to, quality coaching, sport science and sports medicine support, competition immersion and a philosophy of holistic athlete development (Gulbin 2001).

Rationale for Talent Identification

As the aggressive sport systems of Eastern Europe have gradually broken down, the stigma of talent identification as an undesirable practice is fast disappearing. In fact, apart from the use of drugs, which in some of these countries was rampant and cannot be condoned under any circumstances, talent identification was positive in many cases, as young athletes were well catered for both personally and from a sport development viewpoint. According to Bloomfield (1992), the positive features of such programs are as follows:

> Successful talent identification programs begin with an objective-focused data collection phase combined with the subjective coach assessment, or "coach's eye".

- Children are directed toward sports, or particular events, for which they are physically and physiologically best suited. This in turn means that they will probably obtain good results and enjoy their training and participation more.
- Because of the nature of the program, their physical health and general welfare are well looked after.
- They are usually the recipients of specialized coaching, which is well supported by the sports medical team and sometimes by a sport psychologist.
- The administrators of many of these programs are now concerned about the vocational opportunities for the athletes after the conclusion of their competitive career and cater for them with high-quality secondary or tertiary education or vocational training.

Additionally, talent identification programs allow countries to be strategic about where limited resources can be applied. For example, it has been estimated that the federal investment required to win an Olympic medal is approximately AUD 8 million, while that required for an Olympic gold medal is reported to cost AUD 37 million (Hogan and Norton 2000). While medals are not the sole outcome of sport funding, the process of talent identification does encourage resources to be used in a more targeted and efficient way with clear performance objectives and outcomes.

The "Traditional" Talent Identification Process

Talent identification can be either very simple or highly sophisticated. One may find, for example, a school basketball coach recruiting players simply because they are tall for their age. Or a school swimming coach, when walking around the school yard, may observe the way children are standing: If they have naturally large pronated feet, they may be prospectively good breaststroke swimmers, as this physical capacity is needed for an efficient kick. While there is always room for the "coach's eye", sophisticated programs, on the other hand, are highly oriented toward sport science and medicine, with a comprehensive test battery used to screen the young athletes.

Elements for Successful Talent Identification Programs

Ackland (2001, 2006) has written extensively about the talent identification process, which may also be used to monitor the status of athletes throughout the training cycle. The process usually involves five stages, as follows:

1. **Understanding the important aspects for success in competition.** This leads us to ask whether athletes in a particular sport demonstrate distinctiveness in physical capacities that might provide an advantage for sporting performance. If certain sports require a unique physical structure (e.g., height, weight or proportionality), then these factors must be present in the talent identification test battery. Similarly, if an athlete must fit within a team structure (e.g., members of a rowing eight or a sprint kayak team, cyclists in a team pursuit event or members of a synchronized swimming team), then the relevant elements of size and physical capacity for the most optimal team composition must also be assessed. It is also important to keep abreast of any rule or equipment changes within a specific sport, as this may affect the requirements for success at the elite level. For example, rule changes such as legalizing drafting in the cycle leg of the triathlon have led to greater emphasis on an athlete's running capabilities in comparison with the swimming and cycling components (Hausswirth et al. 2001).

2. **Recording a set of data on an athlete.** Once the important parameters that differentiate talented individuals are determined, the coach or sport scientist must set about collecting a valid and reliable set of data for the athlete. The chapters in part II of this book provide discussion on the alternative methods for assessing these characteristics.

3. **Gathering a set of normative or comparative data.** To be effective, these normative data must be of a certain standard; otherwise you may not expose the critical variables to work on with your athlete. Good-quality normative data must be derived from high-performance athletes, measured in recent years and appropriately stratified for player position, event type or weight category (or some combination of these). Quality data on high-performance athletes are difficult to collect and are often jealously guarded by sport institutes and professional teams.

Normative data will rapidly diminish in utility when the sport experiences substantial changes in rules, equipment and athlete preparation. These changes can affect the physical and physiological capacities that provide an advantage for success in the sport or event. Therefore, coaches should always seek to use the most up-to-date set of comparison data in order to create an individual profile. The data must also accurately reflect the status of the athlete and so must be appropriate for the individual's age, gender and category (player position, event or weight category). One should exercise extreme caution with reference to trusting the average values reported in the literature. Too often, we rely on convenient data without regard to their impact on the resulting profile chart.

4. **Using these data to construct a profile of the athlete.** Examples relating to this aspect are provided later in this chapter.
5. **Interpreting the profile to guide the selection process or provide the basis for an ongoing training program.** Similarly, a number of examples of profile interpretation are given at the end of this chapter.

According to Ackland (2006), it is unfortunate that many coaches and sport scientists have not always used the process effectively. Often we observe a huge battery of tests being administered in a talent identification program, without too much forethought having been given to the importance of the selected parameters and use of quality normative data for comparison. In simple terms, one may be better off measuring fewer, but more discriminatory, physical, physiological and psychological variables when applying this model, rather than using the generalized "shotgun approach". Readers should bear this advice in mind throughout the remainder of this chapter.

General Talent Identification Screening

Sport systems or institutions within them that have sophisticated screening programs examine several parameters that usually relate to junior athletes and then construct a general profile of each subject. The following test series illustrates this.

Hereditary Factors There are important hereditary factors that one should consider in the selection of talented athletes, as the evolution of research encompassing familial, twin and molecular studies has substantially contributed to our understanding of this area. For example, early research by Cratty (1960) demonstrated that 24 college-aged male athletes produced comparable performances with their fathers in both the long jump ($r = 0.86$) and the 100 yd sprint ($r = 0.59$) some 34 years earlier. Studies examining the characteristics of identical twins have demonstrated that the heritability estimates for critical athletic traits such as maximal aerobic power, maximal aerobic and anaerobic capacity, maximal mechanical power and muscle force, body mass index, somatotype, maturation, muscle area and fibre distribution are significantly and substantially dependent on genetics (Klissouras 2001). At the molecular level, where modern scientific techniques can identify specific genes and their locations on the chromosomes, scientists have now recorded over 187 specific gene entries that influence and are relevant to fitness and performance (Rankinen et al. 2006). Thus, as genetics can limit sporting prowess, the fundamental goal of the talent identification process should be to try to determine the innate differences that exist within populations.

Time Spent in Sport It is important to establish how long each athlete has trained for his or her sport and how much coaching he or she has received. In some instances, certain children will have been in a particular sport for a much longer time than others and will have already received considerably more specialized coaching. In such cases they may not develop much further with their skills or their cardiorespiratory capacity, for example, if these factors are important selection criteria.

Maturity Maturity is an important factor that must be considered in any talent selection procedure. It is well known that early maturers are often taller, heavier, more powerful and faster than their counterparts during the early to mid teenage years. This has often led to selection biases in sporting competitions grouped by chronological age (Helsen, Van Winckel, and Williams 2005). Many coaches have observed that early maturers

Relative newcomers to serious training for a sport have great potential for further improvements in their skills.

have an advantage in junior sport and often neglect their skill development, whereas less physically mature athletes are forced to develop their skills earlier to perform well. These late developers when fully matured may have definite advantages over the early developers, because their physical capacity levels are eventually matched but their skills may be better.

Physical Capacity Screening

All talent identification programs must be concerned with every facet of the particular sport or event. The athletes' functional capacity generally forms the bulk of many talent identification assessments; however, it is important to reiterate the points made in chapter 1—that performance in sport is multifaceted and must be approached from many perspectives, and that no single physical capacity on its own can be the absolute criterion for success in sport. The coach should be careful not to place too much emphasis on just one or two variables, but should assess the body as a total unit. It is important for the athlete to have several physical capacities that are basic to a particular sport, but young athletes must also have the other essential physiological, psychological and skill characteristics before they can be regarded as talented.

Absolute Size and Proportionality Proportionality can be an important self-selector for various sports and events. Accurate forecasts with relation to individual and team performances can be made on the basis of measurements of height and body mass alone.

The stability of height has not been studied in any detail, but reasonable forecasts can be made on the basis of existing information and the experience of successful coaches. Malina and Bouchard (1991) and Tanner (1989), when discussing the stability of growth, suggested that until the age of 2 or 3 years a child's growth is not particularly constant, but that from 2 to 3 years to the onset of adolescence, height is reasonably stable and predictable. During adolescence, because of the various spurts and plateaus, it is again unstable, but in the immediate postadolescent period becomes stable once more. At any time during a child's life, height or bone lengths are more constant than body mass.

Predictions of height and body mass (weight) have been made reasonably accurately for some time, and Lowery (1978) gave several formulas to estimate height. He also quoted the work of Bayley and Pinneau (Bayer and Bayley 1959), who used skeletal age and tables to forecast growth. Noninvasive prediction of adult stature can also be achieved using formulas and curves established from longitudinal population studies, and are typically accurate within ~3 to 6 cm (1.2-2.4 in.) (Beunen et al. 1997; Sherar et al. 2005). Another way to estimate final height and body mass is to determine a subject's skeletal age from an X ray of the hand and wrist using Greulich and Pyle's atlas (1959), then use growth standards (Tanner 1989) for height and body mass to determine the mature measures. This method has been used by some coaches to forecast final heights and body masses of athletes in situations in which these variables are important factors, either when the athlete is too large for the event, as in gymnastics, or too small, as in the throwing events.

The composite shape of the athlete is a consideration in sports in which overcoming or even promoting aerodynamic drag is vital. Cycling, speedskating and ski jumping, for example, are substantially affected by the resultant shape that the athlete can make with his or her body. While clothing or equipment can help alter body shape, the winter sport discipline of skeleton offers few opportunities to lower frontal surface area as the sliders lie prone on a sled and reach speeds of up to 130 km/h (80.8 miles/h). Wind tunnel testing conducted by the Australian Institute of Sport and Monash University has demonstrated that the shape presented by the shoulders and buttocks can substantially affect aerodynamics (figure 2.1). Since these anatomical features are mostly fixed, consideration ought to be given to composite shape in the identification and selection of athletes when aerodynamic form is important.

The proportions of athletes vary greatly within racial groups and even more so between them, and chapter 6 illustrates the fact that certain individuals have very definite leverage advantages over others. It is important to point out that very little research has been done on the stability of the proportions of the growing child and adolescent. Ackland and Bloomfield (1996) have suggested, however, that various segment breadths remain stable throughout adolescence and can, therefore, be used for predictive purposes. On the other hand, they also found that many segment lengths were unstable during adolescence and suggested that these should not be used as prediction criteria in talent identification programs.

Somatotype Body type is a general physical capacity that can be a useful indicator of future elite performance. The stability during growth of the somatotype has been discussed by Malina and Bouchard (1991), who suggested that ectomorphy was reasonably stable during growth but that mesomorphy and endomorphy in adolescent boys were not as predictable. Because boys experience a marked increase in muscle mass during the adolescent growth spurt as a result of testosterone secretions, it is logical that somatotype variations will occur in this component. However, reasonably accurate forecasts can be made in mesomorphy from around Tanner stage 3.5 to 4 of pubescent development (at approximately 14 to 14.5 years of age), when a boy's musculature rapidly develops. The stability of the somatotype components in girls during adolescence is less well understood, but many coaches suggest that adolescent girls from around

Figure 2.1 Aerodynamic testing of an Olympic skeleton competitor in a wind tunnel.
Photo courtesy of Greg Ford, Monash University.

age 12.5 to 13 years tend to steadily increase their endomorphy rating as they develop and that this is accompanied by a drop in their power/weight ratio. This decrease can be detrimental for them in weight-bearing ballistic sports such as gymnastics, basketball, netball and volleyball in which powerful movements are essential.

> One must exercise care when predicting future adult performance based on adolescent testing because of the varied stability of physical traits during adolescent growth.

The most suitable body types for various sports and events are discussed in detail in chapter 4. To be within one component of the mature body type at any time during the adolescent growth period would indicate that the young athlete is generally "on track" as far as the attainment of an optimal shape is concerned. Well-trained coaches are now aware that special intervention programs using strength and power training or nutrition (or both) can alter the primary somatotype components over a one- to two-year period.

Exceptions to the standard body shape must be carefully considered because coaches and sport scientists report that some high-level performers are outside the standard somatotype for a particular sport or event; some examples are cited in chapter 4. It must be emphasized again that body type is only a general physical capacity and is useful only when combined with several other measures, as it cannot be used as a single criterion in itself.

Body Composition As with somatotype, body composition is only a general indicator of high-level performance. The stability during growth of an individual's body composition was reviewed by Malina and Bouchard (1991), who suggested that body fat was not particularly stable from birth to 5 or 6 years or during adolescence. They also stated, however, that the mark of excess fatness (in adults) appears to be greater for those who have thicker subcutaneous fat measurements during childhood.

The most suitable body composition for young athletes in various sports has not been systematically documented. However, if young athletes are within 1% to 2% body fat for a male or 3% to 4% for a female at around Tanner stage 3.5 to 4 of their pubescent development, or when they are close to their peak height velocity (PHV), then they are "on track" for the optimal body composition that is generally required for the given sport or event. It should be noted, however, that the demands of some sports for low fat levels, for example in field or court sports, are not as great as in other sports such as gymnastics, where power to weight ratios play a significant role in an athlete's performances (see chapter 13).

As with body type, intervention programs that employ strength and power training or dietary modification (or both) can be used to assist the individual to reach accepted body composition levels, for both lean body mass and fat mass in athletes who are reasonably close to the accepted levels.

Posture As with proportionality, posture is an important self-selector for various sports and events within them. Chapter 7 discusses the value of certain types of postures that give some competitors very definite advantages over their opponents. However, the stability of posture during growth has not been examined in a systematic way, and again most of the information has been formulated from anecdotal evidence. The general feeling about posture is that it steadily develops during adolescence and becomes more extreme as the individual ages. Perceptive sport scientists and coaches, however, will be able to identify these developments and forecast with some degree of accuracy what the athlete's ultimate posture will be.

The posture of athletes, like that of the nonathlete population, varies greatly. However, there are several postures that give athletes an immense advantage over their competitors. Coaches need to carefully observe the athlete's posture in order to determine whether or not it is advantageous to the performance. Posture can be partially modified using flexibility and strength training, but this must be done at an early stage in the individual's development because changes are difficult to make during late adolescence and early adulthood.

Strength and Power Strength and explosive power in many sports are essential physical capacities, and performances in sport have improved at a rapid rate since these capacities became important components of the modern training program.

The stability of strength, according to Malina and Bouchard (1991), is reasonably high from year to year, for example from 11 to 12 years of age; but the correlations decline as the age gap widens. These authors reported that correlations between 7 and 12 years and between 12 and 17 years were only low to moderate. One important point they made, however, was that a composite of strength measures tended to be more stable than single variable scores.

Eastern European coaches have suggested for some time that around the PHV period (about 14 years of age) in the adolescent male's growth spurt is a reasonably reliable time to make future forecasts on mature strength levels. This has been done particularly well in the sport of weightlifting and can probably apply to other sports.

The importance of strength and power in the majority of sports is now well accepted, and early identification of high strength and power levels can be very helpful to the coach. With the modern strength and explosive power training equipment and methods that are now available and are outlined in chapters 8 and 9, it is possible to develop a strong and powerful athlete if this training begins during adolescence.

Flexibility Flexibility is a very important physical capacity in the majority of modern sports. Very little research has been done with relation to its stability during growth, and the studies that have been conducted often concern gross flexibility in which several body segments may be involved. Malina and Bouchard (1991) reported data from several sources that used the sit and reach test. These showed that in boys, the scores were stable from 5 to 8 years; they declined to ages 12 to 13 years and then increased again to approximately 18 years. The girls' scores were stable from 5 to 11 years, increased to 14 years, and then gradually flattened out from that time on.

With such variable data currently available, it is difficult to reach any conclusions about the validity of flexibility as an indicator of future talent. Modern flexibility training methods are so good that reasonably high levels can now be reached by most athletes if their programs are started either in preadolescence or in the early stages of adolescence. Chapters 11 and 13 outline these programs in detail for a large range of sports.

Speed Speed of movement is an essential physical capacity for high levels of performance in many sports. Again, very little research has been done in this area except for running speed, which is only one aspect of speed. However, most coaches are aware of whether an athlete is fast or slow either in running speed or in other movements from an early stage, because they are continually comparing their athletes with others of similar age.

Speed training techniques are now well developed, and intervention programs are commonplace in modern training; but young athletes should be at least 10 years of age before they embark on such a program if high levels of performance are to be reached by the time they become senior athletes.

Health Status A medical examination is usually conducted, during which special attention is paid to the musculoskeletal and cardiorespiratory systems. Athletes with medical conditions that may at some time in the future limit their training or participation should be very carefully evaluated. Komadel (1988) identified the diseases listed by the Czechoslovak Ministry for Health as contraindications to competition in sport at the highest level. Many Western sports medicine physicians would not totally agree with all of these, citing asthma and diabetes mellitus as diseases that can now be well controlled by high-level athletes with drugs that are currently available. The majority of sport physicians currently working with elite athletes would, however, generally endorse this list.

> Talent identification programs aimed at postadolescent athletes avoid the trait stability issues inherent in adolescent talent identification programs. These programs can also reduce developmental time frames and can provide a second chance for early-retiring or deselected athletes.

A New Paradigm

The traditional approach in talent identification, or more specifically *talent detection,* has been to undertake mass screening testing in primary and secondary schools with junior-level athletes. While this technique has been successful in producing medal-winning performances at the national and international level (Gulbin 2001; Hahn 1990), the approach can be demanding of time and resources and requires unwavering long-term commitment by the sport. Access to talent via schools alone can be compromised if teachers or schools decline to participate in talent detection programs. Therefore, consideration ought to be given to broader recruitment strategies and alternative talent pools to improve the opportunities and rates of development for athletes (Gulbin 2004).

An Alternative Talent Pool

One talent pool that has been consistently ignored in the area of talent identification has been the postjunior- or senior-level athletes. These athletes are approximately 18 to 28 years old and possess well-developed skills and fitness traits that can be transferable to other sports. Many athletes may have unfulfilled ambitions in sport or have injury or deselection issues that prevent continued participation in their current sport despite their being highly motivated to continue (Cockerill 2005). There are limitations to this technique, however, including the imbalance between donor and recipient sports, the inability of athletes to make the transition into early-specializing sports such as swimming and gymnastics and the difficulty in achieving rapid success in sports with large numbers of participants (Baker and Horton 2004). Nevertheless, consideration of this unique subpopulation presents resourceful opportunities for talent identification practitioners interested in boosting numbers in the later-specializing sports.

Talent transfer instances have usually been undertaken independently, although more systematic programs have been promoted and trialled (Battaerd 2004; Halson et al. 2006). Well-publicized talent transfer examples include the speedskater Eric Heiden, who switched to road cycling and won the U.S. professional road championships after winning five gold medals at the 1980 Winter Olympics, and basketball player Michael Jordan's transfer to professional baseball. Other common talent transfer scenarios within sports have included switches from gymnastics to diving and aerial skiing, from sprinting to bobsled and from surf lifesaving to flat-water canoeing.

Developmental time frames are an important consideration in talent identification programs. Coaches in certain sports may find themselves with a smaller pool of senior athletes than they would like as a result of injury, poor succession planning or retirements after major competitions. Talent identification programs using the talent transfer model can enable a quick injection of talent into training squads, supporting or at times surpassing existing athletes. While theoretical rates of talent development have resulted in the 10-year generalized benchmark of sporting expertise (Ericsson, Krampe, and Tesch-Romer 1993), in practice 28% of Australian high-performance athletes who specialize later in their athletic careers can reach senior national representation in four years or less (Oldenziel, Gagne, and Gulbin 2004).

Assessing physical and physiological data from older athletes virtually eliminates the confounding issue of the maturational differences that exist amongst adolescent athletes (Malina and Bouchard 1991). Reduced extrapolation of data is required to ascertain elite athletic potential provided that diagnostic tests more closely approximate the specific demands of the event. Other advantages of screening older athletes in comparison with younger athletes are that older athletes have an ability to set goals, a greater degree of independence, a training and work ethic and a better understanding of competition demands. These are the unknown qualities that are important for performance but often difficult to measure in traditional talent identification programs. Finally, governments and sporting institutes may have also invested significant financial resources by providing scholarships to athletes yet to achieve the performance returns projected. Reinvesting in publicly funded athletes not only is shrewd, but also provides a second chance for athletes to delay sporting retirement and to achieve success on alternative playing fields.

Talent Transfer Case Studies at the Australian Institute of Sport

In 2002, a group of researchers at the Australian Institute of Sport (AIS) sought to boost senior women's sprint cycling stocks by recruiting noncyclists with explosive leg power characteristics. From over 247 women tested, the top 26 with the best 10 s peak power and 30 s average power were selected to commence a sprint cycling program. The women selected were aged 16 to 29 years and came from sporting backgrounds such as basketball, rock climbing, rowing, netball, surf lifesaving, triathlon, rugby, alpine skiing, water polo and in-line skating. In less than two years, senior achievements from five different athletes were the following:

- National road race—club champion
- National team-sprint—champion
- National road race—series champion
- National team-sprint—silver medal
- Australian Road Cycling—squad member
- Oceania Championships—bronze medal
- Athens Paralympics—gold and bronze medals

> Establishing the value of genetic information and its relationship with athletic performance is a valid and worthwhile scientific pursuit. It is the inappropriate application of scientific knowledge that raises a number of legal, ethical and social issues.

This AIS talent transfer program in sprint cycling was able to demonstrate that athletic cycling novices could rapidly transfer into both sprint and road cycling and achieve success at the national and international level in less than two years.

In 2004, with the recent reintroduction of skeleton into the 2002 Winter Olympics, researchers at the AIS investigated the possibility of identifying women with the potential to qualify for the 2006 Torino Winter Olympics in less than two years. A key to the sport of skeleton is the explosive sprint start over ~30 m (33 yd) distance that accounts for approximately half of the performance variability. As a result, the skeleton event is highly suited to the talent identification process. A talent detection approach was undertaken to identify fast, powerful women who had the capacity to dominate the start component of the event while concurrently learning how to drive the sled. Ten women were selected predominantly on their sprinting prowess and subsequently placed into a talent development program with a dedicated expert coach, excellent competition opportunities and rigorous sport science and medicine support.

The age range for the selected athletes was 18 to 29 years. In their inaugural 2004-2005 sliding season, Melissa Hoar (a two-time world champion in the surf-lifesaving event of beach flags) qualified for the World Championships and achieved an astounding 13th place after only three months in the sport. In the following season, former beach sprinter Michelle Steele produced world-class performances with three top-six finishes in World Cup competition, subsequently qualifying for the Torino Winter Olympics where she finished in 13th place. Thomas Rowland's (1998) prophetic statement regarding the potential of talent identification applies unreservedly to Michelle Steele's phenomenal transfer from beach to ice in only 14 months: "Indeed, it is not difficult to imagine that some unsuspecting individuals will pass their lives unaware they possessed extraordinary talents in events such as luge or fencing or steeplechase" (p. 199).

What Does the Future Hold?

Apart from the influence of environment and hormonal factors, many of our individual physical and physiological characteristics have a genetic basis. For example, the major determinant of our muscle fibre type is via genetics, inherited from our parents, while only a small percentage of fibres can be altered by specialist training (see chapter 8). In recent years, the Human Genome Project has processed billions of consecutive pieces (bases) of raw data to identify sequences that control specialized protein production, including the proteins that are associated with elite athleticism. Already it has been shown that elite athletes generally have different genetic profiles in comparison with control groups for attributes such as speed and endurance. For instance, a functional muscle gene that is required for forceful contractions at high velocity is *ACTN3,* and no female sprint/power athlete (n = 35) or sprint Olympian (n = 32) was found to be deficient in the gene (Yang et al. 2003). Similarly, important regulatory genes such as *EPAS1* (Henderson et al. 2005) and *ACE* (Gayagay et al. 1998) are both reported to enhance cardiorespiratory function and are more likely to be found in elite endurance athletes than nonathletes.

Whilst the early applications of this work are centred on understanding the mechanism of, and correcting the defective genes responsible for, Duchenne muscular dystrophy and other medical conditions, it does not take too much imagination to understand that such techniques can be readily applied to elite sport. For example, the protein *musTRD,* which regulates the growth of muscle and influences muscle hypertrophy, has been identified. It also represses the development of slow-twitch fibres and activates the development of fast-twitch fibres. In a similar vein, the gene sequence $p70^{s6k}$, which increases the growth and stretch of muscle, has been identified (Hardeman 2001).

The utility of genetic profiling for talent identification is readily understood, though it is acknowledged that there are moral and ethical arguments to be considered. Genetic profiling could, for example, be used to predict future strength and power capability in children from an early age or to identify the gene action profiles that will permit optimum training among a group of junior tennis players. However, it is important that any debate regarding genetics and talent identification be based on the understanding that the acquisition of scientific knowledge and a better appreciation of human function is the supreme goal, not the desire to conduct genetic interventions or therapies. We must remain vigilant, however, since misuse of genetic information may lead to unacceptable practices aimed at seeking ergogenic advantage, thereby providing new challenges for the administrators of sport across the world.

In conclusion, the selection of prospective elite athletes using talent identification is a complex problem. Because there is a dearth of knowledge as to how maturity affects the physical, physiological and psychological makeup of the individual, it is often difficult to ascertain whether the results of tests given early in an athlete's life will be useful predictors for the future. With more longitudinal testing of elite groups, the sport scientist and sport physician will be able to predict future success more accurately, and in so doing make talent identification a more reliable method of choosing future sporting talent. A further important point was raised by Hahn (1990), who stated that "talent identification and talent selection programs will not of themselves guarantee the emergence of champions. To realize their potential, the people selected must have regular access to top-level coaching, as well as appropriate facilities and equipment. They must (also) have a clear network of support, and if possible, such additional elements as sport science and sports medicine services" (p. 11).

Profiling

Athletic profiling has been carried out by coaches in an informal way for many years. During the past three decades formal profiling has steadily developed, first in Eastern Europe and more recently in Western Europe, North America and some Commonwealth countries. The purpose of this section is to discuss modern sport profiling and how it can aid athletes to produce their best performances.

As mentioned in chapter 1, all high-level athletes are unique individuals with many physical, physiological and psychological strengths; otherwise they would not have reached the elite level. Athletes also have weaknesses that need to be known by the coach, who will then be able to take remedial action to strengthen them.

Types of Profiling

Two types of profiling can be carried out with athletes: general and specific profiling. The application of either is largely a function of the level of athlete and the point within the season at which the profiling occurs. The interpretation of these testing results and the quality of feedback provided to athletes (either individually or as a group) from the coach or sport scientists are important elements of the profiling process.

General Profiling

General profiling is done in a detrained state and administered at the commencement of a season. The results of a series of tests will give the coach a general profile of the athlete. In order to ascertain the individual's actual status within the group, one should evaluate the results not only against those of other high-level athletes in the same sport or event, but also against those of the athlete's own teammates. The most important comparison will be that which is made against other elite athletes in the same sport or event. However, the individual's status within the team or squad is often of interest to the athlete and is certainly of value to the coach.

When the coach has evaluated the test results with the relevant sport scientists or the sport physician (or both), the season's training schedule with each individual's strengths and weaknesses in mind can be planned. Often, this general training plan will also take into consideration skill weakness, so that several team members ranking lower than their teammates in the various tests will be given an intervention program in addition to their normal training. It is important to keep in mind that general profiling is often more useful for potential elite athletes in the developmental stages while they are still at the national junior or youth level; for the senior international-level athlete it is not as valuable.

Most coaches profile only once each season at this level in order to identify weaknesses. Others have follow-up tests at varying intervals to monitor the general progress of the intervention programs. Agility athletes in "closed" sports, such as gymnasts and divers, are regularly monitored for body composition because power to weight ratios are crucial factors in their performances.

Specific Profiling

Specific profiling is usually done with elite senior athletes in sports in which events are won by very small margins or times, as in aerobic sports such as swimming, rowing, kayaking, running and cycling, where it is important to accurately evaluate the individual's adaptation to the stress of heavy training at regular intervals.

In some programs, tests are done as regularly as every two weeks, but more often a month apart, which fits in well with a 3/1 training cycle. When a major championship is approaching, some coaches request that only the "key" stress adaptation tests be done at more regular intervals, which could be as often as every week. Within sports that are more skill oriented, specific profiling is done approximately every six months, because alterations to the athlete's physical capacities take a longer time to occur.

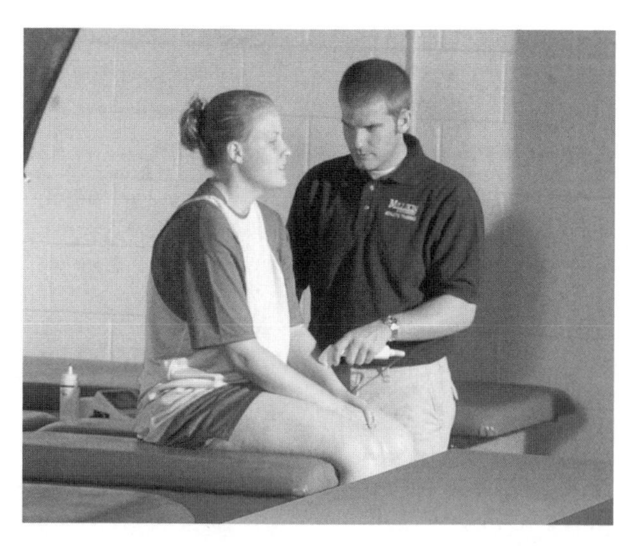

Personal health problems that may affect training and performance can be addressed with regular health checks.

Health Status Checks

At the commencement of each season, the sports medicine physician should conduct a general health evaluation. This can also be followed up in midseason or at other times thought appropriate by the medical and coaching team. These checks give the sport physician and other specialists such as the nutritionist an opportunity to discuss, at both the team and individual levels, various personal health problems the athletes may have. Young athletes in particular feel that such discussions are very valuable.

Sport Science Tests Used in Profiling

Various sports have differing demands; therefore only tests that are highly specific to the given sport should be used. It is pointless to give a pistol shooter or archer complex cardiorespiratory evaluations at regular intervals, because endurance is not an important factor in these events. One should not completely dismiss the idea of such testing, though, as all athletes need a reasonably high physical fitness level to assist them with other more specific skills and mental tasks, even if there is no specific cardiorespiratory component in their event. However, unless there is a definite need for both endurance and skill in an event, it is often a waste of time and money to give a sophisticated test such as a maximal oxygen uptake ($\dot{V}O_2max$) test to a nonendurance athlete.

Selecting the right test for the right situation is the skill of the sport scientist. Table 2.1 summarizes a number of general and specific tests that are undertaken both in the laboratory and on the field to gain insights into the athlete's physical, physiological, psychological and skill capabilities.

Setting Up the Profile

As mentioned previously, profiles are normally constructed with data obtained from other elite athletes within the same sport or event. Until recently it has been difficult to obtain international-level data for comparative purposes, as there was a dearth of such information in published form and because the test protocols sometimes varied from country to country. It is pleasing to note that both these problems are now being overcome and that more reliable data are available for comparative purposes.

To develop a profile when one has the data is relatively simple. Norms must first be constructed based on percentile scores, T-scores, Z-scores or deviations from the mean of these data. For example, table 2.2 gives physical capacity percentile scores for international female tennis players.

A profile sheet can be used to enter the data; then the graph can be constructed from the data. Figure 2.2 shows data from an international-level male rower under *score* and his profile graph under *percentile scores*. The profile should be drawn simply so that athletes can easily understand it, because they should take part in the discussion with the coach and the relevant sport scientist.

Evaluating the Profile

Physical capacity profiles are simple to interpret, and examples of these profiles are given in figures 2.3 and 2.4. These are of athletes who were first tested to set the training agenda, then at a later time retested to show whether the intervention program had improved their physical status. The athlete in figure 2.3 was a female gymnast of linear build with low levels of strength and power but a high level of skill. She had a somatotype that rated 2.5-3.0-3.5 with a percentage fat level of 12.5. A general strength training program over and above her normal training, three times a week for 20 months, was prescribed. During this time her somatotype changed to a 2.0-4.0-3.0 and her percentage body fat was reduced to 9.6. The dotted line on the right-hand side of the original graph shows the other changes that were made during this time.

Table 2.1 Sport Science Tests to Assist in Profiling

PHYSICAL CAPACITY TESTS	
Structural	**Functional**
Height and body mass	Strength and explosive power
Somatotype and body composition	Speed of movement
Proportionality	Flexibility
	Posture
	Balance and agility
PHYSIOLOGICAL TESTS	
Cardiovascular	**Biochemical and haematological**
Aerobic power tests using specific ergometers	Haemoglobin (Hb)
Anaerobic power and capacity tests using specific ergometers	Blood haematocrit
	Lymphocyte and neutrophil count
	Ferritin
	Uric acid/urea
	Creatine phosphokinase (CPK)
	Testosterone/cortisol levels
PSYCHOLOGICAL TESTS	
Mental toughness	Independence
Aggression	Personality: extroversion-introversion
Anxiety	Leadership
Arousal	Mood
Attention	Motivation
Cohesion	Self-concept and self-esteem
SKILL TESTS	
Objective tools	**Subjective tools**
High-speed videography	Coach's eye
Global positioning systems	Various sport-specific challenges
Accelerometers	

The subject in figure 2.4 had relatively low cardiorespiratory fitness, was slightly overweight and was not particularly strong for his body mass. His somatotype was 3.0-4.5-4.0 and percentage body fat 12.9. After an 18-month training period, which included a moderate diet and three strength training sessions each week in addition to his normal training, his physical status and cardiorespiratory fitness level had improved significantly. Examples of this can be seen on the profile graph, which shows his second test, done nine months into his intervention program, and his third test, administered a further nine months later. It is interesting to note that his body type was 2.5-4.5-4.0 at the second test and 2.0-5.0-4.0 at the third, while the body fat measures had changed from

Table 2.2 Percentile Scores of International Female Tennis Players

Variable	10	20	30	40	50	60	70	80	90
					PERCENTILE				
Height (cm)	163.0	164.7	165.7	166.5	167.5	168.5	169.5	171.3	173.0
Body mass (kg)	51.6	53.9	56.5	58.7	60.7	62.7	64.8	67.4	70.9
BODY COMPOSITION									
Body density (g/ml)	1.009	1.022	1.031	1.039	1.047	1.055	1.063	1.070	1.075
Body fat (%)	30.3	28.8	26.2	24.1	22.1	20.1	17.9	15.3	11.8
Triceps skinfold (mm)	17.0	16.1	15.5	15.0	14.5	14.0	13.4	12.8	11.9
Subscapular skinfold (mm)	11.0	10.1	9.5	9.0	8.5	8.0	7.4	6.8	5.9
Suprailiac skinfold (mm)	14.4	12.9	11.8	10.8	10.0	9.1	8.1	7.00	5.5
Abdominal skinfold (mm)	14.8	13.2	11.6	10.7	10.1	9.3	7.9	7.2	5.4
PROPORTIONALITY									
Brachial index	73.2	75.2	76.6	77.8	79.0	80.1	81.8	82.7	84.7
Crural index	98.2	100.2	101.6	102.8	104.0	106.1	108.3	109.7	112.0
Relative sitting height (%)	50.3	51.0	51.6	52.0	52.5	52.9	53.3	53.9	54.6
FLEXIBILITY									
Arm flexion-extension (deg)	182.2	189.7	195.1	199.7	204.0	208.2	215.9	224.2	235.7
Forearm flexion-extension (deg)	124.6	129.9	133.7	137.0	140.0	143.0	146.2	150.0	155.3
Foot dorsi-plantarflexion (deg)	54.7	58.2	60.8	63.0	65.0	67.0	69.1	71.7	75.2
Leg flexion (deg)	107.0	113.2	117.7	121.5	125.0	128.5	132.2	136.7	142.9
STRENGTH									
Grip strength (kgf)	20.9	23.5	25.4	27.1	28.6	30.1	31.7	33.6	36.2
Arm flexion strength (kgf)	14.2	24.8	32.5	39.0	45.0	51.0	57.4	65.1	75.7
Arm extension strength (kgf)	11.8	21.5	28.5	34.5	40.0	45.5	51.4	58.4	68.1
POWER/SPEED									
Jump and reach (cm)	37.3	39.9	41.8	43.5	45.0	46.5	48.1	50.0	52.6
40 m dash (s)	7.1	6.8	6.6	6.4	6.3	6.1	5.9	5.7	5.4
Agility run (s)	18.7	17.8	17.2	16.7	16.2	15.6	15.1	14.5	13.6

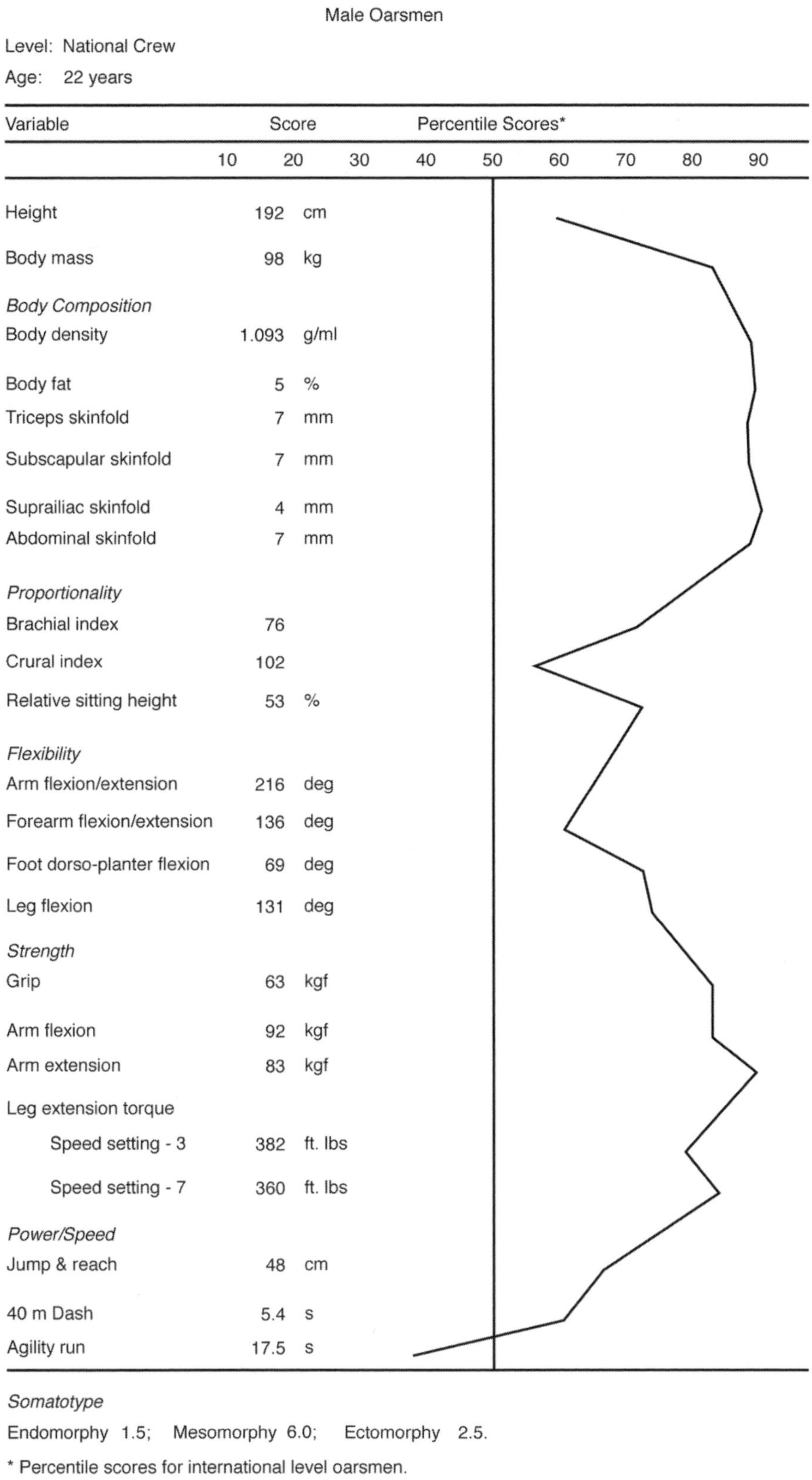

Male Oarsmen

Level: National Crew

Age: 22 years

Variable	Score		Percentile Scores*
			10 20 30 40 50 60 70 80 90
Height	192	cm	
Body mass	98	kg	
Body Composition			
Body density	1.093	g/ml	
Body fat	5	%	
Triceps skinfold	7	mm	
Subscapular skinfold	7	mm	
Suprailiac skinfold	4	mm	
Abdominal skinfold	7	mm	
Proportionality			
Brachial index	76		
Crural index	102		
Relative sitting height	53	%	
Flexibility			
Arm flexion/extension	216	deg	
Forearm flexion/extension	136	deg	
Foot dorso-planter flexion	69	deg	
Leg flexion	131	deg	
Strength			
Grip	63	kgf	
Arm flexion	92	kgf	
Arm extension	83	kgf	
Leg extension torque			
Speed setting - 3	382	ft. lbs	
Speed setting - 7	360	ft. lbs	
Power/Speed			
Jump & reach	48	cm	
40 m Dash	5.4	s	
Agility run	17.5	s	

Somatotype

Endomorphy 1.5; Mesomorphy 6.0; Ectomorphy 2.5.

* Percentile scores for international level oarsmen.

Figure 2.2 A profile of an international-level oarsman.

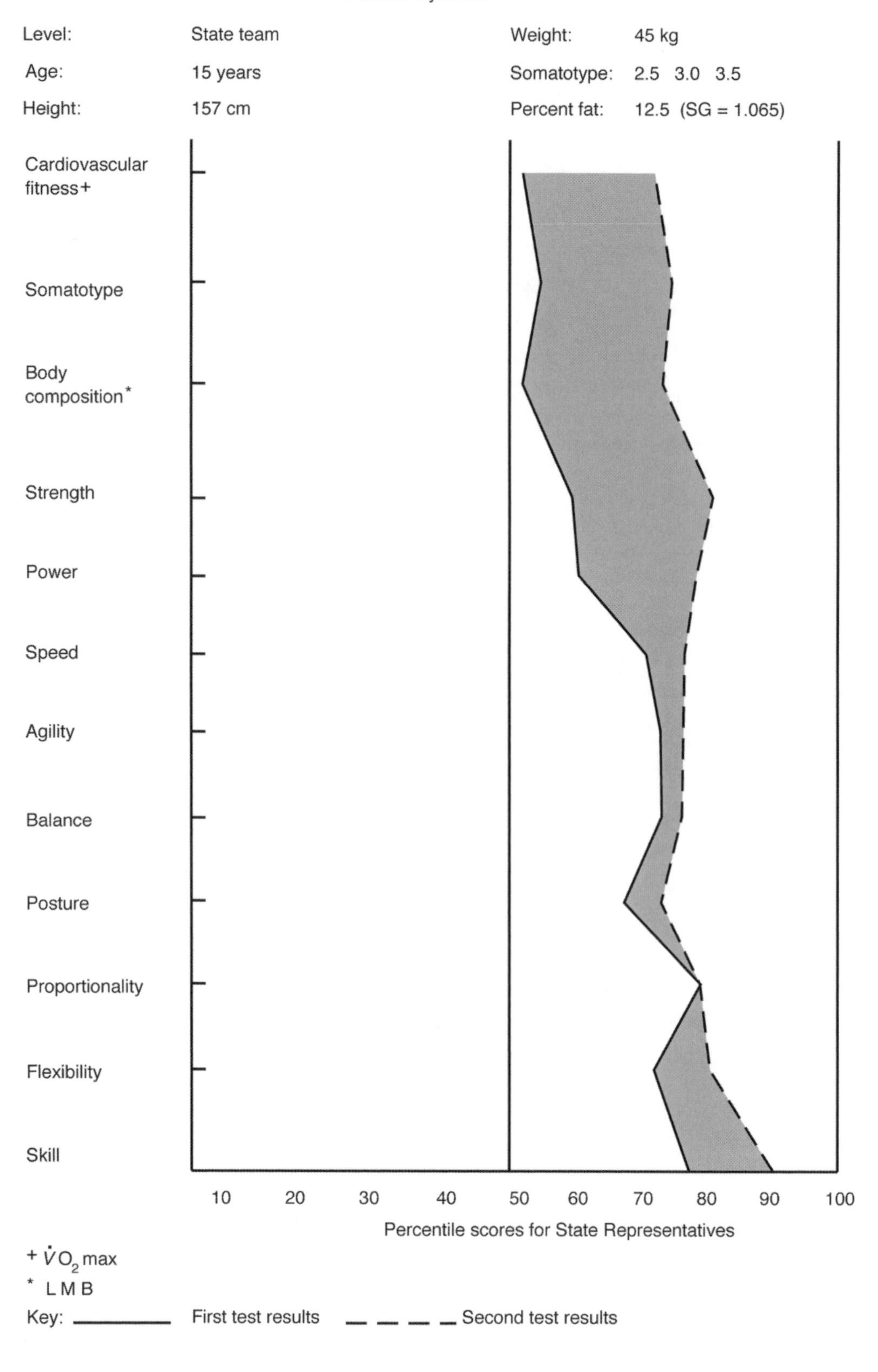

Figure 2.3 The profiles of a female state-level gymnast recorded before and after a 20-month intervention program.

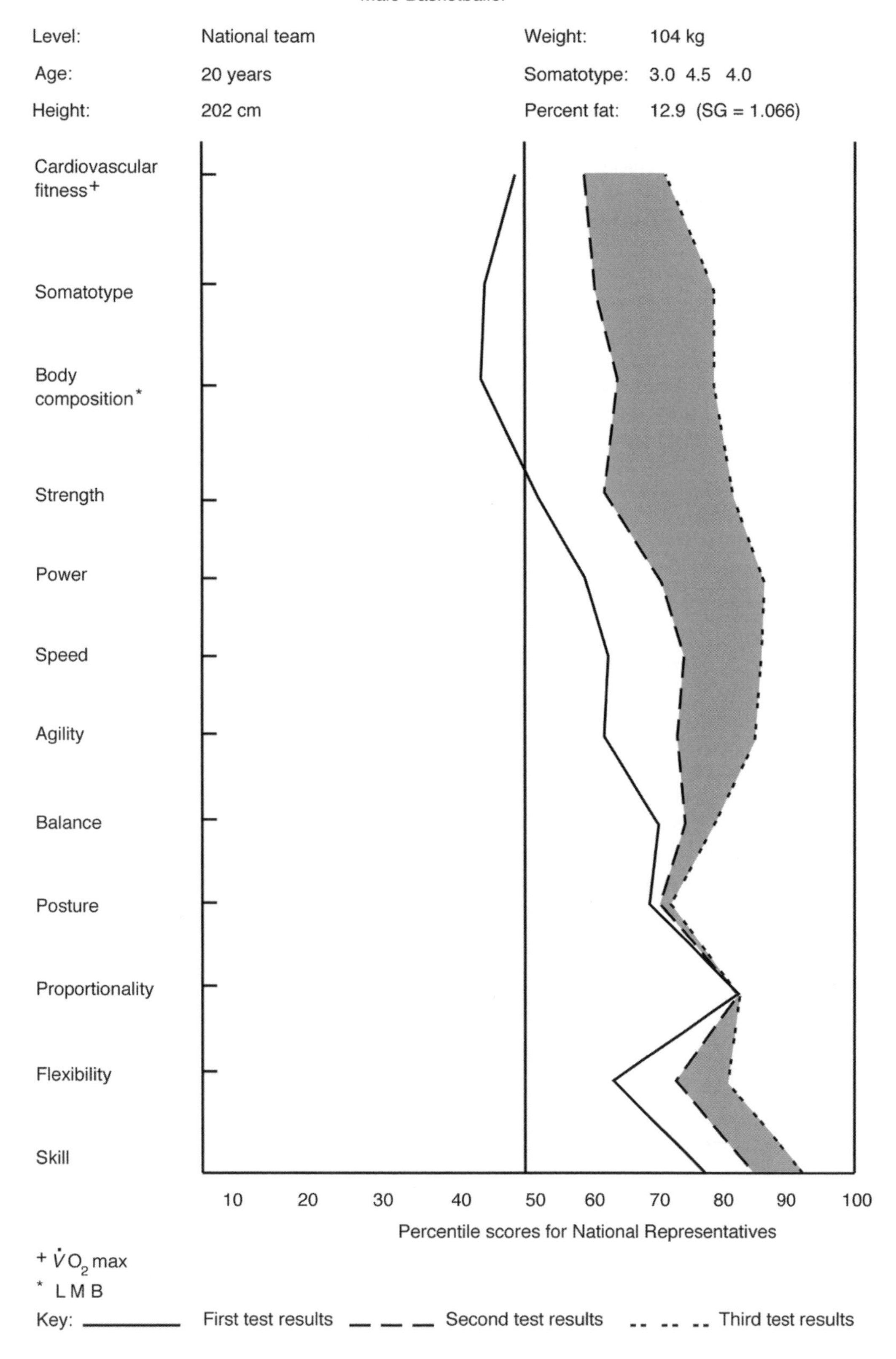

Figure 2.4 The profiles of a male national-level basketball player recorded three times during an 18-month intervention program.

12.9% to 11.0%, and then to 9.7% at the third testing session. The other positive changes can be seen on the right-hand side of the graph.

Program Development

After the profile has been evaluated, the decision should be made as to whether the individual needs an intervention program. Whether one is needed or not, the coach and athlete should confer at this stage, as each team member should know his or her status in comparison with other athletes of a similar level, or the ranking on each variable within the team or squad. In some cases athletes can even take part in the development of their own program, especially if they are mature individuals.

It is important at this point that a formal intervention program be devised that is in a printed form. Targets or goals should be realistic, and the competitor should be clear about what these are and how they can be achieved. The coach should also set the time of the next test session, whether it be three or six months hence or at the beginning of the next season. This gives the athlete a target date to aim for rather than a nebulous period over which the program should be carried out.

Summary

Until recently, profiling has been carried out only randomly in various institutes and centres of excellence throughout the world. Now that more international sport science literature is available with which to make comparisons between athletes, and more test protocols are being internationally standardized, profiling is becoming an important feature of the elite athlete's development.

part II

Applied Anatomy

Assessment and Modification of Physical Capacities

three

Absolute Body Size

Timothy S. Olds, PhD; and Grant R. Tomkinson, PhD

This chapter explores the importance of absolute body size for sporting success, drawing on differences between sporting and nonsporting bodies, historical trends in the size of sporting bodies and differences in body size across performance levels. It concludes by describing basic anthropometric measurements that can be used to quantify aspects of body size.

Morphological optimisation is the process by which the ideal body size and shape for a particular sport are selected or modified.

Principle of Morphological Optimisation

Successful athletes are characterized by a mix of physiological, psychological and anthropometric qualities that varies from sport to sport. Because sport is competitive, these characteristics are optimised, so that only the "fittest" survive at the elite level.

What Is Morphological Optimisation?

Where body size and shape (body morphology) are important, this process is called *morphological optimisation* (Norton et al. 1996). Genetics, training, supplements and drugs, changes to the game space or rules and strategies and even fashion can affect body morphology and modify selection pressures. In Australian football, for example, the game has become much faster with more periods of intense work and longer breaks. These changes, driven by more liberal interchange rules, have favoured bigger players and have also increased injury rates (Norton, Craig, and Olds 1999). The tendency in sport in general has been toward more homogeneous physiques within sports (and within positions and events in any given sport) and toward more heterogeneous physiques between sports.

How Do We Know Whether Absolute Body Size Is Important?

Three types of data index the importance of absolute body size. We can refer to comparisons between sporting and nonsporting bodies, chart historical trends in the size of sporting bodies and compare the bodies of elite and nonelite performers.

Are Athletes Different From the Source Population?

The first test is to examine whether the absolute body size of the sport group is different from that of the population from which it is drawn (the "source" population). The average stature of Australian state- and national-level male heavyweight rowers, for example, is 190.5 cm (75 in.), compared to 178.6 cm (70.3 in.) for 18- to 35-year-old Australian males. (The characteristics of the source population will vary from country to country and from time to time. In the United States, the average stature of 18- to 35-year-old males was 174.1 cm [68.5 in.] in 1988, for example.) Only 0.8% of rowers are shorter than the average young male. We would conclude that height is likely to be important for rowing success.

Divergence in average values is not the only way the characteristics of a sport group can differ from those of the source population. In some cases, the average values are very similar, but the variability is different (typically lower in the sport group). The average stature of elite male teams pursuit cyclists (178.9 cm [70.4 in.]) is almost identical to that of the source population (178.6 cm), but the variance in stature in cyclists (25 cm^2) is less than half that of the source population (61 cm^2). We just don't see very tall or very short teams pursuit cyclists. One possible reason is that for efficient drafting, team members should all be about the same size.

Table 3.1 Mean Stature (cm) and Mass (kg) for Male Rugby Union Players at Various Competitive Levels

Level	Stature (cm)	Mass (kg)	Mesomorphy
Five Nations	184.0	99.3	7.3
Super 12/minor nations	184.0	97.8	6.5
State	184.8	86.6	5.4
Club	180.1	82.1	4.8

Data from: T.S. Olds, 2001. "The evolution of physique in male Rugby Union players in the twentieth century," *Journal of Sports Sciences* 19: 253-262.

Are There Body Size Gradients Across Performance Levels?

If body size is important for performance, we would also expect to see better players exhibit more extreme characteristics and lower variability. Table 3.1 shows the stature and mass for rugby players at different competitive levels (n = 1500). There are increases in stature, mass and mesomorphy across levels. Even at the very top level of a sport, there are often significant body size differences between the "best" and the "rest".

Are There Secular Trends in Body Size?

If absolute body size is an important factor in success, we would expect to see secular trends in body size outrunning those in the source population as sports become more professional and more competitive. The increase in mass in world-class male shot-putters is a striking example of secular trends in absolute body size. Figure 3.1 shows data from the 1928, 1960, 1964, 1972 and 1976 Olympics, as well as some measurements taken on elite non-Olympic performers. In these years, mass increased at a rate of 7.2 kg (15.9 lb) per decade—about seven times the rate of increase in the source population. Other sports, such as American football, show similar, though less marked, trends.

In many sports, body size is increasing at a far greater rate than in the general population.

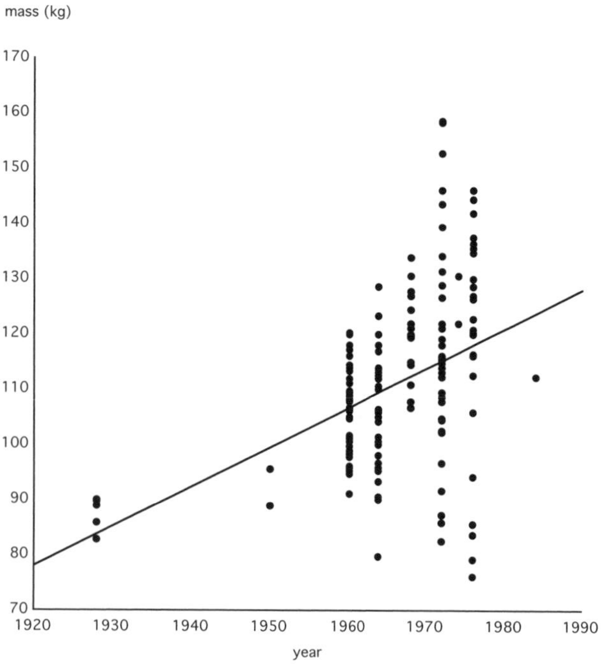

Figure 3.1 Evolution of the mass of Olympic shot-putters.

How Do We Quantify the Importance of Body Size?

Each of these markers of the importance of body size compares the characteristics of one population (the elite sport group) to another: the source population, subelite groups or elite groups of the past. It would be helpful to be able to quantify these differences. Techniques such as discriminant function analysis and logistic regression can be used for this purpose, but a more intuitive method is to use the Overlap Zone (OZ) and Bivariate Overlap Zone (BOZ). Both methods work by calculating the probability that a randomly chosen individual from the source population could be a member of the sport group. Interested readers are referred to Norton and Olds 2000 for a more detailed statistical analysis. An example is shown in figure 3.2, which illustrates the distributions of body mass indexes (BMIs) in Australian Football League (AFL) players, heavyweight boxers and the source population. Note that while the mean values are very similar, the variance is much greater for the source population than for the AFL players. The OZ value is 28.6%, indicating moderate overlap. Compare this to the situation for heavyweight boxers, where both the mean and the variance of the sport group differ greatly from those of the source population. The OZ value is 5.4%.

When comparing bivariate distributions (for example, stature and mass conjointly), a similar procedure is used. An individual is randomly chosen from the source population. The probability that this individual might be a member of the sport group is calculated statistically. An example is outlined in figure 3.3, which shows the 90% density ellipses (i.e., the ellipses containing 90% of all subjects) for the source population and for elite marathoners. Marathoners are shorter and lighter than the source population. The BOZ is 17.8%.

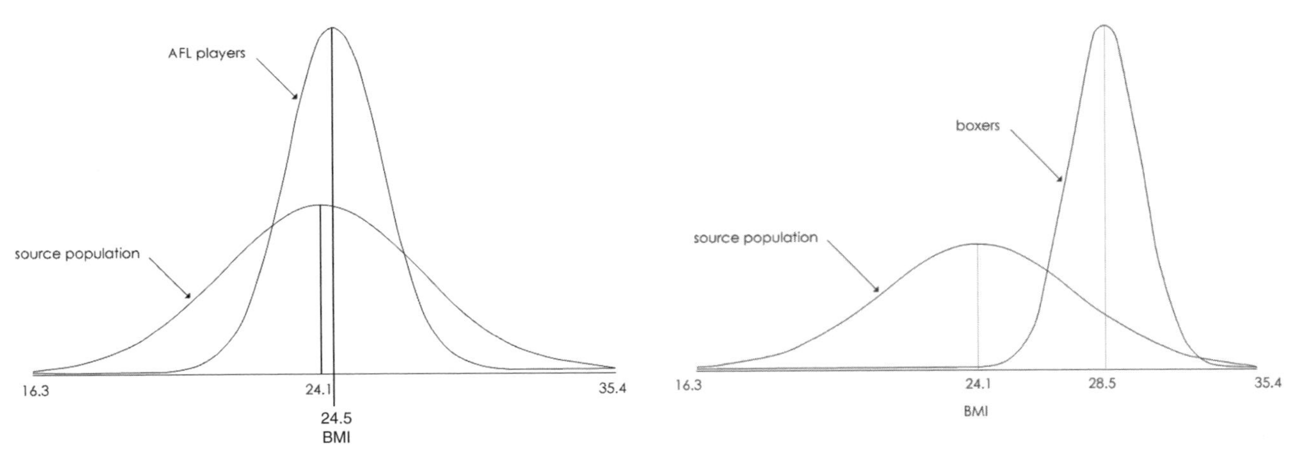

Figure 3.2 Distribution of body mass index (BMI) values in Australian football (AFL) players (left panel) and heavyweight boxers (right panel), compared to the source population of young Australian males. BMI values are plotted on a log scale.

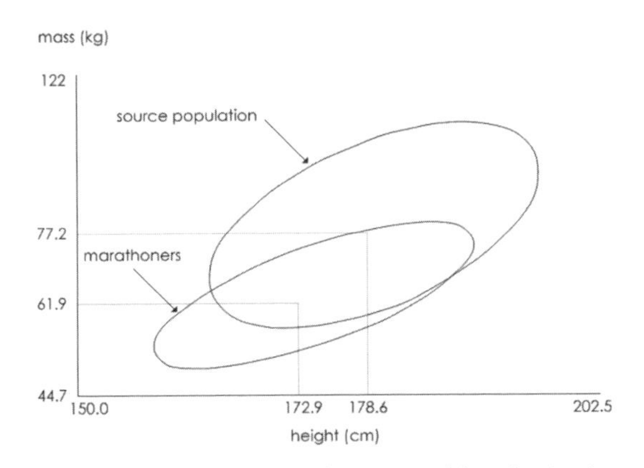

Figure 3.3 Bivariate distributions of height (cm) and mass (kg) for elite marathoners and the source population. The ellipses are 90% density ellipses for both marathoners and the source population. Mass values are plotted on a log scale.

Normative Data on Absolute Body Size in Sportspeople

Table 3.2 shows data on the stature and mass of 12,536 elite adult male athletes from 21 sports. The athletes were all measured after 1980. The OZ values have been calculated for stature and mass separately, and the BOZ for stature and mass conjointly. Mass was log-transformed to normalize the distribution. The antilogs of the mean values are shown for clarity.

Individual data on elite female athletes are much more scarce. Table 3.3 shows data on the stature and mass of 253 adult female athletes from 13 sports, measured since 1980.

Physics of Absolute Body Size

Substantial datasets, coupled with statistics like the OZ and the BOZ, allow us to decide whether absolute body size is important for performance in a given sport. However, they do not explain *why* size is important. This section looks at a number of possible factors: inertia, fluid resistance, angular acceleration and body surface area:mass ratio.

Inertia

Inertia is the tendency of a body to maintain its state of motion, a characteristic that is important in body contact field invasion games such as American football. Inertia is proportional to the mass of the body, so clearly the larger the body size, the greater its resistance to being moved by an opponent. However, a large body mass requires a great deal of power to accelerate rapidly to speed. A 100 kg (220 lb) quarterback will require a power output of at least 1250 W for 2 s to accelerate to 7 m/s. The opposing requirements of rapid acceleration and high inertia may explain why quarterbacks, although very big, are not as big as linemen.

Fluid Resistance

The resistance of a body moving through fluid is proportional to the body's projected frontal area, which in theory is proportional to body surface area (Olds and Olive 1999). Figure 3.4 shows the relationship between projected frontal area (measured by planimetry) and body surface area (calculated using the Dubois equation). Regressions are drawn for three different riding positions: on the hoods, on the drops and with aero bars. The correlations range between 0.78 and 1.00, and the slopes of the regression lines (0.21-0.32) are not significantly different. Larger riders are therefore relatively disadvantaged given equal power output and similar body shape and posture. Similar considerations apply to sports such as downhill skiing and speedskating. Different size relationships apply for uphill riding, where the main resistance is mass (successful Tour de France hill climbers are typically very light), and downhill riding, where greater mass is an advantage.

Table 3.2 Data on 12,536 Male Athletes From 21 Sports

		STATURE (CM)			LN(MASS) (KG)				
Sport	**n**	**Mean**	**SD**	**OZ**	**Mean**	**SD**	**Antilog**	**OZ**	**BOZ**
Source population		**178.6**	**7.7**		**4.339**	**0.148**	**76.7**		
AFL	315	184.9	7.9	39.8	4.417	0.106	82.8	36.2	25.9
Boxing (HW)	31	189.9	3.9	11.0	4.631	0.056	102.6	3.2	1.0
Cycling—off-road	18	178.6	4.6	34.5	4.274	0.055	71.8	21.8	17.2
Cycling—pursuit	24	178.9	5.0	36.7	4.292	0.100	73.1	36.4	26.8
Cycling—road	247	178.1	5.6	39.7	4.227	0.131	68.5	38.8	27.4
Diving	43	169.9	8.4	36.5	4.206	0.150	67.1	40.4	25.7
Hockey	31	179.2	4.6	34.9	4.305	0.054	74.1	21.4	15.2
Kayak	45	181.8	6.8	42.1	4.337	0.124	76.5	45.1	34.0
Marathon	79	172.9	7.6	43.3	4.125	0.108	61.9	19.0	18.1
NBA	1480	200.3	9.2	7.0	4.575	0.130	97.0	18.7	5.4
NFL	7905	187.2	6.7	29.3	4.638	0.173	103.3	16.5	11.3
Rowing (HW)	118	190.5	4.9	13.7	4.471	0.081	87.4	22.1	8.7
RU (backs)	484	179.6	6.1	41.5	4.433	0.111	84.2	34.1	28.1
RU (forwards)	636	187.6	7.6	30.7	4.641	0.103	103.6	7.5	9.5
Running (sprint)	33	179.6	4.4	32.3	4.332	0.094	76.1	36.2	22.8
Running (distance)	43	174.5	6.3	40.5	4.147	0.150	63.2	30.4	23.8
Soccer	283	174.3	6.9	43.7	4.203	0.127	66.9	32.8	28.1
Swimming	255	182.8	6.8	42.5	4.343	0.097	76.9	37.3	28.8
Taekwondo	146	175.3	6.9	43.6	4.240	0.135	69.4	41.9	32.6
Volleyball	130	186.5	9.0	40.2	4.344	0.160	77.0	53.2	29.8
Water polo	190	186.3	6.0	28.1	4.454	0.100	86.0	28.9	17.6

AFL = Australian Football League; BOZ = Bivariate Overlap Zone; HW = heavyweight; NBA = National Basketball Association; NFL = National Football League; OZ = Overlap Zone; RU = rugby union; SD = standard deviation.

The mean and SD for stature and the natural logarithm of mass [i.e., ln(mass)] are shown, as well as the antilogged value of the mean ln(mass). OZ values are shown for stature and mass separately, and BOZ values for stature and mass conjointly.

Angular Acceleration

Some sports, such as diving and gymnastics, require rapid twisting and turning of the body (i.e., high angular acceleration). Angular acceleration is described by the equation

$$\alpha = T / I$$

where α is angular acceleration in rad/s, T is torque in Nm and I is the moment of inertia in kg/ms. Moment of inertia is a measure of the body's resistance to angular motion, which increases with stature when the whole body is rotating lengthwise. Given this, it is not surprising to find that divers (mean stature = 169.9 cm [66.9 in.] for males and 161.2 cm [63.9 in.] for females), gymnasts (169.4 cm [66.7 in.] and 157.0 cm [61.8 in.]) and figure skaters (170.7 cm [67.2 in.] and 157.7 cm [62.1 in.]) are relatively short.

Body Surface Area: Mass Ratio

Smaller body size is usually associated with a larger body surface area (BSA) to mass ratio. The BSA:mass ratio is an index of thermoregulatory ability. Heat production is theoretically proportional to mass (more specifically, muscle mass), and heat loss is proportional to BSA, although subcutaneous fat also plays an insulating role. We would therefore expect to see a high BSA:mass ratio in sports in which heat loss is important, such as distance running. This is indeed the case. The lowest ratios are in discus for women (234.8 cm^2/kg) and in sumo for men (187.6 cm^2/kg). The highest ratios are found in women in the marathon (314.9 cm^2/kg) and in males in distance running (271.2 cm^2/kg). This may result in a selection pressure for smaller body size in sports in which heat loss is important.

Many other size-related factors are likely to affect sport performance. Lever length is likely to be important,

Table 3.3 Data on 253 Female Athletes From 13 Sports

		STATURE (CM)			LN(MASS) (KG)				
Sport	**n**	**Mean**	**SD**	**OZ**	**Mean**	**SD**	**Antilog**	**OZ**	**BOZ**
Source population		**165.6**	**6.8**		**4.081**	**0.139**	**59.2**		
Badminton	14	166.4	2.9	24.7	4.075	0.063	58.9	28.2	14.3
Basketball	32	176.8	7.6	23.8	4.234	0.114	69.0	28.2	18.3
Cricket	12	165.8	6.7	50.8	4.142	0.125	62.9	42.7	43.5
Hockey	26	165.5	6.4	48.1	4.107	0.104	60.8	41.0	30.9
Lacrosse	17	165.2	7.4	52.2	4.097	0.127	60.2	45.6	51.0
Marathon	16	164.3	6.7	48.0	3.924	0.095	50.6	24.0	15.5
Netball	21	173.3	3.8	19.4	4.193	0.074	66.2	22.6	11.9
Running (distance)	23	162.4	4.8	35.2	3.918	0.085	50.3	20.9	17.0
Soccer	10	165.2	5.9	45.9	4.096	0.129	60.1	46.4	44.2
Softball	22	167.4	4.8	38.6	4.096	0.103	60.1	40.8	32.5
Tennis	15	173.9	8.4	38.5	4.136	0.126	62.6	44.3	13.4
Throwing	11	173.7	3.8	17.5	4.287	0.167	72.7	29.6	19.4
Volleyball	34	178.2	6.2	15.1	4.222	0.111	68.2	29.4	14.3

BOZ = Bivariate Overlap Zone; OZ = Overlap Zone; SD = standard deviation.

The mean and SD for stature and the natural logarithm of mass [i.e., ln(mass)] are shown, as well as the antilogged value of the mean ln(mass). OZ values are shown for stature and mass separately, and BOZ values for stature and mass conjointly.

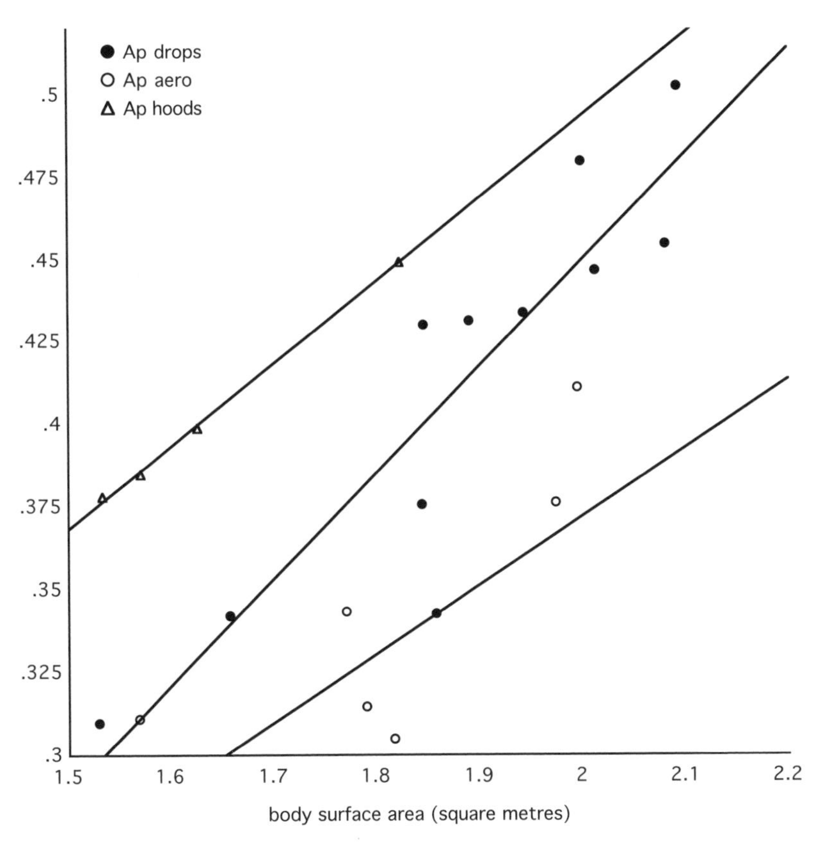

Figure 3.4 Relationship between projected frontal area (m^2) and body surface area (m^2) in 24 cyclists. Separate regression lines are shown for riders riding on the brake handle hoods (triangles), in the drop position (filled circles) and with aero bars (open circles).

because longer levers provide a mechanical advantage for speed but a disadvantage for strength (see chapter 6). Body centre of gravity, which is both relatively and absolutely higher in taller people, will affect sports that require stability, such as wrestling. Height is clearly an advantage in sports in which projectiles are thrown for maximum distance, such as javelin and discus, because projectiles can travel greater horizontal distances before striking the ground if released from a greater height. It is also an advantage when a ball is thrown or struck downward over a barrier such as a net in sports like tennis and volleyball. The reason is that the striker has more options for placing the ball in different parts of the opponent's court.

Secular Trends in Absolute Body Size

Most parents are aware that their children seem to be taller than they themselves were as youngsters and seem to mature earlier. These trends mean that the source population is constantly changing. In addition, the source population—the "market" for sporting bodies—has been continuously expanding in numbers for at least a century.

Secular Trends in the Source Population

Over the last 150 years, there have been substantial secular increases in stature and mass in the source population. Across the developed world, secular increases have averaged about 1 cm (0.4 in.) and 1 kg (2.2 lb) per decade since the mid-19th century (Meredith 1976), with even sharper rises in mass and BMI occurring since about 1980. The secular trend in body size has been accompanied by earlier maturation. Each decade the age of menarche in girls has been falling by about three to four months in many developed countries, and the age at which boys' voices break has been falling at the rate of two months per decade (Himes 1979). At the same time, the world population has been increasing dramatically, from about 1.65 billion in 1900 to over 6 billion today, and modern transport and telecommunications have made the extra billions much more accessible.

These three factors—increasing body size, earlier maturation and larger accessible populations—have greatly expanded the potential source population, accelerating the drive toward more extreme sporting physiques. This is in spite of declining fertility rates, which have reduced the relative proportion of adults in the population. There are a larger number of extreme physiques available to the sport body market, with a declining number of elite sportspeople, as local competitions are swallowed up by global competitions. In Australia, for example, there would have been about 1500 men over 190 cm (74.8 in.) tall in 1900. In the year 2000 there were about 150,000 men over 190 cm. The secular trend may also have the opposite effect, reducing the number of potential athletes. The body size of jockeys, for example, is constrained by the lack of change in the size of racehorses and by the weight-for-age scale. The mass of jockeys remained more or less constant during the 20th century. Between 1900 and 1999, the median mass varied only between 49 and 52 kg (108 and 115 lb).

In addition to the secular trend in absolute body size, there are associated trends in shape. Increases in height have been achieved almost entirely through increases in femur length, so that people today have a smaller sitting height to stature ratio than previously. This tends to reduce the pool of potential elite athletes in sports such as weightlifting, where short levers are advantageous.

The mismatch of supply and demand in the body market drives an expansion of the source population into new countries, to new ethnic groups, across genders and to new age groups. Secular trends can shift the geographical centre of gravity of sport power. The search for jockey-sized adults has also increased the proportion of Hispanic and Asian—and of course female—jockeys. The fact that the size of male marathon runners has barely changed over the last 100 years (figure 3.5) means that fewer and fewer Caucasians are the right size to excel as marathoners. In August 2001, 18 of the fastest 20 male

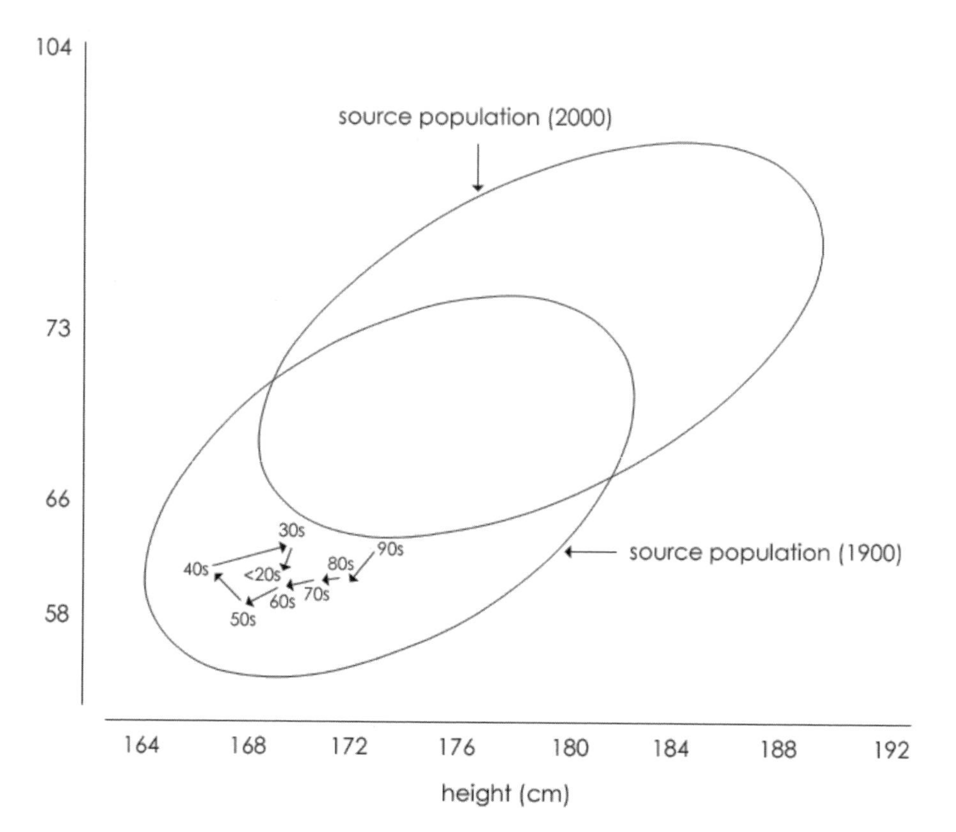

Figure 3.5 Changes in the average stature and mass of male marathon runners in the 20th century. The ellipses are the 67% density ellipses for the source populations (Australian norms) in 1900 and 2000. The arrows represent shifts in the average stature and mass of 668 elite marathoners between 1897 and 1999, through the decades of the 20th century. There has been virtually no change in mass and only a small increase in stature.

marathon times and 13 of the fastest female times were by East Africans and Japanese. Secular trends can equally create age shifts in sport groups. There has been a strong tendency for the height and mass of female gymnasts to decrease over the last 50 years (Norton and Olds 2000). Because it was becoming harder and harder to find gymnasts small enough to compete at the top level, younger and younger gymnasts were recruited into the sport. The average age decreased by about three years over each decade between 1960 and 1990, from about 25 years in 1960 to close to 15 years in 1990.

Three Models of Optimisation

Patterns of change in body size in the sport group must be understood in relation to these changes in the source population. Three patterns of change are possible.

Open-Ended Optimisation

Open-ended optimisation occurs when body size in the sport group becomes more and more extreme in absolute terms (i.e., when athletes get bigger and bigger or, more rarely, smaller and smaller). Figure 3.6 shows an example—increases in the mass of National Football League (NFL) offensive and defensive tackles and quarterbacks between 1920 and 1999. Mass in these positions has increased at a rate of 4.6 kg (10 lb) per decade, about four times as fast as the increase in the source population. For tackles, the BOZ fell from a very low 0.7% in the 1940s to only 0.1% in the 1970s, 1980s and 1990s. Similar patterns occur in the stature of basketballers and the mass of open-class throwers, wrestlers and weightlifters. The logic of open-ended optimisation is obvious, but the demands it places on the "body market" lead to extreme distortions in globalisation of sport, player salaries, poaching and artificial growth. These are discussed in a later section.

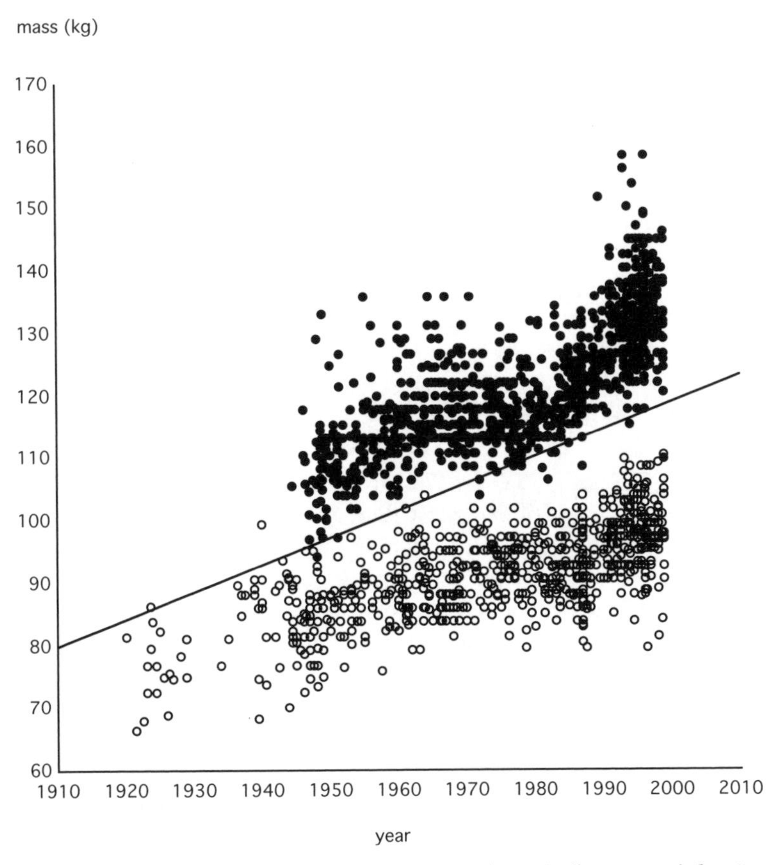

Figure 3.6 Increases in the mass of National Football League defensive and offensive tackles (filled circles; $n = 1045$) and quarterbacks (open circles; $n = 782$) between 1920 and 1999 ($r = 0.62$, $p < 0.0001$).

Relative Optimisation

Relative optimisation occurs when the body size of the sport group may differ from that of the source population but changes over time in step with the source population. The increases in body size we see in the sport group are of a similar magnitude to those in the source population. One example is rugby players. Between 1905 and 1999, the stature of top-level rugby players increased at a rate of 0.9 cm (0.35 in.) per year, which is not different from the rate of increase in the source population (about 1 cm per year). Rugby players are consistently about 0.8 SDs taller than the source population. It is likely that in cases of relative optimisation, competing forces maintain a body size equilibrium. For example, it may be advantageous for rugby players to be tall for line-outs and marks, but also to have a low centre of gravity for scrummaging and breaking out of tackles. Alternatively, very tall players may be more prone to injury.

Absolute Optimisation

Absolute optimisation occurs when the body size of the sport group does not change despite secular changes in the body size of the source population. Figure 3.7 shows an example of absolute optimisation. The stature of international-level male divers did not change between 1950 and 1991, with a mean of 170.8 cm (67.2 in.). Over this period, the stature of the source population increased by about 4 cm (1.6 in.) while that of divers increased by only 0.4 cm (0.16 in.). Absolute optimisation is also found in jockeys and marathon runners (Norton et al. 1996).

Consequences of the Search for Size

The emergence of a globalised market for sporting bodies has had consequences both in the domain of sports and beyond. It has led to geographical shifts in the flow of sporting "labour" and capital, changes in the

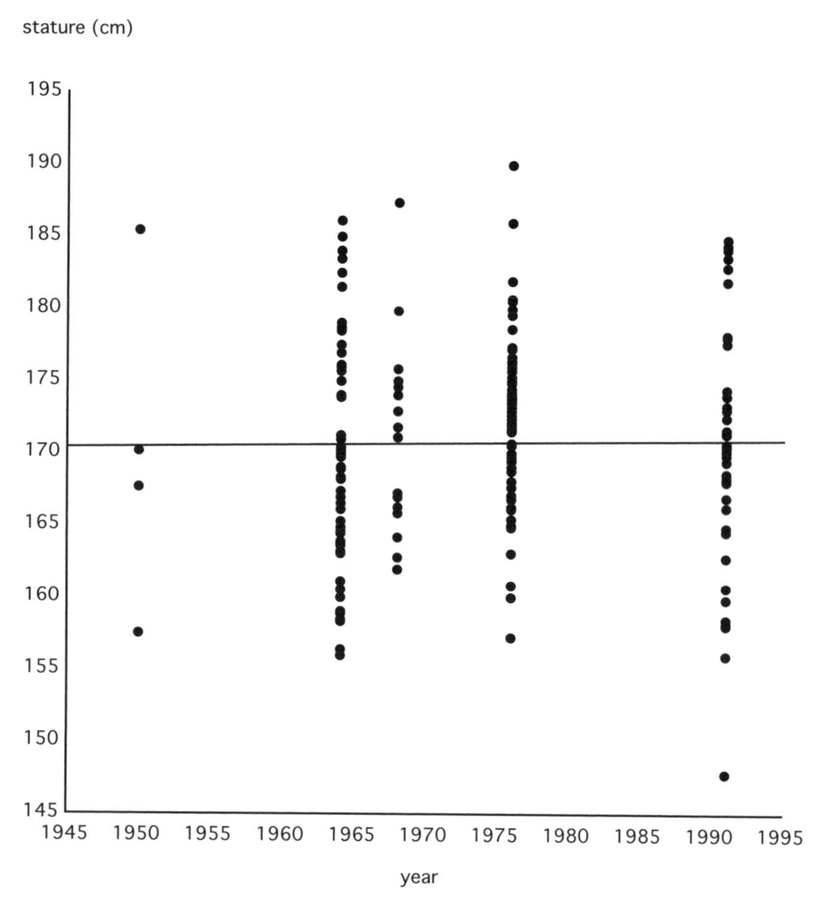

Figure 3.7 Stature of international-level male divers between 1950 and 1991 (n = 180).

way sportspeople work and live and changes in recruitment and training practices.

Specialization and the "Exploding Universe" of Sport Bodies

The tendency in general has been toward more similar physiques within sports but toward very different physiques between sports—an "exploding universe" of sporting physiques. One consequence has been that very few athletes now excel in more than sport. In the history of the Olympics, 39 athletes have won medals in more than one sport, but only four have achieved that feat since 1936. We are unlikely to again see the days when a single athlete can medal in such improbable combinations as shooting and weightlifting (as did the Dane Jensen Viggo in 1896), ski jumping and yachting (Norwegian Thams Tullin in 1924 and 1936) or water polo and fencing (Belgian Boin Victor in 1908, 1912 and 1920).

> The body size and shape of sportspeople are becoming more differentiated: more similar within sports, but more different between sports.

Globalisation of the Body Market and the "Body Drain" in Sport

One consequence of the search for size has been the globalisation of body markets. When local source populations have been exhausted, the net must be cast farther afield. The result is that even "national" teams are often heavily loaded with foreign-born players. In the 1999 Rugby World Cup, for example, there were enough New Zealanders playing for various countries to field two extra teams. Over 40% of the Scottish team were born outside of Scotland. An analysis of the players in the 2002 World Soccer Cup showed that almost 50% of all World Cup players were playing in five wealthy European countries—England, Italy, Germany, Spain and France. Not one of the Cameroon team actually played in Cameroon in 2002. This describes a kind of "body drain" from poor countries to wealthy countries.

The globalisation of sport is even more pronounced in "free" sport markets, such as the National Basketball Association (NBA). In 1999, 4.4% of NBA players were of foreign origin. The foreign-born players were taller (211.1 vs. 200.3 cm [83.1 vs. 78.9 in.]) and heavier (110.6 vs. 99.0 kg [244 vs. 218 lb]) than the American-born players. By 2002, the percentage had risen to 9%. Foreign players represented 5 of the 10 tallest and heaviest NBA players. The 2003 Women's National Basketball Association (WNBA) had 96 American-born and 24 foreign-born players. The international players were significantly taller (188.4 vs. 181.6 cm [74.2 vs. 71.5 in.]) and heavier (78.3 vs. 73.5 kg [172.6 vs. 173 lb]) than the American-born players. Nine of the tallest 12 players were foreign born.

As well as reaching into other countries, sports are also reaching into ethnic groups previously excluded from sport. There were no black sprinters in any men's 100 m Olympic final until 1928; since 1984, almost all finalists have been black. We find similar overrepresentation of Aboriginals in Australian football (who constituted 1% of players in 1950 and 11% in 2005), blacks in the NBA and NFL and Hispanics in American baseball. The search for size here has had a positive influence on breaking down ethnic divisions.

Sport Salaries

As required absolute body size becomes more extreme, it has become necessary to pay much larger salaries to attract individuals with extreme physiques from other careers and from other sports. Since 1975, player salaries in the four major U.S. sports have increased from about five times the male median salary to 15 to 40 times the median male salary. Similar trends, though somewhat less extreme, have occurred in rugby union and Australian football.

Size also translates into sport earnings in "winner-take-all" markets. In the NBA and NFL, an extra 1 cm of stature may translate into an extra $35,000 to $70,000 in career earnings due to increased career longevity; 1 kg would equate to $15,000 to $30,000, and 1 BMI unit to $90,000 to $100,000. It is not surprising, then, that there is substantial investment in anthropometric talent identification and in artificial growth enhancement.

Poaching

In the future, we are likely to see more aggressive intersport poaching, like the Australian rugby union's campaign to recruit players from Rugby League. The Australian Institute of Sport recently launched a program to win the gold medal in women's skeleton, a kind of bobsled event being reintroduced into the Winter Olympics after a long absence. The Australian Institute of Sport deliberately targeted beach sprinters from surf lifesaving as having the right physical and physiological characteristics. We are likely to see an acceleration of the "body drain" from poor sports to rich sports.

Kinanthropometric Assessment

The quantification of human morphology is a fundamental requirement for the assessment and monitoring of high-performance athletes. One serious concern that has plagued scientists in this field for many years has been the lack of a universally adopted protocol for kinanthropometric assessment. Many different landmarking protocols and measurement procedures are currently in use around the world. The techniques adopted in this chapter are consistent with those prescribed by the International Society for the Advancement of Kinanthropometry (ISAK 2001) and supported by the Laboratory Standards Assistance Scheme of the Australian Sports Commission (adopted for use in Australia in 1993).

Anatomical Conventions

A description of body landmarks and other sites for measurement requires a standard reference system. This avoids confusion, especially when the body is oriented in a posture other than standing. The relative location of parts of the body may be accurately defined using the following terminology for a subject standing in the anatomical position, as shown in figure 3.8.

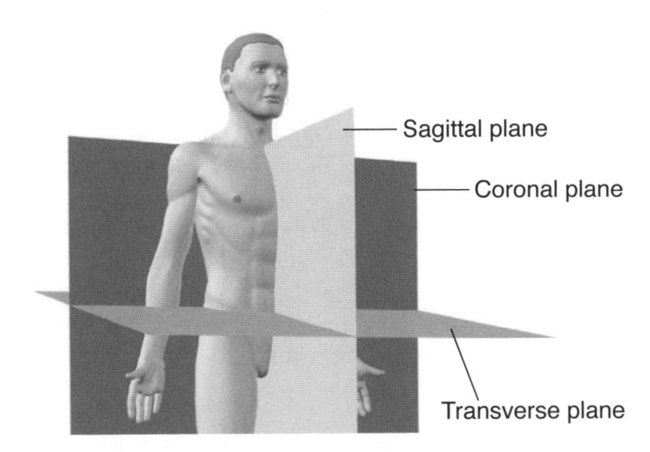

Figure 3.8 The planes and aspects of the human body in the anatomical position.

- **Sagittal plane—**a vertical plane that divides the body into left and right portions. The midsagittal plane is that which runs centrally through the body to create equal left and right parts. The term "medial" refers to relative locations that are closer to this midsagittal plane, as opposed to "lateral", which means away from this plane.
- **Coronal plane—**a vertical plane that is perpendicular to the sagittal plane and that divides the body into front and rear portions. The term "anterior" refers to relative locations that are closer to the front of the body, while "posterior" means toward the rear.
- **Transverse plane—**a horizontal plane that is perpendicular to both sagittal and coronal planes and that divides the body into upper and lower portions. The term "superior" refers to relative locations toward the top of the head, while "inferior" means toward the soles of the feet.

With respect to limbs, which may be oriented in many positions in relation to the trunk, it is often useful to employ the terms "proximal" and "distal". "Proximal" refers to relative locations toward the trunk, while "distal" means toward the extremity.

Landmarks

The International Society for the Advancement of Kinanthropometry recommends that where possible, measurements be taken on the right-hand side of the body while the subject is in a relaxed standing posture. All dimensions are defined relative to landmarks. Landmarks are usually identified by palpation and marked with a fine- to medium-point nonpermanent pen or pencil. Sites should be rechecked after their initial marking, since movement of the skin over the skeleton may alter the relative position of the mark when pressure is released.

The following does not constitute an exhaustive list of landmarks, but identifies those required for the measurements generally used in sport science and medicine. Each landmark is shown with its corresponding number and definition in the ISAK system. The landmarks and methods for their location are described next and shown in figure 3.9. For a more complete description of landmark locations and methods, the reader is referred to ISAK 2001.

1. Acromiale—the point at the most superior and lateral border of the acromion process when the subject stands erect with arms relaxed and hanging vertically
2. Radiale—the point at the most proximal and lateral border of the head of the radius
3. Midacromiale-radiale—the point equidistant from the acromiale and radiale landmarks
4. Triceps skinfold site—the posterior part of the triceps, in the midline, at the level of the midacromiale-radiale landmark
5. Biceps skinfold site—the most anterior part of the biceps at the level of the midacromiale-radiale landmark
6. Stylion—the most distal point on the lateral margin of the styloid process of the radius
7. Midstylion—the midpoint, on the palmar surface of the wrist, of the horizontal line at the level of the stylion
8. Subscapulare—the undermost tip of the inferior angle of the scapula
9. Subscapular skinfold site—the site 2 cm (0.8 in.) along a line running laterally and obliquely downward from the subscapulare at a 45° angle
10. Mesosternale—the midpoint of the corpus sterni at the level of the centre of the articulation of the fourth rib with the sternum
11. Iliocristale—the point on the iliac crest where a line drawn from the midaxilla (middle of the armpit), on the longitudinal axis of the body, meets the ilium
12. Iliac crest skinfold site—the site at the centre of the skinfold raised immediately above the iliocristale
13. Iliospinale—the most inferior or undermost part of the tip of the anterior superior iliac spine (ASIS)
14. Supraspinale skinfold site—the site at the intersection of two lines: (1) the line from the iliospinale to the anterior axillary border and (2) the horizontal line at the level of the iliocristale
15. Abdominal skinfold site—the site 5 cm (2 in.) to the right of the midpoint of the navel
16. Trochanterion—the most superior aspect of the greater trochanter of the femur
17. Tibiale laterale—the most superior aspect on the lateral border of the head of the tibia
18. Midtrochanterion–tibiale laterale—the point equidistant from the trochanterion and the tibiale laterale
19. Medial calf skinfold site—the site on the most medial aspect of the calf at the level of the maximal girth
20. Front thigh skinfold site—the site equidistant from the inguinal fold and the anterior patella, on the midline of the thigh
21. Tibiale mediale—the most proximal aspect on the medial border of the head of the tibia
22. Sphyrion tibiale—the most distal tip of the medial malleolus of the tibia
23. Akropodion—the most anterior point of the foot, which may be the first or second digit
24. Anterior patella—the most anterior and superior margin of the anterior surface of the patella when the subject is seated and the knee bent at a right angle
25. Dactylion—the tip of the middle (third) finger
26. Glabella—midpoint between the brow ridges
27. Gluteal fold—the crease at the junction of the gluteal region and the posterior thigh
28. Inguinal fold—the crease at the angle of the trunk and the anterior thigh
29. Orbitale—the lower bony margin of the eye socket
30. Pternion—the most posterior point of the calcaneus
31. Tragion—the notch superior to the tragus of the ear
32. Vertex—the most superior point on the skull when the head is held in the Frankfort plane (i.e., when the orbitale and tragion are horizontally aligned).

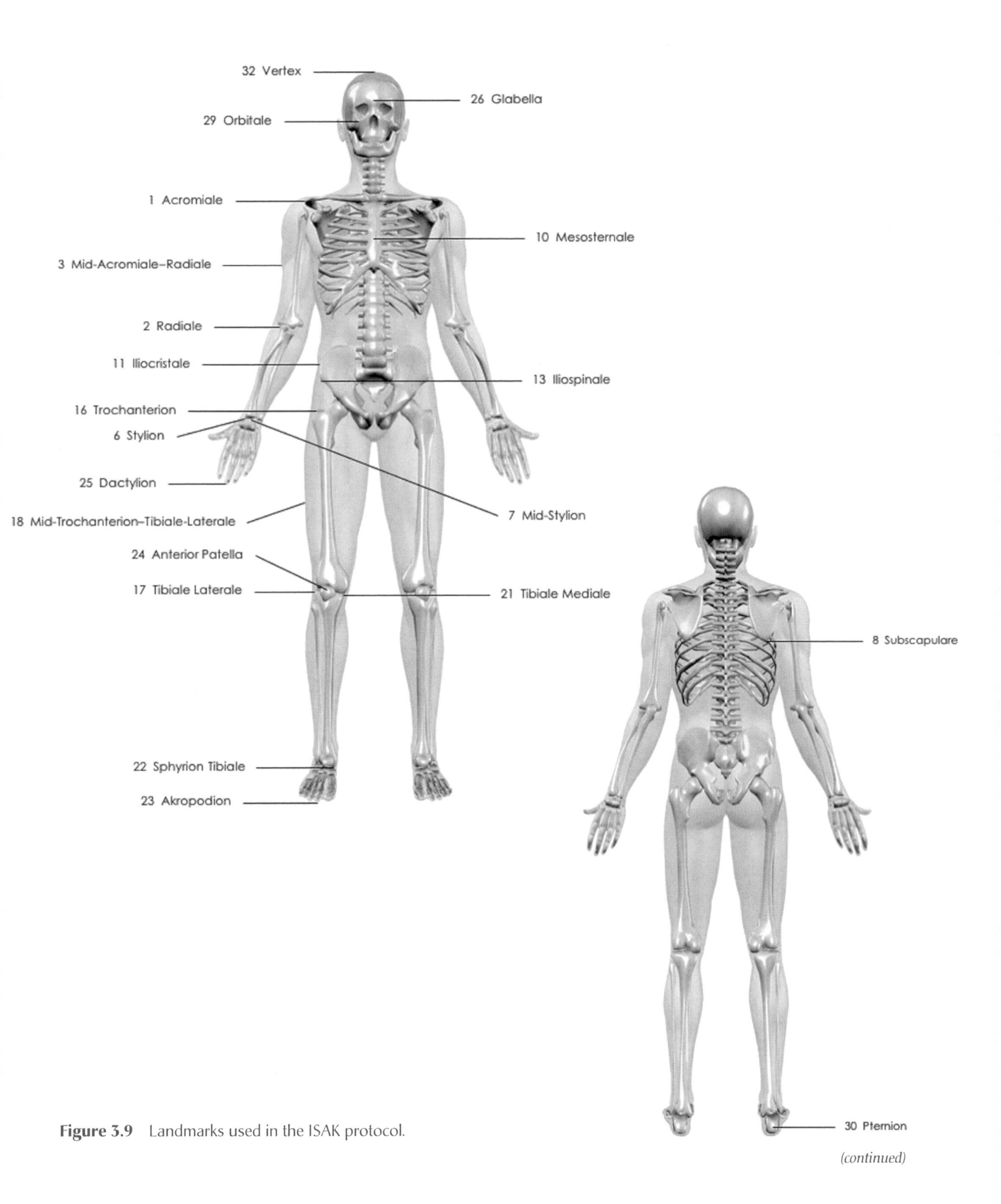

Figure 3.9 Landmarks used in the ISAK protocol.

(continued)

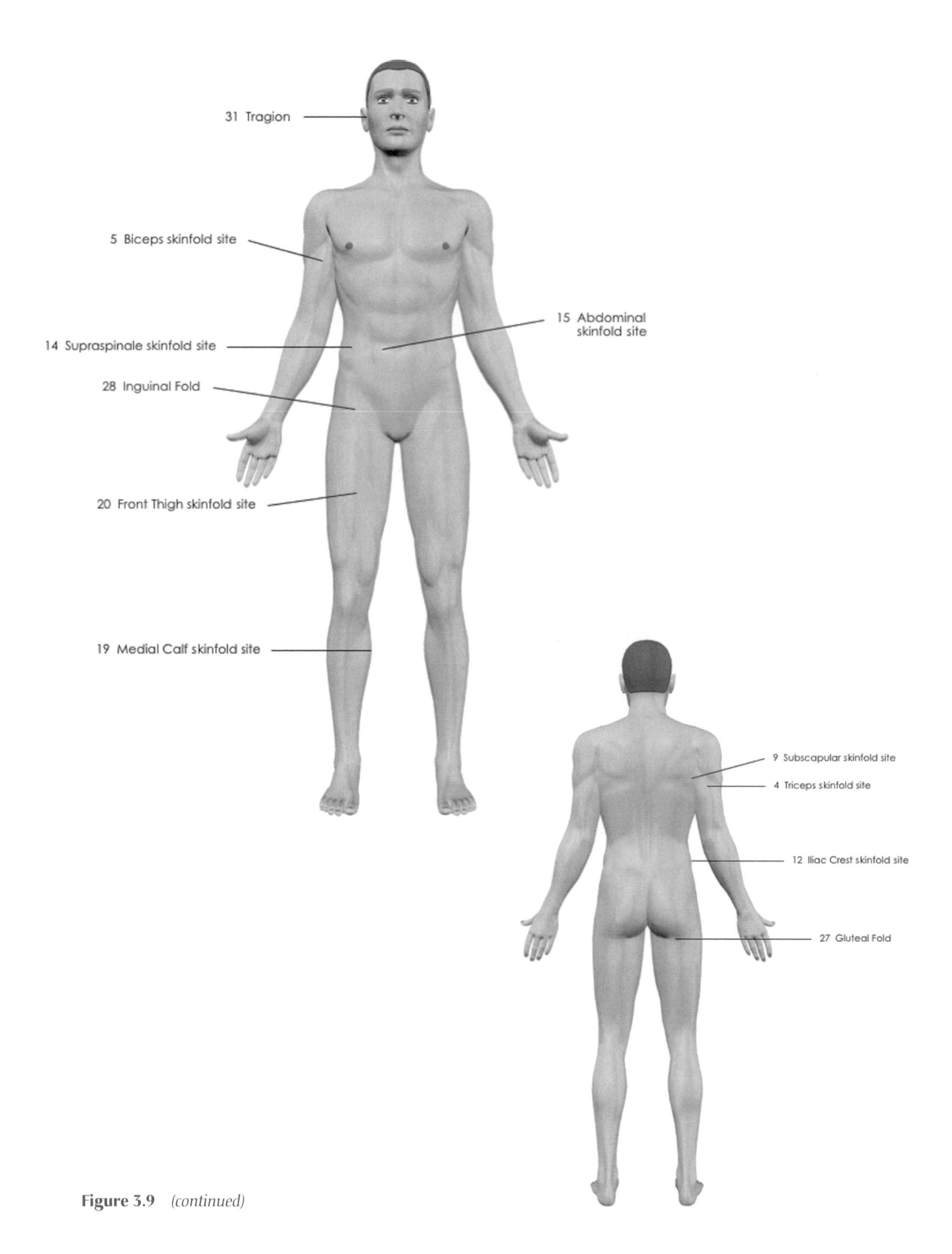

Figure 3.9 *(continued)*

Equipment

The following equipment would be regarded as the minimum requirement for a complete anthropometric assessment of athletes. Full descriptions of these instruments and their relative merits may be found in ISAK 2001. Different makes are available; those in parentheses are recommended by ISAK.

- Length assessment—stadiometer, anthropometer (Siber-Hegner GPM), segmometer (Rosscraft)
- Body mass assessment—a beam-type balance scale or high-quality load cell scale with a range of 0 to 150 kg and calibrated to 0.1 kg
- Breadth assessment—anthropometer (Siber-Hegner GPM), small sliding caliper (Rosscraft or Mitutoyo), wide-spreading caliper (Siber-Hegner GPM or Rosscraft)
- Girth assessment—anthropometric tape (Lufkin W606PM)
- Skinfolds—skinfold calipers (Harpenden or Slim Guide)
- Other equipment—anthropometric box (a 50 × 40 × 30 cm [20 × 16 × 12 in.] box is extremely useful for anthropometric measurement)

Measurements

For all of the following measurements, subjects should wear minimal clothing and no footwear. Where time permits, measurements should be made twice. If two skinfold measurements differ by more than 5%, or other measurements by more than 1%, a third measurement should be made. Ideally, each group of measures (e.g., all skinfolds) should be attempted once, then the whole group measured again. In this way the measurer is less likely to be influenced by previous scores. The mean of two assessments, or the median of three measurements, should be retained as the measured value. The following list of dimensions uses the ISAK numbering system.

Basic Measures

"Basic" measurements refer to three simple measurements that should be part of any standard anthropometric battery: mass, stature and sitting height.

1. **Body mass.** Check that the scale is at the zero mark, then direct the subject to stand in the centre of the platform. The mass is recorded to the nearest 0.01 kg.

2. **Stature.** Direct the subject to stand erect with heels, buttocks and shoulders pressed against the upright portion of the stadiometer. The heels should be touching, and the arms should hang by the sides in a natural manner. Lower the headboard so that it touches the Vertex. Place your thumbs at the level of the orbitale and index fingers at the level of the tragion and align the subject's head in the Frankfort plane. Place your remaining fingers under the mastoid process of the temporal bone. Instruct the subject to "look straight ahead and take a deep breath", and gently but firmly stretch the vertebral column. Read the score from the stadiometer scale while stretching the subject.

3. **Sitting height.** The subject sits on the anthropometric box at the base of the stadiometer, with the thighs projected horizontally (legs may be crossed to achieve this position) with the buttocks and shoulders pressed against the upright portion of the stadiometer. Align the head in the Frankfort plane and follow the instructions for standing height for this measurement. Subtract the height of the box from the resulting score to obtain sitting height.

Skinfolds

Unless otherwise indicated, the subject assumes a relaxed standing position. When taking skinfolds, hold the caliper in the right hand, and lift and hold the fold with the thumb and index finger of the left hand at the marked site. The skinfold should not include muscle. It is held throughout the measurement. The caliper must be applied at right angles to the fold so that the pressure plate is 1 cm (0.4 in.) from the left thumb and index finger. A reading is taken 2 s after the application of the caliper, and the score recorded to the nearest 0.1 mm (Harpenden caliper) or 0.5 mm (Slim Guide caliper). Skinfold sites are shown in figure 3.10.

4. **Triceps skinfold.** A fold parallel to the long axis of the arm is raised at the triceps skinfold site (figure 3.11).

5. **Subscapular skinfold.** A fold is raised at the subscapular skinfold site. The line of the skinfold is determined by the natural fold of the skin.

6. **Biceps skinfold.** A fold parallel to the long axis of the arm is raised at the biceps skinfold site.

7. **Iliac crest skinfold.** A fold is raised at the iliac crest skinfold site. The line of the skinfold generally runs slightly downward posterior-anterior. The subject's right arm should be abducted or placed across the trunk.

8. **Supraspinale skinfold.** A fold is raised at the supraspinale skinfold site. The fold runs medially downward at an angle of about 45°.

9. **Abdominal skinfold.** A vertical fold is raised at the abdominal skinfold site.

10. **Front thigh skinfold.** The fold is raised at the front thigh skinfold site, and runs parallel to the long axis of the thigh. The subject sits at the front edge of the anthropometry box with the torso erect and arms hanging by the side. Because this

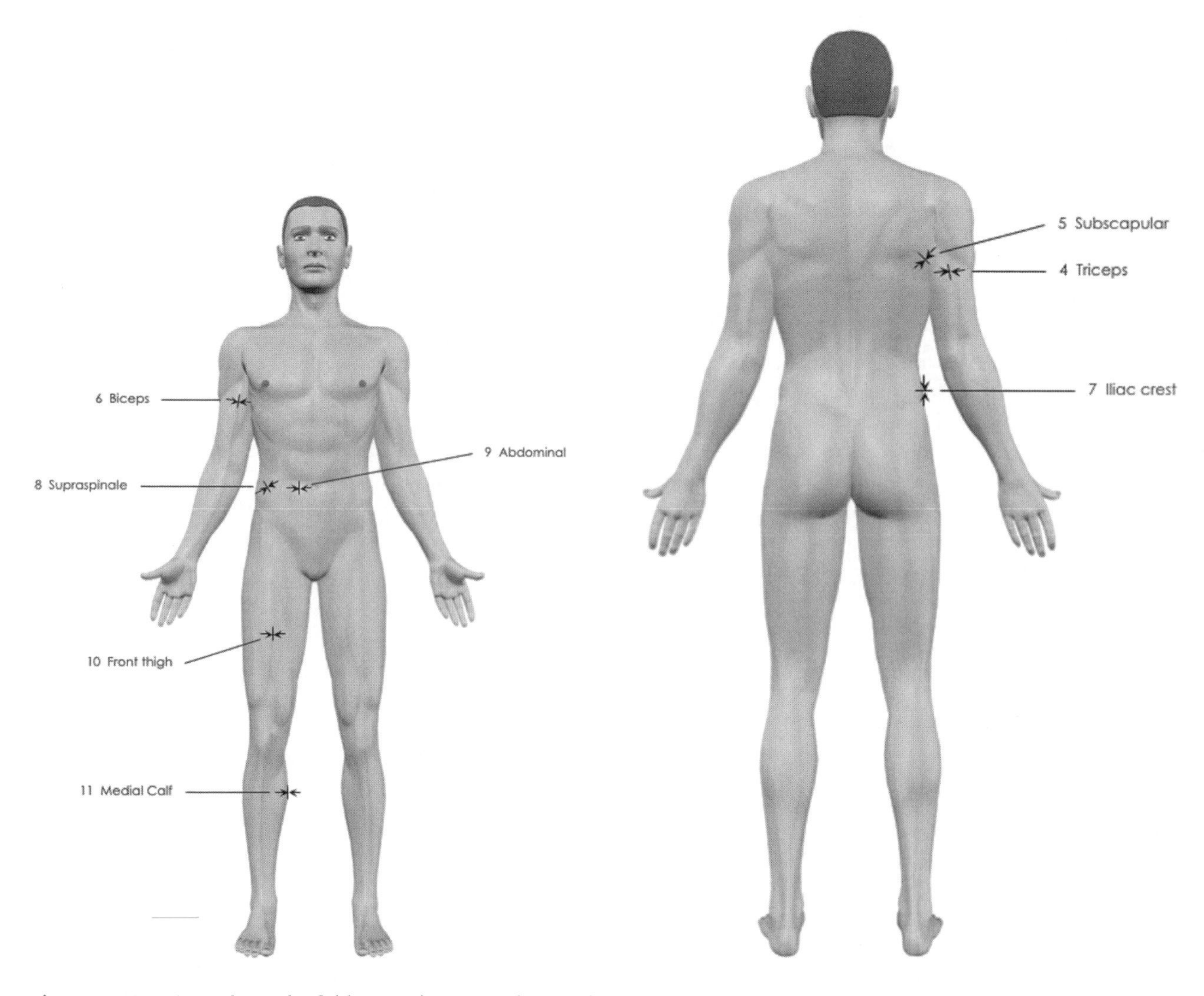

Figure 3.10 Sites where skinfolds are taken according to the ISAK protocol.

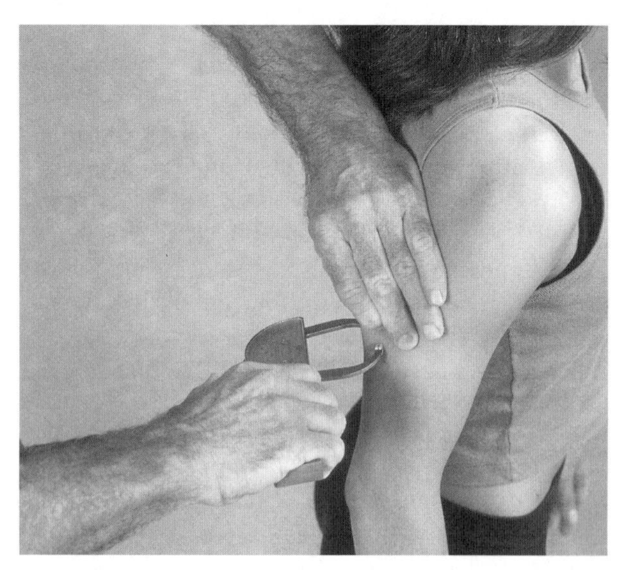

Figure 3.11 Taking the triceps skinfold.

can be a difficult skinfold to take, the subject may assist by lifting with both hands the underside of the thigh. An assistant may also be asked to assist by raising the fold with both hands about 6 cm (2.4 in.) either side of the landmark.

11. **Medial calf skinfold.** The subject places his or her right foot on the box, with the right knee bent at about 90°. A vertical fold is then lifted at the medial calf skinfold site.

Girths

When taking girths, hold the tape perpendicular to the long axis of the limb or torso. Where breathing will affect the size of the measurement, measurements should be taken at the end of a normal expiration (end tidal). Care should be taken to minimize indentation of the skin. The locations of girth measurement sites are shown in figure 3.12.

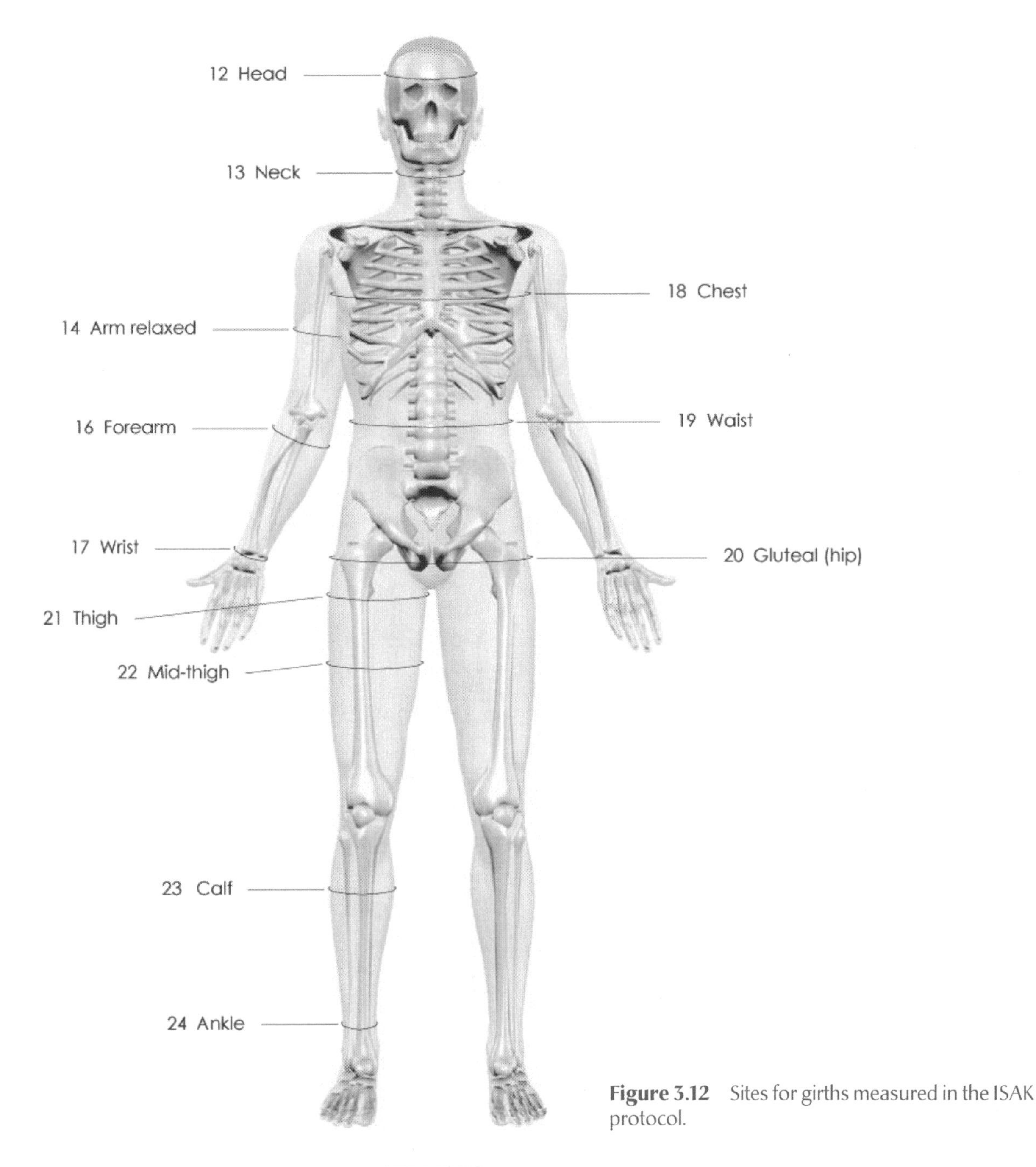

Figure 3.12 Sites for girths measured in the ISAK protocol.

12. **Head girth.** With the subject seated and the head in the Frankfort plane, place the tape horizontally around the head immediately superior to the glabella (figure 3.13). Apply firm pressure to flatten the hair.

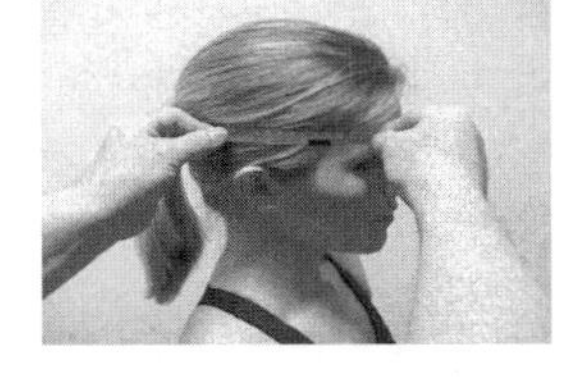

Figure 3.13 Measurement of head girth.

13. **Neck girth.** The subject remains seated with the head in the Frankfort plane while the tape is placed around the neck, perpendicular to the long axis and at a level immediately superior to the thyroid cartilage ("Adam's apple").

14. **Arm girth relaxed.** The girth is measured at the level of the midacromiale-radiale. The subject assumes a relaxed position with the arm hanging by the side.

15. **Arm girth flexed and tensed.** The subject's right arm is raised anteriorly to the horizontal with the forearm supinated and flexed at between 45° and 90° to the arm. The girth is measured at the level of the peak of the contracted biceps.

16. **Forearm girth.** The measurement is taken at the maximum girth of the forearm distal to the humeral epicondyles.

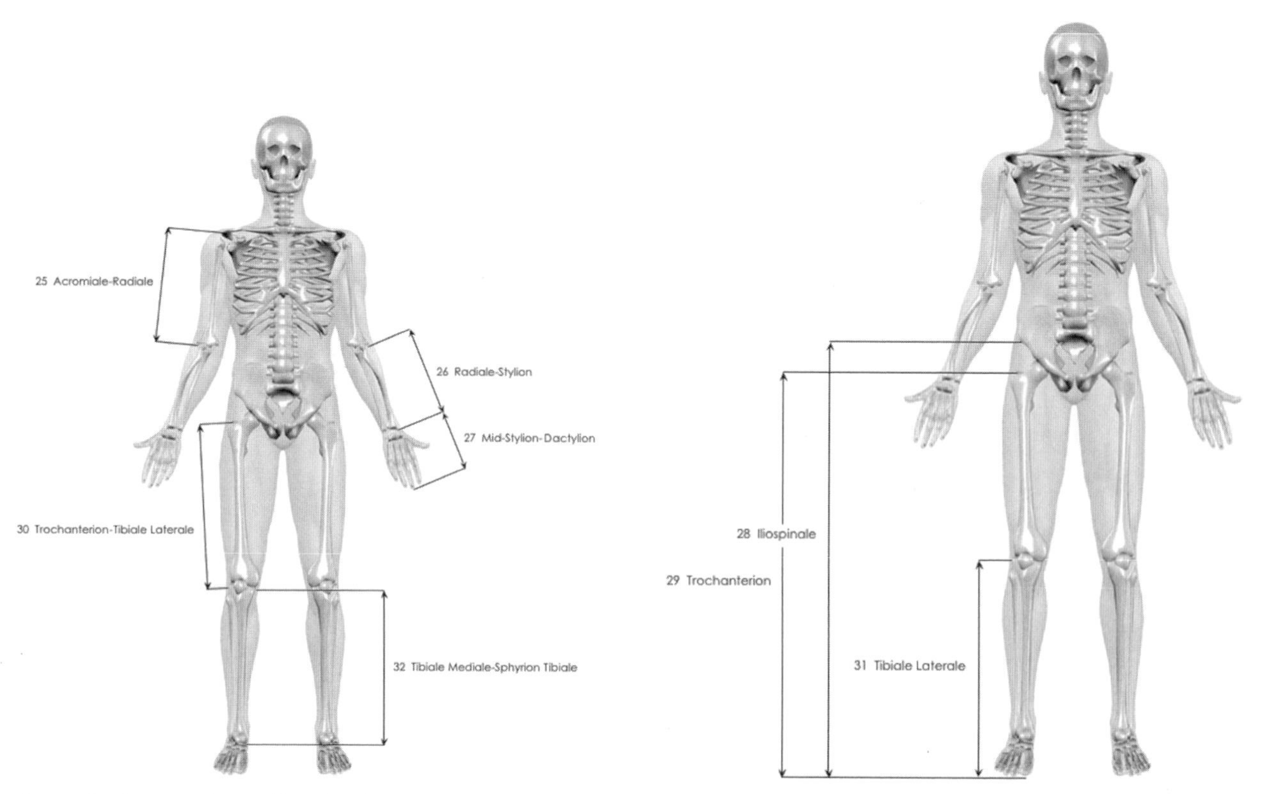

Figure 3.14 Lengths and heights in the ISAK protocol. Note that some lengths are shown on the left-hand side of the body for convenience, but all measurements are taken on the right-hand side where possible.

17. **Wrist girth.** The girth is the minimum wrist girth distal to the styloid processes, with the forearm supinated and the hand relaxed.

18. **Chest girth.** The girth is taken at the level of the mesosternale. The subject initially raises the arms and then lowers them as the measurement is taken.

19. **Waist girth.** This is the minimum girth between the lower costal (10th rib) border and the iliac crest. The subject folds the arms across the chest.

20. **Gluteal (hip) girth.** The subject should stand erect with feet together and gluteal muscles relaxed. The girth is taken at the level of the greatest posterior protuberance of the buttocks.

21. **Thigh girth.** The subject stands erect with the feet slightly apart and weight equally distributed. Pass the tape around the leg and slide up using a cross-handed technique to a horizontal position 2 cm (0.8 in.) below the gluteal fold.

22. **Midthigh girth.** Place the tape horizontally around the thigh at the level of the midtrochanterion–tibiale laterale landmark. The subject stands as for the thigh girth measurement.

23. **Calf girth.** This is the girth at the medial calf skinfold site. The subject should stand on a box with weight evenly distributed.

24. **Ankle girth.** The minimum girth of the ankle superior to the sphyrion tibiale. The subject should stand on a box with weight evenly distributed.

Lengths and Heights

A large sliding caliper or segmometer is used to measure lengths (figure 3.14). The way to use these instruments is to place the points of the instrument on each of the two landmarks that define a length. That is, a length is a point-to-point measurement.

25. **Acromiale-radiale length (arm length).** This is the linear distance from the acromiale to the radiale. The right forearm should be pronated. Anchor the end pointer of the segmometer to the acromiale and move the housing pointer to the radiale as shown in figure 3.15.

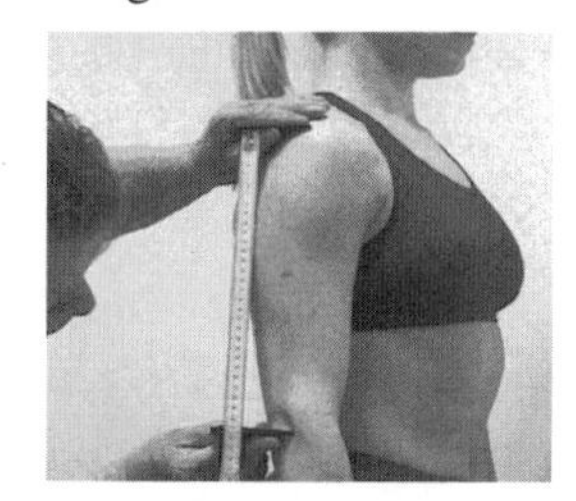

Figure 3.15 Measurement of the acromiale-radiale length using a segmometer.

26. **Radiale-stylion length (forearm length).** The subject stands as already described. This length is the linear distance from the radiale to the stylion. Anchor the end pointer to the radiale and move the housing pointer to the stylion.

27. **Midstylion-dactylion length (hand length).** This is

the linear distance from the midstylion to the dactylion. Anchor the end pointer at the midstylion mark and move the housing pointer to the dactylion.

28. **Iliospinale height (lower limb length 1).** The subject stands with feet together facing the box, so that the toes are placed in the cut-out portion of the box. The height from the top of the box to the iliospinale is measured. Add the height of the box to this score to obtain iliospinale height.

29. **Trochanterion height (lower limb length 2).** The subject stands with feet together and the lateral aspect of the right leg against the box. The height from the top of the box to the trochanterion is measured. Add the height of the box to this score to obtain trochanterion height.

30. **Trochanterion–tibiale laterale length (thigh length).** The subject stands with feet together and arms folded across the chest. This length is the linear distance from the trochanterion to the tibiale laterale.

31. **Tibiale laterale height (leg length 1).** With the subject standing on the box, measure the vertical distance from the top of the box to the tibiale laterale.

32. **Tibiale mediale–sphyrion tibiale length (leg length 2).** The subject is seated with the right ankle resting on the left knee. The length is the linear distance from the tibiale mediale to the sphyrion tibiale.

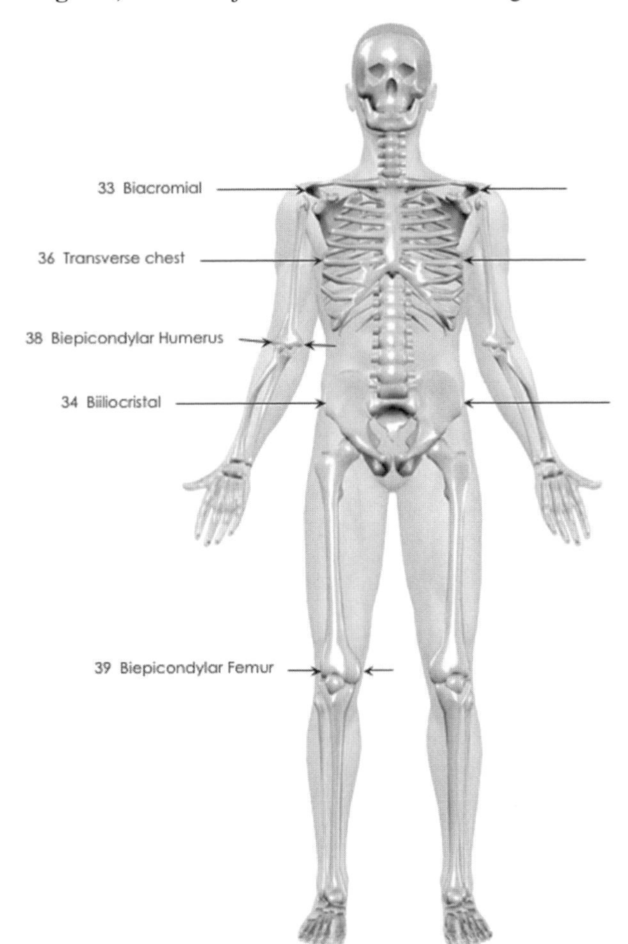

Figure 3.16 Breadths taken in the ISAK protocol.

Breadths

Breadths are measured with small and large sliding calipers. The caliper lies on the back of the hands while the thumbs rest against the inside edge of the caliper branches and the extended index fingers along the outside edges of the branches (figure 3.16).

33. **Biacromial breadth.** This is the distance between the most lateral points on the acromion processes. Standing behind the subject, the measurer holds the caliper branches at an angle of about 30° pointing upward. Firm pressure is applied to the acromion process during the measure (figure 3.17).

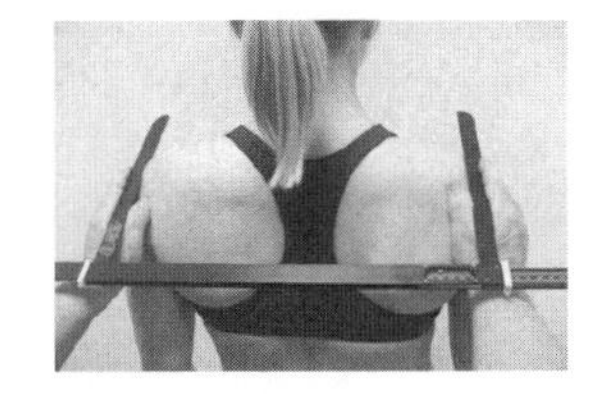

Figure 3.17 Measurement of the biacromial breadth.

34. **Biiliocristal breadth.** The subject stands with arms across the chest. While standing in front of the subject and with the pointers angled upward at 45°, measure the distance between the left and right iliocristale. Firm pressure is applied during the measurement.

35. **Foot length.** The subject stands on the box. Foot length is the perpendicular distance between the coronal planes passing through the akropodion and the pternion.

36. **Transverse chest breadth.** This is the distance between the most lateral aspects of the thorax when the top of the body of the caliper is positioned at the level of the mesosternale, and the branches are angled downward at 30°. Care must be taken to avoid including the latissimus dorsi or pectoralis major muscles.

37. **Anterior-posterior chest depth.** The subject assumes a seated position with the torso erect and the palms resting on the thighs. The depth is the linear distance between the mesosternale and the spinous process of the vertebra at the horizontal level of the mesosternale. Measurement is taken at "end tidal".

38. **Biepicondylar humerus breadth.** This is the linear distance between the medial and lateral epicondyles of the humerus. The subject stands with the right arm raised anteriorly to the horizontal and the elbow flexed at right angles (figure 3.18). Apply strong pressure.

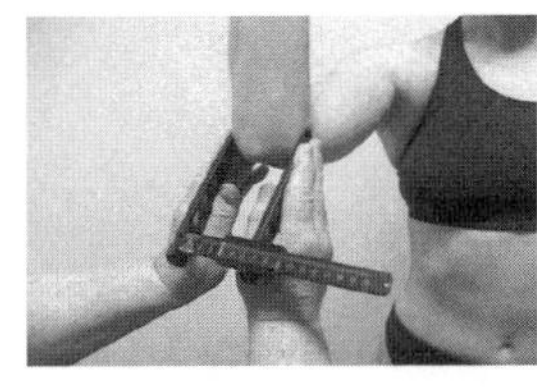

Figure 3.18 Measurement of the biepicondylar humerus breadth.

39. **Biepicondylar femur breadth.** This is the linear distance between the medial and lateral epicondyles of the

femur. The subject is seated with the knee bent at 90°. Apply strong pressure.

40. **Arm span (not a standard ISAK dimension).** Direct the subject to face the wall with the arms extended horizontally so that the left dactylion touches a corner and the right stretches out maximally. Mark the position of the right dactylion and measure the distance from the corner.

Precision and Accuracy

Prior to any testing, the precision and accuracy of each anthropometrist should be quantified. Precision refers to test-retest repeatability of a measurement, while accuracy refers to how well an anthropometrist's measurements compare to those of a "gold standard" criterion anthropometrist (usually ISAK Level 4). The statistic used to quantify both precision and accuracy is the technical error of measurement (TEM), which is given by the formula

$$\text{TEM} = \sqrt{(\Sigma d^2 / 2n)}$$

where d is the difference between the first and second measurements (for precision) or between the anthropometrist's measurement and the criterion measurement (for accuracy), and n is the number of subjects or pairs of measurements. TEMs should be calculated for each site, preferably using 20 or more subjects similar to those who are to be involved in testing. TEMs are more commonly expressed as a percentage (%TEM), such that

$$\%\text{TEM} = 100\ (\text{TEM} / M_{overall})$$

where $M_{overall}$ is the average of all measurements (that is, both test and retest measurements for precision, and both the anthropometrist's and criterion measurements for accuracy). In general, %TEMs of about 5% for skinfolds and about 1% for other measurements are very satisfactory.

Summary

While many things contribute to sporting success, absolute body size is in many cases a critical factor. The clearly "abnormal" body size of many performers, the sharp differences between elite and nonelite performers and the rapidly changing size of sporting bodies over time are all indexes of the importance of size. There is now a global market for sporting bodies that is rapidly changing the face of sport worldwide. Standardized anthropometric assessment by trained and experienced practitioners should be a part of the routine assessment of athletes, both for talent identification and for monitoring purposes.

four

Somatotype in Sport

J.E. Lindsay Carter, PhD; and Timothy R. Ackland, PhD

The scientific measurement of human physique and body types has occurred mostly during the past century; however, human biologists and physicians have been interested in the field for many centuries. Hippocrates formally classified two fundamental body types in the 5th century BC. The first he called *habitus phthisicus,* which was characterized by a long, thin body dominated by a vertical dimension, and the second *habitus apoplecticus,* whose main physical characteristic was a short, thick body that was strong in the horizontal dimension. After Hippocrates, many scientists developed categorical classifications ranging from two to five "types" (Tucker and Lessa 1940). These "discrete categories" were very general, and not everyone could be placed accurately in them.

Development of Somatotype Rating Methods

The somatotype is a quantified expression or description of the present morphological conformation of a person. It consists of a three-numeral rating, for example, **3½-5-1.** The three numerals are always recorded in the same order, each describing the value of a particular component of physique (Duquet and Carter 2008).

The Sheldon Method

The word *somatotype* was coined by William H. Sheldon, who developed a new method to classify body types, using front, back and side view photographs and measurements on the photos of 4000 male subjects (Sheldon, Stevens, and Tucker 1940). His method, which was based on the presence of three extreme body characteristics, became the basis for the systems that were to develop in the latter part of the 20th century.

Sheldon rejected the classical typologies and rated physiques numerically from 1 to 7 on three components, which he called endomorphy, mesomorphy and ectomorphy. Every person was a mixture of the components that Sheldon claimed were derived from the embryonic tissue layers of the body. Sheldon, Dupertuis, and McDermott (1954) expanded the method and provided photographs of 18- to 65-year-old males in an "atlas" that displayed photographs of 1175 men representing 88 somatotypes. A method was set out in which photographs of the subject were compared with those in the atlas. Sheldon constructed age-adjusted weight scales to reflect his view, as stated in the 1940 book, that the somatotype was a *genotype* and therefore did not change throughout adulthood. The main objections to Sheldon's methods were the insistence that the somatotype was permanent and did not change and the limitations of the 7-point scales.

The Parnell Method

Richard Parnell (1958) was interested in the relationship between physique and behaviour and developed a more objective method than that of Sheldon for both males and females. He used anthropometry and constructed the M4 Deviation Chart on which to record these measures, thus making the ratings more objective and eliminating the need for a photograph. Parnell also raised the question of the phenotype versus the genotype. However, his scales were age adjusted to match Sheldon's concept of permanence of the somatotype.

The Heath-Carter Method

The current method of somatotyping is that of Heath and Carter (1967) and Carter and Heath (1990), who refined both the Sheldon and Parnell methods to derive a more objective assessment of the body type. These investigators made the component scales from ½ to 16+ for endomorphy, ½ to 12+ for mesomorphy, and ½ to 9+ for ectomorphy. That is, the scales were open-ended and could go beyond the upper ratings if warranted by the data. This turned out to be an important addition to the method, because it had become obvious that among various subpopulations around the world, there are many cases of obese individuals who rate well over 7 in endomorphy (figure 4.1*a*). Furthermore, highly mesomorphic athletes, especially those competing in gymnastics, weight- and powerlifting, as well as in the sport of bodybuilding, are rated well over 7 in mesomorphy (figure 4.1*b*). Similarly, the Nilotes of northeastern Africa included some individuals who were more than 7 in ectomorphy (figure 4.1*c*).

> It is well known by coaches that various body types can be altered over time by intervention programs such as dietary or strength training regimens; therefore the phenotype rating is a more realistic criterion for the field of sport science.

Another important modification in the Heath-Carter method was that the rating was of the present somatotype—the phenotype rating. This enabled ratings to be made from childhood to old age using the same criteria. Because the phenotype represents the body at any given moment in time, human biologists have more recently accepted it as a more valid representation of the somatotype.

In the Heath-Carter method the somatotype components were redefined as follows:

- Endomorphy—the relative adiposity of a physique
- Mesomorphy—the relative musculoskeletal robustness of a physique
- Ectomorphy—the relative linearity or slenderness of a physique

The definitions of endomorphy and mesomorphy reflect the observed shape and composition of the body, and the anatomical model of body composition. The observed adiposity, muscle and bone are rated relative to stature, and ectomorphy is rated according to the height-weight ratio (HWR). Ratings can be made by three methods:

- The *photoscopic somatotype,* based on the standardized somatotype photograph (front, side and

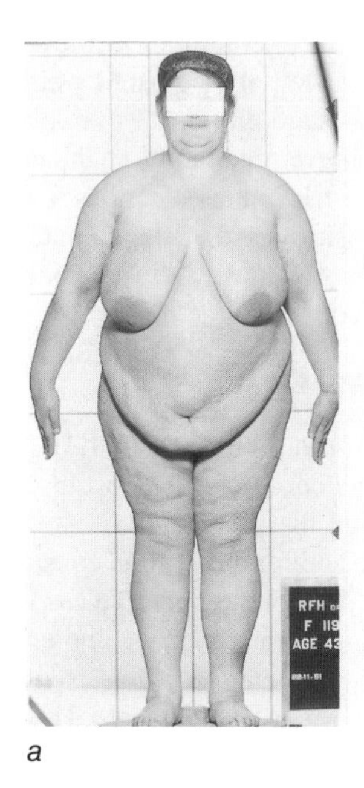

a

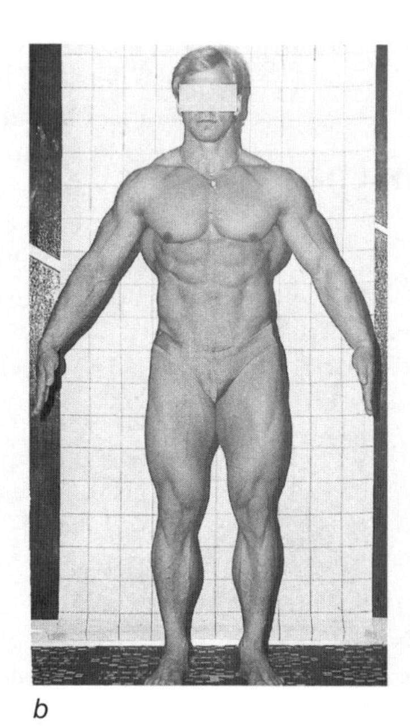

b

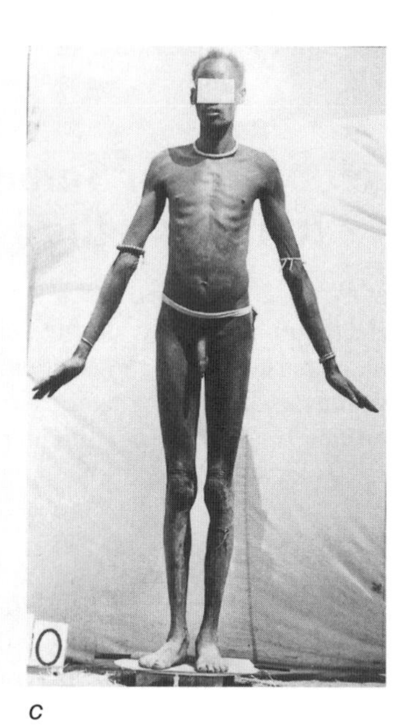

c

Figure 4.1 Examples of extreme somatotypes. *(a)* Endomorph, **14-4-½**; *(b)* mesomorph, **1-10½-1**; and *(c)* ectomorph, **1½-2-8**.

HEATH-CARTER SOMATOTYPE RATING FORM

Name: ____________ Age: ________ Sex: ______ Date: ________

Skinfolds (mm)
- triceps =
- subscapular =
- supraspinale =
- Sum 3 skinfolds =
- $\times \left(\frac{170.18}{ht=}\right)$ = ☐
- calf =

HEIGHT CORRECTED SUM OF 3 SKINFOLDS (mm)

Upper limit	10.9	14.9	18.9	22.9	26.9	31.2	35.8	40.7	46.2	52.2	58.7	65.7	73.2	81.2	89.7	98.9	108.9	119.7	131.2	143.7	157.2	171.9	187.9	204.0
Mid-point	9.0	13.0	17.0	21.0	25.0	29.0	33.5	38.0	43.5	49.0	55.5	62.0	69.5	77.0	85.5	94.0	104.0	114.0	125.5	137.0	150.5	164.0	180.0	196.0
Lower limit	7.0	11.0	15.0	19.0	23.0	27.0	31.3	35.9	40.8	46.3	52.3	58.8	65.8	73.3	81.3	89.8	99.0	109.0	119.8	131.3	143.8	157.3	172.0	188.0
Endomorphy	½	1	1½	2	2½	3	3½	4	4½	5	5½	6	6½	7	7½	8	8½	9	9½	10	10½	11	11½	12

Height (cm) =	139.7	143.5	147.3	151.1	154.9	158.8	162.6	166.4	170.2	174.0	177.8	181.6	185.4	189.2	193.0	196.9	200.7	204.5	208.3	212.1	215.9	219.7	223.5	227.3
Humerus breadth (cm) =	5.19	5.34	5.49	5.64	5.78	5.93	6.07	6.22	6.37	6.51	6.65	6.80	6.95	7.09	7.24	7.38	7.53	7.67	7.82	7.97	8.11	8.25	8.40	8.55
Femur breadth (cm) =	7.41	7.62	7.83	8.04	8.24	8.45	8.66	8.87	9.08	9.28	9.49	9.70	9.91	10.12	10.33	10.53	10.74	10.95	11.16	11.36	11.57	11.78	11.99	12.21
Biceps girth - T[1] (cm) =	23.7	24.4	25.0	25.7	26.3	27.0	27.7	28.3	29.0	29.7	30.3	31.0	31.6	32.2	33.0	33.6	34.3	35.0	35.6	36.3	37.0	37.6	38.3	39.0
Calf girth - C[2] (cm) =	27.7	28.5	29.3	30.1	30.8	31.6	32.4	33.2	33.9	34.7	35.5	36.3	37.1	37.8	38.6	39.4	40.2	41.0	41.7	42.5	43.3	44.1	44.9	45.6

Mesomorphy	½	1	1½	2	2½	3	3½	4	4½	5	5½	6	6½	7	7½	8	8½	9

Body mass (kg) =

$HT/\sqrt[3]{body\ mass}$ = ☐

Upper Limit	39.65	40.74	41.43	42.13	42.82	43.48	44.18	44.84	45.53	46.23	46.92	47.58	48.25	48.94	49.63	50.33	50.99	51.68
Mid-point	and	40.20	41.09	41.79	42.48	43.14	43.84	44.50	45.19	45.89	46.32	47.24	47.94	48.60	49.29	49.99	50.68	51.34
Lower limit	below	39.66	40.75	41.44	42.14	42.83	43.49	44.19	44.85	45.54	46.24	46.93	47.59	48.26	48.95	49.64	50.34	51.00
Ectomorphy	½	1	1½	2	2½	3	3½	4	4½	5	5½	6	6½	7	7½	8	8½	9

	Endomorphy	Mesomorphy	Ectomorphy
Anthropometric Somatotype			

Rater: ____________

1 Biceps girth corrected for fat by subtracting triceps skinfold

2 Calf girth corrected for fat by subtracting calf skinfold

Figure 4.2 Heath-Carter somatotype rating form.

From J.E.L. Carter and B. Heath, 1990, *Somatotyping: development and applications* (New York: Cambridge University Press). Adapted with the permission of Cambridge University Press.

back views), a table of possible somatotypes for a given HWR and photoscopic criteria

- The *anthropometric somatotype,* based on the somatotype rating form or equations
- The *criterion method,* from a combination of the anthropometric plus photoscopic methods

In the absence of criterion raters, most somatotypers use the anthropometric method, but they often take photographs as a record of the physique of the subjects (Carter and Heath 1990). The photoscopic, criterion and anthropometric ratings are made to the nearest half-unit (½). Decimal ratings for individuals are of spurious accuracy. Ratings from equations may be made to a tenth of a unit (0.1-0.9) for statistical purposes. Mean somatotypes in the text are in bold characters to the nearest half-unit to draw attention to the three-number unity of the whole somatotype rating.

Heath and Carter (1967) developed a somatotype rating form on which to record the anthropometry (figure 4.2). The resulting somatotypes were then plotted on a somatochart (figure 4.3). When somatotypes are placed on the somatochart they are called somatoplots. The somatoplots can be assigned descriptive names according to component dominance as shown in figure 4.4.

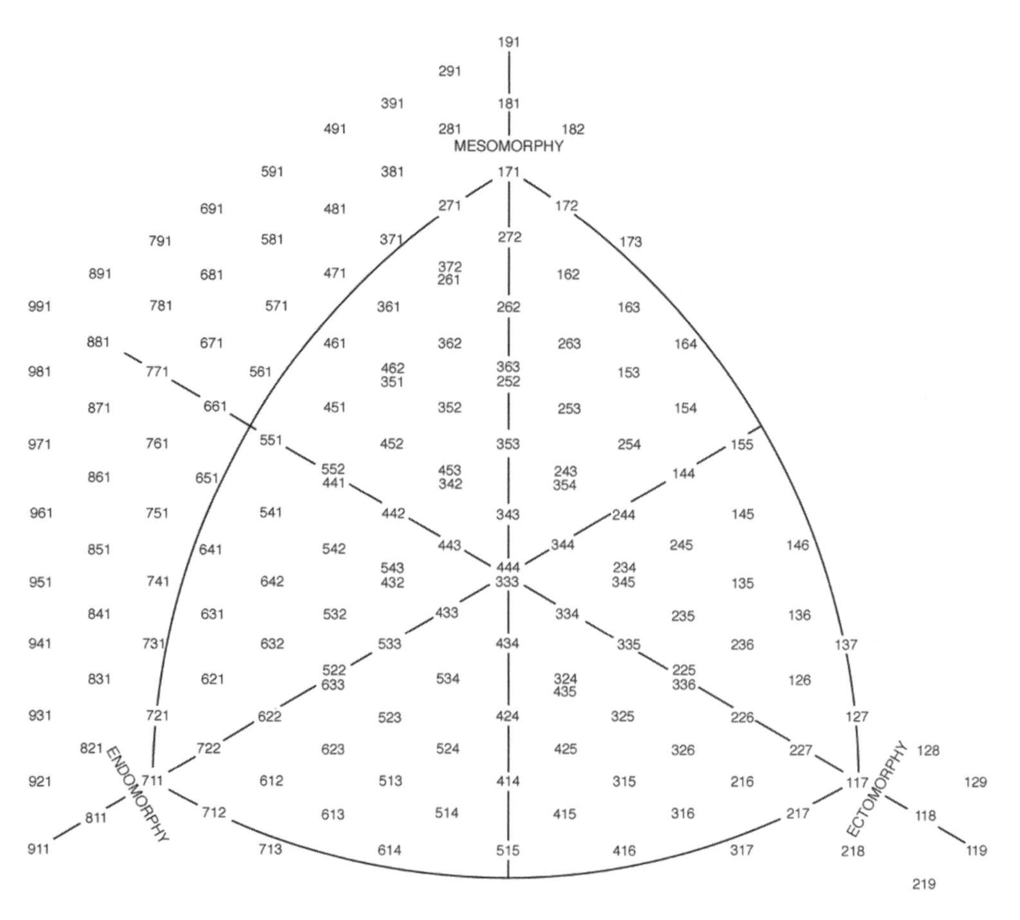

Figure 4.3 Somatochart.

From J.E.L. Carter and B. Heath, 1990, *Somatotyping: development and applications* (New York: Cambridge University Press). Adapted with the permission of Cambridge University Press.

The Heath-Carter Anthropometric Method

The Heath-Carter anthropometric somatotype (Carter and Heath 1990) may be calculated using either a rating form, or equations (Duquet and Carter 2008) or a computer program (Goulding 2002). This section illustrates the rating form method, since it provides a "first principles" view of the assessment process. Similarly, the location of a particular somatotype on the somatochart may be made by hand, or X,Y coordinates may be generated for subsequent plotting (Carter and Marfell-Jones 1994; Goulding 2002).

Anthropometry

The 10 measures needed to calculate the somatotype should be taken using the methods prescribed by the International Society for the Advancement of Kinanthropometry (ISAK 2001)—see chapter 3. Skinfolds, breadths and girths are taken on the right side of the body. Record the following information on the rating form in figure 4.5: skinfolds—triceps, subscapular, supraspinale and calf skinfolds; bone breadths—biepicondylar humerus and femur breadths; girths—flexed arm and calf girths; stretch stature; body mass.

Rating Procedure

The procedure for calculating each somatotype component involves several specific steps. In this section we describe the procedures used to rate endomorphy, mesomorphy and ectomorphy in an athlete.

Endomorphy

There are three steps for calculating endomorphy.

- Sum the three skinfolds (triceps, subscapular and supraspinale).
- Apply the dimensional scaling factor (sum 3 skinfolds × 170.18/stature).
- Circle the closest value in the table and then circle the rating value for that column.

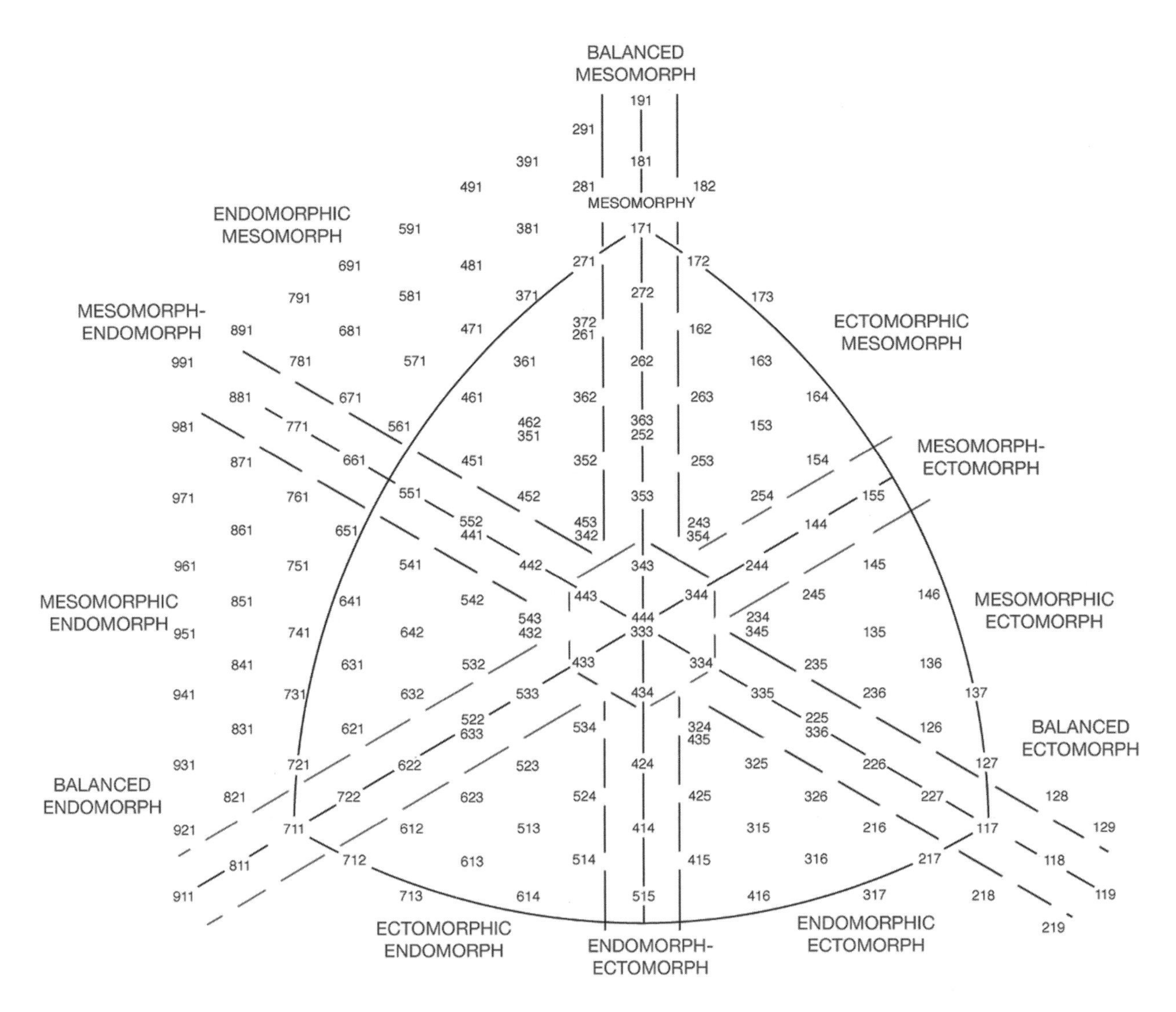

Figure 4.4 Thirteen somatotype categories.

Adapted, by permission, from K. Norton and T. Olds, 1996, *Anthropometrica* (Sydney, Australia: UNSW Press).

Mesomorphy

The following steps are used to calculate mesomorphy.

- Mark the subject's height on the height scale and then, in the row below, circle the numeral closest to this value.
- For each bone breadth, circle the figure in the appropriate row that is nearest to the measurement value.
- Subtract triceps skinfold (T) in centimeters from biceps girth, and calf skinfold (C) in centimeters from calf girth, before circling the figure in the appropriate row that is nearest to the calculated value.
- Considering the breadth and girth rows only, calculate the algebraic sum of column deviations (D) from the circled height column: positive deviations to the right, negative deviations to the left of the height column (see, for example, the calculations in figure 4.5, where D = +9 + –2 = +7).
- Considering the height measurement, a correction (K) is necessary when the subject's height is not identical to the circled height numeral. As shown in table 4.1, the correction factor is negative if the subject's height is greater than the circled value, positive if less than the circled value.
- Calculate mesomorphy rating using the following formula: Mesomorphy = D / 8 + K + 4.0.
- Circle the closest rating value, being conservative if the calculation falls midway between values.

HEATH-CARTER SOMATOTYPE RATING FORM

Name: B. A. Age: 20 Sex: M Date: 25-10-93

Skinfolds (mm)
- triceps = 6.4
- subscapular = 7.1
- supraspinale = 4.6
- Sum 3 skinfolds = 18.1
- × (170.18 / Ht= 178.3) = 17.3
- calf = 5.2

HEIGHT CORRECTED SUM OF 3 SKINFOLDS (mm)

Upper limit	10.9	14.9	18.9	22.9	26.9	31.2	35.8	40.7	46.2	52.2	58.7	65.7	73.2	81.2	89.7	98.9	108.9	119.7	131.2	143.7	157.2	171.9	187.9	204.0
Mid-point	9.0	13.0	(17.0)	21.0	25.0	29.0	33.5	38.0	43.5	49.0	55.5	62.0	69.5	77.0	85.5	94.0	104.0	114.0	125.5	137.0	150.5	164.0	180.0	196.0
Lower limit	7.0	11.0	15.0	19.0	23.0	27.0	31.3	35.9	40.8	46.3	52.3	58.8	65.8	73.3	81.3	89.8	99.0	109.0	119.8	131.3	143.8	157.3	172.0	188.0
Endomorphy	½	1	(1½)	2	2½	3	3½	4	4½	5	5½	6	6½	7	7½	8	8½	9	9½	10	10½	11	11½	12

- Height (cm) = 178.3
- Humerus breadth (cm) = 7.2
- Femur breadth (cm) = 9.75
- Biceps girth - T[1] (cm) = 33.1
- Calf girth - C[2] (cm) = 33.8

139.7	143.5	147.3	151.1	154.9	158.8	162.6	166.4	170.2	174.0	(177.8)	181.6	185.4	189.2	193.0	196.9	200.7	204.5	208.3	212.1	215.9	219.7	223.5	227.3
5.19	5.34	5.49	5.64	5.78	5.93	6.07	6.22	6.37	6.51	.65	6.80	6.95	7.09	(7.24)	7.38	7.53	7.67	7.82	7.97	8.11	8.25	8.40	8.55
7.41	7.62	7.83	8.04	8.24	8.45	8.66	8.87	9.08	9.28	.49	(9.70)	9.91	10.12	10.33	10.53	10.74	10.95	11.16	11.36	11.57	11.78	11.99	12.21
23.7	24.4	25.0	25.7	26.3	27.0	27.7	28.3	29.0	29.7	3[illegible].3	31.0	31.6	32.2	(33.0)	33.6	34.3	35.0	35.6	36.3	37.0	37.6	38.3	39.0
27.7	28.5	29.3	30.1	30.8	31.6	32.4	33.2	(33.9)	34.7	3[illegible].5	36.3	37.1	37.8	38.6	39.4	40.2	41.0	41.7	42.5	43.3	44.1	44.9	45.6

Mesomorphy: ½ 1 1½ 2 2½ 3 3½ 4 4½ (5) 5½ 6 6½ 7 7½ 8 8½ 9

D = +7/8 = .875 − .07 = .805 + 4.000 = 4.805

- Body mass (kg) = 69.2
- HT/∛body mass = 43.4

Upper Limit	39.65	40.74	41.43	42.13	42.82	(43.48)	44.18	44.84	45.53	46.23	46.92	47.58	48.25	48.94	49.63	50.33	50.99	51.68
Mid-point	and	40.20	41.09	41.79	42.48	43.14	43.84	44.50	45.19	45.89	46.32	47.24	47.94	48.60	49.29	49.99	50.68	51.34
Lower limit	below	39.66	40.75	41.44	42.14	42.83	43.49	44.19	44.85	45.54	46.24	46.93	47.59	48.26	48.95	49.64	50.34	51.00
Ectomorphy	½	1	1½	2	2½	(3)	3½	4	4½	5	5½	6	6½	7	7½	8	8½	9

Anthropometric Somatotype	Endomorphy	Mesomorphy	Ectomorphy
	1½	5	3

Rater: T.A.

1 Biceps girth corrected for fat by subtracting triceps skinfold

2 Calf girth corrected for fat by subtracting calf skinfold

Figure 4.5 Calculation of the anthropometric somatotype using the rating form.

Ectomorphy

Two steps are needed to calculate ectomorphy.

- Calculate the ratio of height to the cubed root of body mass and then record this value.
- Circle the closest numeral on the scale, then circle the rating value corresponding to that column.

The resulting somatotype as shown in figure 4.5 is **1½-5-3,** an ecto-mesomorph. This somatotype would be plotted between the **1-5-3** and the **2-5-3** somatotypes on the somatochart (figure 4.3). When children or small adults are being measured, the mesomorphy columns may be extended to the left (downward), or they may be extended to the right (upward) for extremely large subjects.

Somatotype components and the X,Y coordinates for plotting can be calculated by equations as follows:

- Endomorphy = $-0.7182 + 0.1451 \times \text{sumSF} - 0.00068 \times \text{sumSF}^2 + 0.0000014 \times \text{sumSF}^3$, where sumSF = (sum of triceps, subscapular and supraspinale skinfolds) × (170.18 / height in centimeters). This is called height-corrected endomorphy.

Table 4.1 Height Correction Values (K) for Mesomorphy Assessment

Difference in height (mm)	K	Difference in height (mm)	K
1	0.01	11	0.14
2	0.03	12	0.16
3	0.04	13	0.17
4	0.05	14	0.18
5	0.07	15	0.20
6	0.08	16	0.21
7	0.09	17	0.22
8	0.11	18	0.24
9	0.12	19	0.25
10	0.13		

- Mesomorphy = $0.858 \times$ humerus breadth $+ 0.601 \times$ femur breadth $+ 0.188 \times$ corrected flexed and tensed arm girth $+ 0.161 \times$ corrected calf girth $-$ height $\times 0.131 + 4.5$.
- Ectomorphy = $0.732 \times$ HWR (if HWR is greater than or equal to 40.75); or if HWR is greater than

40.75 and less than 38.25, then ectomorphy = 0.463 × HWR – 17.63; or if HWR is equal to or less than 38.25, then ectomorphy = 0.1.

The X,Y coordinates for plotting on the two-dimensional somatochart are as follows:

$$X = \text{Ectomorphy} - \text{Endomorphy}$$

$$Y = 2 \times \text{Mesomorphy} - (\text{Endomorphy} + \text{Ectomorphy})$$

For more details regarding somatotype ratings, see Carter and Heath 1990, Carter 1996 or Duquet and Carter 2008.

Somatotype and Sport Performance

The somatotype is a general physical characteristic and is only one indicator of an athlete's suitability to perform at a high level. One must combine it with other variables like body composition, proportionality, strength and power when evaluating the physique of an athlete.

Somatotype and High-Level Performance

Physique plays an important role in the self-selection of individuals for competitive sport. There is also a considerable amount of information in the sport science literature on the suitability of various somatotypes not only for particular sports, but also for specific events or positions within many sports (Carter 2003). As a large number of factors are involved in the physical makeup of a champion athlete, there is not necessarily one perfect somatotype for a particular sport or event within that sport. Much of the literature gives mean somatotype ratings of high-level athletes; however, one should remember that there will be a range of somatotypes that will cluster around the mean in each group. Figure 4.6 illustrates this point, as it demonstrates the spread of somatotypes among divers who competed in the World Championships in 1991.

Differences Between the Sexes

The extent to which males and females differ with respect to some external characteristic is referred to as sexual dimorphism. In this context, the comparisons are within

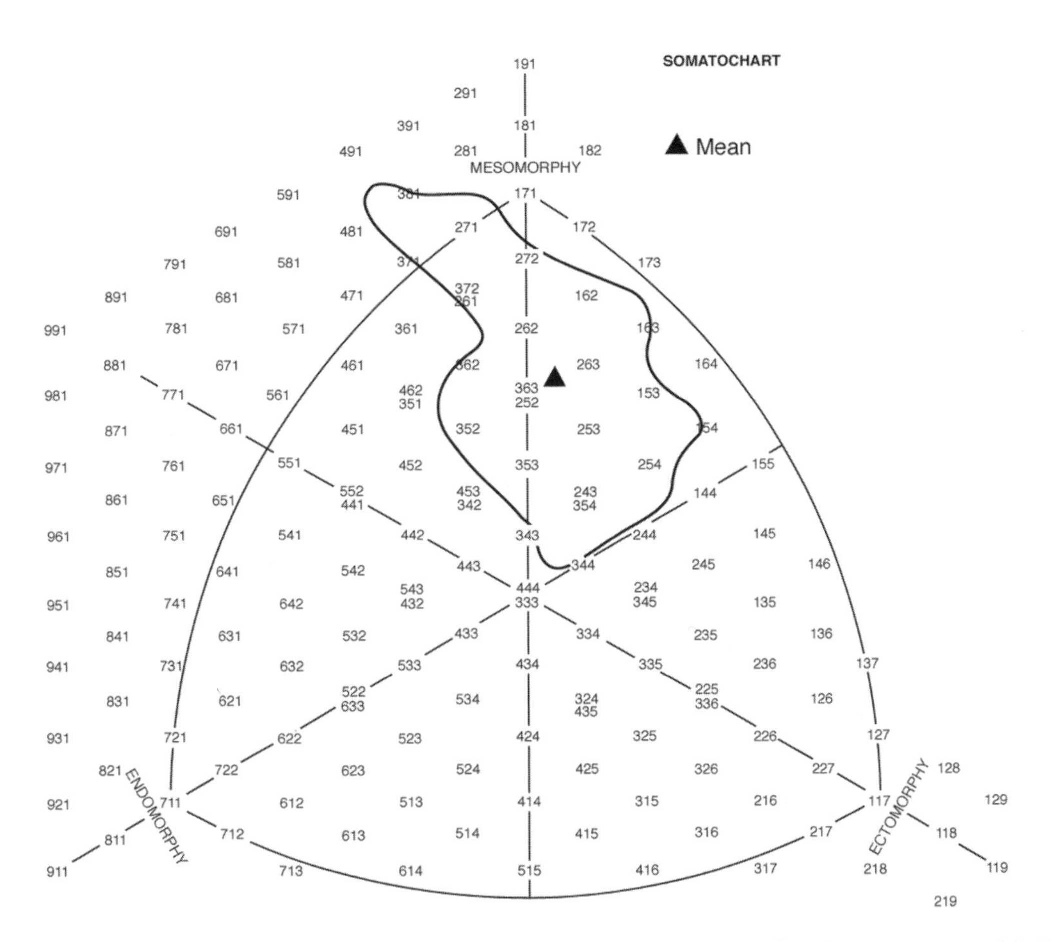

Figure 4.6 The somatotype distribution of 43 male divers at the World Championships in 1991. Mean = **2.0-5.3-2.4**.

Adapted, by permission, from J.E.L. Carter and M. Marfell-Jones, 1994. Somatotypes. In *Kinanthropometry in aquatic sports,* edited by J.E.L. Carter and T. Ackland (Champaign, IL: Human Kinetics Publishers).

sport and event. The majority of data collected before the 1960s on high-level athletes were on males; however, in recent decades this deficiency has been steadily redressed. In sports in which sexual dimorphism has been studied, the data indicate that high-level female athletes are generally more endomorphic and less mesomorphic than their male counterparts but are similar in ectomorphy. These differences are illustrated in the somatocharts in the section on sports in this chapter. The following is a rough guide that has been used in the past by sport scientists in sports for which a reasonable number of data have been gathered:

- In sprint running, the female is usually about 1.0 unit more in endomorphy and 1.0 to 1.5 less in mesomorphy, with ectomorphy at a similar level. In middle distance running, the female is generally about 0.5 more in endomorphy and 1.0 less in mesomorphy, with ectomorphy again similar.
- The range of differences between male and female swimmers is the same. However, when greater body bulk or explosive power or both are needed in sports such as field games or judo, females are up to 2.0 units higher in endomorphy and 2.0 lower in mesomorphy than males. Ectomorphy is generally similar, but may vary by 0.5 in some cases.

Carter and Heath (1990, p. 289) suggested that more intensive training at the top level does not significantly modify these differences and that "this may help to account for the continued differences in the performance of trained male and female athletes".

Differences Between Athletes and Nonathletes

High-level male and female athletes are generally higher in the mesomorphic sector of the somatochart when compared to individuals from the normal population, who are generally more endomorphic as well as more randomly spread around the somatochart (especially in the high endomorphy and low mesomorphy areas). For example, male athletes tend to be much more mesomorphic than a reference male population, with a mean somatotype of approximately **2-5-2½** compared to **3½-4½-2,** while female athletes have a mean somatotype of approximately **3-4-3** compared to a reference population with a mean of **5-3-2.**

Athlete Somatotype Changes Over Time

Carter and Heath (1990) stated that somatotypes of athletes in some sports over the past 60 years had undergone various changes. In this period male swimmers became less endomorphic and field games athletes developed more mesomorphy, while high jumpers and 400 m runners became less mesomorphic and more ectomorphic. However, track sprinters showed very little change but tended toward more mesomorphy. Furthermore, a shift was noted by Štěpnička (1986) toward higher mesomorphy in cross-country skiing, judo and basketball, sports in which power is an important element.

These changes are probably the result of several factors:

- The first would be self-selection for the sport, with more individuals with suitable body types selecting themselves for various sports or events. Athletes are also coming from a considerably larger population interested in sport participation than was the case before the 1970s. Further, various races and ethnic groups who have some very special physical features that suit particular sports are now competing regularly in international meets (e.g., African male and female distance runners).
- The second factor relates to the very intensive endurance training that has been carried out during the past several decades. Depending on the sport, many endurance athletes now spend from 4 to 5 h training each day and therefore expend a great deal of energy. This reduces their endomorphy while at the same time increasing their ectomorphy in some sports such as distance running, or their mesomorphy in such sports as swimming, rowing and canoeing, which involve very intensive anaerobic-aerobic training.
- The third reason is that sport scientists have been able to tailor athletes' nutritional needs to suit them personally. In some cases they need "force-feeding" to gain bulk or more endomorphy, and in other sports they need to reduce adipose tissue in order to "make the weight" and to compete at an optimal performance weight with a very low fat mass (FM), thus increasing their ectomorphy.
- The fourth factor is the universal use of intensive strength and explosive power training through which competitors now increase their lean body mass (LBM), as well as their mesomorphy rating.

Desirable Body Types for High-Level Performance

"Athletes who have or acquire the optimal physique for an event are more likely to succeed than those who lack the general characteristics" (Carter 1985, 115). Before evaluating differences in somatotypes among sports, it is important to note that we should expect some differences between individual and team sports. The "tasks" in some events, such as shot put and high jump, are quite specific and different from each other, and so are the successful physiques. In team sports, players have to be proficient in a variety of skills and require a physique that is suited to accomplishing all of them (e.g., power forwards in basketball, soccer midfielders). Norton and Olds (1996, p. 352) called this the process of morphological optimisation, which is "the process whereby the physical demands of a sport lead to the selection of body types (structure and composition) best suited to that sport".

> The tasks in some events, such as shot put and high jump, are quite specific and different from each other, and so are the successful physiques. In team sports, players have to be proficient in a variety of skills and require a physique that is suited to accomplish all of these (e.g., power forwards in basketball, soccer midfielders).

The following text and table 4.2 summarize updated data from studies of elite athletes in Olympic or World Championships. Reference is also made to pertinent national or other high-level samples for which the results help us understand somatotype patterns and possible selection influences at different levels of competition. Some of the rounded (½-unit) means in table 4.2 are derived from several studies in a sport as summarized by Carter and Heath (1990); therefore the overall sample size is not always given. In other studies referenced in the table, the sample size and calculated mean are given where known. In the following sections mean somatotypes are plotted using selected means from table 4.2. Although the scatter of somatotypes for each sport is not shown, the distances between somatotype means give an estimate of differences between sports and samples.

Table 4.2 Mean Somatotypes of High-Level Sportsmen and Sportswomen

Sport and source	N	Males	N	Females
RACQUET SPORTS				
Tennis[1]	61	2-4½-3	69	3½-3½-3
Tennis—All-Africa Games 1995[5]			8	4.3-3.6-2.2
Badminton[2,3]	7	2.5-4.6-3.2	6	4.1-4.4-2.5
Squash and Racquetball[1,3]	19	2½-5-3	6	3.4-4.0-2.8
AQUATIC SPORTS				
Swimming[4]	231	2-5-3	170	3-4-3
Swimming—All-Africa Games 1995[5]			15	3.5-3.8-2.8
Water polo[4]	190	2½-5½-2½	109	3½-4-3
Rowing[1]		2½-5½-2½		3-4-3
Rowing—World Junior Lightweight 1997[6]	383	2.3-4.7-3.0	222	3.4-3.7-2.6
Rowing—Olympics 2000[7]				
Lightweight rowers	56	1.4-4.4-3.4	14	2.0-3.4-3.4
Open rowers	153	1.9-5.0-2.5	73	2.8-3.9-2.6
Rowing—Pan American Games 1986[8]				
Lightweight	20	1.8-5.1-2.9	13	2.4-3.9-3.3
Canoeing[1]		2-5½-2½	12	3-4½-2½
Canoeing—Olympics 2000[7]				
Sprint paddlers	50	1.6-5.0-2.2	20	2.4-4.6-2.3
GYMNASTICS AND POWER SPORTS				
Gymnastics[1]		1½-6-2		2-4-3
Gymnastics—World Championships 1987[9]	165	1.5-5.6-2.1	201	1.8-3.7-3.1
Diving[4]	43	2.0-5.3-2.4	39	2.8-3.8-2.8
Skating[1]	19	1½-5-3	27	2½-4-3
Weightlifting[1]				
(<60 kg)		1½-7-1		
(60-79.9 kg)		2-7-1		
(80-99.9 kg)		2½-8-½		
(>100 kg)		5-9-½		
Bodybuilding[1]		2-8½-1		2½-5-2½

(continued)

Table 4.2 *(continued)*

Sport and source	N	Males	N	Females
TRACK AND FIELD, CYCLING, TRIATHLON				
Track and field[1]				
Sprinting, running, hurdles		1½-5-3		2½-4-3
400 m, 400 m hurdles		1½-4½-3½		2-3½-3½
800 m, 1500 m		1½-4½-3½		2-3½-3½
5000 m, 10,000 m		1½-4-3½		
Marathon		1½-4½-3½		5½-5½-1½
Shot, discus, hammer		3-7-1		2½-3-4
High, long, triple jump		1½-4-3½		
Distance runners—All-Africa Games 1995 and 2004[10]	17	1.4-3.4-4.2	11	2.2-2.6-4.0
Middle distance	16	1.6-2.9-4.3	15	2.0-2.3-4.6
Long distance	16	1.4-3.9-3.9	8	2.1-3.4-3.3
Marathon				
Cycling[1]				
Track		2-5½-2½		
Road		1½-4½-3		
Triathlon—World Championships 1997[11]				
Senior	20	1.9-4.2-3.0	18	2.8-3.6-3.0
Junior	29	2.3-4.2-3.1	20	3.5-3.6-2.9
MOBILE FIELD SPORTS				
Field hockey[1,3]		2½-4½-2½	17	3.5-4.5-2.2
Soccer[1,3]		2½-5-2½	11	4.2-4.6-2.2
Soccer—Copa America 1994[12]				
Goalkeepers	15	2.6-5.5-1.9		
Other players	95	2.0-5.3-2.2	17	4.1-4.5-2.4
Lacrosse[2,3]	26	2.9-5.4-2.5		
CONTACT FIELD SPORTS				
Rugby[1]		3-6-2	56	4.5-4.5-1.8
Rugby—South Africa Provincial 1995[13]				
Forwards	161	4.5-6.6-1.4		
Backs	188	3.1-5.8-1.9		
Rugby—South Africa schoolboys[14]				
Primary school backs	110	2.6-4.3-3.2		
Primary school forwards	127	3.0-4.4-2.9		
Secondary school backs	170	2.9-4.7-2.1		
Secondary school forwards	199	3.3-5.5-1.9		
Australian football[2]	23	2.1-5.7-2.5		
American football[1]				
Linemen		5-7½-1		
Backs		3-5½-1½		
SET FIELD SPORTS				
Baseball[1] (males only)		2½-5½-2		
Softball[1] (females only)			22	3.8-4.3-2.7
Cricket[3]			12	4.9-4.4-2.0
Cricket—South Africa club[15]	88	3.8-4.6-2.4		
Golf[1]		4-5-2		4-4-2½

COURT SPORTS				
Basketball[1]		2-4½-3½		4-4-3
Basketball—World Championships 1994[16]				
Guards			64	2.9-3.9-2.6
Forwards			57	2.8-3.5-3.2
Centres			57	3.2-3.1-3.4
Netball[3]			7	3.0-3.8-3.3
Volleyball[1]		2.5-4.5-3.5		3.5-4.0-3.0
Volleyball—USA teams 1988[17]	11	1.9-4.7-3.6	15	2.8-3.3-3.6
MARTIAL ARTS				
Judo[1]		2-6½-1½		4-4-2
Wrestling[1]				
(<60 kg)		1½-5½-2½		
(60-79.9 kg)		2-6½-1½		
(80-99.9 kg)		2½-7-1		
(>100 kg)		4-7½-1		
Boxing[1]				
(<60 kg)		1½-5-3		
(60-79.9 kg)		2-5½-2½		
(80-99.9 kg)		2½-6-2		

[1]Rounded (½-unit) means derived from several samples in Carter and Heath (1990). For some samples, exact means are given; [2]Withers, Craig, and Norton (1986); [3]Withers et al. (1987); [4]Carter and Marfell-Jones (1994); [5]De Ridder et al., "Kinanthropometry in African Sports: Somatotype" (2000); [6]Claessens et al. (2001, 2002); [7]Ackland et al. (2001); [8]De Rose et al. (1989); [9]Claessens et al. (1991); [10]De Ridder et al., "Kinanthropometry in African Sports: Somatotype" (2000) and De Ridder et al. (2001); [11]Landers et al. (1999); [12]Carter et al. (1998); [13]Wilders and De Ridder (2001) [14]De Ridder (1992); [15]De Ridder and Peens (2000); [16]Carter et al. (2005); [17]Carter et al. (1994).

Racquet Sports: Tennis, Badminton, Squash, Racquetball

Competitors in racquet sports have variable body types. The mean somatotype for male tennis players is **2-4½-3,** while females are **3½-3½-3.** For high-level female tennis players the somatoplots cover a fairly wide circular area, with most means (of the various groups) just to the left of the centre of the somatochart. Eight female tennis players at the All-Africa Games in 1995 had a meso-endomorphic somatotype of **4.3-3.6-2.2.** All were on or to the left of the mesomorphic axis on the somatochart. Male badminton players have very similar body types to male tennis players; however, squash players of both sexes were higher in mesomorphy than tennis players (figure 4.7).

Aquatic Sports

Two sports using specific enclosed pools (swimming and water polo) and two using open water (rowing and canoeing) are reviewed.

Swimming

For swimmers measured at the 1991 World Swimming Championships, the male somatotype was **2-5-3,** which placed these athletes in the ecto-mesomorphic category of the somatochart. This group was characterized by a low level of adiposity, moderate to high musculoskeletal robustness and moderate linearity. The female swimmers were rated at **3-4-3.** Most studies on high-level swimmers during the past 20 years have revealed very similar body types, particularly for those individuals competing in freestyle, butterfly and breaststroke events. Backstrokers, however, have tended to be a little more ectomorphic than swimmers competing in the other strokes. Female swimmers at the All-Africa Games in 1995 had a somatotype of **3.5-3.8-2.8,** which is similar to that for the other samples mentioned (figure 4.7).

Water Polo

Male elite water polo players at the 1991 World Swimming Championships had a mean of **2½-5½-2½,** or balanced mesomorphy. Data from Hebbelinck, Carter,

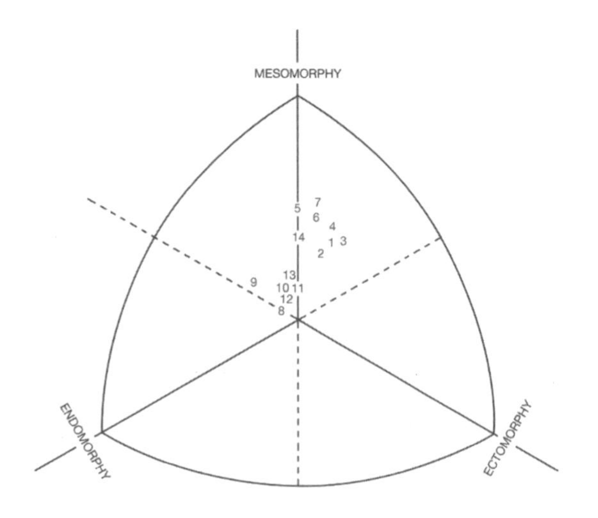

Males:

1. Tennis (2-4½-3)
2. Badminton (2.5-4.6-3.2)
3. Squash (2½-5-3)
4. Swimming (2-5-3)
5. Water polo (2½-5½-2½)
6. Rowing (1.9-5.0-2.5)
7. Canoeing (1.6-5.0-2.2)

Females:

8. Tennis (3½-3½-3)
9. Badminton (4.1-4.4-2.5)
10. Squash (3.4-4.0-2.8)
11. Swimming (3-4-3)
12. Water polo (3½-4-3)
13. Rowing (2.8-3.9-2.6)
14. Canoeing (2.4-4.6-2.3)

Figure 4.7 Mean somatoplots for racquet and aquatic sports.

and De Garay (1975) on the 1968 Olympic water polo players placed them in the endo-mesomorphic **(3-5½-2½)** category on the somatochart. The more intense training required of players since the 1960s is the most likely reason the endomorphy rating has been reduced by ½ unit to 2½ units. Female players at these championships had a mean rating of **3½-4-3,** which is on the border between the central and endo-mesomorphic somatotype categories, with mesomorphy being the dominant feature of their somatotype. There were some differences according to playing position for both males and females (figure 4.7).

Rowing

Male heavyweight rowers at the international level have a somatotype of **2½-5½-2½,** which is a high balanced mesomorphic rating. This is expected because of the requirement for both strength and power to pull the boat through the water, while rowers also need a high level of cardiorespiratory endurance because the event continues for 5 to 6 min. Female rowers at the elite level have central or balanced mesomorphic somatotypes, with a mean rating of **3-4-3,** which is almost identical to that of swimmers (Carter, Aubry, and Sleet 1982; Hebbelinck et al. 1980). However, they are generally taller and heavier than swimmers (figure 4.7).

Claessens and colleagues (2001, 2002) somatotyped male and female junior rowers at the World Championships in 1997. The males had a mean of **2.3-4.7-3.0,** and the finalists had lower endomorphy (2.2 vs. 2.4) than nonfinalists; but there were no differences between scullers and sweep oarsmen. The females had a mean of **3.4-3.7-2.6;** the finalists and nonfinalists did not differ, but the scullers were lower than sweep rowers on endomorphy (3.2 vs. 3.6). Thus the junior rowers were generally lower in mesomorphy (especially males) compared to heavyweight rowers from the 1968 and 1976 Olympics.

At the 2000 Olympics, Ackland and colleagues (2001) studied lightweight male and female rowers. Their respective somatotypes were **1.4-4.4-3.4** and **2.0-3.4-3.4.** These means are slightly lower in mesomorphy than those of lightweight competitors at the Pan American Games in 1986 (De Rose et al. 1989). The male mean was **1.8-5.1-2.9,** and the female mean was **2.4-3.9-3.3.** The mean for female open-class rowers at the 2000 Olympics was **2.8-3.9-2.6,** and for males the mean was **1.9-5.0-2.5.** These comparisons suggest that there are differences between juniors, lightweight and open-class rowers for both males and females. The more recent studies suggest lower endomorphy in both males and females at the Olympic level.

Canoeing

When the data from Olympic male (Carter 1984) and high-level Eastern European canoeists are combined (Mészáros and Mohácsi 1982; Štěpnička 1974), these athletes have a mean somatotype rating of **2-5½-2½.** This is an identical somatotype to that reported previously for international oarsmen, which is probably attributable to the similar demands of canoeing and rowing. Female canoeists, on the other hand, have a mean rating of **3-4½-2½** on data taken from competitors in the 1968 and 1976 Olympic Games (Carter 1984). It is interesting to note that although male rowers and canoeists have almost identical body types, female canoeists are more mesomorphic and less ectomorphic than female rowers. This is confirmed by the somatotypes of sprint paddlers at the 2000 Olympics, showing that the males had a mean of **1.5-5.0-2.2** (similar to male Olympic rowers) and the females of **2.4-4.6-2.3** (different from female Olympic rowers) (figure 4.7).

Gymnastics and Power Sports

Gymnastics, diving and skating, as well as weightlifting and bodybuilding, are reviewed in this section.

Gymnastics

The majority of high-level male gymnasts are either balanced mesomorphs or ecto-mesomorphs with a mean rating of **1½-6-2.** They are mature in comparison to female competitors and are very powerful athletes, with heavy musculature and little adiposity. However, modern female gymnasts at the highest level are mostly pre- or midadolescent, which makes them younger than most other international athletes except perhaps for some swimmers. These athletes are ecto-mesomorphic or central types, with a mean rating of approximately **2-4-3;** however, older competitors have often been up to 1.0 unit higher in endomorphy. Carter and Heath (1990) suggested that body type was a very important factor from an early age in gymnastics; therefore it is important in talent identification for coaches to select their future gymnasts carefully (figure 4.8).

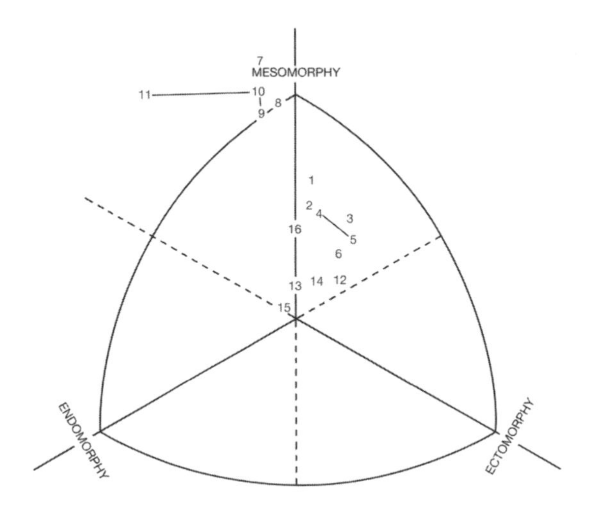

Males:

1. Gymnastics (1.5-5.6-2.1)
2. Diving (2.0-5.3-2.4)
3. Skating (1½-5-3)
4. Cycling – track (2-5½-2½)
5. Cycling – road (1½-4½-3)
6. Triathlon (1.9-4.2-3.0)
7. Body building (2-8½-1)
8. Weight lifting (<60 kg) (1½-7-1)
9. Weight lifting (60-79.9 kg) (2-7-1)
10. Weight lifting (80-99.9 kg) (2½-8-1)
11. Weight lifting (>100 kg) (5-9-½)

Females:

12. Gymnastics (1.8-3.7-3.1)
13. Diving (2.8-3.8-2.8)
14. Skating (1½-5-3)
15. Triathlon (3.5-3.6-2.9)
16. Body building (2½-5-2½)

Figure 4.8 Mean somatoplots for gymnastics and power sports, cycling and triathlon. Lines connect events in the same sport.

Claessens and colleagues (1991) reported mean somatotypes of **1.5-5.6-2.1** for males at the World Championships at Rotterdam in 1987. The males were mostly balanced mesomorphs (58%) and ecto-mesomorphs (38%). The mean for females was **1.8-3.7-3.1,** with most classified as ectomorph-mesomorphs (61%) and central somatotypes (19%). The somatotypes of 13- to 18-year-old gymnasts were compared with reference groups of Flemish girls of similar ages. The young gymnasts had mean somatotypes clustered around **2-4-3½**, and the reference group means were in the central and balanced endomorphy categories around the **4-3-3** somatotype. This demonstrates that only a small, select group of female somatotypes from the general population are suited to, or can acquire the somatotypes needed for, success at the top level of the sport.

In contrast, seven female gymnasts (age = 14.3 years) at the All-Africa Games in 1995 had a mean of **2.2-3.2-4.6,** decidedly more ectomorphic than most other samples. This suggests that most of these gymnasts do not have somatotypes that are likely to be successful at the world level.

Diving

It is commonly believed that divers have almost identical body types to gymnasts, but this is not as clear when comparisons are made at the highest level. In a sample of 82 divers assessed at the 1991 World Championships, the mean somatotype of male divers was **2.0-5.3-2.4,** which classified them as balanced mesomorphs; when they were compared with gymnasts they had slightly more endomorphy, less mesomorphy and more ectomorphy. Female divers from the same championships were central somatotypes with a mean of **2.8-3.8-2.8.** They were very similar to the gymnasts in mesomorphy and ectomorphy but were 1.0 unit higher in endomorphy (figure 4.8).

Skating

Male skaters who compete in figure skating and ice dancing have high levels of mesomorphy, are reasonably high in ectomorphy and are low in endomorphy, with a rating of **1½-5-3** (figure 4.8). When they are compared with other gymnastics types, notably gymnasts and divers, they are lower in mesomorphy but higher in ectomorphy. Conversely, female skaters have a mean rating of **2½-4-3,** which gives them an almost identical body type to their counterparts in gymnastics and diving (Ross et al. 1977).

Weightlifting

Using combined data from three Olympic Games, Carter (1984) found that the higher the weight class, the higher the endomorphy and mesomorphy ratings but the lower the ectomorphy component. All weightlifters require large muscle mass, which gives them very high ratings in mesomorphy. They are among the most mesomorphic of all the high-level athletes, with lifters over 80 kg (176 lb) being the least ectomorphic. The combined mean of this group was **2½-7½-1** (figure 4.8).

Bodybuilding

Bodybuilders are of interest because their sport emphasizes extremely high mesomorphy and low endomorphy. The average somatotype of male bodybuilders is about **2-8½-1** (Carter and Heath 1990). The somatotype of amateur bodybuilders is **1½-8½-1** (balanced extreme mesomorphy), while the top competitors range from 10 to 12 in mesomorphy (Borms et al. 1986). A recent rating of the champion bodybuilder, Ron Coleman, was **1-13-1.** While some athletes in other sports, notably shot put, hammer throw, higher classes in weightlifting and wrestling and American football have very high mesomorphy ratings, these are often accompanied by ratings in endomorphy from 3 to 6 or more, so that the mesomorphy is not so obvious as it is in bodybuilders. Female bodybuilders have somatotypes that cluster around **2½-5½-2½,** or balanced mesomorphy, with the top competitors probably around 6 or 6½ in mesomorphy (figure 4.8). A recent study of 11 female Brazilian bodybuilders (Cyrino et al. 2002) resulted in a somatotype of **1.6-4.6-2.5,** slightly more ecto-mesomorphic than for those in Carter and Heath (1990).

Track and Field, Cycling, Triathlon

In this section we review somatotype data for track and field along with cycling and triathlon.

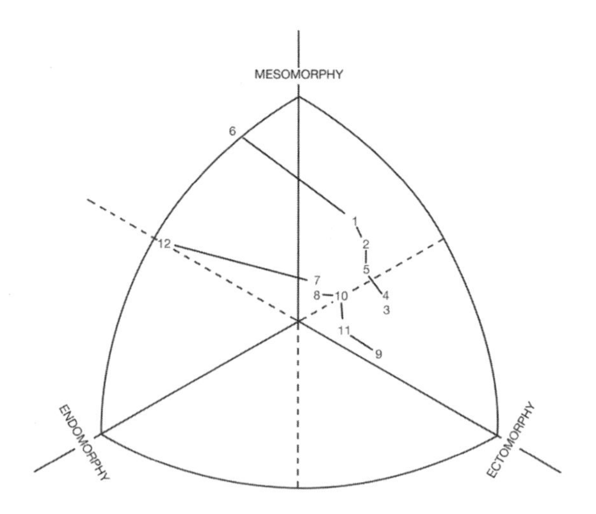

Males:

1. Sprinting (1½-5-3)
2. 400 m + 400 m hurdles (1½-4½-3½)
3. Middle & long distance (1.5-3.2-4.2)
4. Marathon (1.4-3.9-3.9)
5. High, long and triple jumping (1½-4-3½)
6. Shot, discus and hammer (3-7-1)

Females:

7. Sprinting (2½-4-3)
8. 400 m (2-3½-3½)
9. Middle & long distance (2.1-2.4-4.6)
10. Marathon (2.1-3.4-3.3)
11. High, long & triple jumping (2½-3-4)
12. Shot and discus (5½-5½-1)

Figure 4.9 Mean somatoplots for track and field. Lines connect events in the same sport.

Track and Field

Athletes from three Olympic Games (n = 452) were studied by Carter (1984), who plotted the somatotypes of male athletes in 11 track and field events (figure 4.9). The mean somatoplots were almost on a straight line, running from extreme endo-mesomorphy for the throwers to ectomorphy-mesomorphy for the distance runners. The only group that did not fit this pattern were marathon runners, who were slightly more mesomorphic and less ectomorphic than 5000 and 10,000 m runners. Carter (1984) obtained relatively similar results for female Olympic athletes to those for males, with the means of the female athletes approximately parallel to those of the males but lower on the somatochart.

For athletes in middle distance, long distance and marathon from the All-Africa Games in 1995, the male runners were extremely low in endomorphy, with the majority of somatotypes in the meso-ectomorphic category. The marathoners were more mesomorphic (3.9 vs. 3.4 and 2.9) than middle and long distance runners, respectively. The female runners were also mostly meso-ectomorphic, with the long distance runners more ectomorphic than marathoners and the latter more mesomorphic than both middle and long distance runners. Sexual dimorphism was observed, with the males being slightly lower on endomorphy and higher on mesomorphy than females. The males and females were similar in ectomorphy. The slightly higher mesomorphy in marathoners of both sexes is somewhat surprising, but perhaps a more robust musculoskeletal system is an advantage to sustain the heavy training and hill climbing required in undulating courses.

Recently, Underhay and colleagues (2005) combined data for long distance runners (3, 5, 10 km) from the 1994 and the 2005 All-Africa Games and found means of **1.6-2.7-4.2** for males (n = 21) and **2.0-2.7-4.0** for females (n = 19). Males were significantly lower in endomorphy than the females. In a study of elite 10 to 21 km South African male distance runners (n = 23), Semple, Neverling, and Roussouw (2003) found a mean somatotype of **1.5-3.4-4.0,** which is higher in mesomorphy than observed in the other studies. These studies of African athletes who are currently the best in the world suggest that these runners are more ectomorphic and less mesomorphic than those in previous studies. In these samples about 90% were classified as "black".

Cycling

In the past it was common for cyclists to compete both on the track and on the road. However, the last two decades have seen specialization occur, with two distinct body types emerging (figure 4.8). Male track cyclists are a powerful group with a somatotype of **2-5½-2½,** while road racers **(1½-4½-3½)** have lower mesomorphy and endomorphy ratings and are higher in ectomorphy (White et al. 1982). The latter rating is quite understandable for an endurance cyclist, who mostly needs endurance and at times needs a reasonable amount of power in this event. This will often be the case during hill climbs or short sprints, in which a tactical advantage is required.

Triathlon

A study of 16 highly trained "international" female triathletes in 1990 revealed that they had a mean somatotype of **3.1-4.3-2.6,** and they were generally closer in somatotype to Olympic swimmers than middle distance runners (no comparisons with cyclists were available) (Leake and Carter 1991). In a study of male South African (in 1993) and USA triathletes (in 1987), no differences in somatotype were reported between the two samples, and the combined mean was **1.7-4.3-3.1** (Travill, Carter, and Dolan 1994). This mean rating did not differ from that of 15 World Championship 1500 m swimmers **(1.8-4.5-3.4)** or from a calculated Olympic model of combined swimmers and cyclists **(1.5-4.6-3.3)** (figure 4.8).

The somatotypes of World Championship senior and junior triathletes were reported by Landers and colleagues (1999). The male juniors **(2.3-4.2-3.1)** were slightly more endomorphic than the seniors **(1.9-4.2-3.0)**. Also, the junior females **(3.5-3.6-2.9)** were more endomorphic than the senior females **(2.8-3.6-3.0).** The senior males were very similar to those in the study by Travill and colleagues (1994). Six South African triathletes had a mean of **2.9-4.7-2.7** (Wildschutt et al. 2002). About half were balanced mesomorphs and the other half were evenly divided between endo- and ecto-mesomorphs. The "best 3" **(2.1-4.3-3.2)** were more ectomorphic and less endomorphic and mesomorphic than the "rest" **(3.6-5.0-2.0)** and were also closer to the Caucasians in Travill and colleagues' study.

Mobile Field Sports

Male competitors in mobile field sports are low in both endomorphy and ectomorphy, with reasonably high levels of mesomorphy, enabling them to perform with speed and agility (figure 4.10). Female competitors in this group are high in mesomorphy with moderately high levels of endomorphy, but are low in ectomorphy.

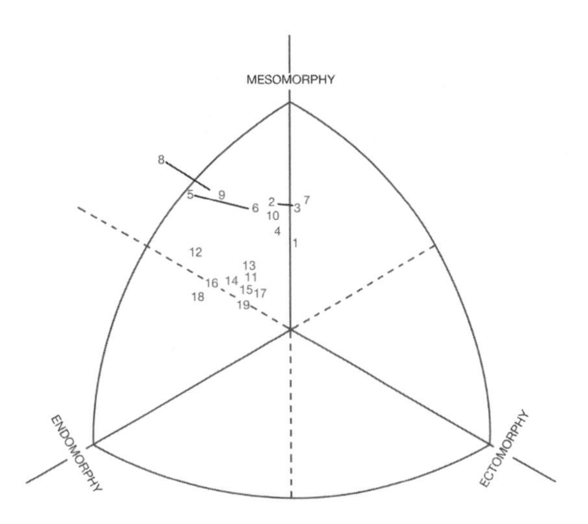

Males:

1. Field hockey (2½-4½-2½)
2. Soccer goalkeepers (2.6-5.5-1.9)
3. Soccer other positions (2.0-5.3-2.2)
4. Lacrosse (2.9-5.4-2.5)
5. Rugby forwards (4.5-6.6-1.4)
6. Rugby backs (3.1-5.8-1.9)
7. Australian rules (2.1-5.7-2.5)
8. American football – linemen (5-7½-1)
9. American football – backs (3-5½-1)
10. Baseball (2½-5½-2)
11. Cricket (3.8-4.6-2.4)
12. Golf (4-5-2)

Females:

13. Field Hockey (3.5-4.5-2.2)
14. Soccer (4.2-4.6-2.2)
15. Lacrosse (4.1-4.5-2.4)
16. Rugby (4.5-4.5-1.8)
17. Softball (3.8-4.3-2.7)
18. Cricket (4.9-4.4-2.0)
19. Golf (4-4-2½)

Figure 4.10 Mean somatoplots for mobile, contact and set field sports. Lines connect positions in the same sport.

Field Hockey

For field hockey, the average male somatotype from several studies was **2½-4½-2½,** but there were differences in mesomorphy and ectomorphy among countries. Female hockey players are endo-mesomorphs with a mean of **3½-4½-2½.**

Soccer

Six national men's teams were somatotyped at the 1995 Copa America in Uruguay. The 15 goalkeepers **(2.6-5.5-1.9)** were more endomorphic than the 95 other players **(2.0-5.3-2.2)**, but there were no differences among playing positions or by team or team place when goalkeepers were excluded. Most players were in the balanced mesomorphy category, with others in the endo-mesomorphy or ecto-mesomorphy category. The Brazilian 1994 World Cup team had a mean somatotype of **2.1-4.4-2.0,** which suggests that they were slightly less mesomorphic than the Copa America teams. The mean somatotypes of other national teams show considerable variation, suggesting that the physiques best suited for soccer can be different according to country. For example, mesomorphy ranged from 3.2 for the India team to 5.9 for the CSSR team (Carter and Heath 1990). Data are limited for female soccer players, but the mean somatotype for South Australian female players was **4.2-4.6-2.2** (Withers et al. 1987).

Lacrosse

The somatotypes of male **(2.9-5.4-2.5)** and female **(4.1-4.5-2.4)** South Australian lacrosse players are very similar to those of soccer players from the same province.

Contact Field Sports

Individuals competing in contact field sports tend to be basically high in mesomorphy with some degree of endomorphy, depending on their position on the field (figure 4.10). This type of athlete must sustain "hard knocks" for prolonged periods of time, and thus these competitors are generally low in ectomorphy. Contact sport players need to be fast and powerful; and if they are not basically mesomorphic, they need special skills with high levels of agility, speed and dynamic balance.

Rugby

The mean somatotype of rugby players is around **3-6-2;** however, forwards tend to be endo-mesomorphic, and backs tend toward balanced mesomorphs (Carlson et al. 1994). In both position groups in recent years there has been a slight increase in mesomorphy with a similar drop in endomorphy. Anecdotal evidence attributes these minor shifts to more strenuous endurance and more intensive strength training.

Kieffer and colleagues (2000) found that the USA national team (n = 65) tight forwards **(3.7-7.1-0.9)** were more endo-mesomorphic than loose forwards **(2.5-5.9-1.9)** and backs **(2.4-5.9-1.8)**. Moreover, South African provincial players (n = 349) have increased in mesomorphy over the past two decades (Wilders and De Ridder 2001). There were some differences in playing positions among the forwards and backs, whose mean somatotypes were **4.5-6.6-1.4** and **3.1-5.8-1.9,** respectively. Similar findings were obtained by Rienzi and colleagues (2003) for 246 male players from 28 countries competing in the 1996-1997 Rugby Sevens in Uruguay. The forwards **(2.9-6.1-1.3)** were more endomorphic and mesomorphic and less ectomorphic than backs **(2.4-5.7-1.5)**. Players in the best teams were less endomorphic and more mesomorphic than those in the worst teams, but forwards and backs in the best teams did not differ in somatotype. A possible reason for the latter finding is that much more open-field mobility is needed among the forwards in the Rugby Sevens compared to the regular 15-player teams.

The development in somatotypes with age of young backs and forwards can be seen in the study of South African schoolboys by De Ridder (1992). The differences in somatotypes of 237 primary (age = 13.1 years) and 367 secondary (age = 18.1 years) school rugby players by position were small in the primary school sample (**2.6-4.3-3.2** for backs and **3.0-4.4-2.9** for forwards), but more marked in the secondary school sample (**2.9-4.7-2.1** for backs and **3.3-5.5-1.9** for forwards). These latter somatotypes are approaching those of the successful provincial-level players and illustrate the influence of morphological optimisation, age and training on the somatotype.

Australian Football

Australian football is a very mobile game with a large field to be covered by only 18 players, and therefore these athletes need a high level of cardiorespiratory endurance. Although there is heavy contact from time to time, this is not as sustained as in rugby or American football. The rounded mean somatotype for Australian football is **2-5½-2½,** which reflects the dynamic nature of the game. However, it should be understood that some specialist players, such as ball gatherers (midfield players), who need high levels of speed, are short and mesomorphic, whereas others such as ball markers (set position players), who need jumping and catching skills, are tall and more ecto-mesomorphic. Therefore, the spread of body types in Australian football is quite diverse.

American Football (Professional)

American football is a highly specialized and controlled contact sport in which player tasks are clearly defined. Linemen, whose task it is to impede or stop the progress of the opposing team members, have high levels of endo-mesomorphy and have an approximate mean somatotype rating of **5-7½-1.** Backfield players, on the other hand, who are responsible for the majority of the ball carrying, passing and defensive tackling, need to be strong, fast and agile, with an approximate somatotype rating of **3-5½-1½.**

Set Field Sports: Baseball, Cricket, Golf

Baseball, cricket and golf are generally played in set field positions and therefore do not have the cardiorespiratory requirements of the more mobile field sports. These sports have very special skill requirements, and running speed is also important in baseball and cricket. This gives rise to a wide spread of body types in these sports in comparison to many others (figure 4.10). However, the past 20 years have seen a slight rise in their mesomorphy component, as these athletes have become more concerned about the development of explosive power, which basically involves strength and power training (see chapters 8 and 9).

Male cricket players at the national club championships in South Africa were compared within five player positions. The means were quite similar except for wicket-keepers, who were mesomorph-ectomorphs **(2.8-3.7-3.4)**. The overall mean was **3.8-4.6-2.4,** with other positions being mostly endo-mesomorphs and endomorph-mesomorphs. Endomorphy means ranged from 2.7 (wicket-keepers) to 4.2 (spin bowlers); mesomorphy means ranged from 3.7 (wicket-keepers) to 4.5 (spin bowlers); and ectomorphy means ranged from 2.2 (spin bowlers and all-rounders) to 3.4 (wicket-keepers).

Court Sports

The three court sports, basketball, netball and volleyball, can be played indoors or outdoors.

Basketball

The majority of early data on elite male basketballers were gathered on Olympic and amateur players in Europe and North America. The mean somatotype of these athletes was **2½-4½-3½.** The spread of these players classified them as ecto-mesomorphs, ectomorph-mesomorphs and meso-ectomorphs, which seems perfectly logical when one takes into consideration the different playing positions on the court. For female basketball players, there was large somatotype variation, partly due to the different functions of the playing positions. In adult samples, the mean female somatotypes were to the left on the somatochart, around a **4-4-3** somatotype (figure 4.11).

However, somatotypes of female basketball players from the 1994 World Championships (n = 168 from 14 countries) showed that the guards **(2.9-3.9-2.6)** were more mesomorphic than centres **(3.2-3.1-3.4)** and less ectomorphic than both centres and forwards **(2.8-3.5-3.2)**. In addition, when the top four (T4) and the bottom four (B4) teams were compared by position, the T4 guards were more ectomorphic than the B4 guards (2.9 vs. 2.1); the T4 forwards were less mesomorphic (3.0 vs. 3.7) and more ectomorphic (3.7 vs. 2.7) than B4 forwards, but there were no differences between the centres (T4 vs. B4) on any component. At the All-Africa Games in 1995, 11 female basketball players had a somatotype of **3.1-3.4-2.7,** a central somatotype (De Ridder et al., "Kinanthropometry in African Sports: Somatotypes" 2000). This is similar to the overall mean of **3-3½-3** at the 1994 World Championships.

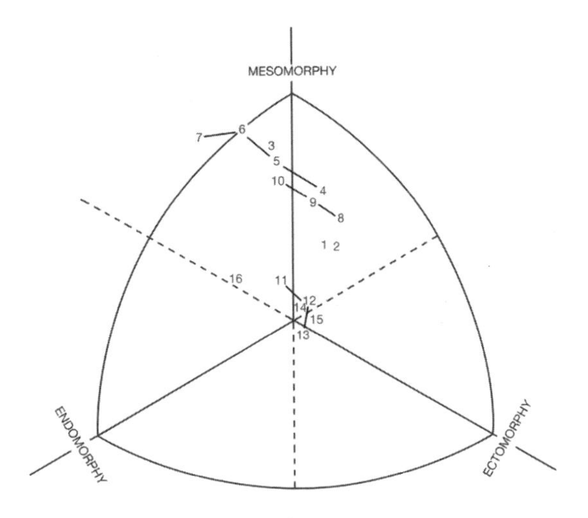

Males:

1. Basketball		(2½-4½-3)
2. Volleyball		(1.9-4.7-3.6)
3. Judo		(2-6½-1½)
4. Wrestling	(<60 kg)	(1½-5½-2½)
5. Wrestling	(60-79.9 kg)	(2-6½-1½)
6. Wrestling	(80-99.9 kg)	(2½-7-1)
7. Wrestling	(>100 kg)	(4-7½-1)
8. Boxing	(<60 kg)	(1½-5-3)
9. Boxing	(60-79.9 kg)	(2-5½-2½)
10. Boxing	(<100 kg)	(2½-6-2)

Females:

11. Basketball – guards	(2.9-3.9-2.9)
12. Basketball – forwards	(2.8-3.5-3.2)
13. Basketball – centers	(3.2-3.1-3.4)
14. Netball	(3.0-3.8-3.3)
15. Volleyball	(2.8-3.3-3.6)
16. Judo	(4-4-2)

Figure 4.11 Mean somatoplots for court sports and martial arts. Lines connect positions or events in the same sport.

Netball

Netball is a similar game to basketball, played mainly by females and limited to British Commonwealth or associated countries (figure 4.11). The mean somatotype of players in South Australia was **3.0-3.8-3.3,** with a spread on the somatochart that accounted for the various player positions. When the Australian netball players are compared with World Championship female basketballers, they are similar in endomorphy and mesomorphy to the guards but are higher in ectomorphy. Female netball players (n = 48) at the All-Africa Games in 1995 had a mean of **3.8-3.7-2.4;** their somatotypes were distributed around the ectomorphic axis with ratings from 1 to 5, and with more in and around the endomorph-mesomorph category (De Ridder et al., "Kinanthropometry in African Sports: Somatotypes" 2000).

Volleyball

The distribution of male volleyball players covers the range from endo-mesomorphy to meso-ectomorphy, with the majority of the somatotypes being ecto-mesomorphic. The mean rating for high-level players is **2½-4½-3½,** which is similar to that for international basketballers. Female volleyball players are reasonably varied in their body types, with an approximate mean of **3½-4-3,** which places them in the central region of the somatochart (figure 4.11).

Data from the USA national teams in 1988 show a slightly different somatotype distribution, with both the male **(1.9-4.7-3.6)** and female players **(2.8-3.3-3.6)** less endomorphic, and the females less mesomorphic and more ectomorphic, than in previous studies. There is more variation in mesomorphy and ectomorphy than in endomorphy for both men and women.

Martial Arts

Those athletes who compete in the various martial arts are highly mesomorphic, mainly because strength, power, speed and agility are fundamental to all styles (figure 4.11). However, they do vary in both their endomorphy and ectomorphy ratings according to the weight divisions in which they compete. This was illustrated in comparisons among combat sports by Kang (2001), who somatotyped 176 Korean national-level players selected from four different sports and divided into four weight groups. The mean somatotype components were significantly different among the sport events. The means by sport were taekwondo, **2.4-4.5-2.7;** judo, **3.1-6.7-1.4;** boxing, **3.0-5.0-2.5;** and wrestling, **2.8-6.1-1.6.** The means by weight class were <60 kg, **2.4-5.0-2.5;** 65 to 72.9 kg, **2.5-5.3-2.2;** 73 to 80.9 kg, **3.1-5.8-1.8;** and >81 kg, **3.6-6.1-1.5.** Regardless of sport, the somatotypes were more endo-mesomorphic with increased weight class. The results of this study are similar to those from studies of Montreal Olympic weight-classified sport participants (Carter, Aubry, and Sleet 1982).

Judo

Male judoists at the elite level are predominantly endo-mesomorphs; the remainder are mainly in the balanced mesomorph region of the somatochart. Their mean is approximately **2½-6½-1½;** however, if classified by weight, those athletes in the heavyweight classes often rate higher in both mesomorphy and endomorphy and lower in ectomorphy than lighter judoists. On the other hand, females who compete in judo at a high level distribute themselves mainly in the mesomorph-endomorph and central regions, with a mean of **4-4-2** (figure 4.11).

Seven female judoists at the All-Africa Games in 1995, mostly in lighter weight classes, had a mean of **3.0-4.1-2.4;** all were slightly endo-mesomorphic except for one, who was ectomorphic (De Ridder et al., "Kinanthropometry in African Sports: Somatotypes" 2000). In a comparison of top class Belgian male judoists (n = 24) and karateka (n = 24), Claessens and colleagues (1986) found that the somatotypes differed on all components—judoists were **1.9-5.8-2.0** and the karateka were **2.6-5.2-2.6.** The somatotypes of a South American selection of male (n = 20) judoists whose weight classes ranged from 60 to 142 kg were around **2½-5-1½** in the lighter weight classes and approximately **7-7-½** in the heavier weight classes (Gavilan and Godoy 1994). The female (n = 18) judoists in weight classes from 52.5 to 97.5 kg had somatotypes that ranged from **4-3-2** (lightest) to **6-5-1** (heaviest). For both sexes the athletes became more endomorphic and mesomorphic, as well as less ectomorphic, as weight class increased.

Wrestling and Boxing

As with other combative sports, endomorphy and mesomorphy increase and ectomorphy decreases in the higher weight classes in wrestling. The lighter wrestlers tend to be balanced mesomorphs and the heavier ones endo-mesomorphs. Their mean is **2½-6½-1½,** but they range from **1½-5½-2½** in the under 60 kg class to **4-7½-1** in the heavyweight class (Carter 1984). The mean somatotype for boxers is approximately **2½-5½-2½,** with a range from **1½-5-3** for lighter competitors to **2½-6-2** for heavier boxers (figure 4.11).

Exceptions to the Standard Body Shape and Size

It is not uncommon to find athletes at the state or national level, and occasionally at the international level, whose physiques do not appear to be suited to the sport or a specialized event within a sport (Ackland and Bloomfield 1995). One must decide when not to interfere with an individual's body build, and Peter Snell from New Zealand provided a good example of this phenomenon (figure 4.12). Snell was an outstanding middle distance runner, winning gold medals at Tokyo in 1964 for the 800 and 1500 m events, despite the fact that his muscular physique **(1½-6-2)** appeared more suited to the 200 m distance. However, with a maximal oxygen uptake ($\dot{V}O_2$max) of 5.502 L/min (73.3 ml·min^{-1}·kg^{-1}), Snell possessed the circulatory capacity to successfully compete in the middle distance events (Pyke and Watson 1978), even though he was carrying much more body mass than his competitors.

When one carefully examines participants in every sport, it is possible to find these "exceptions to the rule" because of the many factors that make up a championship performance, and it is obvious that some champions will succeed despite what appears to be an unsuitable body

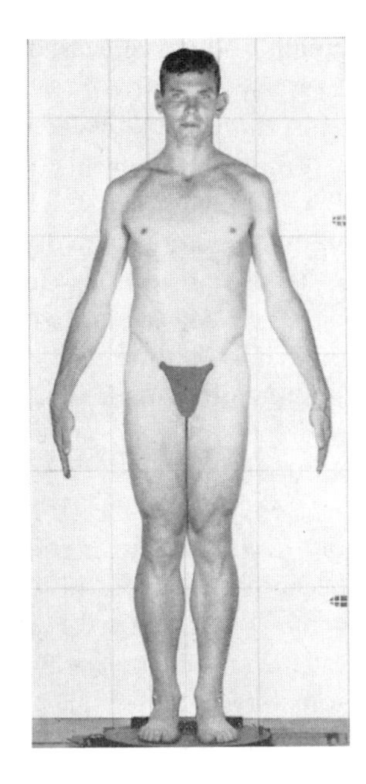

Figure 4.12 Peter Snell, somatotype = **1½-6-2**; an exception to the typical somatotype of an 800 to 1500 m runner.

type. With this in mind, coaches should not place too much pressure on their athletes with stressful intervention programs, because somatotype is only one of the general physical capacities needed by athletes.

Sport Selection and Somatotype

The term "self-selection" for various sports and events has often been used, because physically talented children find that they are better at some sports than others. Intelligent coaches also recognize that various physical capacities are important in certain sports and endeavour to identify children who possess these attributes. It is clear that an individual with a highly suitable somatotype will probably do well in a particular sport. Of the three somatotype components, mesomorphy is the most important because strength, explosive power, speed and agility are highly related to this characteristic.

Somatotyping has now reached the stage where there are enough data on junior and senior athletes to enable coaches or sport scientists to predict, with some degree of accuracy, the most appropriate sport or event for their athletes. One must always be aware, however, that predictions of this type are not foolproof, and if too much emphasis is placed on this variable the athlete and coach may be disappointed. The following statement by Carter and Heath (1990, 287) sums this up well: "The guidance of coaches, aware of the varying patterns of somatotypic development, is crucial in helping young athletes to discover their aptitudes for particular sports. With the somatotypes of successful athletes for models, the objective is to predict the most likely adult somatotypes, and to estimate the influence of appropriate nutrition and training in modifying the somatotype for optimal performance in the chosen sport".

Among male athletes, the **2-5-3** somatotype and those around it, and for females the **3-4-3** and those around it, seem to have the "best" all-round physiques for success in a variety of sports. These athletes may be excellent performers in one or several sports if they acquire the appropriate skills and other capacities. Two all-round physiques are shown in figure 4.13.

Somatotype Modification

Guidance with diet and exercise can help to modify body type within the limits of the individual's genotype. Athletes should know their somatotype and where they are on the somatochart in relation to other high-level performers in their sport. If they are within 1.0 unit of the mean of the elite group on all of the components, very little modification is needed. However, if they are outside the limits of the distribution, an attempt can be made to move them closer to that mean. To modify the body type of the individual, a personal intervention program should be developed in order to make appropriate changes in one or more of the components. This can be achieved in the following ways:

- Endomorphy: Athletes may wish to either increase or decrease this component in order to move closer to the mean of their group. Large amounts of complex carbohydrates and some additional protein are needed to increase fat mass. An athlete who wishes to reduce this rating should follow a low-kilojoule diet in conjunction with performing endurance training (see also chapter 5).

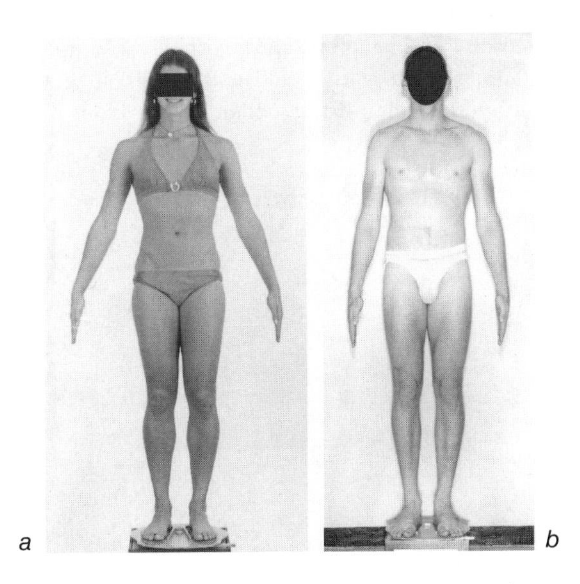

Figure 4.13 Two "all-round" physiques. *(a)* Female somatotype = **3½-4-3**; *(b)* male somatotype = **2½-5-3½**.

- Mesomorphy: In the majority of cases athletes will wish to increase their mesomorphy rating. They can do this by using a low-repetition, high-resistance strength and power training program as outlined in chapters 8 and 9. Competitors who wish to reduce their mesomorphy rating will find this very difficult to achieve, because testosterone secretions determine the rating in this component to a great extent. However, if no strength and power training are carried out and a low-kilojoule diet is accompanied by light endurance training, it may be possible to achieve a slightly lower rating.
- Ectomorphy: This component is changed when endomorphy or mesomorphy or both are altered, or with changes in height or weight.

Changes in Body Type During Growth

The question whether one can accurately predict the final adult somatotype from ratings made during childhood and adolescence has not been fully answered, mainly because too few longitudinal studies have been completed. However, both Carter and Heath (1990) and Malina and Bouchard (1991) have suggested that there are general trends that the somatotype usually follows. In general, young boys move from endo-mesomorphy toward balanced mesomorphy, then tend to decrease in mesomorphy, moving slightly toward ectomorphy in midadolescence. After this point they generally move back toward mesomorphy as their muscle mass increases. Girls move from endo-mesomorphy toward central somatotypes, thereby decreasing their mesomorphy in adolescence. They then move toward endo-mesomorphy and balanced endomorphy-mesomorphy in late adolescence. Carter and Heath (1990) suggested that individuals may differ considerably from these trends and that genetic factors as well as nutrition and exercise also play important roles.

A thorough knowledge of these trends can be important in talent identification for selected sports, because the informed coach will be better able to predict what the ultimate body type will generally be like. Experienced coaches and teachers in the areas of gymnastics and dance have been able to make quite accurate predictions, especially with females, while many coaches of males in sports that require body bulk, explosive power, agility and speed in their athletes have also made reasonably accurate forecasts.

Body type is therefore a variable that must be taken into account in any high-quality junior training program, because its modification can have a significant effect on some athletes. Several examples are available in the literature to show that at least two, and in some cases three, somatotype components have been changed, particularly in males for whom mesomorphy was an important factor in their success.

Summary

Somatotype, although a general physical characteristic, is very important to athletes when almost all the other variables that make up a high-level performance are equal. The review in this chapter shows that to be successful in a sport or event at the top level, athletes need certain somatotypes. There are large differences in somatotype between many sports, and often within a sport depending on player position or weight category. Several procedures can be used to modify the individual's somatotype, but the most widely used are endurance, strength and power training and nutrition. For the young athlete, it is also important for the coach to take maturation into account, because it is during this period that growth hormones and training can have a significant influence on body type.

five

Body Composition in Sport

Deborah A. Kerr, PhD; and Arthur D. Stewart, PhD

Assessment of body composition in sport is undertaken to determine the physique of the athlete and to monitor the effect of modifications to the athlete's training and dietary intake. When used correctly, body composition assessment is a valuable tool to assist the athlete in achieving peak performance. The data provided can quantify the tissues that propel or impede performance and profile morphological change over time in parallel with alterations in training. There is, however, the potential for misuse of body composition data. For instance, it is not uncommon for coaches to set arbitrary cutoffs for athletes' body fat levels that are anecdotal and not based on any scientific evidence linking performance with body composition. Therefore, it is critical that the sport scientist be able to advise athletes and coaches on the relationship between body composition and sport performance. Additionally, body composition is not just about taking the measurement, but also about understanding the impact that this may have on the athlete. It is important to understand that body composition is just one factor influencing sport performance. This chapter outlines the factors that affect body composition, describes the methods of assessing body composition and explains how to interpret body composition data.

Factors Affecting Body Composition

Important factors that affect body composition include genetics, growth, aging and nutrition. Genetic factors interact with nutrition and lifestyle factors in complex ways that are not completely understood. Both growth and aging are accompanied by changes in adipose tissue, muscle tissue and bone mass.

Genetics

Body composition is influenced by complex interactions between genetic and environmental factors. Identifying how the genetics of an individual interacts with environmental factors, such as nutrition and lifestyle, is difficult. Opinions have differed on the importance of genetics in obesity, but this may be due to the lack of a consistent approach in studying this area (Bouchard and Perusse 1993). Study designs include twin, family and adoption studies with body composition assessed by body mass index (BMI) or several skinfold sites. With use of these approaches, heritability has been estimated to be from very low to as high as 90% for BMI (Bouchard et al. 1988). But with most of these studies, it is difficult to separate the effects of shared environment from the genetic influences. Some researchers have attempted to control for these factors by comparing the BMI of twins reared apart with that of those reared together (Stunkard et al. 1990). The heritability appears to vary from between 40% and 70% for the BMI. The assumption is that the twins reared apart were not placed in a similar environment and that intrauterine factors had little influence. There exists, however, evidence that such intrauterine factors as poor nutrition may result in a low birth weight, which may predispose the newborn to increased risk of cardiovascular disease, diabetes and hypertension later in life. This theory—known as the intrauterine-programming hypothesis—remains hotly debated amongst the scientific community (Poulter 2001), with criticisms including the inability to sort out the confounding effects of genes and other unknown factors affecting the growth of the foetus.

Another approach to teasing out the gene–environment interactions, proposed by Bouchard and colleagues (1988), was to study the response of monozygotic (MZ) twins to overfeeding and overexercising. Since MZ twins share the same genes, any differences are assumed to be a result of nongenetic factors. The ability to gain weight in response to overfeeding appears to be under genetic control, as does the response to overexercising. Genetic variability in body composition (Bouchard 1997) means that some athletes may experience more difficulty in maintaining lower body fat levels than others. The implications of Bouchard's studies for athletes are twofold: When in a situation of energy surplus (for example, overeating or inactivity due to injury), some may gain body fat more rapidly than others and then may find it more difficult to lose body fat. Although there is currently no way to identify those athletes who may have more difficulty in losing body fat, one should consider this factor when dealing with athletes, especially if there is a family history of overweight and obesity.

Body composition is influenced by complex interactions between genetic and environmental factors, which include nutrition, exercise and other aspects of lifestyle.

Growth

Puberty is a time of rapid changes in size, shape and body composition for both sexes. The onset and timing of puberty, however, can vary considerably between individuals. This difference in maturation has important implications for sporting ability; the late-maturing boy will not be as strong as early-maturing boys, as the spurt in strength will follow the spurt in height (Ackland, Blanksby, and Bloomfield 1994). In girls, late maturation is an advantage in sports in which low body mass and narrow hips assist movement, as in gymnastics, ballet and distance running (Ackland, Elliott, and Richards 2003).

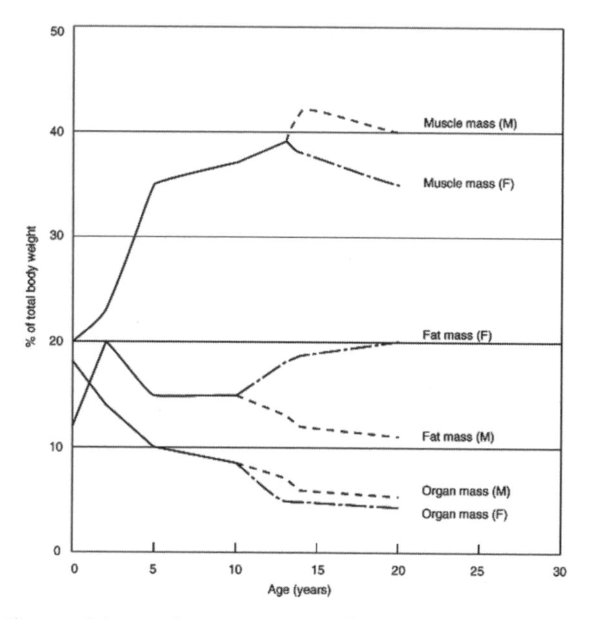

Figure 5.1 Body composition changes during growth: (M) = males, (F) = females.

Changes in Adipose Tissue During Growth

Typical growth in adipose tissue (figure 5.1) includes a period of rapid increase in the first year and a half of life, in which the tissue reaches a maximum level of approximately 20% of total body mass. This percentage decreases during childhood in response to the body's increased energy requirements for growth and movement, then reaches a plateau just prior to adolescence. It is important to understand that the absolute amount of adipose tissue actually increases during this period (Marshall 1978) but relative to the change in total body mass, fat decreases proportionally.

During pubertal growth the gain in adipose tissue drops markedly for both males and females in response to an increase in energy requirements. The absolute level of adipose tissue may even decline for males between the ages of 15 and 17 years (Marshall 1978). This growth phenomenon has the effect of sharply lowering the proportion of adipose tissue in relation to total mass (figure 5.1).

Following the adolescent growth spurt in size, the typical development of the adipose tissue compartment differs for males and females. While absolute tissue gains are reported for both sexes, the increasing deposition of fat among females is much more marked than for males. As a result, the percentage of adipose tissue with respect to total mass increases sharply for females, whereas among males the influence of a rapidly developing musculoskeletal system predominates so that no net change in the proportion of adipose tissue occurs. The absolute amount of adipose tissue appears to decline for adolescent males (He et al. 2004). Following puberty, the increased deposition of fat is much more marked in females than in males. This can be stressful for young female athletes and can affect their body image. It is important for sport scientists and coaches to understand that increased fat deposition occurs with maturation and not to place unrealistic expectations on female athletes to achieve low body fat levels. Body image issues are more common at this time; so when undertaking any assessment of body composition, the sport scientist needs to be sensitive to the impact that measuring may have on young athletes. One should consider ensuring a private location for taking the measurements and, where possible, using a same-gender anthropometrist. The incidence of eating disorders increases around the time of puberty. Although eating disorders are more common in females, young males are becoming much more concerned about their body image and should be treated with equal sensitivity.

Changes in Muscle Tissue During Growth

Composite data reported by Holliday (1978) showed that muscle mass increased from 20% of total body mass in the newborn up to 35% at 5 years of age (figure 5.1). The major increase in muscle mass occurs particularly from 1.5 to 5 years and corresponds to a shift in energy requirements from growth to greater activity levels.

A second growth spurt in muscle mass occurs during the adolescent growth period and is consistent with this time of increased activity and food consumption. Males under the influence of testosterone show a more marked increase in muscle mass than females. Muscle mass reaches a level greater than 40% of total body mass in adult males (healthy nonathletes) compared with a maximum of 39% in females by the end of adolescence. However, this value declines at full maturity for females to about 37% due to the rapid increase in adipose tissue growth at this time. Malina (1978) argued that these adult values are somewhat deflated since they are derived from elderly cadaver dissections. He suggested that the true proportion of muscle mass to total body mass would be more in the order of 45% for the adult male.

Changes in Bone Mass During Growth

The longitudinal growth of the human skeleton has usually occurred by early adulthood. That is, the process of ossification of long and short bones is completed, and no more increase in bone length will occur after the epiphyseal growth plates have closed. Bone will, however, continue to adapt to the mechanical stresses placed upon it by changing in width, bone mineral density and internal architecture. The time around puberty is critical for bone mineral accrual. About 26% of adult bone is accumulated during the two years around the period of peak bone accrual (Bailey et al. 1999). Therefore, puberty is an important time for optimising bone health, with sufficient calcium and physical activity the most important positive influences.

The skeleton is designed to adapt to the mechanical forces of daily weight bearing. Exercise has positive effects on bone mineral accrual in the growing skeleton (MacKelvie et al. 2003), especially if commenced prior to puberty (Bass et al. 1998) and if the activity involves high impacts as in jumping (Iuliano-Burns et al. 2003). For optimal bone health, a range of high-impact weight-bearing activities such as team and racquet sports and jumping activities should be maintained throughout childhood. To optimise bone health and other health benefits from exercise, participation in physical activity and sport throughout childhood should be actively encouraged (MacKelvie, Khan, and McKay 2002).

Aging

Aging affects total amounts of adipose tissue, with fat levels typically rising in both men and women throughout the middle years. Skeletal muscle mass and bone mass show age-related declines that are reversible to varying degrees with resistance training, even in later life.

Changes in Adipose Tissue With Aging

After the attainment of adult status when the energy demands for growth have diminished, environmental factors such as nutrition and exercise play a most important role. In both males and females, total mass gradually increases into middle age, most probably because total fat also increases, although this may not necessarily be portrayed in the skinfolds because much of the gain is in visceral fat. Generally, among adult women, skinfold thicknesses steadily increase up to age 60 and then decline again (Wessel et al. 1963). For the adult male, the proportion of body fat has been reported to remain fairly constant; however, a redistribution of fat has been observed (Norris, Lundy, and Shock 1963) whereby relative gains were noted for the trunk region. Since the time of that study, the global obesity epidemic has dramatically altered the picture for Western populations, and it is clear that fat levels appear to rise in both men and women throughout middle age.

Changes in Muscle With Aging

Age-related functional decline occurs in skeletal muscle mass; this is known as sarcopenia (Blumberg 1994). As much as one-third of muscle mass can be lost over three decades after the age of 50 years (Borst 2004). The aetiology of the skeletal muscle loss with aging is unclear, but factors such as declining physical activity, altered protein synthesis and turnover and reductions in insulin-like growth factor 1 (IGF-1) have been suggested (Carmeli, Coleman, and Reznick 2002). Exercise, in particular progressive resistance training, is known to cause muscular hypertrophy in the trained muscles. The ability to gain muscle strength appears to be present even late in life, as Fiatarone and colleagues (1994) have shown in a group of elderly subjects. The nursing home patients, who ranged from 72 to 98 years of age, demonstrated increases in muscle strength of 113%. Muscle protein turnover does appear to slow with aging compared with that in younger subjects (Welle, Thornton, and Statt 1995). However, Rennie and Tipton (2000), upon reviewing the literature on exercise and protein metabolism, concluded that the basal rate of protein turnover is reduced by inactivity and therefore normalized by exercise training.

Changes in Bone Mass With Aging

Bone is continually being broken down and rebuilt in a process known as remodelling, under the regulation of circulating hormones and local growth factors. The bone remodelling cycle takes between four and six months to complete in adults (Epstein 1988) and consists of five successive events: quiescence, activation, resorption, reversal and formation (Raisz 1999). Following resorption of a packet of bone by the osteoclast, new bone is laid down by the osteoblast. When bone resorption exceeds formation, bone loss occurs; if prolonged, this can lead to osteoporosis and increased fracture risk.

Osteoporosis is a condition of low bone mass associated with greater bone fragility and increased risk of fractures (World Health Organization 1994). Clinically, osteoporosis is defined in terms of the bone mineral density (BMD) that is below the age-adjusted reference range. An individual is considered osteoporotic if his or her BMD is 2.5 standard deviations (SD) or more below the young adult mean for bone density. Osteopenia is a condition of low bone mass in which the BMD is more than 1 SD below the young adult mean but less than 2.5 SD below this value. Sport osteopenia refers to low bone mass, seen most commonly in female athletes. The risk of developing osteoporosis and subsequent fractures is largely determined by the peak bone mass, achieved in early adulthood. Up to 60% to 80% of the variability in bone density has been attributed to genetic factors (Nguyen et al. 1998).

Studies of exercise effects on bone density in adults suggest that weight-bearing exercise (e.g., jogging, walking, running and gymnastics) has a positive effect on bone mass (Prince et al. 1995). However, more favourable effects on the skeleton have been found with resistance training in both premenopausal (Lohman et al. 1995; Snow-Harter et al. 1992) and postmenopausal women (Kerr et al. 1996, 2001; Nelson et al. 1994). In a unilateral exercise study in postmenopausal women, Kerr and colleagues (1996) compared two resistance training regimens that differed only in the number of repetitions of the weight lifted. The exercise effects on the skeleton were specific to the site of exercise loading and dependent on the weight lifted. Randomised controlled trials of exercise in men are lacking, or have been too short in duration to allow definitive conclusions. In two studies involving male subjects, the intervention was only for 16 weeks (Menkes et al. 1993; Ryan et al. 1994). As the bone remodelling cycle takes four to six months to complete, these studies are not of sufficient duration to detect bone density changes. Men and women of all ages should be encouraged to participate regularly in either weight-bearing activity or progressive resistance training. The American College of Sports Medicine (Kohrt et al. 2004) recommends weight-bearing activity three to five times per week, as well as resistance exercise two to three times per week, for bone health. Activities to improve balance and prevent falls should also be included for elderly men and women.

Nutrition

As a way of improving performance, athletes are often interested in losing body fat or gaining muscle bulk—commonly referred to as *bulking up*. An individual is in energy balance when energy expenditure (largely from physical activity) matches the energy intake from food. When this occurs, body weight is maintained at a constant level. For people to lose body weight, energy expenditure must exceed the energy intake from food. Alternatively, for them to gain weight, energy intake must exceed energy expenditure. Therefore, through manipulation of both energy expenditure and food intake, changes in body composition will occur.

An individual is in energy balance when energy expenditure (largely from physical activity) matches the energy intake from food. When this occurs, body weight is maintained at a constant level. For people to lose body weight, energy expenditure must exceed the energy intake from food.

The components of energy expenditure are the basal metabolic rate (BMR), the *thermic effect* of food and the energy expended in physical activity. These factors together ultimately determine an athlete's energy requirements to maintain energy balance. The type of activity undertaken, the size of the athlete, gender and quantity of muscle an athlete possesses influence energy requirements for a sport. Energy imbalance, leading to weight gain, can occur when athletes taper off their training or decrease activity because of injury without adjusting their food intake. If athletes continue to consume the same energy from food, weight gain will occur. Understanding energy balance is important for athletes hoping to obtain optimum sporting performance.

Body Composition Assessment

Methods of assessing body composition include three criterion and research methods: densitometry, air displacement plethysmography and dual-energy X-ray absorptiometry. Field methods include anthropometry, used by sport scientists primarily to assess changes in body composition in athletes in response to interventions; bioelectric impedance analysis; and indexes of height and weight.

Criterion and Research Methods

Densitometry, one of the most frequently used methods of ascertaining body composition, involves submersing the individual, determining the net underwater weight and then measuring the quantity of air remaining in the lungs following a maximal expiration. Air displacement plethysmography and dual-energy X-ray absorptiometry are methods that have been developed more recently.

Densitometry

The capacity for humans to float freely in water can be attributed to the density of all constituent tissues together with the buoyancy provided by entrapped air. A person with a high fat mass is commonly observed to float more readily than someone with a more muscular build. This observation exemplifies the principles of densitometry, which were first identified by Archimedes (287-212 BC). Densitometry was viewed as the criterion method of determining body composition until the 1970s.

Densitometry requires the body to be divided into a simple two-compartment model: fat-free mass (FFM) and fat mass (FM). Because fat (defined anatomically as ether-extractable lipid) is the only body constituent with a specific gravity of less than 1.0, its quantity can be estimated using several key assumptions. The density of FM is assumed to be a constant 0.90 g/ml, and FFM is assumed to have a uniform density of 1.10 g/ml. Although the former assumption is reasonable, for the latter to be true it is assumed that the constituent tissues of the FFM have constant densities and are in constant proportions. With these assumptions, measuring density involves calculating total body mass and volume and relating their ratio to % fat (Brozek 1960; Siri 1961). In assessing the relative underwater (hydrostatic) weight, the observer realizes the pivotal role of body gas and the effect of exhaling to differing extents. As a consequence, densitometric assessment relies on both determining the hydrostatic weight and estimating residual volume.

Task 1: Determining Net Hydrostatic Weight A tank of water similar to that shown in figure 5.2 is required for the measurement of net hydrostatic weight. For patient comfort and hygiene, the water should be filtered, chlorinated and heated to approximately 30° C and have a submersible seat suspended from a solid foundation above the tank. A force-measuring device (preferably a load cell or strain gauge, but otherwise an autopsy scale) is fitted to the top of the chair in series with the overhead point of attachment for the recording of weight.

The protocol listed in table 5.1 together with the water density data in table 5.2 should be adopted for the measurement of gross hydrostatic weight (Wgross). After this value has been determined, the following formulas may be used to calculate body density (Db).

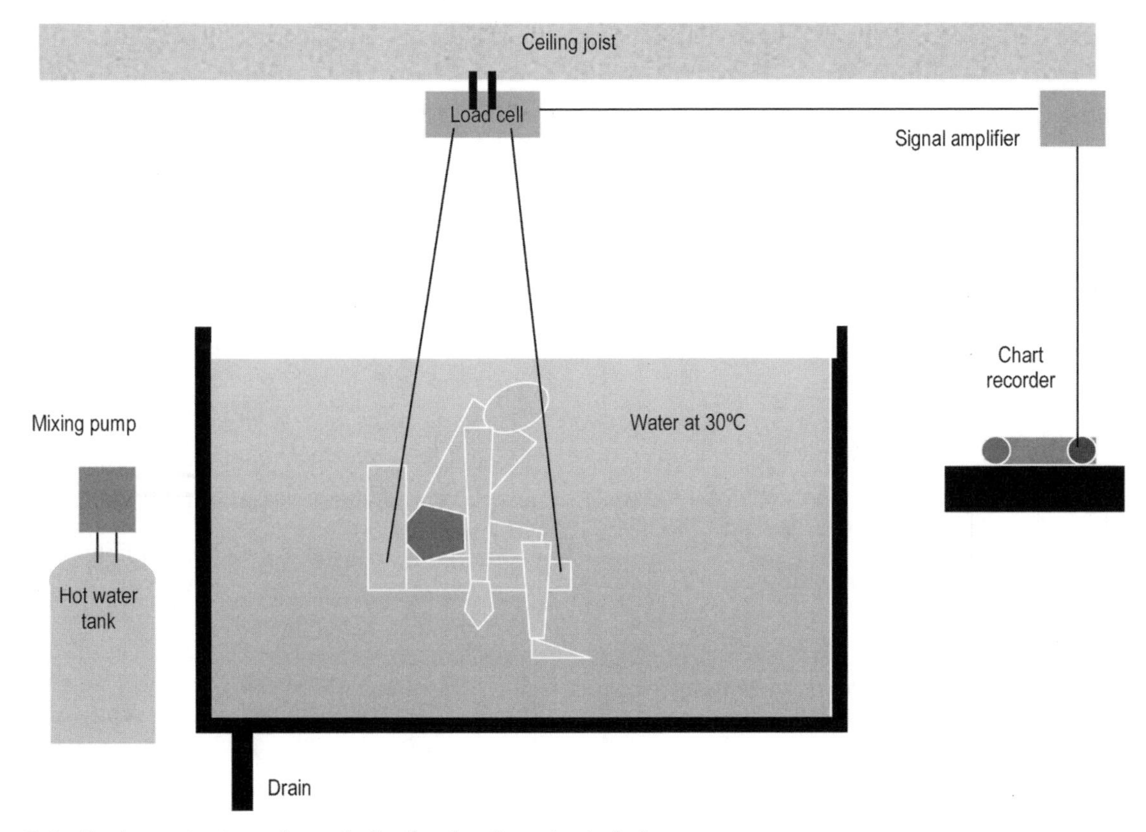

Figure 5.2 Equipment setup schematic for the densitometry technique.

Table 5.1 Protocol for the Measurement of Hydrostatic Weight

Procedure	Rationale or annotation
Record body weight in air (Wa) to the nearest 0.01 kg.	Subject is to wear a only light swimsuit that does not entrap any air.
Turn filtration system off.	This avoids turbulent currents that may affect recordings.
Record water temperature and determine water density (Dw).	Variations in water density influence body weight in water.
Direct the subject to enter the tank as gently as possible and stand on the floor of the tank away from the seat. Record the weight of the seat (Wseat) to the nearest 0.01 kg.	The added buoyancy given to the seat with the subject in the tank must be taken into account during recording (Wseat).
Instruct the subject to sit on the seat and then, with lungs full of air, to practice submerging completely and responding to prearranged signals to surface.	Three practice trials with air-filled lungs help to instil confidence in the subject.
Instruct the subject to exhale forcefully and completely while submerging, then hold a steady position until the signal to surface is given (~5 s). Record the weight of the subject plus seat to the nearest 0.01 kg.	Maximal exhalation is the critical factor in this technique.
Repeat the procedure a further nine times.	Subjects learn to expel more air with each additional trial.
The average of the last three trials is used to provide the submerged weight of the subject plus seat (Wgross).	This is done provided that all three trials are perceived as valid.

Table 5.2 Density of Water at Different Temperatures

Temperature (°C)	Density of water (g/ml)
25	0.997
31	0.995
35	0.994
38	0.993

From:

$$Db = Wa / V,$$

it follows that

$$V = Wa - Wnet / Dw - (RV + C)$$

where:

Wa = weight of the body in air (kg)

Wnet = gross weight in water (Wgross) – weight of the seat (Wseat)

Dw = density of water (g/ml)

RV = residual volume

C = a constant volume that represents the entrapped air in the intestinal tract (~100 ml)

Task 2: Determining the Volume of Entrapped Air Residual volume is the quantity of air remaining in the lungs following a maximal expiration. A constant value for RV can be predicted from sex, age and stature, but such predictions do not represent athletes who may have atypical thoracic breadth and depth or compliance of the chest wall. For these reasons it is recommended that RV be measured directly by a closed circuit nitrogen-oxygen dilution method (Wilmore and Behnke 1969). This involves breathing pure oxygen within a closed and nitrogen-free spirometry system until nitrogen equilibrium is reached. Then, RV may be calculated through measurement of the variation in nitrogen concentration and the volumes of gases involved at the commencement and cessation of the test, using the following equation. Note that the concentration of N_2 at the mouth during exhaling is assumed to accurately reflect the alveolar concentration of N_2.

$$RV = \frac{VO_2\,(E\,N_2 - I\,N_2)}{Ai\,N_2 - Af\,N_2} - DS \times \text{BTPS factor}$$

where:

RV = residual volume

VO_2 = initial volume of O_2 in the breathing bag

E N_2 = percentage of N_2 at equilibrium

I N_2 = percentage of N_2 initially in the breathing bag

Ai N_2 = percentage of N_2 initially in alveolar air during breathing of room air

Af N_2 = percentage of N_2 in alveolar air at the end of the test

DS = dead space of mouthpiece, valve and sensing element of the N_2 analyser

BTPS factor = wedge spirometer conversion factor

Because hydrostatic pressure affects lung volumes, it is advisable to measure RV with the subject in the tank rather than in air. Repeating the exact position achieved at the moment of underwater weight recording may be impossible, and most laboratories measure subjects seated or standing in the tank with the head out of the water. Predicted % fat is affected greatly by lung volume (a 100 ml variation is equivalent to about 0.7% fat) but less so by errors in the recorded weight (a 50 g difference is equivalent to about 0.5% fat), while variations in total mass and water temperature introduce a negligible effect. Common practice is to add a constant 100 ml to the RV estimation to approximate the volume of entrapped gas in the gastrointestinal tract. A study measured enteric air (Bedell et al. 1956) to be 115 ± 125 ml, and it is possible that higher values could be anticipated in athletes who might consume greater food quantities than normal, nonathletic subjects.

Common Methodological Problems In practice, densitometry presents problems for those who are not water confident or who fail to exhale completely. Common practice is to record at least five or more trials, which can itself be problematic. Although unlikely in the case of athletes, some subjects may experience positive buoyancy when fully submerged, in which case a weighted belt may be placed around the abdomen and its weight subsequently subtracted from Wgross.

Data Treatment With the assumptions and calculations presented, density can be related to % fat via either of these equations:

$$\% \text{ fat} = [(4.95 / d) - 4.5] \times 100 \quad \text{(Siri 1961)}$$

$$\% \text{ fat} = [(4.57 / d) - 4.142] \times 100 \quad \text{(Brozek et al. 1963)}$$

A large number of predictive equations rely on densitometric assessment for their criterion measurement (Durnin and Womersley 1974; Withers et al., "Body Density of Male Athletes" 1987, "Body Density of South Australian Females" 1987). Many generalized equations in widespread use have failed to provide accurate estimates of the measured body density for athletes, suggesting that the basic assumptions of constant density and proportions of tissues composing the FFM may be violated. Data from seven female and six male unembalmed cadavers (Clarys, Martin, and Drinkwater 1984; Martin et al. 1985) highlighted a range of muscle mass from 41.9% to 59.4%, bone mass from 16.3% to 25.7% and other lean tissues from 24.0% to 32.4% of total body mass. This, together with variability in bone density (Martin et al. 1985), may explain the apparent absurdity of professional football players being predicted to have –12% fat (Adams et al. 1982) despite their having observable superficial adipose tissue. These players were highly strength-trained Afro-Caribbean individuals whose FFM density was probably about 1.17 g/ml.

Compromised skeletal health involving lowered bone mineral content (BMC) or density (BMD) can also invalidate the assumption of a constant FFM density. Bone loss arising from the female athlete triad of disorders (amenorrhoea, eating disorder and osteoporosis) may be relatively rapid and possibly irreversible (Yeager et al. 1993).

In the light of such evidence, it is now generally accepted that the accurate measurement of body density should be used to monitor changes in the composition of an individual rather than to attempt to predict % fat. Increasingly common is the useful role that densitometry can play as part of a higher-level model of body composition using three or four components. In this scenario, body density is measured in conjunction with assessments of bone mineral and body water, effectively subdividing the body into four principal constituents (FM, BMC, water and fat-free soft tissue). A recent well-controlled four-compartment study investigating a heterogeneous sample of males and females, whites and blacks and athletes and nonathletes concluded that there was no positive relationship between either BMC or BMD and the mineral fraction of the FFM or the density of the FFM (Evans et al. 2001). Group mean errors of estimating % fat from densitometry have been predicted to be between 2% and 5% of body mass in athletes, and the causes of the deviations appear complex and not simply related to differences in muscularity (Prior et al. 2001).

Air Displacement Plethysmography

The development of air displacement plethysmography (ADP) is new, and this method has been sufficiently reliable for commercial viability only in recent years. The BOD POD body composition system (Life Measurement Inc., Concord, CA) is becoming established in nutritional and related studies. Air displacement plethysmography utilizes a similar theoretical model to densitometry in assessing body volume and relating the resultant density to % fat via a two-compartment model.

In ADP, the subject sits inside a sealed chamber of known volume. This is linked to a second chamber, also of a known volume, via a sealed diaphragm. The second chamber is perturbed via a computer signal, which leads to pressure fluctuations resulting from the volume changes. These pressure fluctuations are used to determine the volume of the test chamber, both with and without the subject inside. Because of the different behaviour of air compressed under isothermal and adiabatic conditions, scaling constants are applied to account for the combined effects of clothing, hair and body surface area (the latter is still predicted from DuBois and DuBois 1916). These scaling constants relate surface area to stature and mass raised to appropriate powers but ignore body density, composition, size or proportions. In practice, subjects wear swimwear and a swim cap. Although lung volumes can be estimated, they are preferably measured from pressure–volume differences induced by tidal breathing. Once such a breathing pattern is established, the airway is occluded at midexhalation for approximately 3 s, and the subject performs a gentle puffing manoeuvre that induces pressure changes; these are detected and related to volume by the application of simple gas laws. Compliance with the procedure is addressed via calculation of a score, referred to as the "figure of merit" (Dempster and Aitkens 1995), relating airway pressure to chamber pressure. Values of <1 are considered the criterion for accepting the validity of the thoracic gas measurement.

Dempster and Aitkens (1995) provide validation data between 25 and 150 L, representing the normal range for all but the largest athletes. Within-day reproducibility for ADP has been shown to be slightly better than for densitometry (1.7 vs. 2.3 %CV [coefficient of variation]; McCrory et al. 1995). Predicted % fat in 69 collegiate football players was lower by ADP than by dual-energy X-ray absorptiometry (DXA) or a three-compartment model using fat, bone mineral and residual mass (Collins et al. 1999). Factors including mixed ethnicity, the lack of a forced expiratory manoeuvre in ADP or the DXA software's not "expecting" such large muscular subjects could explain the observations. By contrast, a study comparing body composition of 47 "Euro-American" female athletes and 24 "controls" using ADP and DXA did not show significant differences (Ballard, Fafara, and Vukovich 2004), and the authors concluded ADP to be reliable and valid for measuring female athletes.

Studies using ADP with athletes are becoming more common, but those published to date have shown differences from densitometry (both under- and overestimating predicted % fat). The violation of key assumptions regarding the FFM in athletes remains a significant issue for both methods. Although ADP is likely to supersede densitometry in the future, further validation work in different athletic samples is necessary before it can be universally endorsed as a replacement for densitometry. A further criticism of ADP is related to the sensitivity to air currents in adjacent rooms, which may invalidate the calibration procedure, so appropriate location for such a device is essential.

Dual-Energy X-Ray Absorptiometry

Developed relatively recently as a means of determining body composition, DXA uses an X-ray source to quantify both total and regional body constituents, rapidly and with minimal subject discomfort.

Development of Dual-Energy X-Ray Absorptiometry

Dual-energy X-ray absorptiometry is a relatively new method for assessing body composition. Originally developed as a scanning system for investigating bone, DXA originated from photon absorptiometry using gamma-emitting radioisotopes, which, having passed through an absorbing substance, attenuate according to the atomic number of the molecules in their path (Cameron and Sorenson 1963). When photons of two energies are used, there is a steeper attenuation at the lower energy, and the ratio of the attenuation at the two energies (the *R-value*) is a measure of tissue composition. While fat is principally carbon, hydrogen and oxygen, lean soft tissue contains several other elements—especially sodium, chlorine, potassium and phosphorus.

Since the late 1980s, radionuclide sources have been replaced by DXA, using the same principles but with an X-ray source. In practice, the subject lies on a scanning table beneath which the source emits X rays in a linear (pencil beam) or fan (array beam) arrangement. Strong and weak beams follow the same path and are detected by a moving arm above the subject, synchronized to the X-ray emission from the source (figure 5.3). The principles of DXA are summarized comprehensively by Pietrobelli and colleagues (1996). R-values for the molecular constituents of the human body can be calculated independently from direct measurement. Lipids show values of 1.20 to 1.22; protein has a value of 1.29, glycogen of 1.30, intracellular fluid of 1.39 and bone mineral of 2.86.

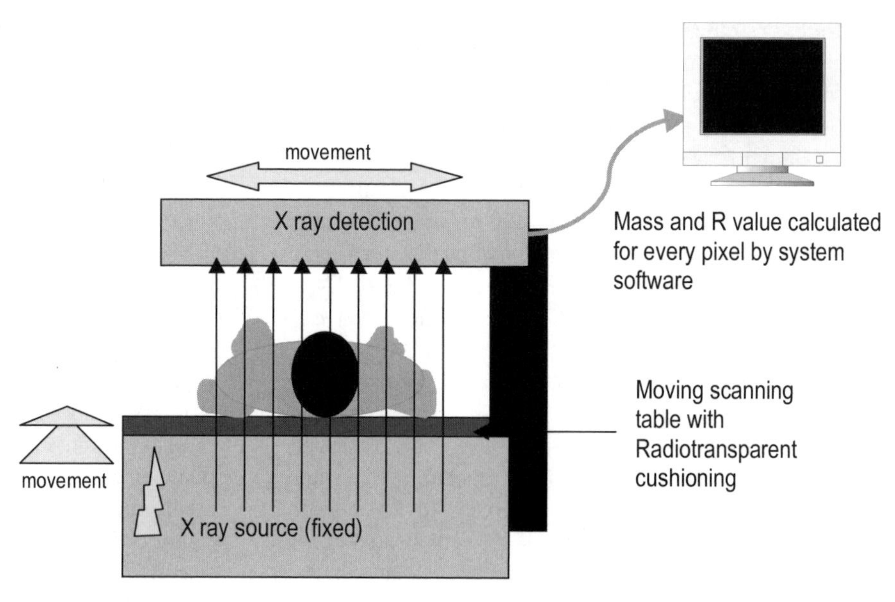

Figure 5.3 Typical dual-energy X-ray absorptiometry scanner equipment schematic and subject setup.

Basically, DXA makes a map of the R-values of each pixel in a scan area. Via system software, which is specific to each manufacturer, calculations are made to discern bone from nonbone and then to subdivide the nonbone pixels into fat and fat-free soft tissue. The end result is a regional map of the body, with the composition of BMC, fat-free soft tissue (FFST) and fat. Bone mineral density is a criterion for assessing bone disease such as osteoporosis and is an *areal density* (units = g/cm^2), where BMC (g) is divided by bone pixel area (cm^2).

Various assumptions govern the calculation of the composition as just outlined. Firstly, it is assumed that the R-value characteristics of individuals do not vary significantly. However, more crucial is the determination of the soft tissue component of those pixels that contain bone. These are estimated according to the ratio of fat to FFST in neighbouring nonbone pixels. This estimation relies on an assumed fat distribution model that calculates the composition predicted by the gradient of differing "soft tissue shells" and projected into the bone "shadow" (Nord and Payne 1995). Although perhaps 40% of a scan may contain bone pixels, this estimation can be done reliably in subjects who are of near-average composition. A third assumption is that the R-value is unaffected by the tissue thickness. While DXA can be calibrated for different body thicknesses, in practice, subjects whose anterior-posterior distance exceeds 25 cm (10 in.) may yield problematic results (Jebb, Goldberg, and Elia 1993).

Practical Issues Practical issues relating to DXA scanning data include the necessary regulations governing the administration of radioactive substances. Because X rays can increase the relative risk of developing cancer, it is imperative that X-ray dosage be minimized in any circumstance. In comparison to the substantial doses associated with conventional X rays, a whole-body DXA scan involves considerably less—normally about one to two days' background radiation, or the equivalent of that experienced in one transatlantic flight. Even so, to comply with local ethical procedures, it is common to have women of childbearing age undergo a pregnancy test before being scanned. Further practicalities include the dimensions of the scan table, which is normally about 185 cm (72.8 in.) long. Athletes significantly taller than this can be scanned in two sections and the scan data merged (Prior et al. 1997). Scan width may be less of an issue, but experience suggests that muscular athletes need to have their arms constrained to stay within the scan area. Most athletes are happy to be scanned in minimal sport apparel, which DXA software adds to the soft tissue calculation. Watches, jewellery and any metal items must be removed before the scan.

The utility of DXA for assessing athletes lies in its ability to quantify both total and regional body composition. It can do this with minimal subject discomfort; and with modern array beam devices having scans that typically last 4 min, it is far more rapid than alternative reference methods.

Precision and Accuracy Dual-energy X-ray absorptiometry has good precision compared with other techniques. Short-term precision is normally evaluated via repeated scans on several subjects and expressed as a % coefficient of variation (the SD divided by the mean, expressed as a percentage; i.e., %CV). Typical %CV values are ~3% for % fat, 0.9% for BMC and BMD and ~0.8% for total FFST (Fuller, Laskey, and Elia 1992). Lower precision errors for fat have been reported by Mazess and colleagues (1990), whose subjects had greater adiposity. Higher precision errors are associated with regional areas for all tissues. Part of this is attributable to the subject's assuming a different location and posture within the scan area after repositioning, which can affect where the regional boundaries are placed (e.g., between the torso and arms) during analysis.

Dual-energy X-ray absorptiometry has been shown to be accurate for measuring pigs (of similar body composition to humans) through comparison with subsequent chemical analysis. Mitchell, Conway, and Potts (1996) showed agreement within 0.4% fat for both methods using 48 pigs. Svendsen and colleagues (1993) found the in vivo accuracy of DXA in pigs to have a standard error of the estimate of 2.9% despite incompleteness of the homogenisation of carcasses. In theory, accuracy should be better in humans, for whom the fat distribution model has been designed.

Studies With Athletes Since DXA offers a rapid, convenient three-compartment method, the authors of several studies have developed predictive models of body composition in athletes using DXA as the reference method (De Lorenzo et al. 2000; Stewart and Hannan, "Prediction" 2000). In addition, regional fat and fat-free composition using DXA has enabled quantification of tissue distribution in different athletic groups (Stewart and Hannan, "Sub-Regional Tissue" 2000) and supported a hierarchical view of regional fat loss during exercise (Nindl et al. 1996). Increasingly, muscle mass is being predicted using limb FFST in DXA, with an assumed constant ratio of muscle on the limbs compared to the torso of 3:1 (Hansen et al. 1993). This ratio was derived from cadaver studies, but specific physical activity is likely to alter the regional distribution of muscle and varies significantly between male and female athletes (Stewart 2003). Caution is also necessary in interpreting pre- and postseason changes in regional and total FFST. Muscle, glycogen and water would all be detected similarly by DXA; and in many cases, all would contribute to the actual morphological changes taking place. With the advantages offered by DXA, scanners are more commonly used to assess athletes, so it is important to remember that results remain specific to each device and software version. Cross-validation of results from different scanners remains an objective for future research.

When taking measurements it is important to be sensitive to the potential psychological impact of assessment on the athlete. Many athletes are preoccupied with their body composition, especially skinfolds, and can be sensitive to comments from coaches, parents and peers.

Field Methods

Anthropometry is a field method that involves the measurement of body composition through use of such techniques as skinfolds, girths, height, weight, limb lengths and bone breadths. Bioelectric impedance analysis is based on the differing electrical properties of fat and lean tissues of the body; height–weight indexes such as body mass index are used to predict mean values for a population rather than individual adiposity.

Anthropometry

Sport scientists are primarily interested in assessing changes in body composition in athletes in response to training and dietary interventions. The choice of method for assessing physique depends largely on available resources, testing conditions and the application of the results either for individual reporting or for a research outcome. Anthropometry is the most common field method used; it is portable, noninvasive and inexpensive and includes techniques such as skinfold measurements, girths, height, weight, limb lengths and bone breadths (see chapter 3). However, to minimize measurement error, the anthropometrist needs to have undergone accreditation in anthropometry and to be willing to follow a standard protocol (International Society for the Advancement of Kinanthropometry).

Frequently, an estimation of the subcutaneous adipose tissue mass, or body fat as it is commonly termed, is all that is required. Skinfold measures are used routinely in elite sport programs to provide an estimate of the subcutaneous adipose tissue mass. A minimum of six skinfold sites, but ideally eight (triceps, subscapular, biceps, iliac crest, supraspinale, abdominal, front thigh and medial calf), are taken.

Interpreting Anthropometric Data When interpreting anthropometric data it is important to recognize the variability in physique between athletes. The physique and level of adiposity that equate to optimal performance in one athlete may not be the same for another. Genetic variability in body composition (Bouchard 1997) means that some athletes can maintain a low level of adiposity without having to restrict their energy intake, whereas others may have more difficulty in maintaining a lower body fat.

When taking measurements it is important to be sensitive to the potential psychological impact of anthropometric assessment on the athlete. Many athletes are preoccupied with their body composition, especially skinfolds, and can be sensitive to comments from coaches, parents and peers. The anthropometrist needs to ensure privacy when taking the measurements and to monitor an athlete's response to either the result or the measurement process itself. Some athletes are overly concerned about the results, which could be an indication of body image disturbances or a more serious underlying eating disorder. This can show up in the athlete's asking to have the measurements taken more often. In general, monthly or bimonthly body composition assessment is sufficient to monitor change, as there is unlikely to be a real change in the measurement at more frequent intervals. Individual data should be kept confidential and not discussed or displayed publicly. When discussing the results with athletes, one should always interpret the data with reference to how they relate to improving or limiting athletic performance.

Assessment of Adipose Tissue (Body Fat) Using Skinfolds A skinfold caliper measures the compressed thickness of a double layer of skin and the underlying subcutaneous adipose tissue. Therefore, skinfold calipers do not measure visceral fat, located within the abdomen. It is common for skinfold values to be included as independent variables in formulas aimed at predicting body mass proportions. These prediction strategies, however, most commonly use criterion measures derived from densitometry. Thus, the limitations of both densitometry and skinfold methodologies are incorporated in these regression equations.

Most regression equations for the prediction of the proportion of FM using skinfolds are derived using multiple linear regression analyses and therefore are population specific. That is, they may be applied, with due caution, to individuals who are similar in age, gender and body type to the criterion population (Norton 1996). To predict body fat from skinfold measures requires the acceptance of certain assumptions. A major source of error is skinfold compressibility, which causes a decrease in the reading after the initial application of the skinfold caliper to the fold (Martin et al. 1985). Skinfold compressibility varies considerably between and within individuals and at different skinfold sites on the same person. The important implication, however, is that two individuals may have identical skinfold values but very large differences in uncompressed adipose tissue thickness. Norton (1996) has summarized the available prediction equations and has provided details of the population source. These body composition prediction strategies are therefore of little clinical or scientific value; and it is now commonly believed that the skinfold measures should be used directly for comparison without modification, with the exception of some dimensional scaling. The individual skinfold scores may be summed and the result used for comparative purposes over time or between populations.

Comparison to Normative Data Using the O-Scale System The O-scale system (Ward et al. 1989) provides a method of comparing individual skinfold results with a normative database from more than 20,000 observations categorized by age and gender. Individual adiposity ratings are determined from nine standard intervals (stanines) that provide divisions at the percentile equivalents of P4, 11, 23, 40, 60, 77, 89 and 96. Unlike other systems that provide arbitrary labels to define obesity, the O-scale system presents an unbiased description of adiposity and physique in comparison with a healthy standard.

The essential measures to be recorded in this system are age, gender, height, body mass and six skinfolds (triceps, subscapular, supraspinale, abdominal, front thigh and medial calf). From these data the adiposity (A) and proportional weight (W) ratings may be calculated as shown next.

Initially, the arithmetic sum of the six skinfolds (S6SF) is dimensionally scaled to account for individuals of varying size using the equation

$$pS6SF = S6SF\ (170.18 / Ht)$$

where:

pS6SF = the proportional sum of six skinfolds (mm)

Ht = the subject's height (cm)

The A rating is then determined by reference to the normative data shown in tables 5.3 and 5.4 (Ward et al. 1989).

Table 5.3 The O-Scale Adiposity Ratings for Male Subjects

	STANINE THRESHOLD VALUES								
Age (years)	1	2	3	4	5	6	7	8	9
6	43.0*	47.4	57.4	63.0	70.0	80.9	92.7	121.0	
7	40.2	44.6	51.2	59.0	70.9	83.0	99.5	131.0	
8	41.2	45.7	50.7	56.8	65.4	77.6	99.5	137.9	
9	43.6	47.1	50.9	55.9	64.2	77.7	105.2	172.4	
10	45.1	47.1	53.7	59.1	65.4	83.7	129.1	183.2	
11	41.5	45.1	50.8	58.4	68.3	90.9	154.7	193.2	
12	37.6	43.1	47.0	53.4	65.7	89.3	126.6	188.9	
13	34.8	40.2	44.9	51.7	62.7	86.1	116.4	166.5	
14	34.7	37.2	43.4	49.3	57.3	70.9	103.5	146.1	
15	33.5	35.7	42.1	47.0	55.9	69.0	100.8	146.1	
16	32.3	35.4	40.4	44.6	53.3	63.1	79.4	126.7	
17	32.3	35.4	39.5	44.7	53.3	62.4	79.4	107.8	
18-19	31.5	34.3	41.7	47.6	57.0	70.3	87.3	109.3	
20-25	35.0	40.9	48.1	57.8	71.5	89.0	109.0	130.0	
25-30	38.3	45.5	54.5	66.8	81.8	99.5	119.3	144.0	
30-35	41.9	49.8	60.3	72.2	87.3	103.9	121.3	145.5	
35-40	43.9	53.0	62.3	73.9	88.1	102.5	121.9	143.0	
40-45	46.0	53.9	64.2	74.6	87.5	102.5	121.0	142.5	
45-50	44.7	55.2	64.8	76.3	90.5	106.8	123.4	147.0	
50-55	47.2	56.3	66.3	75.7	87.8	105.0	121.0	140.0	
55-60	46.9	56.8	65.8	76.4	87.5	101.1	115.9	136.0	
60-65	47.3	53.9	64.8	74.5	87.2	98.3	116.8	134.3	
65-70	43.0	53.0	60.5	71.6	84.3	92.9	104.8	121.5	

*Proportional sum of six skinfolds (mm).

Courtesy of Bill Ross, rosscraft@shaw.ca.

Table 5.4 The O-Scale Adiposity Ratings for Female Subjects

	STANINE THRESHOLD VALUES								
Age (years)	1	2	3	4	5	6	7	8	9
6	46.8*	56.1	61.7	69.5	77.9	96.7	128.6	144.0	
7	44.3	47.4	60.2	68.3	76.1	91.8	113.2	140.0	
8	43.7	49.2	63.9	69.8	81.4	94.5	111.7	143.2	
9	45.4	53.4	66.1	73.2	87.7	98.6	111.7	143.3	
10	49.2	59.6	67.6	78.6	98.3	109.7	143.2	173.5	
11	51.9	56.4	66.5	75.6	96.4	108.8	150.0	173.4	
12	53.0	59.3	66.5	77.8	98.7	111.4	153.0	175.6	
13	46.7	56.9	67.9	77.4	97.7	114.9	153.0	165.5	
14	46.7	60.9	69.0	81.9	99.6	113.4	147.4	164.8	
15	49.4	62.6	72.4	85.4	99.6	113.2	145.3	162.1	
16	53.8	65.0	76.2	90.3	101.1	112.0	142.4	158.1	
17	62.1	69.4	78.3	92.8	106.5	117.6	141.4	156.4	
18-19	63.4	70.5	78.5	90.2	103.4	118.2	135.9	155.7	
20-25	64.0	72.5	81.2	92.0	104.2	118.9	138.0	164.0	
25-30	65.2	74.1	82.2	93.0	107.9	122.9	141.0	169.2	
30-35	64.1	72.0	81.9	94.6	108.0	126.0	144.3	172.2	
35-40	64.5	73.9	85.5	97.9	112.1	131.7	148.0	178.4	
40-45	69.5	80.5	90.3	102.4	120.7	140.9	161.1	187.3	
45-50	72.5	83.2	97.7	110.5	125.7	141.8	165.1	194.0	
50-55	70.0	84.5	96.2	112.5	127.8	144.8	168.3	196.5	
55-60	46.9	90.1	102.6	115.7	130.5	152.8	169.9	198.2	
60-65	78.3	85.3	96.8	114.6	130.6	146.4	166.0	194.0	
65-70	74.3	84.8	97.0	110.4	130.7	140.7	153.4	164.6	

*Proportional sum of six skinfolds (mm).

Courtesy of Bill Ross, rosscraft@shaw.ca.

The W rating is determined by geometrically scaling the subject's weight to a common height in order to produce a proportional weight (pWt) as follows:

$$pWt = Wt\ (170.18 / Ht)^3$$

where:

Wt = the subject's weight (kg)

Ht = the subject's height (cm)

The W rating is then determined by reference to normative data shown in tables 5.5 and 5.6 (Ward et al. 1989).

When used in combination, the A and W scales provide a significant description of the physique and composition of the individual. The A rating may be regarded as a fatness rating with respect to the population, while the pS6SF score may be used for intraindividual comparisons over time. The W rating is not a fatness rating, but rather one of ponderosity, and together with the A rating may be used to indicate musculoskeletal development.

Bioelectric Impedance Analysis

Bioelectric impedance analysis (BIA) machines are gaining popularity due to their low cost and ease of use and the minimal training required. As a result, these machines can now be found in most gyms and even in household bathrooms. It is important to understand the limitations of BIA and to be aware that the results must be interpreted with caution.

Bioelectric impedance analysis is based on the differing electrical properties of fat and lean tissues of the body:

Table 5.5 The O-Scale Proportional Weight Ratings for Male Subjects

	STANINE THRESHOLD VALUES								
Age (years)	**1**	**2**	**3**	**4**	**5**	**6**	**7**	**8**	**9**
6	55.2*	56.8	59.9	62.6	64.8	66.7	69.6	73.9	
7	49.5	55.1	56.7	59.8	63.2	35.2	67.5	69.3	
8	49.8	54.2	55.8	57.9	60.5	63.4	66.7	67.8	
9	49.4	53.3	55.1	57.4	59.7	62.5	66.1	69.1	
10	50.1	53.1	54.3	59.5	59.5	66.8	66.0	71.9	
11	48.1	50.4	53.5	55.8	59.6	63.3	70.2	75.7	
12	46.3	50.6	52.8	54.9	58.3	62.2	67.3	74.4	
13	46.2	48.8	51.4	54.2	57.2	61.6	67.0	73.2	
14	46.6	48.8	51.3	54.2	57.3	60.8	64.5	71.3	
15	46.8	49.2	51.4	54.3	57.5	61.2	66.8	71.7	
16	47.1	49.8	52.7	55.3	57.3	61.4	66.8	71.7	
17	47.9	50.8	53.5	56.3	59.3	62.4	67.5	71.8	
18-19	49.5	52.8	56.4	59.0	62.5	64.5	67.8	70.8	
20-25	51.3	54.8	57.8	61.8	65.6	69.4	74.6	80.1	
25-30	53.1	56.2	59.8	63.2	67.5	71.4	76.4	84.3	
30-35	53.8	57.7	61.2	64.6	68.7	73.2	78.3	85.2	
35-40	55.2	58.6	61.8	65.4	69.7	73.8	79.0	86.2	
40-45	55.6	59.1	62.7	66.4	69.7	73.8	78.9	86.0	
45-50	55.6	59.6	63.5	66.8	70.8	75.0	79.7	86.8	
50-55	55.9	59.9	63.4	66.6	70.7	74.8	79.6	86.3	
55-60	56.6	60.4	63.5	66.7	71.3	76.1	80.7	87.8	
60-65	55.9	60.3	63.3	66.3	70.5	74.8	79.8	87.3	
65-70	53.0	57.5	62.1	66.5	69.5	73.9	77.8	81.3	

*Proportional weight (kg).

Courtesy of Bill Ross, rosscraft@shaw.ca.

Body fluids are highly conductive, and fat and bone are not (Segal et al. 1985). An estimate of FFM is calculated after normalization for height. The FM is derived from the total body mass by subtraction of the estimated FFM. The leg-to-leg Tanita BIA (Tanita Corporation, Arlington Heights, IL) is gaining popularity as it includes a weighing scale. Similarly to skinfold equations, BIA equations require validation against a criterion method such as DXA and are therefore population specific.

Factors that affect the recorded electrical resistance in the BIA technique have been outlined by Baumgartner (1996). Variations in diet, hydration, ethnicity and disease states affect the body's electrolyte balance, which in turn influences the FM estimate (Malina 1987). In athletes, it is important to control for testing conditions such as hydration (Segal 1996). To date, most studies have evaluated the precision and accuracy of BIA under standard conditions of normal hydration, which is not always possible in the athletic setting (Clark et al. 2004). An additional issue with BIA is the lack of comprehensive, normal-range data on a variety of sports. In the sporting setting, BIA in its current form is of limited value to the athlete.

Indexes of Height and Weight

Weight–height indexes have been used for many years in an attempt to determine the "ideal weight" for an individual. The best known of these is the BMI (weight [kg]/height [m]^2). All indexes provide a measure of ponderosity, which is not the same as measuring adiposity. For an individual of any given stature, body mass will vary according to the amount and density of lean body

Table 5.6 The O-Scale Proportional Weight Ratings for Female Subjects

	STANINE THRESHOLD VALUES								
Age (years)	**1**	**2**	**3**	**4**	**5**	**6**	**7**	**8**	**9**
6	53.1*	54.4	57.4	60.2	63.8	66.7	71.3	72.9	
7	51.3	53.8	56.2	57.6	60.8	64.1	68.9	72.8	
8	51.7	54.3	55.8	57.3	59.8	62.7	66.6	71.6	
9	49.9	52.0	54.4	56.6	59.7	63.2	67.7	72.2	
10	47.6	51.2	53.2	55.8	60.0	63.7	71.1	75.8	
11	46.6	49.3	52.0	53.8	58.2	65.0	70.7	74.7	
12	46.2	49.2	51.8	54.8	59.6	63.9	72.8	80.2	
13	46.0	49.8	52.2	56.3	59.9	65.3	71.8	77.0	
14	46.3	50.2	53.3	56.7	60.3	64.8	71.8	78.0	
15	47.2	50.3	54.2	57.2	60.5	64.3	71.0	76.3	
16	47.3	52.2	55.3	57.7	60.8	63.8	70.8	75.0	
17	49.0	52.8	55.8	58.4	61.6	64.4	70.0	75.3	
18-19	51.8	54.8	57.5	60.4	63.5	66.8	71.0	77.8	
20-25	52.2	55.2	57.6	60.8	64.2	68.3	72.9	80.0	
25-30	52.5	55.2	57.7	61.0	64.8	68.9	74.8	83.0	
30-35	52.3	55.3	58.5	61.5	64.8	69.1	74.8	84.5	
35-40	53.1	56.2	58.8	62.4	66.3	70.7	76.7	88.0	
40-45	54.4	57.6	60.8	63.8	68.1	73.2	80.2	89.2	
45-50	55.2	58.7	62.0	65.2	69.8	74.6	82.3	91.8	
50-55	54.2	57.8	62.2	65.3	69.6	74.3	82.7	93.0	
55-60	55.5	59.1	62.5	66.8	72.8	78.1	84.4	95.5	
60-65	56.3	59.0	63.8	67.4	71.9	77.5	85.4	93.5	
65-70	53.3	58.7	65.3	69.2	74.8	78.8	84.3	91.7	

*Proportional weight (kg).

Courtesy of Bill Ross, rosscraft@shaw.ca.

mass as well as the adipose tissue mass. The BMI does not distinguish the body composition or structure of individuals, so misclassification is a problem especially in sportspeople with a muscular physique (Ross et al. 1988).

Indexes of weight and height have been universally employed by scientists and clinicians for their ease of measurement, especially in epidemiological research. The use of indexes such as the BMI in epidemiological studies may be justified to predict mean values for a population, but not to predict individual adiposity. International cutoffs (table 5.7) devised by the International Obesity Taskforce (www.iotf.org), if applied to athletes, would place substantial numbers of elite athletes into overweight or obese categories.

Body Composition and Sport Performance

Levels of body fat are important in certain sports, such as running and jumping sports, and in certain positions in team sports. It is essential to remember, however, that body composition is only one factor among many that contribute to performance.

Link Between Body Composition and Sport Performance

Body composition for certain sports is linked with performance, with some sports being less tolerant of composition diversity than others. In sports in which the

Table 5.7 Classification of Overweight in Adults According to the BMI

Classification	BMI (kg/m^2)	Risk of comorbidities
Underweight	<18.5	Low (but risk of other clinical problems increased)
Normal range	18.5-24.9	Average
Overweight	≥25	
Preobese	25.0-29.9	Increased
Obese class I	30.0-34.9	Moderate
Obese class II	35.0-39.9	Severe
Obese class III	≥40.0	Very severe

Obesity is classified as BMI (body mass index) ≥ 30 kg/m^2.

Information from the International Obesity Taskforce (www.iotf.org).

body weight must be transported, a lean physique and minimal body fat can give a competitive advantage (Tittel 1978). These sports include distance running and jumping sports; for these, the emphasis is on assessing body fat with skinfolds. In team sports, the position played can demand a certain physique for optimum performance. For example, positions that require the player to run will demand a lower body fat than set position players or goalkeepers. In other sports that require athletes to "make weight", such as lightweight rowing and combative sports, a low level of body fat is also desirable (Wilmore and Costill 1988). Body composition is only one of many factors that will determine sport performance. Therefore, athletes can still achieve competence in their chosen sports without having the optimal physique for the given sport.

> Body composition is linked to performance, with some sports being less tolerant of composition diversity than others. For sports in which the body weight must be transported, a lean physique and minimal body fat can give a competitive advantage.

Sport-Specific Differences in Body Composition

Since the 1960s, interest in body composition has flourished as coaches and athletes have observed that success demands not only a particular physique but also a certain ratio of muscle mass to fat mass. However, extreme caution is required prior to wholesale application of this information because of the great variation in data collection and treatment methods. A lack of standardization of techniques and sampling sites in the past, together with the use of a variety of data treatment strategies, has meant that it is not possible to pool some of the results for profiling and comparative purposes with any real accuracy.

Published values for skinfolds of elite athletes should be used as a guide only and not to determine a skinfold "cutoff" for an individual athlete. Such definitive values do not account for individual genetic variability and may not be applicable to recreational athletes. It is also important to explain to an athlete who does not meet elite skinfold values that body composition is only one factor contributing to performance. For some athletes, achieving the "ideal" skinfold sum may require a severe restriction of energy intake to the detriment of performance and may never be achievable. Chapter 13 provides further discussion on differences in body composition for various sports. Tables 5.8 and 5.9 display data for various national and international athletes by sporting category.

Nutritional Changes to Modify Body Composition

Athletes often wish to modify their body composition in an attempt to gain a competitive advantage, either through gaining weight and "bulking up", reducing body fat or (in weight-category sports) making the lowest weight category possible. It is essential for coaches and athletes to know when such changes are appropriate and to understand which nutritional strategies are safe and effective.

Gaining Weight and Bulking Up

Strength and power athletes consume very-high-protein diets in an attempt to promote muscle hypertrophy, or to "bulk up" in the common parlance. Although protein intake is important, many athletes consume excessive amounts of protein, especially protein powders, and neglect other aspects of their training or nutrition. By far the most powerful stimulus for muscle hypertrophy is

Table 5.8 Normative Data for International- and National-Level Female Athletes

Sport	Level	Position/Event	Number of subjects	SKINFOLD SUM (MM)* Mean	Range
Athletics[1]	National	SASI jumps	4	61.1 ± 12.7	41.7-72.8
		SASI throws	9	95.3 ± 49.4	53.0-203.7
		SASI sprint	7	60.3 ± 11.9	45.1-83.9
		SASI middle distance	20	59.2 ± 19.6	37.4-110.6
		SASI long distance	6	51.3 ± 8.8	40.4-68.3
		Scotland distance runners[5]	10	57.9 ± 14.9	32.2-72.6
Basketball[2]	International	Guard	64	76.6 ± 22.2	36.4-143.5
		Forward	65	76.0 ± 20.1	40.9-131.7
		Centre	47	88.0 ± 21.1	45.7-146.8
Cricket[1]	National		27	90.8 ± 19.7	55.9-141.1
Cycling, road[1]	National		32	61.9 ± 12.0	33.8-89.5
Diving[3]	International		39	65.6 ± 17.0	32.1-114.3
Gymnastics[1]		SASI elite	68	37.9 ± 6.1	27.4-57.6
Hockey[1]		SASI senior	57	87.4 ± 18.5	48.1-140.3
Netball[1]		SA senior	33	83.4 ± 17.3	51.5-124.0
Rowing[1]		SASI lightweight	24	73.4 ± 13.4	55.5-105.2
		SASI heavyweight	30	87.5 ± 17.8	60.7-119.4
Triathlon[4]	International		19	62.8 ± 13.4	40.3-98.4
Swimming[3]	International		170	72.6 ± 19.6	37.9-147.1
Synchronized swimming[3]	International		137	81.7 ± 22.1	37.5-145.8
Volleyball[1]		SASI senior	29	90.5 ± 25.1	35.8-147.1
Water polo[3]	International		109	89.8 ± 23.8	39.7-151.6

*Sum of seven skinfolds (unless otherwise indicated) = triceps, subscapular, biceps, supraspinale, abdominal, front thigh, medial calf.

[1]Adapted with permission from the South Australian Sports Institute (SASI) and published previously by Woolford et al. 1993; [2]Ackland, Schreiner, and Kerr 1997; [3]Note: Sum of six skinfolds from Carter and Ackland (1994) = triceps, subscapular, supraspinale, abdominal, front thigh, medial calf; [4]Note: Sum of eight skinfolds from Ackland et al. (1998) = triceps, subscapular, biceps, iliac crest, supraspinale, abdominal, front thigh, medial calf; [5]Sum of eight skinfolds (all ISAK sites).

Unpublished data from A.D. Stewart, 1999. University of Edinburgh. Data collection 1996-1998.

Table 5.9 Normative Data for International- and National-Level Male Athletes

Sport	Level	Position/Event	Number of subjects	SKINFOLD SUM (MM)* Mean	Range
Athletics[1]	State	SASI pole	3	46.8 ± 0.3	46.4-47.1
		SASI sprint	4	56.1 ± 2.2	53.9-58.3
		SASI middle distance	9	38.6 ± 12.0	25.8-68.2
		SASI long distance	4	49.8 ± 6.4	41.3-56.4
Australian Rules football[1]	National	Under 17 years	20	67.2 ± 6.9	44.7-104.1
Boxing[1]	State		13	57.5 ± 17.7	34.2-95.2
Cricket[1]	National		22	77.8 ± 23.0	52.3-135.2
Cycling[1]	State	Road	24	58.1 ± 11.9	42.9-85.0
	Scotland national	Road and time trial	16	69.0 ± 17.2	44.2-101.6
	National	Track	83	53.9 ± 12.7	26.4-85.3
Diving[2]	International		43	45.9 ± 11.4	28.0-79.7
Gymnastics[1]	State	SASI elite	41	41.6 ± 7.2	27.5-59.1
Hockey[1]	State	Under 21 squad	22	59.4 ± 17.0	38.7-107.2
Kayaking[1]	State	SASI senior	64	58.0 ± 14.0	37.4-96.7
Orienteering (and fell running)[3]	National	Scotland	12	60.6 ± 16.7	42.9-96.3
Racquet sports	Area and national	Scotland	10	76.1 ± 27.6	42.7-121.9
Rowing[1]	State	SASI lightweight	27	45.2 ± 6.5	35.8-65.1
	State	SASI heavyweight	18	66.9 ± 18.0	46.1-111.8
	Scotland area and national	Mixed light- and heavyweight[3]	15	70.2 ± 12.5	50.7-91.1
Rugby union[1]	State	SASI senior	58	92.2 ± 32.9	50.6-223.2
	International	Scotland team[3]	11	93.3 ± 30.3	62.0-147.8
Triathlon[4]	International		19	48.3 ± 10.2	36.8-85.9
	Area and national	Scotland[3]	10	63.7 ± 12.6	34.4-77.3
Swimming[2]	International		231	45.8 ± 9.5	26.6-99.9
Volleyball[1]	State	SASI senior	17	56.8 ± 13.2	36.9-79.6
Weightlifting[1]	State	SASI squad	47	74.9 ± 34.4	33.9-190.2
Water polo[2]	International		190	62.5 ± 17.7	27.9-112.1

*Sum of seven skinfolds (unless otherwise indicated) = triceps, subscapular, biceps, supraspinale, abdominal, front thigh, medial calf.

[1]Adapted with permission from the South Australian Sports Institute (SASI) and published previously by Woolford et al. 1993; [2]Note: Sum of six skinfolds from Carter and Ackland (1994) = triceps, subscapular, supraspinale, abdominal, front thigh, medial calf; [3]Note: Sum of eight skinfolds from Ackland et al. (1998) = triceps, subscapular, biceps, iliac crest, supraspinale, abdominal, front thigh, medial calf; [4]Note: Sum of eight skinfolds (all ISAK sites).

Unpublished data from A.D. Stewart, 1999. University of Edinburgh. Data collection 1996-1998.

resistance training (see chapter 8). The role of nutrition is to ensure that the athlete is in positive energy balance and is meeting his or her protein requirements to allow weight gain to occur. However, a common problem for athletes in sports that also have a high aerobic requirement for training, such as rowing and football, is being able to consume sufficient food to meet their energy needs. In this situation, the volume of training may need to be adjusted by the coach if hypertrophy is the main goal. Monitoring body composition (body weight, skinfolds, girths and corrected girths) is important to determine whether or not energy intake is sufficient for the training load.

Athletes and coaches can have unrealistic expectations about the rate at which an athlete can gain weight. For an athlete who increased his or her energy intake by 2100 kJ/day, the expected weight gain would be around 0.25 to 0.5 kg/week (0.55-1.1 lb/week). It is important to understand that for teenage males, the strength spurt occurs only after peak growth in height has been experienced. Therefore "bulking up" is more easily achieved once growth has slowed.

The amount of protein required for resistance-trained athletes may be as high as 1.8 g/kg body weight per day, particularly when training is first begun (Lemon 2000). These estimated requirements are approximately double the amount recommended for a nonathlete, but most athletes easily meet these requirements from their diet. This may not be the case for athletes on restricted intakes or those who are trying to make weight. These athletes may require specialist input from a sport dietician. Once athletes reach a steady state in their resistance training, the requirements are estimated to be 1 to 1.2 g/kg body weight per day. Highly trained endurance athletes' requirements may be around 1.6 g/kg body weight per day (Tarnopolsky 2004). Athletes are encouraged to snack between meals rather than relying on just three meals per day. Liquid meal replacement drinks or homemade high-protein milk drinks with added skim milk powder are practical and low-cost ways of increasing both the protein and carbohydrate intake.

Protein and carbohydrate intake, both immediately before and after resistance training, can enhance protein synthesis, especially for those beginning a resistance training program (Tarnopolsky et al. 2001; Tipton et al. 2004). Therefore, athletes looking to maximize the benefits of the resistance training session should consume food or drink containing protein and carbohydrate either before or after the session. A simple way to do this is to consume a small carton of yoghurt or milk with a piece of fruit or a cereal bar. Athletes need guidance in adopting sound nutritional practices rather than eating just anything or thinking that the answer lies simply in taking protein powders and dietary supplements.

Reducing Body Fat

Most commonly, athletes will need to lose body fat after having taken time off from training due to injury or after the off-season. It is not uncommon for athletes to return to training having gained a considerable amount of body fat. Some athletes find it difficult to adjust their energy intake when activity levels are reduced. The most common factor contributing to higher body fat levels in athletes is overconsumption, especially of foods and drinks with high energy density such as high-fat foods and alcohol. Athletes can react to weight gain by attempting drastic weight loss. Some "just want to get the weight off", not realizing that it will take some time to achieve this safely and without causing performance decrements.

Popular Weight Loss Diets

Athletes are prone to trying popular weight loss diets just as members of the public are; and with the rise in obesity rates, the popularity of these diets is set to continue. Popular diets are not always safe for athletes and should not be undertaken without appropriate assessment by a sport dietician. Athletes are susceptible to these diets if they think such regimens will help them lose body fat quickly, as is often claimed. Anecdotal evidence, or testimonials from other athletes or celebrities, make these diets very appealing. In some cases, coaches suggest a diet that they themselves or a family member may have tried. This places athletes in a difficult situation, as they are often keen to follow the coach's advice whether or not it is correct. A structured weight loss program for an athlete is a fine balance between ensuring that glycogen stores are replete and restricting intake enough to allow some weight loss to occur. An assessment by a sport dietician is the optimal approach to identifying the best possible strategies to assist with weight loss. Ideally, weight loss should be undertaken early in the season; close to competition, it should be avoided, as the performance decrements are likely to be far greater than any competitive advan-

Athletes are prone to trying popular weight loss diets; however, these are not always safe for athletes and should not be undertaken without appropriate assessment by a sport dietician. A structured weight loss program for an athlete is a fine balance between ensuring that glycogen stores are replete and restricting intake enough to allow some weight loss to occur.

tage conferred by body fat loss. Although this may seen obvious, athletes have been known to attempt weight loss even a week before a major competition in the mistaken belief that this will assist their performance.

Effective Weight Loss for Athletes

By far the most consistent evidence for losing weight or body fat effectively is that for an energy-restricted diet. Therefore, to lose body fat, it is necessary to create an energy deficit in the diet. This means either eating a little less food or increasing the energy output from exercise (or a combination of the two). For many athletes there are limited options for altering their training program to increase the aerobic output, but this is an approach that can be discussed with the coach and exercise physiologist.

To identify possible dietary changes, the first step is often to ask athletes to record their food intake for three to four days. Details of the type of food, how it was prepared and portion sizes should be included. The act of recording food intake is known to alter athletes' eating behaviour so that they eat less food, but it also serves to increase their awareness of what they are eating. This can be an important first step in modifying eating behaviour and identifying possible changes. Many young athletes eat high-fat takeaway foods or snacks too often, or consume excessive amounts of soft drinks and cordials without realizing that these are sources high in energy.

Often athletes state that they are eating healthy foods but still not able to lose weight. A common reason is that they are eating too much food. Portion sizes, especially of food eaten away from the home, have increased over the past few decades. Athletes may need to adjust their portion sizes to assist with weight loss.

Making Weight

In weight-category sports, athletes attempt to gain a competitive advantage by making the lowest weight category possible. These sports include the combative sports such as judo, wrestling, boxing, weightlifting and lightweight rowing. Failing to make weight can be disastrous for an athlete who has trained for an event, but unfortunately does happen. Therefore, when working with the weight-category athlete it is important to understand the rules of competition, especially the weigh-in procedures, such as the length of time between weigh-in and competition and the frequency of weigh-ins. For example, does the athlete need to weigh in each morning of competition? This is an important consideration in advising the athlete regarding hydration and nutritional strategies during competition.

Those working with athletes who need to make weight should look at the whole picture, including the sport, the competition rules and the individual athlete. As a first step, the assessment of body composition, in particular body fat, can be useful in identifying whether the desired weight category is realistic. This decision needs to be based, firstly, on the athlete's body composition and size. If an athlete already has low body fat as assessed by the skinfold sum (at least six and preferably eight sites), then significant weight loss can be safely achieved only by loss of muscle mass. It is important that the sport scientist or dietician discuss this issue with the coach and athlete before weight loss is undertaken, as loss of muscle mass can compromise strength and endurance capacity. Many athletes will attempt to "make weight" too close to competition, not allowing sufficient time for a slow rate of weight loss. This increases both the physiological and psychological stress on the athlete and can lead to dangerous weight loss practices, such as dehydration or severe dieting. An energy deficit of 2000 kJ/day would be expected to result in a weight loss of 0.5 kg/week (1.1 lb/week), which is generally equivalent to an approximately 5 mm skinfold sum (across seven sites) loss per week. An eating plan can be worked out by the sport dietician with the athlete to maximize carbohydrate intake while still ensuring a slight energy deficit.

Making weight is a highly complex issue that needs to be carefully managed with the medical team, including a sport dietician, exercise physiologist, sport psychologist and sport doctor. For example, the exercise physiologist can advise the athlete on appropriate amounts of additional aerobic exercise to assist in weight loss. It is important that the medical team work together with the athlete to ensure safe and effective practices for making weight. This also includes adopting a sensible approach to after-competition habits. If the precompetition deprivation and dieting are followed by a period of overeating, rapid regain of the weight lost can occur. Major competitions such as the World Championships or the Olympics may also coincide with increased alcohol consumption and time off training. This can lead to rapid weight gain, with the result that it is more difficult to make weight the next time. Just as athletes set goals for making weight, they also need to set themselves realistic goals to deal with the postcompetition phase to avoid finding themselves 6 to 8 kg (13-17.6 lb) heavier at the start of the next preseason.

Summary

Interaction of training and diet alters body composition for the athlete in a complex way. Genetic differences ultimately mean that the responses of two similar individuals to the same regimen are likely to be different, so guidelines for one may not be appropriate for the other. The role of body composition assessment is, therefore, to profile each athlete over time, matching performance against composition data. Criterion or multicomponent methods may assist the accuracy of tissue composition predictions using portable techniques; but they are not necessarily the "Holy Grail", and further research is necessary before normative values can be developed. Where normative data already exist for anthropometric profiles of athletes, they need to be used systematically by sport scientists and collaboratively with allied health professionals. With this dialogue, problems of making weight or unrealistic training goals, which can jeopardize health, are avoided and performance can be optimised.

six

Proportionality

Timothy R. Ackland, PhD; and J. Hans de Ridder, PhD

Human proportionality has been observed for thousands of years, but it was not until the 5th century BC that a Greek named Polykleitos sculpted Doryphoros, the "Spearbearer" (figure 6.1), which was representative of the ideal body form with the proportions of a champion athlete. This figure was then used as a proportional model by sculptors for several centuries. During the Renaissance, Alberti, da Vinci and Vesalius all studied human proportionality and in some cases used linear measures of human proportions for their sculptures and drawings.

It was not until 1887, when Sargent made various observations with relation to the physical build of athletes, that anyone documented the effect of body proportions on sport performance. During the early 20th century, Amar (1920), Kohlrausch (1929), Arnold (1931) and Boardman (1933) measured a large number of athletes and made some very perceptive observations about their suitability for various sports and events. However, it was Cureton in 1951 who did the first definitive work in which a precise knowledge of anatomy was applied to sport, when he reported on his study of 58 Olympic place-getters and world champions in the sports of swimming, track and field and gymnastics.

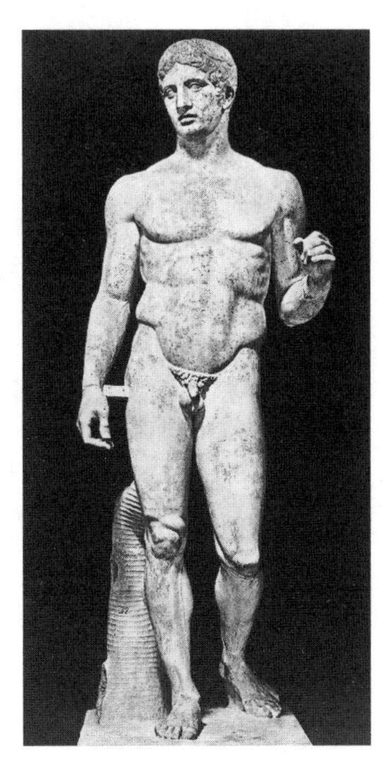

Figure 6.1 A sculpture of the "Spearbearer" by Polykleitos.

Significance of Proportionality Modification in Sport

Even the casual observer will have noted that human body proportions vary greatly from person to person. These variations play an important part in the self-selection process for various sports and events, and it is obvious that there is little that can be done to alter some anatomical body proportions. With this limitation in mind, it is up to coaches to modify an athlete's technique when his or her proportions are not suitable for various sport skills, by either shortening or lengthening the various levers of the body in order to obtain the desirable outcome. This is the essence of high-quality coaching. How to achieve such modification is fully discussed later in this chapter.

Effect of Growth on Proportionality

During the growth period, every individual undergoes proportionality changes to a greater or lesser extent; it is important for coaches to realize this because there will be times when allowances will need to be made for such changes. The following information will enable the coach to understand the growth process better from a proportionality perspective.

• Figure 6.2 illustrates the changes that occur during the growth of various parts of the body in both males and females. The head is well advanced at birth, while the trunk is reasonably developed and the arms and legs lag well behind.

The essence of high-quality coaching lies in coaches' catering for each individual and in their ability to formulate the most efficient technique for their athletes.

• From birth, girls are usually more advanced than boys in relation to height. Their height spurt commences at approximately 10.5 years of age and reaches its peak height velocity (PHV) at approximately 12 years of age, with boys following girls by roughly two years (Tanner 1989). Bone maturity in boys compared to girls is also approximately two years delayed.

• According to Malina and Bouchard (1991, 260), "Maximum growth is obtained first by the tibia and then the femur, followed by the fibula and then the bones of the upper extremity. Maximum growth of stature occurs, on the average, more or less at the same time as maximum growth of the humerus and radius". The authors further stated that "in early adolescence a youngster has relatively long legs, because the bones of the lower extremity experience their growth spurts earlier than those of the upper extremity". Many coaches have observed this "long-legged" phenomenon and realize that they simply have to wait, in their expectation of fully integrated and coordinated movement, for the trunk to develop before the athlete has balanced proportions.

These changes in proportionality with growth present special challenges for talent identification programs in which a particular athlete structure has been specified as providing an advantage for competition. Ackland and Bloomfield (1996) studied the stability of human proportions throughout adolescent growth to determine whether adult proportionality could be predicted from preadolescent structures. Their results demonstrated that bone breadths and upper limb and trunk longitudinal proportions remained stable through the adolescent period. In contrast, proportionality of the lower limbs and component segments changed markedly from pre- to postadolescence, which raised some doubts as to their suitability for inclusion in talent identification programs involving young children.

Proportionality Assessment

For many years observers have made individual comparisons by viewing the raw anthropometric scores of one athlete in relation to those of another, or of one team in relation to those of another. If coaches are very aware of the significance of the raw measures, for example when comparing the heights and weights of rowing crews or those of rugby packs, then this method of comparison is of value. However, this practice has its limitations when the aim is to determine the magnitude of differences between individuals, groups or genders or in data over time.

Human size and proportions are generally assessed by the use of anthropometry or medical imaging technologies such as magnetic resonance imaging and computed

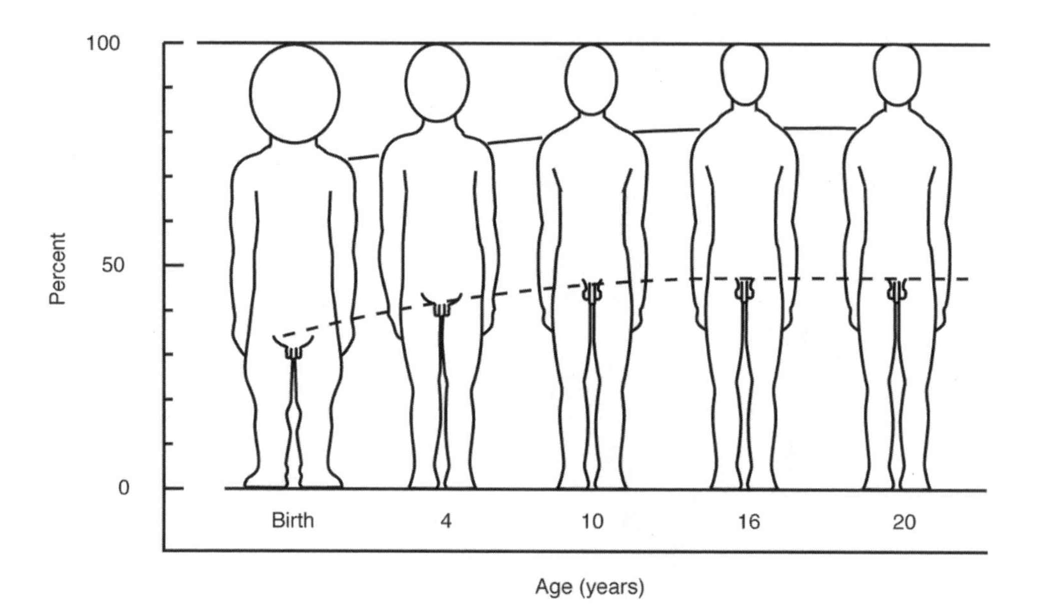

Figure 6.2 General changes in body proportions with age.

Adapted, by permission, from A. Hills, 1991, *Physical growth and development of children and adolescents* (Brisbane: Queensland University of Technology).

tomography scan, as well as laser body surface scanners (see chapter 3). The most common measures are those that assess the lengths, breadths, girths and volumes of body segments. Anthropometric variables are often expressed as indexes to allow a more meaningful description of proportionality. For example, the relationship of the length of the leg (foreleg or lower leg) to the thigh, that is, the crural index, can be calculated in the following way:

$$\text{Crural index} = \frac{\text{(Lower) leg length} \times 100}{\text{Thigh length}}$$

Similarly, the brachial index, which demonstrates the length of the forearm in relation to the arm (upper arm), may be calculated as follows:

$$\text{Brachial index} = \frac{\text{Forearm length} \times 100}{\text{Arm length}}$$

An obvious problem with indexes is that one can never be certain, for example, whether a high brachial index score is the result of a particularly long forearm or a short upper arm.

Two methods have been put forward to address this problem. The first is the somatogram, initially proposed by Behnke, Guttentag, and Brodsky (1959) and later modified to a form described by Behnke and Wilmore (1974). A second method, known as the unisex phantom stratagem, was devised by Ross and Wilson (1974) as a metaphorical model for assessing human proportionality.

The Somatogram

The somatogram strategy provides a graphical representation of the body's proportions using girth measurements. Typically, data for 11 girths are required (figure 6.3) together with k values for each individual measure that are assigned for the reference man or reference woman. Each measured girth (g) is divided by the appropriate k value (table 6.1) to obtain the d value (d = g / k). Then, the sum of the g values divided by the sum of the k values provides a reference score (D; i.e., $\Sigma g / \Sigma k = D$), which is used to create each positional plot on the somatogram. If the anthropometric proportions of an individual conform to the mean of the population, then all the values will fall on the central or zero line. One should note, however, that this system does not account for variations in body size.

For example, a value for the percentage deviation from D for the biceps girth in figure 6.3 is calculated as follows:

$$\begin{aligned} d\text{ (biceps)} &= g\text{ (biceps)} / k\text{ (biceps)} \\ &= 37.5 / 5.29 \\ &= 7.09 \end{aligned}$$

$$\begin{aligned} D &= \Sigma g / \Sigma k \\ &= 6.51 \end{aligned}$$

Therefore,

$$\begin{aligned} \%\text{ deviation from D} &= (7.09 - 6.51) / 6.51 \times 100 \\ &= 8.9\ \%. \end{aligned}$$

Similar values are calculated for the other girths, and the somatogram is created. In this example, the shape of the subject (a male weightlifter) is clearly abnormal compared with the proportions of the reference man. This young weightlifter has proportionally larger shoulder and chest girths, as well as biceps and forearm girths, than the reference man.

Table 6.1 Girth Measurement *K* Values for a Reference Man and Reference Woman Aged 20 to 24 Years

Girth measure	*K* (reference man)	*K* (reference woman)
Shoulder	18.47	17.51
Chest	15.30	14.85
Abdominal	13.07	12.90
Hips	15.57	16.93
Thigh	9.13	10.05
Biceps	5.29	4.80
Forearm	4.47	4.15
Wrist	2.88	2.73
Knee	6.10	6.27
Calf	5.97	6.13
Ankle	3.75	3.70

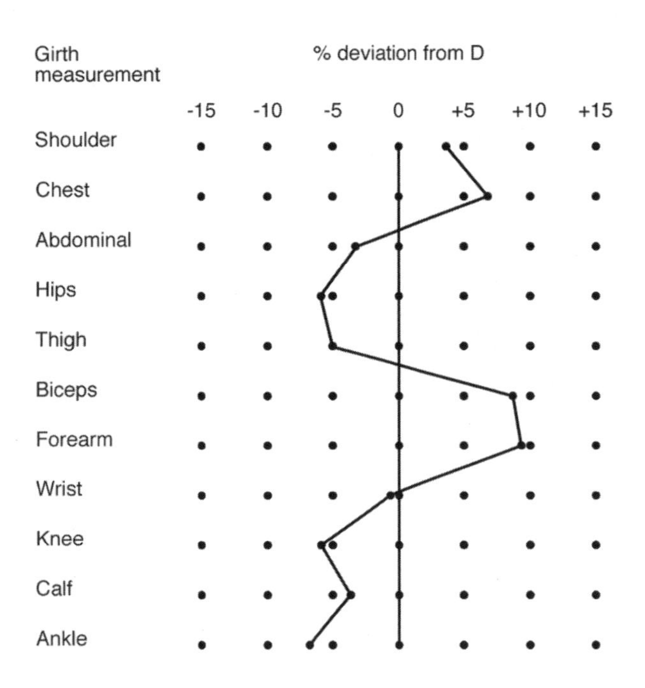

Figure 6.3 Somatogram of a male weightlifter aged 30 years.

The Unisex Phantom

Ross and Marfell-Jones (1991) described the phantom as a calculation device, not a normative system, based on a hypothetical unisex reference human with defined mean (P) and standard deviation (S) values for over 100 lengths, breadths, girths, skinfold thicknesses and fractional masses. In this system, raw data are compared with phantom values, and the resulting deviations, in the form of Z-scores, are the basis for analysis. The phantom stratagem has been used to compare the growth of individuals over time, as well as to compare data for individuals with that of specific groups.

The original data were derived primarily from a collation of anthropometric information by Garrett and Kennedy (1971), whereupon reported mean values for males and females were geometrically adjusted to a standard stature. The resulting phantom P and S values are distributed normally, unimodally and symmetrically. Selected phantom specifications for anthropometric variables are presented in tables 6.2 and 6.3, and phantom Z-scores for an individual may be calculated using the following formula:

$$Z = 1 / S (V (170.18 / H)^d - P)$$

where:

Z = the phantom Z-score

S = the phantom standard deviation for variable V

V = the subject's score on variable V

H = the subject's stature

d = a dimensional exponent that is consistent for a geometrical similarity system; d = 1 for lengths, breadths, girths and skinfolds; 2 for all areas; 3 for masses and volumes

P = the phantom mean score for variable V

Table 6.2 Phantom Specifications for Selected Body Mass, Length and Breadth Variables

Variable	P^*	S^{**}
Body mass (kg)	64.58	8.60
Standing height (cm)	170.18	6.29
Sitting height (cm)	89.92	4.50
DIRECT LENGTHS (CM)		
Acromiale-radiale (arm)	32.53	1.77
Radiale-stylion (forearm)	24.57	1.37
Midstylion-dactylion (hand)	18.85	0.85
Acromiale-dactylion	75.95	3.64
Iliospinale height	94.11	4.71
Trochanterion height	86.40	4.32
Trochanterion–tibiale laterale (thigh)	41.37	2.48
Tibiale laterale height	44.82	2.56
Tibiale mediale–sphyrion	36.81	2.10
Akropodion-pternion (foot)	25.50	1.16
Arm span	172.35	7.41
BREADTHS (CM)		
Biacromial	38.04	1.92
Transverse chest	27.92	1.74
Biiliocristal	28.84	1.75
Biepicondylar humerus	6.48	0.35
Bistyloid wrist	5.21	0.28
Hand	8.28	0.50
Biepicondylar femur	9.52	0.48
Anterior-posterior chest	17.50	1.38

P^* = phantom mean value.

S^{**} = phantom standard deviation value.

Table 6.3 Phantom Specifications for Selected Girth and Skinfold Variables

Variable	P^*	S^{**}
GIRTHS (CM)		
Head	56.00	1.44
Neck	34.91	1.73
Relaxed arm	26.89	2.33
Flexed arm	29.41	2.37
Forearm	25.13	1.41
Wrist	16.35	0.72
Chest	87.86	5.18
Waist	16.35	0.72
Hip	94.67	5.58
Thigh	55.82	4.23
Calf	35.25	2.30
Ankle	21.71	1.33
SKINFOLDS (MM)		
Triceps	15.40	4.50
Biceps	8.00	2.00
Subscapular	17.20	5.10
Iliac crest	22.40	6.80
Supraspinale	15.40	4.50
Abdominal	25.40	7.80
Front thigh	27.00	8.30
Medial calf	16.00	4.70

P^* = phantom mean value.

S^{**} = phantom standard deviation value.

Adapted from T.R. Ackland and J. Bloomfield, 1995, Functional anatomy. In *Textbook of science and medicine in sport,* 2nd ed., edited by J. Bloomfield, P. Fricker and K. Fitch (Melbourne, Australia: Blackwell Scientific Publications).

Figure 6.4 Proportionality differences between aquatic sportsmen and sportswomen.

Using data collected at the World Swimming Championships in Perth in 1991, Carter and Ackland (1994) reported proportionality differences between male and female competitors (figure 6.4). These differences in proportion are shown for length, breadth, girth and skinfold measures, with Z-score data plotted for each gender group separately. Upon inspection of this figure, clear differences can be identified not only between gender groups, but also between the phantom mean score and scores for aquatic athletes in general (see especially the skinfold scores).

In a second example, male water polo players from the Kinanthropometry in Aquatic Sports study (Carter and Ackland 1994) who played in the specialized positions of centre forward (CF) and goalkeeper (GK) are compared directly with the offensive and defensive wing players (OTH). In contrast to use of the phantom mean Z-scores as the basis for plotting, as in figure 6.4, the data for OTH in figure 6.5 are used to create the zero line, and Z-score differences are then plotted for the specialized positions. With respect to the data in figure 6.5, Ross and colleagues

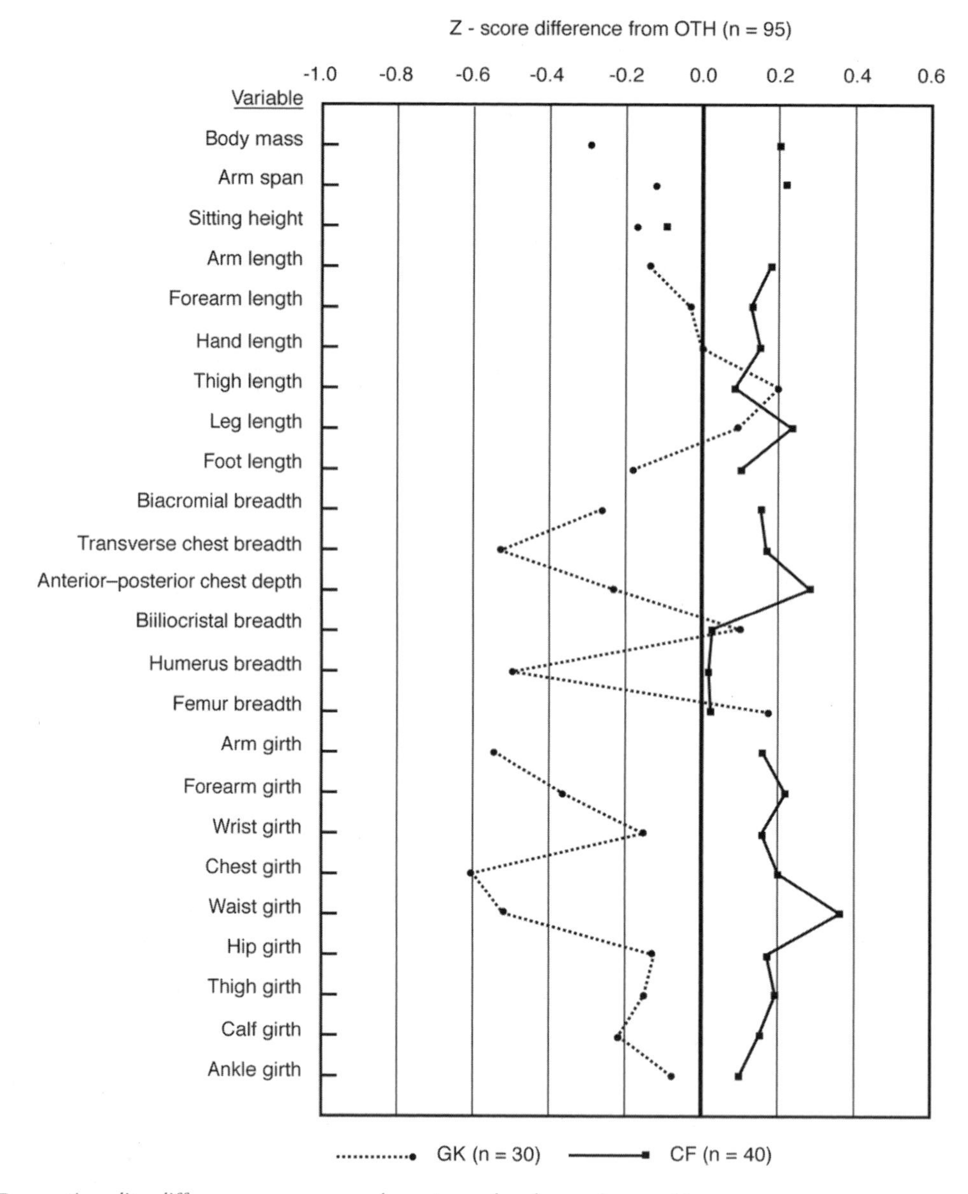

Figure 6.5 Proportionality differences among male water polo players by position.

(1994) reported that CF were proportionally heavier than OTH and had proportionally larger arm spans, leg lengths and waist girths. The GK, however, were proportionally lighter and had a shorter sitting height than OTH. Furthermore, although GK were taller than OTH in absolute dimensions, they were proportionally smaller in biacromial and transverse chest breadth and possessed smaller arm, forearm, chest and waist girths.

Proportionality Applied to Sport Performance

The basic laws of physics as they relate to leverage play an important part in sport, and bone lengths can be either an advantage or a disadvantage depending on the physical demands of the sport in which the individual competes. These lengths are absolute when the individual has reached full maturity and cannot be altered with training.

Lever Lengths

In some sports, such as weightlifting, athletes with short levers have an advantage over those who possess long levers because the weight needs to be lifted through a shorter distance (Hart, Ward, and Mayhew 1991), and a shorter limb enables the load or resistance to be located closer to the axis of rotation. In other sports such as diving or gymnastics, in which the body rotates rapidly in a given distance, short levers are also an advantage to the performer in that they reduce the segment or whole-body moment of inertia (or both). On the other hand, if an athlete requires a long, powerful stroke, as in swimming, canoeing or rowing, then a longer lever, provided it is accompanied by the muscular power to propel it, offers an advantage. The same point can also be made about sports in which hitting, bowling or throwing is important. For example, velocity generation for a tennis serve, fast bowling in cricket, a volleyball spike or a baseball pitch is higher for long-levered athletes, if they have the muscle power to rotate the longer segments. In a study on elite male javelin throwers, it was concluded that these athletes possess exceptionally long upper extremities, which is much to their advantage in this sport (Kruger et al. 2005). The reason is that the velocity of the end of the throwing segment is directly related to the product of segment angular velocity and segment length.

Insertion Point

Although the gross bone length is usually referred to as the lever, this is only a general concept in sport to differentiate athletes with long or short levers from one another. The insertion point of the tendon as it attaches to the bone is also an important determinant of the lever's effectiveness. If this point is farther away from the joint axis of rotation, then it will positively affect the muscle's mechanical advantage, making the athlete stronger or more powerful or both. Many coaches will have observed that two athletes may be similar in size, body shape, lever lengths and muscle mass yet one is more powerful than the other. This is usually due to the tendon insertion position in combination with muscle fibre type (see also chapters 8 and 9).

Trunk Versus Extremity Indexes

In general, individuals with long extremities and relatively short trunks are physically weak types, while people with short extremities and long trunks are usually powerful types. For decades, coaches have made such observations in order to assess the strength, power and speed potential of their athletes and have mostly found them a useful guide. However, although the idea is generally accurate, another dimension needs to be added to it that would enable coaches to make more accurate forecasts of their athletes' potential. This additional factor relates to the indexes of the trunk and the upper and lower extremities, as well as indexes of body segments comprising the extremities themselves. This is discussed fully in the next section.

Proportionality Characteristics of Athletes

One sometimes hears the statement "Athletes are born and not made"; and in the case of proportionality or "lever advantage", this is largely correct. It is very clear to those who have studied various types of athletes that some people are greatly advantaged by their body segment lengths. However, one should not overestimate the salience of this physical capacity, because there are many other important factors that go to make up an optimal performance. If one uses a high jumper to illustrate this point, it is clear that as well as having the optimal height and body shape, that is, long lower limbs compared to the trunk, and long lower legs in comparison to the thighs, the high jumper needs the following characteristics:

- Sufficient muscle mass with a high proportion of fast-twitch muscle fibre and tendon insertion points that give a greater mechanical advantage
- A high level of skill that will enable the jumper to coordinate his or her body segments to "smoothly" clear the bar
- Psychological control, which will help the athlete block extraneous information and focus on each jump in order to attain the best result

As noted in chapter 1, all sport performances consist of a multiplicity of variables; therefore coaches should not stress any one as being much more important than others. Nevertheless, the athlete's proportions are vitally important in ballistic events in which explosive power and speed are necessary.

Racquet Sports (Tennis, Badminton, Squash, Racquetball)

Being agility athletes, racquet sport players have variable proportions because of the multifaceted demands of their games. No definitive research has been carried out on these athletes; however, coaches state that they are a variable group of individuals. The shots used in the racquet sports are executed with the upper extremities, and therefore knowledgeable coaches can help a player to compensate for an inefficient lever system. For example, the forearm can be flexed at the elbow to facilitate greater control during volleying, while it must be almost fully extended to enhance service technique. Two hands may be used for ground strokes to compensate for a lack of strength; but if the athlete is powerful enough, then it may be best to use one hand (particularly when being forced to reach for a return) because a longer lever, provided enough force can be applied to the stroke, will increase reach and racquet head velocity. Furthermore, while longer-levered athletes may structure their basic game on high-velocity shots, shorter players must be agile and fast around the court in order to compensate for their lack of power because of their shorter levers.

Aquatic Sports

Advantageous human proportions have been studied for competitors in aquatic sports. In this section we focus attention on the sports of swimming, water polo, rowing and canoeing.

Swimming

Swimmers at the highest levels of competition have been increasing in height and weight for the past 30 years (Ackland, Mazza, and Carter 1994; Carter 1984). Elite-level swimmers are generally heavier and taller and have a more robust upper body (figure 6.6), as well as larger feet, than lower-level competitors. Even among top-level swimmers there are special characteristics that differentiate them from one event to another. If a comparison is made between sprint and middle or long distance swimmers, the differences are quite marked. For example, sprint swimmers have a higher brachial index than middle and long distance swimmers because of their longer forearms and shorter upper arms (Bloomfield and Sigerseth 1965). Furthermore, Ackland (1999) reported significantly longer segment proportions for arm span, lower limb length and foot length among sprint swimmers (50 and 100 m events) compared to those in long distance events. Within strokes (Ackland, Mazza, and Carter 1994), the freestyle and backstroke swimmers are taller than competitors in the other strokes and have longer limbs, while butterfly swimmers were found to have longer trunks than the others. Breaststroke swimmers, on the other hand, were shorter, very robust and powerful in the trunk region.

Water Polo

Both male and female world-class water polo players (Ackland, Mazza, and Carter 1994) are tall and well built, but there are significant proportionality differences within this sport when players in the various positions are compared with one another. Centre forwards and centre backs are larger and more robust than the other players, mainly because their greater size enables them to use their bodies very effectively in both attack and defence. Goalkeepers, on the other hand, are tall, less bulky and longer-limbed compared to the other players. They have a high skeletal mass, with lower upper body girths, and are basically more ectomorphic. Ackland, Mazza, and Carter (1994, 145) stated that this attribute gives them a "reduced upper limb inertia which would facilitate relative quickness of movement to protect the goal".

Figure 6.6 High-level swimmers are tall and have powerful trunks as portrayed by this world sprint swimming champion.

Rowing and Canoeing

High-level rowers are taller and heavier than the population average (open men—193.3 cm, 94.3 kg [76.1 in., 208 lb]; open women—180.8 cm, 76.6 kg [71.2 in., 169 lb]), and are almost 1.5 and 1.0 standard scores greater in body mass, respectively, than average when scaled to a common stature (Kerr et al. 2006). Carter's (1984) statistics on oarsmen from four Olympic Games show a continuing trend with increases in both height and body mass. Reporting on data collected at the 2000 Olympic Games in Sydney, Kerr and colleagues (2006) showed that open and lightweight rowers possessed very similar proportional segment lengths, but that the open rowers had significantly greater scores for proportional girths and breadths in the upper body and legs. The authors suggested that this was primarily due to the weight restrictions imposed for lightweight competitors. Of particular note, however, were the distinctly small proportional hip dimensions for open rowers (both male and female), which implies that their equipment places some constraint on this parameter.

Canoeists display similar developmental trends to rowers, in both height and body mass, but are not as tall or as heavy as oarsmen (Carter 1984; de Garay, Levine, and Carter 1974). Proportionality profiles for male and female paddlers from the 2000 Olympic Games were similar (Ackland et al. 2003), but paddlers do not have unusually long arms with respect to stature as one might expect. However, canoe and kayak paddlers do possess greater than average proportional thigh length, shoulder and chest breadths and upper body girths whilst also having very narrow hips (male paddlers) and very low proportional skinfold scores.

Gymnastics, Diving and Power Sports

Gymnasts are relatively short and light (Carter 1984), with long bodies and short legs, giving them a low lower limb/trunk ratio. They also possess a low crural index (Cureton 1951). Divers are taller than gymnasts but have similar trunk and leg ratios (Ackland, Mazza, and Carter 1994; Cureton 1951; de Garay, Levine, and Carter 1974). Male and female world-class divers are a reasonably homogeneous group; however, 10 m platform divers are less robust in physique than springboard divers, with relatively longer lower extremities. Ackland, Mazza, and Carter (1994, 144) stated that this characteristic is advantageous to the 10 m diver because the "decreased moments of inertia are afforded by smaller absolute and proportional limb girths and segment breadths and these may provide an advantage in the performance of aerial manoeuvres". It is also possible that a more knifelike entry can be made by a more linear diver. When the top 10 m divers were compared with the other competitors in the 1991 World Championships, they were found to be leaner; however, no other major differences were apparent. Weightlifters (figure 6.7) have similar proportions to throwers, possessing powerful arms and shoulders. They also have long trunks with short, thick and powerful legs, that is, a low lower limb/trunk ratio; and they generally have a low crural index (Carter 1984; Cureton 1951; Tanner 1964).

Track, Field and Cycling Sports

Athletes in track, field and cycling have probably been scrutinized most closely with respect to limb proportions. In this section we focus on the sprint, middle and long distance runners, as well as the jumpers and throwers.

Sprinters

Sprinters are relatively short and muscular (especially in the region of the buttocks and the thighs) compared to middle distance runners, but of medium height when compared to other track and field athletes. They have normal trunk lengths with short lower limbs, that is, a low lower limb/trunk ratio (Bloomfield 1979; Tanner 1964). Dintiman and Ward (1988) stated that the champion male sprinter approaches 5.0 steps/s at full pace while females average 4.5 steps/s. Such rapid leg movements can be made only by a relatively short limb (a shorter lever generally has a lower moment of inertia or resistance to movement than a longer one), which gives a greater "ground strike rate" and thus more propulsive force to the sprinter (see also chapter 10). It should also be noted that the crural index of sprinters is average. High hurdlers (figure 6.8) in many respects resemble sprinters, but are taller and possess longer legs (Cureton 1951; Tanner 1964), with proportions similar to those of 400 m runners.

Middle, Long Distance and Marathon Runners

Middle distance (MD) runners are tall, linear and long-legged with a normal-length trunk, that is, with a high lower limb/trunk ratio, and an average crural index.

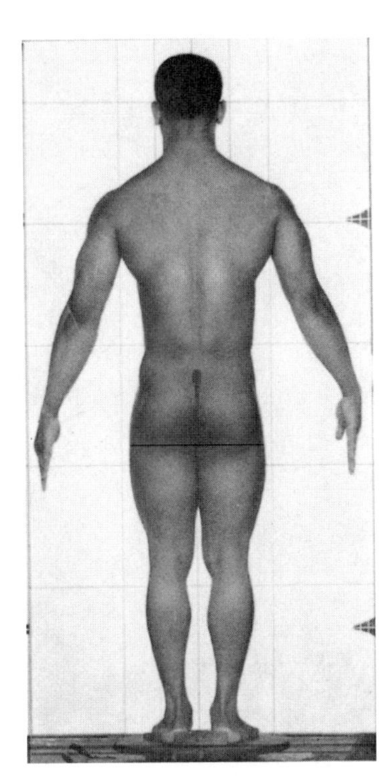

Figure 6.7 The proportions of a weightlifter.

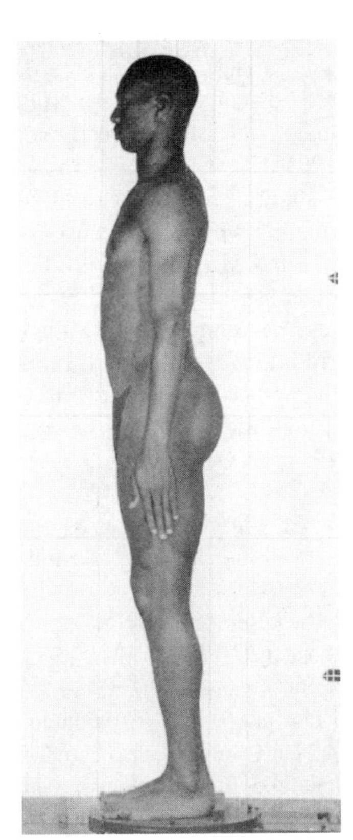

Figure 6.8 High hurdlers resemble sprinters but are taller and possess longer legs.

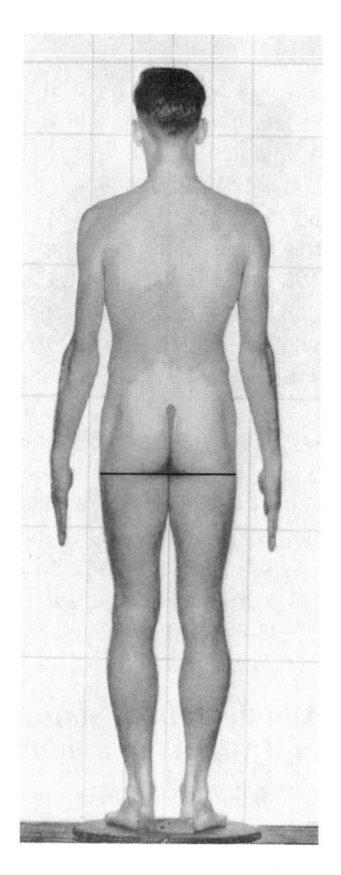

Figure 6.9 The proportions of an elite marathon runner.

This contrasts with long distance (LD) and marathon (M) runners, who become progressively shorter as the race distance lengthens (figure 6.9). The data of elite male MD, LD and M runners who participated in the 1995 All-Africa Games in Zimbabwe were compared in a study by De Ridder and colleagues (2000). The MD runners were significantly heavier and taller than the LD and M runners, with the M runners older than the other two groups. The LD runners also had significantly less percentage muscle than the M and MD runners. Additionally, LD and M runners have short lower limbs in comparison to their trunks, that is, a low lower limb/trunk ratio, and below-average crural indexes (Cureton 1951; de Garay, Levine, and Carter 1974; De Ridder et al. 2000; Tanner 1964).

Elite female African M runners tended to be older and heavier than both the MD and the LD runners (De Ridder et al. 2001) whereas a trend was shown for LD runners to be taller than both MD and M runners, though none of these differences were statistically significant. Underhay and colleagues (2005) compared elite male and female LD runners who participated in the 1995 and 2003 All-Africa Games. As expected, male athletes were heavier and taller than the women, with significantly longer upper extremities. The women had a greater percentage body fat than the men, while the latter group possessed larger muscle and skeletal masses; but most of these gender differences in absolute size were not observed when participants were scaled to a common stature (De Ridder et al. 2005). A similar result was also reported for MD runners (De Ridder et al. 2003).

Jumpers

Athletes who take part in jumping events, particularly the high jump and the triple jump, need to be tall and have long lower limbs relative to their trunk lengths (figure 6.10); that is, they should have a high lower limb/trunk ratio (Bloomfield 1979; Cureton 1951; Eiben 1972; Tanner 1964). They also need a high crural index (Cureton 1951).

Throwers

Throwers are tall and heavy with powerful shoulders and arms, and they are becoming gradually larger each Olympiad (Carter 1984; Cureton 1951; Kruger et al. 2005). Their legs and trunks are of normal length for their height; however, many of them have extremely long, thick arms, especially the discus and javelin throwers (figure 6.11). Javelin throwers, though more linear in shape, also have large proportional muscle and skeleton mass, with broad chests and exceptionally long upper extremities (Kruger et al. 2005).

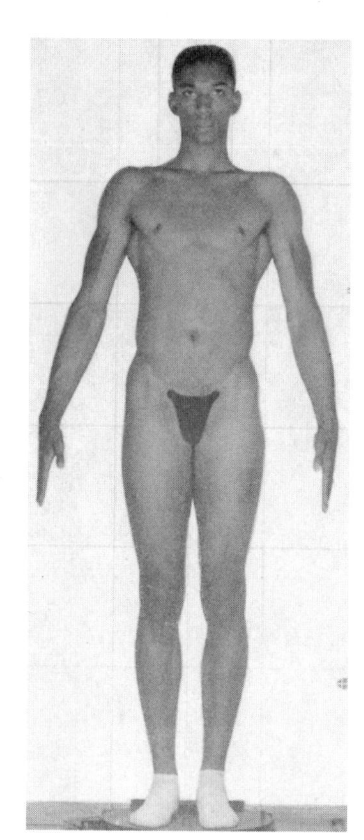

Figure 6.10 The proportions of a former Olympic high jump champion.

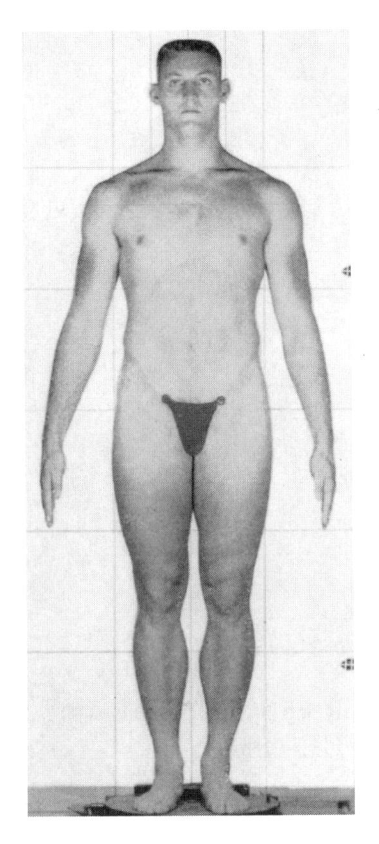

Figure 6.11 An elite discus thrower with typically long arms.

Cyclists

Like many other athletes, cyclists are steadily increasing their height and weight (Carter 1984). Unpublished data from Olds (personal communication) suggests this trend for stature, from 1928 to 2000, to be in the order of +1.3, +2.0 and +0.8 cm (+0.5, +0.8, and +0.3 in.) per decade for road, 4000 m and sprint cyclists, respectively. Similarly, the trend for mass was reported as 0.0, +1.7 (3.7 lb) and +1.3 kg (2.9 lb) per decade, respectively. It is well known that track cyclists are more robust and powerful than road cyclists, but together they form a reasonably homogeneous group as far as their other proportions are concerned. Anecdotal evidence from coaches, one should also note, indicates that high-level track cyclists have short thighs, giving them a high crural index (CI ≈ 1.26 for track sprint cyclists) that increases their mechanical advantage about the hip joint during pedalling.

Mobile Field Sports (Field Hockey, Soccer, Lacrosse)

Individuals in mobile field sports have differing proportions because of the multifaceted demands of their games. As with racquet sport players, little research has been done on proportionality in mobile field sports, though the intelligent coach can use the existing data from agility athletes and sprinters. Such data can be easily applied to athletes in the mobile field sports, who need speed or power and agility in the various specialized positions.

Gomes and Mazza (1998) reported proportionality data for 110 players from six South American national football teams. They concluded that players from different positions do not differ from each other in terms of proportional segment lengths, girths or breadths. Goalkeepers were an exception; they were proportionally heavier, with greater muscle mass and higher skinfold values, than other players.

Contact Field Sports (Rugby, Australian Football, American Football)

Athletes in contact field sports, because they must be powerful, agile and fast, also need to be classified according to the position they occupy on the field. The winger or backfield player, who needs to be very fast, should have similar proportions to a track sprinter, while a forward in rugby or a lineman in American football will need proportions, bulk and agility similar to those of the field games thrower (figure 6.12). Coaches in all games must determine

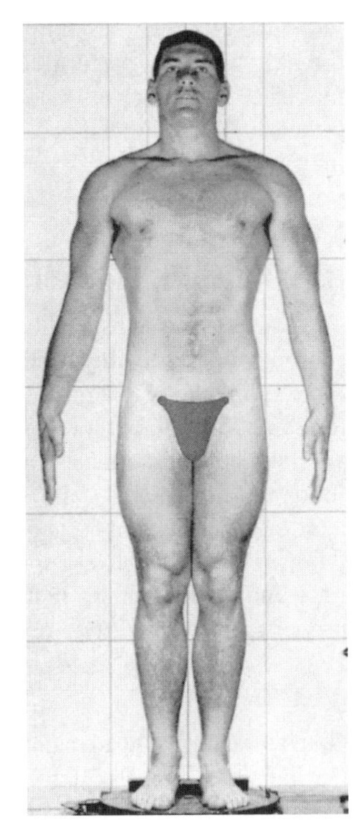

Figure 6.12 The proportions and body bulk needed for a forward in contact football.

the skills that are appropriate for each specialized position in contact sports at the elite level, then identify the proportionality characteristics that will best suit these skills. This is the very essence of intelligent coaching.

Elite rugby players, like athletes in many professional sports, continue to evolve in physical size and shape. Data from 1349 Springbok rugby players, who competed from 1896 to 2004 (n = 664 forwards and 685 backs), were used to map the evolution of the body size and shape. Body mass for the total group increased at a rate of 2.03 kg (4.48 lb) per decade (forwards = 2.47 kg [5.45 lb] per decade; backs = 1.45 kg [3.2 lb] per decade). For stature, the total group increased at 0.84 cm (0.3 in.) per decade over the 108-year period (forwards = 0.94 cm [0.4 in.] per decade; backs = 0.69 cm [0.3 in.] per decade). These rates of increase among elite rugby players were well above those for the general population of young males, and have been increasing at a rate three to four times faster in the last 25 years compared with the rest of the century (Meyer et al. 2005).

Set Field Sports (Baseball, Cricket, Golf)

As with the racquet sport and the mobile field sport groups, the proportionality of cricketers, baseball players and golfers is variable. Intelligent coaches can compensate for an inefficient lever system in these athletes, since many of the skills they use are "set" or "closed", with little or no forward body motion taking place while the skill is being performed. For example, technique modifications in a golf swing, fast bowling technique or baseball pitching may be made to compensate for an inefficient lever system.

Court Sports (Basketball, Netball, Volleyball)

The games of basketball, netball and volleyball are agility sports that partially rely on leaping skills. To do well in these sports the player must be extremely agile and able to jump, so special proportions are needed. These players must be tall and must have long upper limbs, lower limbs and trunks, as well as a high crural index, that is, long lower legs in comparison to the length of their thighs. Thus, elite players in these sports often resemble track and field jumping athletes in both absolute size and proportionality characteristics.

Martial Arts

Noted for their ability to maintain a stable base of support, martial arts exponents often exhibit unique proportionality characteristics that give them a distinct mechanical advantage. Here we focus on the sports of wrestling, judo and boxing in particular.

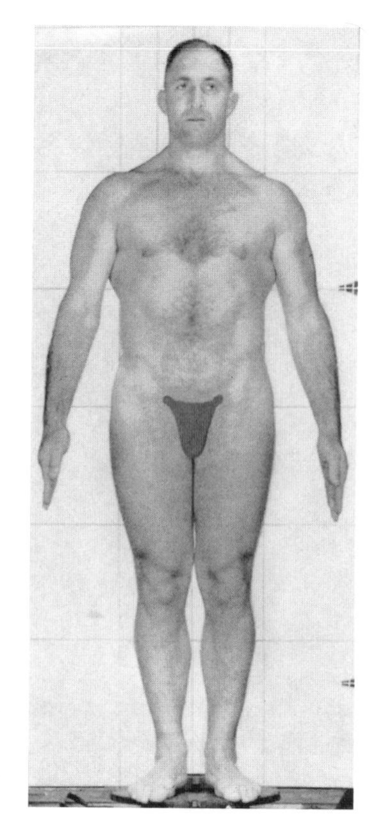

Figure 6.13 The bulk and proportions needed for an international heavyweight wrestler.

Wrestling and Judo

Individuals in wrestling and judo have powerful shoulders and arms, with long bodies and short lower extremities, that is, a low lower limb/trunk ratio. They often have heavy legs and possess a low crural index (figure 6.13). All these features combine to give them a low centre of gravity that makes it difficult to force them off balance (Cureton 1951; Tanner 1964).

Boxing

Athletes in boxing are variable in their proportions and do not have the same basic lever system as individuals in the grappling sports. Because there is a high degree of variability in their proportions, especially in the weight classes, it is up to the coach to compensate by modifying the fighter's technique if necessary.

Racial Characteristics

Physical anthropologists and coaches have observed for some time that there are basic differences in the proportionality characteristics of the major races of the world. Africans (currently in or originally from Africa) have longer upper and lower extremities than Europeans (currently in or originally from Europe), while Asians (currently in or originally from the southeast and western Asian regions) have shorter extremities than both Africans and Europeans. It is interesting to note that Europeans cover a greater range of proportions than either of the other groups. In many cases they have proportionality requirements that suit certain sports admirably, but at the extremes of the range there are fewer individuals with the optimal lever systems for some specialized sports. They therefore find it difficult to compete against persons of other races, who have larger numbers in their population with more suitable proportions.

More research has been done comparing African Americans and European Americans because of the large populations of both in the United States of America and because the two racial groups compete in many sports together. Metheny (1939) found that African Americans exceeded European Americans in body mass, arm length, forearm length, lower limb length, leg length, shoulder breadth, chest depth and width, neck girth and limb girths when they were normalized for stature. Cureton (1951) made similar observations and further reported the work of Codwell, who compared the vertical jumping ability of high school boys in the two groups. He found the African Americans to be superior on this test, and one could only assume that their long lower limbs and legs were important factors in this result.

Field and Court Sports

In the majority of the field and court sports, many variables contribute to high performances; and although proportionality is one, this can be partially compensated for by intelligent coaching to modify the individual's skills. Therefore, except in relation to very specialized positions, racial proportionality differences should not affect performances in these sports.

Track and Field Sports

Tanner (1964), in his definitive work on Olympic athletes, demonstrated conclusively that Africans have longer arms and legs (relative to their stature), narrower hips and more slender calves than Europeans competing in the same event (figures 6.14 and 6.15). He also stated that in sprinting, the African's lighter calves produce "a lower moment of inertia in the leg, and this would permit a more rapid recovery movement, that is a faster acceleration forwards, of the trailing leg" (107). Tanner (1964) further suggested that the original East African was more successful in the middle distance and long distance races while the West African performed very well in the sprints. However, he did not point out the reason for this difference, which is that the tall, slim Nilotic East Africans possess optimal physical proportions for middle distance and longer events and that the shorter, muscular West African is more suited to the power (sprint) events.

Gymnastics and Power Sports

Many coaches have pointed out that individuals with long trunks and short upper and lower limbs do well in gymnastics, and we are now seeing more gymnasts and divers coming from Asian countries. It would appear that Asians are well suited to these events. Tanner (1964) also suggested that the proportions of this group are well suited to power sports, especially weightlifting;

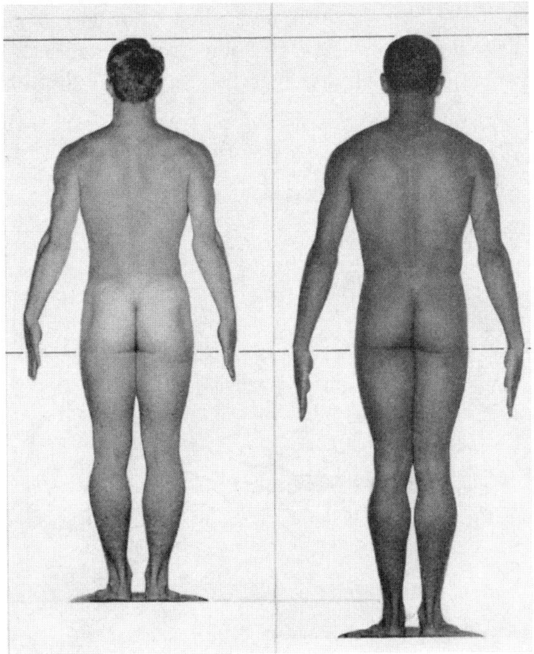

Figure 6.14 A comparison of a European and an African sprinter who have identical trunk lengths.

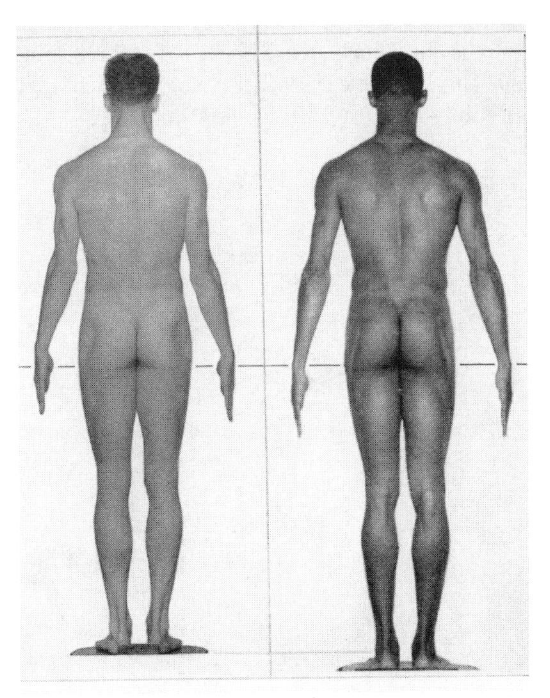

Figure 6.15 A comparison of a European and an African American 400 m runner who have identical trunk lengths.

successful weightlifters of all races have upper and lower limbs that are strikingly short.

Body Modification

The specific proportionality requirements for various athletic events have already been discussed in this chapter, and they clearly demonstrate the important role of this characteristic in high-level performance. As a general rule, human proportions cannot be modified in the same manner as other physical capacities using a simple intervention program, because the mature athlete's bone lengths are absolute and cannot be changed. As a result, proportions are significant parameters in the process of self-selection for various sports and events.

> As a general rule, human proportions cannot be modified in the same way as other physical capacities using a simple intervention program, because the mature athlete's bone lengths are absolute and cannot be changed. Thus they are significant parameters in the self-selection for various sports and events.

As a matter of interest, some athletes' bodies have been modified either accidentally or by design. However, on both moral and ethical grounds, deliberate changes are not recommended under any circumstances. Examples include the following:

- Growth plate compaction: Anecdotal evidence suggests that preadolescent and early-adolescent weightlifters were given very heavy weights to lift over prolonged periods of time in the 1970s and 1980s in some countries in Eastern Europe. It has been reported that this practice compacted the epiphyseal growth plates in the legs, inhibiting the long bone growth and thus shortening the legs. Furthermore, prolonged heavy weight-bearing exercise serves to develop thick, dense bones with a high mineral content. This being the case, a mechanical advantage and a more solid base of support would be obtained.
- Tendon insertion: Anecdotal evidence also suggests that powerlifters who have ruptured a muscle by pulling the tendon off a bone during various lifts, particularly a deadlift, have become stronger in that region of the body when the tendon was reattached to the bone a little farther away from its previous position.

Technique Modification

The only acceptable way to effectively change lever lengths is by a process known as technique modification, in which the coach can play a significant role. By altering the individual's technique, it may be possible to modify the lever system, thereby enabling the athlete to perform the skill in a more mechanically efficient way.

Technique modification to suit each individual athlete is the basis of good coaching. The skilled coach must know when the individual's proportions are unsuitable and which changes will be needed to improve his or her performance. The following examples from the various sport groups illustrate this point.

- A tennis player (figure 6.16) may have the majority of the motor skills and the psychological profile needed to perform well in this sport but be of tall stature with a relatively weak musculature. In addition to instituting a strength training intervention program that will no doubt assist the overall performance, the coach may need to modify the player's volleys and ground strokes. The player can flex the forearm at the elbow, thus shortening

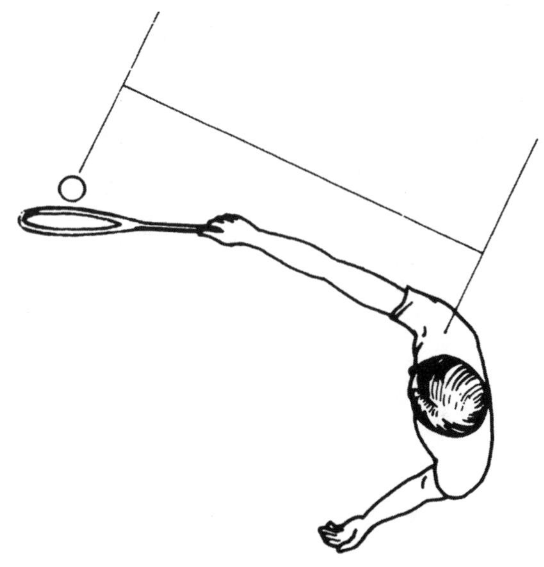

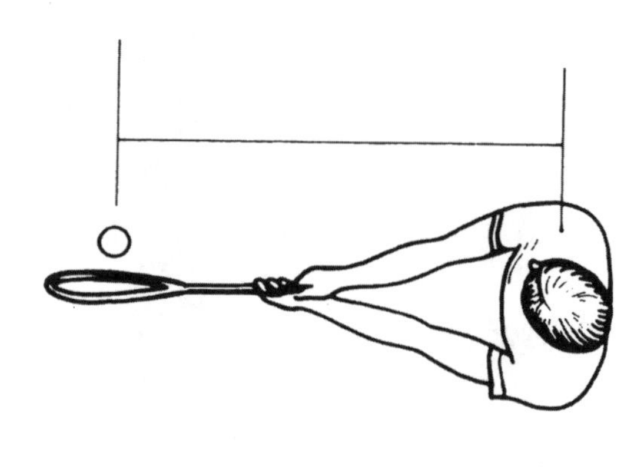

Figure 6.16 Shortening the effective lever using a two-handed backhand in tennis.

Reprinted, by permission, from T.R. Ackland and J. Bloomfield, 1992, Functional anatomy. In *Textbook of science and medicine in sport*, 2nd ed., edited by J. Bloomfield, P. Fricker and K. Fitch (Melbourne, Australia: Blackwell Scientific Publications), 10.

the lever to produce a more powerful forehand ground stroke based on racquet speed rather than lever length. The player may also adopt a double-handed backhand technique, which not only shortens the striking lever but also may facilitate greater stability and a better technique to hit a topspin shot (Ackland and Bloomfield 1995; Bloomfield 1979).

- A swimmer (figure 6.17) may have a large number of the variables essential for a champion but lack the proportions and strength needed for efficient freestyle propulsion. In addition to increasing strength, the swimmer can shorten the propulsive lever by flexing the forearm more than is normal for most swimmers (Bloomfield 1979).
- A golfer with long lower limbs should set up to the ball with a wider stance than a shorter-limbed player in order to prevent swaying laterally "past the ball", which causes the club face to be slightly open at impact and may result in a push-slice.

Figure 6.17 Shortening the effective lever in freestyle swimming.

Reprinted, by permission, from T.R. Ackland and J. Bloomfield, 1992, Functional anatomy. In *Textbook of science and medicine in sport*, 2nd ed., edited by J. Bloomfield, P. Fricker and K. Fitch (Melbourne, Australia: Blackwell Scientific Publications), 10.

Summary

Human proportions cannot normally be modified using a physical intervention program because individual bone lengths are finite and, under normal circumstances, cannot be changed. Proportionality is a self-selector for various sports and events, and some athletes are born with proportions that are highly suited to some sports but not at all suited to others (table 6.4). If an athlete has many of the physical characteristics that are suitable for a particular sport but lacks the leverage capacity to excel, the intelligent coach can modify his or her technique to partially overcome this physical disadvantage.

Table 6.4 Suitability of Various Body Proportions to Particular Sports

	PROPORTIONALITY		
Parameter	**Low**	**Average**	**High**
Stature	Gymnasts Divers Figure skaters Marathon runners	Sprint runners Long distance runners Middle distance swimmers	Middle distance runners Jumpers Throwers Sprint swimmers Basketball players Volleyball players Netball players
Relative lower limb length	Sprint runners Long distance runners Gymnasts Divers Wrestlers Weightlifters	Throwers Swimmers	Middle distance runners Jumpers Basketball players Volleyball players Netball players Rowers
Crural index	Long distance runners Gymnasts Heavyweight wrestlers Weightlifters	Sprint runners Middle distance runners Middle distance swimmers	Jumpers Basketball players Volleyball players Cyclists
Brachial index	Weightlifters Wrestlers	Middle distance swimmers Long distance swimmers	Throwers Water polo players Sprint swimmers Kayak paddlers

seven

Posture

Peter Hamer, PhD; and John Bloomfield, PhD

Some coaches may not be aware that an individual's posture can be very advantageous in many sports. In fact, after proportionality, posture is probably the most important self-selector for various sports and events.

Posture is unique to every individual; no two people have identical postures, although some are very similar. The determinants of an individual's posture are linked to the structure and size of bones, the position of the bony landmarks, injury and disease, static and dynamic living habits and the person's psychological state.

Good posture, both static and dynamic, is important for aesthetic appearance, but more importantly it is essential if the body is to function with an economy of effort. Poor posture can lead to fatigue, pain, muscular tension and poor muscle tone, as well as the sagging of some parts of the body and low self-esteem.

Evolution and the Development of Posture

The evolution of man from quadrupedal to bipedal hominid was accomplished through many adaptations of the musculoskeletal system over millions of years (Krogman 1951; Napier 1967). The four-legged animal possesses a skeletal system similar to that of a bridge, with an arched backbone to support the internal structures and with the legs acting as stanchions to support it. When primates slowly moved toward an upright posture, the advantages of such a system were lost, as the body was solely supported by the hind legs. The following structural changes and their ramifications occurred during this evolutionary period.

> The adoption of an upright posture for bipedal locomotion has provided a unique challenge to the structure and function of the human body.

- The vertebrae had to adapt to vertical weight-bearing stress, and this was achieved by changing from a curved C-shaped vertebral arch to an S-shaped one. The primary thoracic curvature therefore still exists, but other curvatures have developed.
- The erect posture places an extra burden on the pelvis, which now has to support the entire weight of the upper body. With erect standing, the whole structure was tilted upward and thus additional weight was placed on the pelvic basin. As this occurred, the bones in the pelvis changed shape and now resemble a basin that supports the intestines and some organs.
- The foot has changed shape to become less of a grasping appendage and more of a weight bearer. This has occurred via a shortening of the toes and a lengthening of the remainder of the foot; these changes place considerable stress on the arches, sometimes causing problems such as excessive pronation of the feet and functional flat feet.
- To permit the bending and twisting movements of the human spine, the vertebrae have changed shape so that they resemble a partial wedge. This shape, although very good for mobility, has weakened the vertebral column, especially in the lumbar region, where the discs may herniate with overstress. Furthermore, the lumbosacral joint must support the total weight of the upper body; and modifications to the articulation, due to the new alignment of bones, have created an area of instability if the lumbar spine is overstressed.

Changes in Posture During Growth

At birth the infant has two primary vertebral curves. The major curve is in the thoracic region, while the minor one is in the area of the

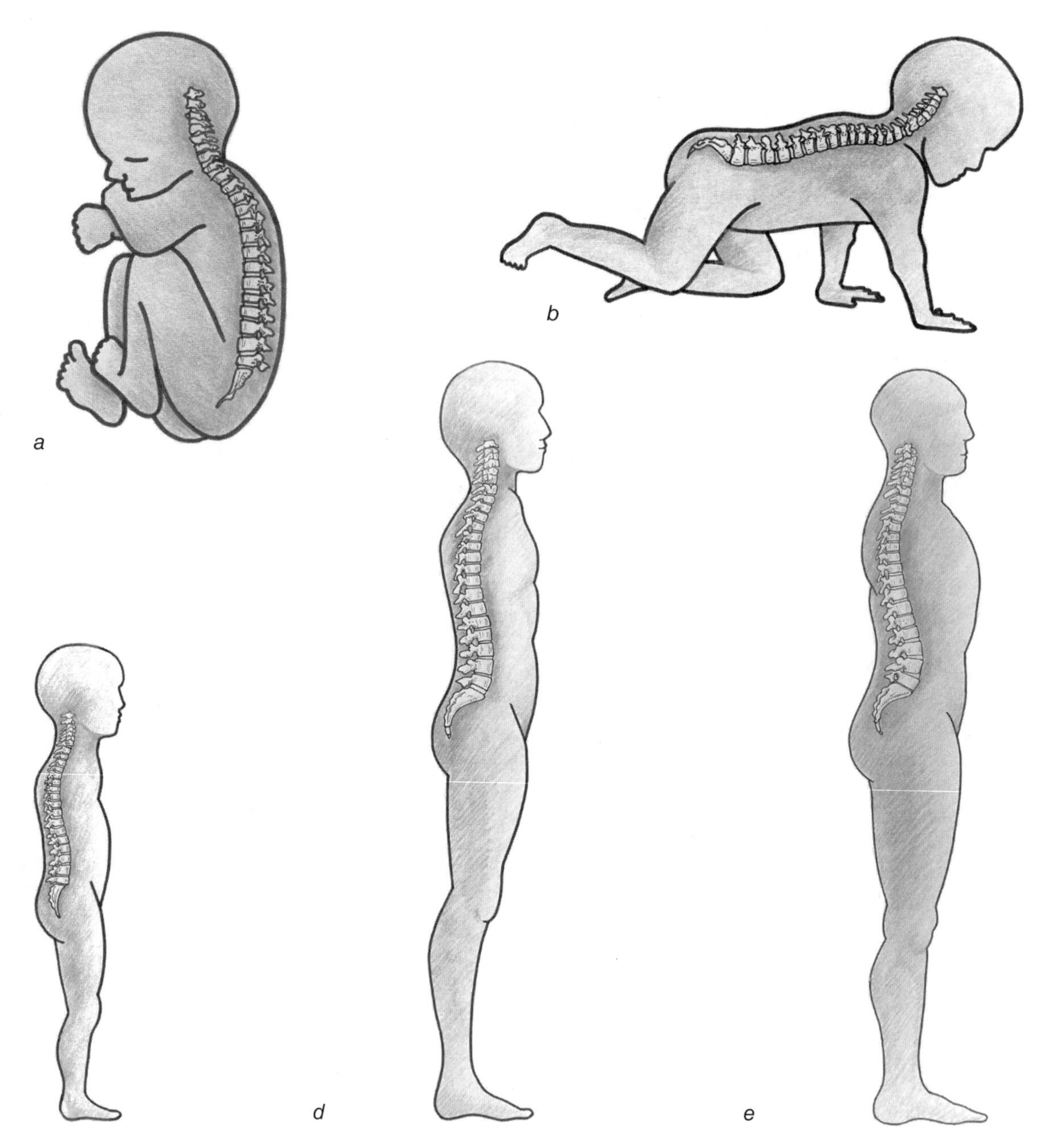

Figure 7.1 Changes in the spinal curvatures from birth to adulthood: *(a)* baby; *(b)* infant; *(c)* 3 years old; *(d)* 6 years old; *(e)* adult.

sacrum. At around 6 months of age, a secondary curve develops in the cervical area and is the result of the infant's holding his or her head up. When the child stands upright, the lumbar curve starts to develop; and by the time the individual reaches 6 years of age there are two definite primary curves and two secondary curves (figure 7.1).

When the baby is born, the legs are flexed and the feet are inverted, but as the child stands and the legs develop, the feet become everted. An 18-month-old child has "bowlegs" when standing, but by 3 years of age will develop "knock knees". In most cases by 6 to 7 years of age the legs will have straightened (figure 7.2) and arches in the feet will have been formed.

Maintenance of Posture

The maintenance of posture in an upright position utilizes the mechanical properties of joints, ligaments, tendons and muscles as well as an automated system of reflexes that are smoothly coordinated within the nervous system, requiring no voluntary (conscious) effort. The reflex that plays the major part in this process is known as the myotatic reflex, which responds to stretch and is described in detail in chapter 11.

The mechanism that enables humans to assume various postures, and maintain them, works in the following way. If a joint starts to flex, the initial slack is taken up by

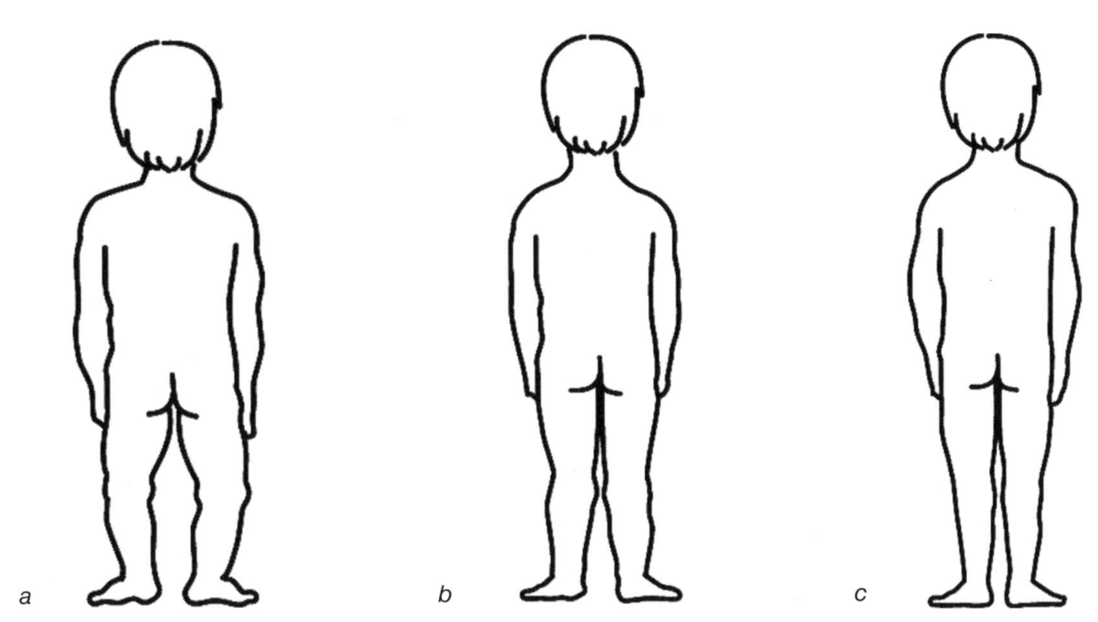

Figure 7.2 The development of leg posture in the child: *(a)* 18 months old; *(b)* 3 years old; *(c)* 6 years old.
Adapted from D. Sinclair, 1973, *Human growth after birth* (London: Oxford University Press).

the tendons crossing the joint, which may offer enough tension and elasticity to resist and then restore the joint position. At the same time the muscle spindles of the muscles controlling the joint are stretched and impulses from these are generated that go to the spinal cord, where synapses are made with motor neurons. These motor neurons then innervate the fibres of the controlling muscles so that they contract and re-extend the joint, thus restoring it to its normal position. The reflex operates only in the presence of facilitation by the vestibular nucleus in the medulla through the spinal cord but is regulated through the extrapyramidal system. The combination of the elasticity of muscles and tendons and the reflexive response of the muscles and nerves enables very smooth and coordinated muscle contractions to maintain normal posture. There are also other important reflexes that assist the maintenance of posture; of these, the myotatic reflex is the least complex. Posture also needs to be considered for positioning of the body prior to any movement taking place; this is important in giving stability to the body during movement

Advantages of Good Posture

Posture can be defined as the relative arrangement of body parts or segments, but generally it is the term used to describe the way a person stands. When we consider good or poor posture, the bones should represent a series of links connected by joints being held together by muscles and ligaments. If these links are arranged in a vertical plane so that the line of gravity passes through the centre of each joint (figure 7.3), then the least stress will be placed on the muscles and ligaments. Good posture therefore is a state of muscular and skeletal balance that protects the supporting structures of the body against progressive deformity or injury. Poor posture, on the other hand, is the faulty relationship of the various segments of the body such that the line of gravity no longer

Figure 7.3 In "good posture", a vertical line should pass through the anterior portion of the ear and then through the centre of each joint of the lower extremity.

passes through the centre of each joint (figure 7.4). This type of posture makes it more difficult to maintain efficient balance over the base of support and causes habitual sagging, which can permanently stretch some muscle groups and shorten others, producing increased stress on supporting structures. The advantage of having good posture is that the least use of energy occurs when the vertical line of gravity falls through the supporting column of bones. In this situation the body requires minimal use of its automated responses in adjusting position to counter the forces of gravity. Good posture, therefore, is both mechanically functional and economical.

Good posture offers efficiency of movement and minimized energy utilization.

Postural Diversity Within Individuals

Postural diversity can occur within individuals because of disharmony between different regions of the body. Sheldon, Stevens, and Tucker (1940) termed this phenomenon "dysplasia" and related it to the person's somatotype by suggesting that a subject could "be in proportion" in one region of the body and yet be quite disproportionate in one or several other segments.

Dysplasic characteristics are seen in many people, but this does not always mean that they have poor posture, because they have in some cases adjusted naturally so that their general postural alignment is quite good. Figure 7.5 gives examples of individuals with dysplasic characteristics such as small hips and a large head, or heavy legs and buttocks or large breasts. These subjects certainly have dysplasia, but their posture is generally good.

Posture and Its Relationship to Somatotype

There is a strong postural relationship with body type (refer to chapter 4), especially among ectomorphs and endomorphs. The ectomorph has more postural deformities than the other groups, especially as these relate to the vertebral column. Such defects as a poked or forward head, abducted scapulae or round shoulders, kyphosis or round back, lordosis or hollow back and scoliosis

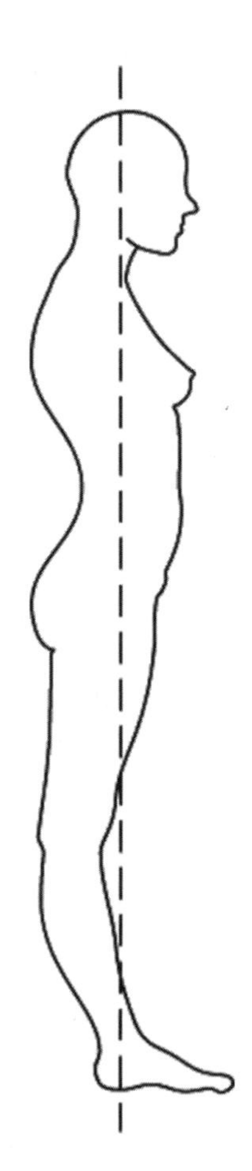

Figure 7.4 If any segment of the body deviates from its vertical alignment, its weight must be counterbalanced by the deviation of another segment in the opposite direction. Also note the severe genu recurvatum (hyperextension) at the knee joint.

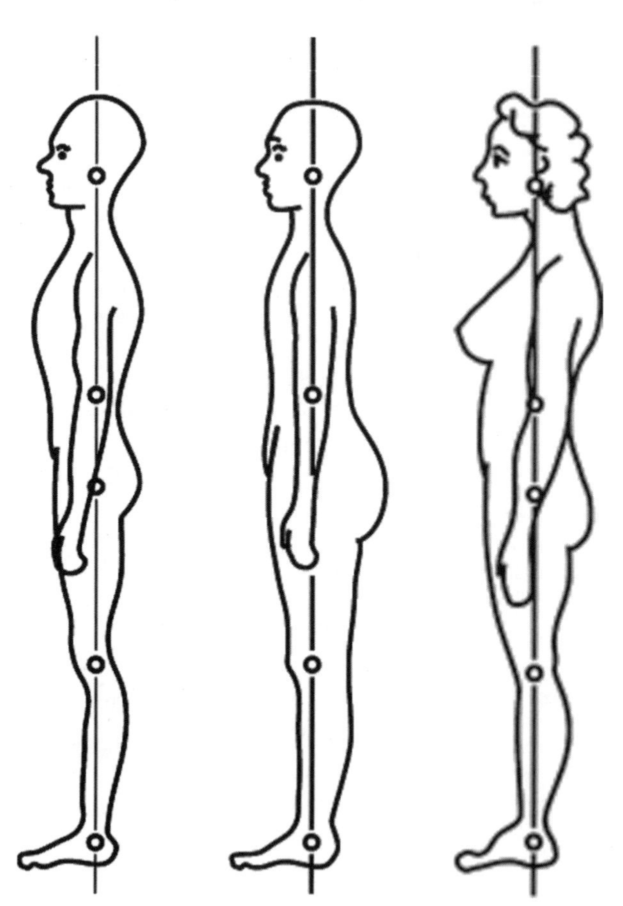

Figure 7.5 Examples of individuals with dysplasic characteristics.

or lateral curvature of the spine are common with primary ectomorphs; and at times two of these problems can be combined. Endomorphs suffer mainly from leg deformities due to the added burden of additional body mass; and such problems as genu valgum or "knock knees", flat feet and everted or duck feet are common. Mesomorphs are generally free from major postural defects but may develop minor problems as they grow older, especially if they increase their body mass.

Postures are commonly associated with body type—ectomorph, endomorph and mesomorph.

Postural Defects

Whereas posture is considered an important self-selector for various sports and events, the coach needs to be aware of the nature of variations to posture and of postural defects that may give an athlete a mechanical advantage or disadvantage or potentially have implications for injury.

Causes of Postural Defects

Ackland and Bloomfield (1992) identified several factors that cause postural defects, some of which are genetic and others of which have environmental genesis:

- **Injury:** When a bone, ligament or muscle injury occurs, it may weaken the support normally provided to the total framework. Therefore, as long as the condition is present, good posture may not be attainable.
- **Disease:** Diseases often weaken bones and muscles or cause loss of joint stability, thus upsetting posture. Examples of such diseases include arthritis and osteoporosis.
- **Skeletal imbalance:** The most familiar imbalance of skeletal lengths is seen in the lower limbs, and in extreme cases this causes a lateral pelvic tilt and may result in the development of scoliosis. However, more subtle skeletal differences such as the location of the acetabulum (figure 7.6) and length of the clavicle provide equal potential for defective posture.
- **Habit:** People acquire postural habits by repeating the same body alignment on many occasions, as when leaning over a desk or slouching in a chair. If body segments are held out of alignment for extended periods of time, the surrounding musculature rests in a lengthened or shortened position. This can give rise to "adaptive shortening" or "stretch weakness" or both (Kendall et al. 2005). Habitual postures associated with an individual's mental attitude over a prolonged period of time and the wearing of high-heeled shoes, which shift the centre of gravity forward, are other examples of variables that may cause postural defects.

Interrelationships of Postural Defects

Although postural defects are treated segmentally, it must be stressed that these abnormalities are usually associated with other changes within the body. Normally the downward gravitational pull on any part of the body is borne by the segment below; but if any segment deviates from its vertical alignment, its weight must be counterbalanced by the deviation of another segment in the opposite direction. Therefore, postural defects must be seen from a total-body perspective. Figure 7.4 illustrates this phenomenon in the following way. The subject, by standing in a tense position, increases the pelvic tilt so that the pelvis rotates forward on the femur, carrying the lumbar spine forward and with it the body's centre of gravity. To compensate for this position, two additional actions occur. First the legs tend to adopt a hyperextended position (genu recurvatum), while the upper part of the body is thrust backward, thus increasing the lumbar and thoracic curvatures.

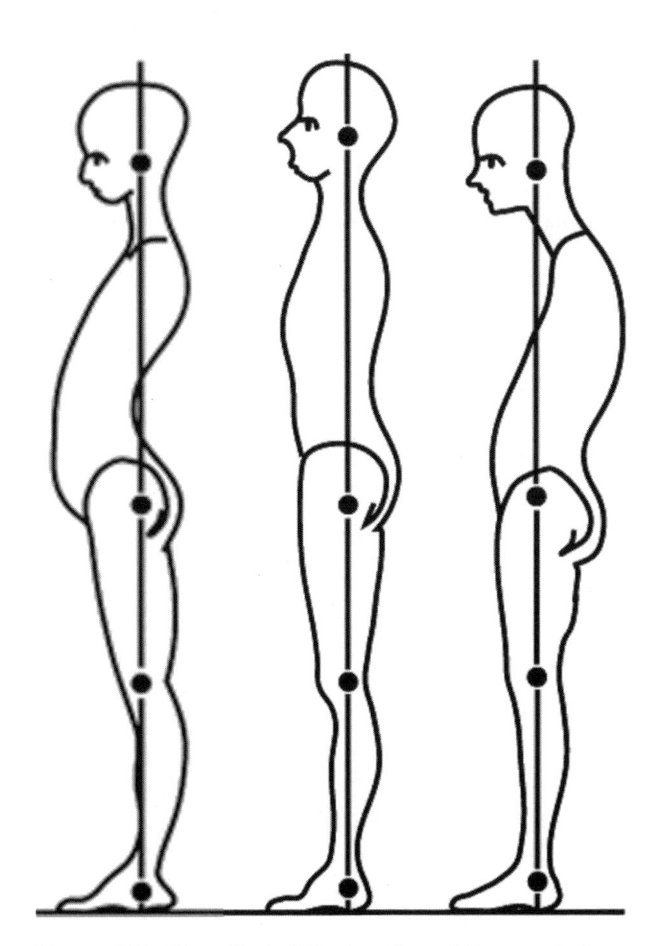

Figure 7.6 The effect of the location of the acetabulum on back posture.

Specific Postural Defects

The defects we discuss next range from those that are only just visible to the trained eye to those that are extreme. Individuals with minor defects may not need a corrective program because their bodies have gradually adapted to these defects; however, those with moderate to severe postural deviations will need various levels of remediation and in extreme cases surgery.

Anteroposterior Defects

It is easiest to see anteroposterior postural defects when one views the athlete from the side (a lateral view), as these postural positions occur in the sagittal plane of the body.

Poked or Forward Head Forward head is an alignment defect in which the neck is extended on a rounded upper back (kyphosis) and the head is partially tilted forward and upward (to keep the eyes level) (Kendall et al. 2005). It is often associated with kyphosis of the upper back and abducted scapulae (round shoulders). To correct this alignment, the posterior muscles of the neck must be actively stretched by flattening of the neck curve, and the anterior muscles should be strengthened to hold this postural alignment. The commonly associated kyphosis of the upper back should also be corrected—importantly, often as the first step.

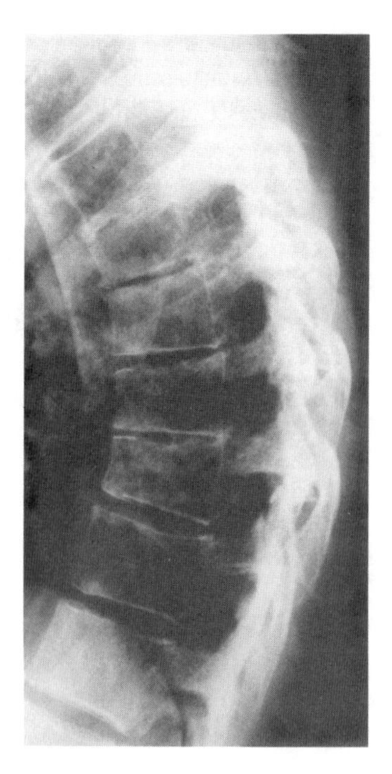

Figure 7.7 A radiograph of severe kyphosis of the thoracic spine depicting vertebral wedging.

Abducted Scapulae (Round Shoulders) Round shoulders is a postural alignment that occurs when the scapulae assume an abducted and often elevated position because the middle and lower fibres of the trapezius muscles are weakened or lengthened, or both weakened and lengthened. The medial borders of the scapulae may protrude from the individual's back. To correct this condition, the adductors of the scapulae need to be strengthened and the pectoralis minor and upper trapezius muscles need to be lengthened (Kendall et al. 2005).

Kyphosis (Round Back, Often Associated With Scheuermann's Disease) Kyphosis increases the convexity of the thoracic curve and is caused by wedging of the thoracic vertebrae (figure 7.7). Treatment in mild cases consists of exercises designed to stretch the upper anterior thoracic region (e.g., pectoralis) and strengthen the muscles of the posterior thoracic area (e.g., middle and lower fibres of trapezius). If the condition is severe, a brace should be considered in order to produce extension of the thoracic spine and decreased lordosis in the lumbar spine (Kendall et al. 2005; Watson 1992).

Lordosis (Hollow Back) Lordotic posture is characterized by an exaggerated lumbar curve, usually caused by a pelvis that tilts too far forward (anterior pelvic tilt; APT). In this condition, the abdominal muscles become stretched and weakened and need to be strengthened along with the extensor muscles of the thigh (gluteal muscles), while the erector spinae and the flexor muscles of the thigh should be stretched (Ackland and Bloomfield 1992; Kendall et al. 2005).

Visceral Ptosis (Protruding Abdomen) Visceral ptosis, characterized by the sagging of the abdominal organs, often accompanies lordosis. The downward drag upon the mesenteries occurs when passive tension in the abdominal wall is not sufficient to hold them in place. Body fat reduction, postural education and exercises that strengthen the abdominal muscles are necessary to alleviate this condition (Hills 1991; Rasch and Burke 1978).

Kypholordosis Kypholordosis is a combination of kyphosis and lordosis and as a result places a great deal of stress on the trunk, because the antigravity muscles are forced to contract vigorously in order to balance the body segments. Individuals with this condition are often in a state of chronic fatigue. Exercises recommended for both kyphosis and lordosis should be carried out to alleviate this condition.

Genu Recurvatum (Leg Hyperextension) Genu recurvatum is characterized by a backward curve of the legs, which creates an unstable knee joint for agility sports (figure 7.4). The causes of this condition are related to the structure of the femur, the tibia and the cruciate ligaments in the knee joint; therefore only general strengthening exercises of the musculature surrounding the knee joint should be carried out, along with education to avoid knee hyperextension. It is important to keep the hamstring muscle group almost as strong as the quadriceps (Hills 1991).

Although everyone is different, common postural alignments can be classified. While many of these are structural, there are postural defects that can be prevented or accommodated. Posture should be assessed to determine the advisability of prevention or management to decrease the possibility of injury or to optimise performance.

Lateral Defects

It is easiest to see the following postural defects when one views the athlete from behind (a posterior view) or from the front (an anterior view). These postural positions occur in the frontal (or coronal) plane of the body.

Scoliosis The common defect known as scoliosis manifests itself in a lateral curvature of the thoracolumbar spine and in many severe cases is accompanied by a longitudinal rotation of the vertebrae (Rasch and Burke 1978). Scoliosis usually begins with a C-shaped curve (functional scoliosis), but over a period of time a righting reflex creates a reversal of the "C" at the upper spinal levels, producing an S-shaped curve (structural scoliosis). Uneven leg lengths, muscle imbalance and ligament lengthening can cause this defect.

Functional scoliosis, if identified early, can be corrected or greatly improved through use of an orthotic device in a shoe, which will increase the person's functional leg length. Education to avoid postural habits that increase the C-curve and exercises that develop general flexibility in the thoracic and lumbar regions of the spine are necessary for individuals with functional scoliosis. In cases of structural scoliosis, medical assistance must be sought.

Genu Varum (Bowlegs) and Genu Valgum (Knock Knees) Genu varum and genu valgum are postural alignments that are genetic, and individuals may need medical attention early in life if they appear to be serious, as there may be knee joint irregularities or partial deformities of the femur or tibia bones. Apart from advising general strengthening exercises for the various muscle groups of the leg, the coach can do little about these defects if they appear to be extreme.

Tibial Torsion ("Pigeon Toes") Often referred to as *inverted feet,* tibial torsion is characterized by internal rotation at the hip joints. This in turn causes the knees to inwardly rotate ("crossed knees" or "squinting patellae") so that the feet become inverted. Although it is a structural defect, tibial torsion can be alleviated via stretching of the medial rotators of the hip joint and strengthening of the lateral rotators (Rasch and Burke 1978).

Pronated Feet ("Duck Feet") Pronated feet is a defect also known as *everted feet* and is characterized by a protruding medial malleolus and pseudo flat feet caused by the rolling inward of the ankles. The best treatment for this defect is the use of an orthotic device from an early age and a series of exercises involving toe flexion, foot plantarflexion and supination (Rasch and Burke 1978).

Flat Feet There are several classifications of flat feet, as follows:

- **True flat feet *(pes planus):*** This is the most significant of the foot defects in which the longitudinal arch is flat. Many people with flat feet experience no symptoms. However, for others the condition may be accompanied by discomfort and interfere with the foot's normal function. In these cases, medical or podiatric consultation (or both) should be sought.
- **Functional flat feet:** This is a defect caused by weakened and stretched muscles, ligaments and fascia in the foot. If not corrected, it can distort the mechanical relationships in the ankle, knee, hip joints and lumbar spine. Medical or podiatric consultation or both should also be sought for this defect.
- **Flexible flat feet:** This condition is characterized by a loss of the arches of the feet during weight bearing; when there is no weight on the feet, they appear normal. It is not regarded as pathological unless it interferes with normal function or becomes painful, in which case medical or podiatric consultation or both should be sought.
- **False flat feet:** This is not a true postural defect but a condition that results from the presence of a fat pad on the plantar surface of the feet. From time to time one hears of elite athletes who have this condition yet are able to function quite normally with it (Rasch and Burke 1978).

Static and Dynamic Posture

Posture is static when a person is in equilibrium or motionless. In sport science, we are much more interested in dynamic posture, that is, the posture of an individual in motion. Generally, there is a high positive correlation between static and dynamic posture—a phenomenon that has been observed by high-level coaches for many years. However, a further item of interest when one compares static and dynamic posture relates to injuries that can be the result of postural defects, as we briefly discuss next.

Injuries Resulting From Static Postural Defects

Researchers have found both strong and poor associations between static postural alignment and injuries, and this topic is one of continuing interest. For example, Lun and coauthors (2004) found that except for patellofemoral pain syndrome, the incidence of lower limb injury did not relate to static biomechanical alignment measurements of the lower limbs. However, previous research by Lorenzton (1988), reported in a study of injured runners, showed that 40% of them had a variety of postural defects, muscle weakness and imbalance or decreased flexibility. Malalignment problems of several types were involved.

- Pronated feet or flat feet, causing excessive pronation during running, resulted in injury.
- During the running cycle, it is necessary to have the correct alignment of the foot and the leg. Runners who did not possess this characteristic, or who had eversion of the heel, predisposed themselves to injury.
- Runners with flat feet (pes planus) were likely to develop a further depressed longitudinal arch without eversion, while those with high arches (pes cavus) suffered from injuries attributed to excessive motion of the subtalar joint.
- Individuals with an increased Q-angle (i.e., a measure obtained by connecting the central point of the patella with the anterior superior iliac spine and the tibial tuberosity) or genu valgum (knock knees) experienced injuries to the patellofemoral joint and the patella itself. Athletes with genu varum (bowlegs) also predisposed themselves to injuries in the patella region as well as to iliotibial band friction syndrome.
- Athletes with leg length discrepancies accompanied by a pelvic tilt developed trochanteric bursitis and iliotibial band friction syndrome, as well as intervertebral compression on the concave side of the lateral lumbar curve.

Not everyone with one or more of these postural defects will incur injury, and some malalignments will have various effects on individuals involved in different sports; but coaches should be aware of the potential increased risk of injury among athletes with these defects. Assessment of how posture changes with the dynamics of skilled movement and with fatigue or changes in environmental influences is also an important part of analysis of sporting performance. From the simplest slow-motion video playback of performance to high-speed 3D motion capture (see chapter 15), coaches, biomechanists and sports medicine practitioners can work with athletes in the prevention of injury and optimisation of performance. Astute observation can often save an athlete from developing a chronic and debilitating injury that could have been avoided.

Prevention of Postural Defects

In order to prevent the occurrence of minor postural defects, most athletes should, as part of their flexibility and strength training programs, carry out a proactive exercise routine designed to assist in the maintenance of good posture. Many postural defects can develop from the overuse of one or several regions of the body and can cause physical discomfort and injury in the more mature athlete. Figure 7.8 shows the overdevelopment of the left side of the upper body of a high-performance left-handed fast bowler in cricket and the accompanying scoliosis that developed over several years. This can occur with unilateral athletes if it is not guarded against, while bilateral athletes such as swimmers, cyclists, wrestlers and boxers can develop various anterior-posterior curvatures—the most common of these being round shoulders resulting from their intensive sport-specific training programs. There are, however, some postural characteristics that, if not too extreme, are a definite advantage to the athlete; these are explained in the section on posture modification later in this chapter.

Posture Assessment

Static posture is usually assessed subjectively in the standing position using a rating chart as a guide for the observer. The form shown in figure 7.9 is based on the New York Posture Rating Test (Adams, Daniel, and

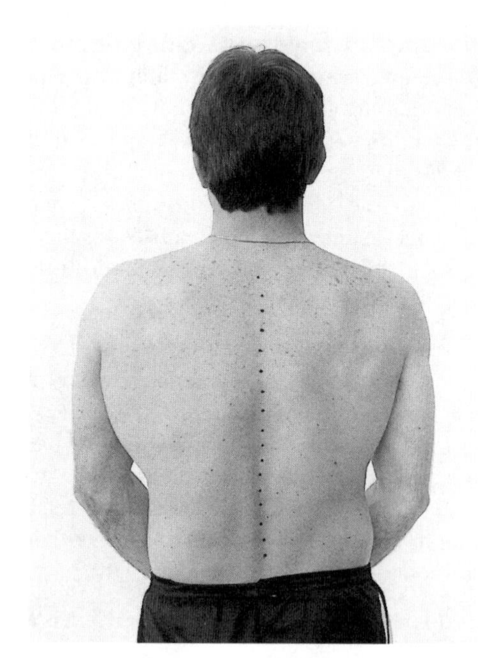

Figure 7.8 Overdevelopment of the left side of the upper body with an accompanying scoliosis in a high-level cricket bowler.

Rullman 1975) and may be used as a simple yet effective screening tool. A trained observer examines the alignment of body segments when viewed from the posterior and lateral perspectives. Observers may use a plumb bob, or one of the many clear grid screens commonly referred to as a posture grid to aid in their subjective evaluations.

The subject stands erect in a natural posture (i.e., not rigidly in the anatomical position), approximately 3 m (3.3 yd) in front of the observer. For each of the 13 posture areas shown in figure 7.9, the subject is given a score from 1 to 5, based on the sketches on the rating form. Only whole numbers are permitted. With respect to the posture of the longitudinal arch of the foot, the subject is required to step onto a chalkboard after the feet have been moistened. The imprint of the feet is then assessed and scored for this item. More objective tests that focus on a particular postural deformity, rather than mass screening, include medical imaging techniques using radiography and computerized tomography. In addition, special photographic techniques such as Moiré topology have

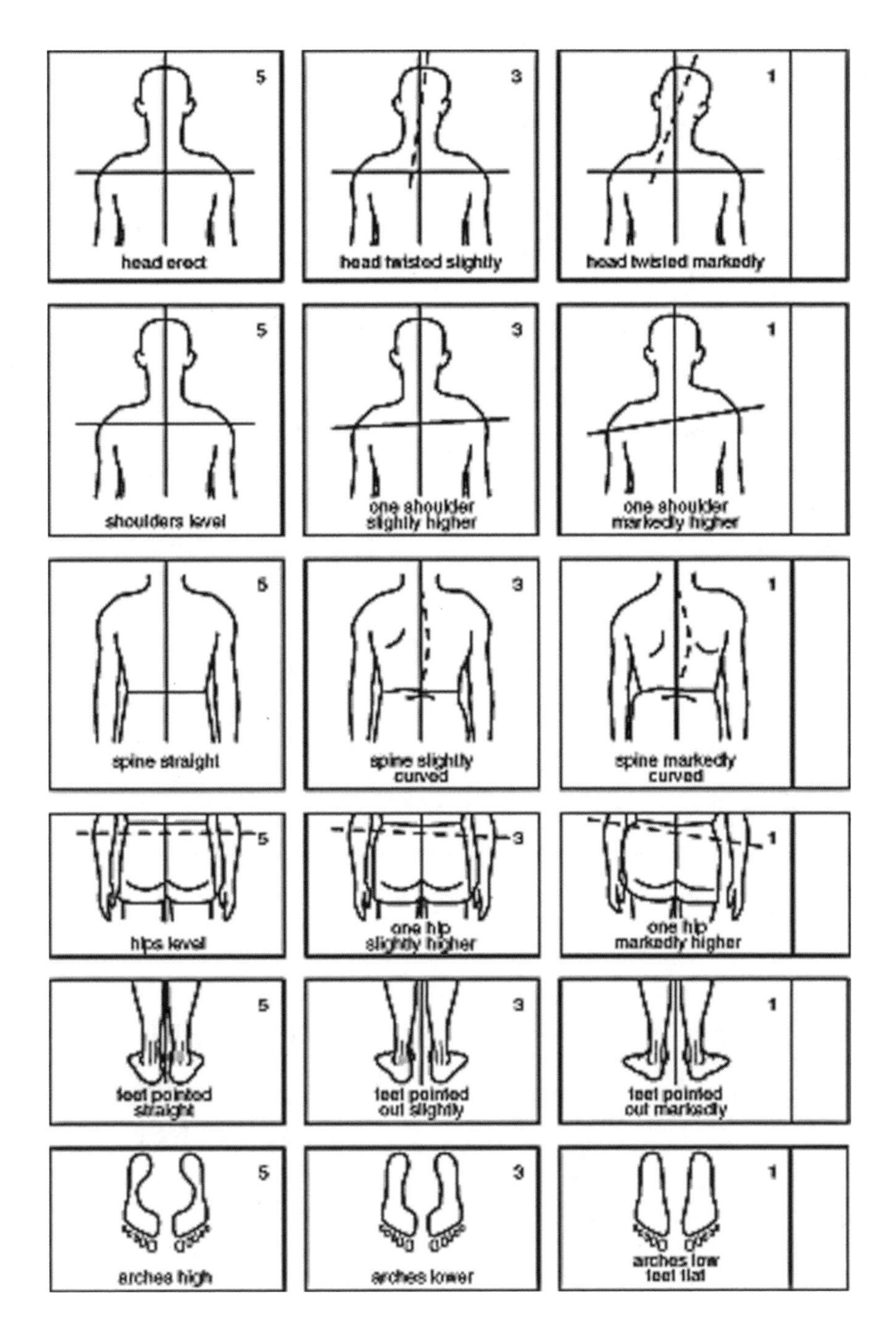

Figure 7.9 Posture screening test.
Based on NY Posture Rating.

(continued)

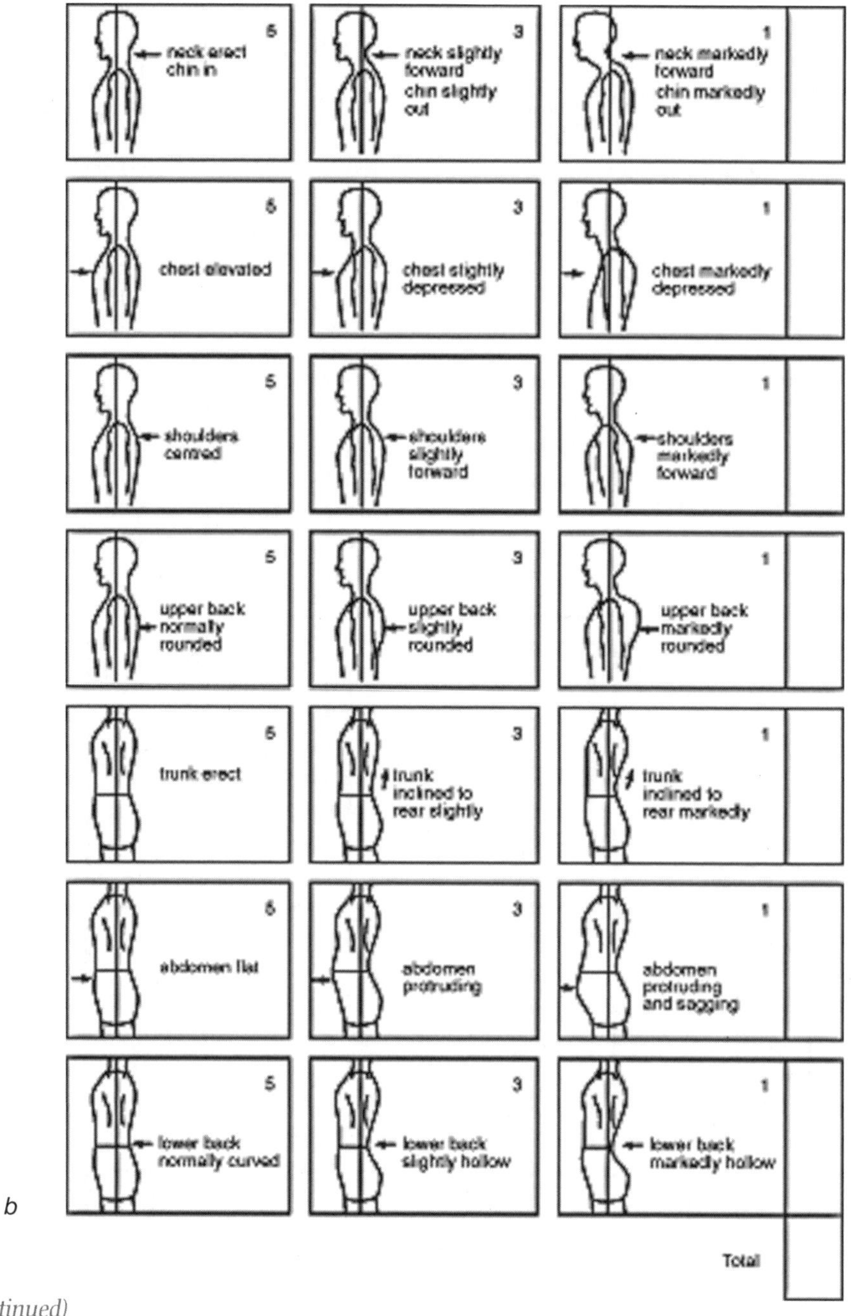

Figure 7.9 *(continued)*

been developed for the accurate assessment of scoliosis and other spinal postural disorders.

Instrumentation and methodologies for measuring dynamic posture are now becoming more readily available, having been developed in association with the biomechanical techniques of high-speed 2D and 3D video analysis and electrogoniometry, which are discussed in chapters 14 and 15. In the field of occupational biomechanics, several quasistatic techniques have been used to quantify dynamic postures in the work environment; these have been reviewed by Chaffin and Andersson (1984).

Desirable Postures for High-Level Sport Performance

There are various minor postural deviations that are well suited to different sports and events, because the alignment of the bones and the muscles covering them produces a mechanical advantage with respect to speed, power, balance or some combination of these (Bloomfield 1979). Posture has a marked effect on performance, but little research has been carried out on its advantages or disadvantages. Various postural phenomena associated with some major sport groups have been identified by coaches, who have made the observations we review next.

Racquet Sports (Tennis, Badminton, Squash, Racquetball)

Competitors in racquet sports have variable postures; however, those with inverted feet (pigeon toes) have a speed advantage over a short distance, as on small courts, because they automatically take short steps that are usually very rapid. There has been some debate as to why this postural characteristic enables them to move fast over a limited distance; the most often cited theory is that tibial torsion tends to shorten the hamstring muscle group, preventing the individual from "striding out" and taking long steps. This characteristic is of little value to athletes who wish to move very fast over any distance more than 15 to 20 m (16.4 to 22 yd) because to attain very high speeds one needs both a fast stride rate and a reasonably long stride. Furthermore, it is also thought that cutting down the player's stride length improves dynamic balance because there is more ground contact while the player is moving.

It is also important to reinforce the point made earlier in this chapter relating to unilateral athletes; many racquet sport players fit into this category, especially mesomorphic males. Such players should have compensatory strength and flexibility training on the opposite side of the body to the preferred side in order to prevent muscle overdevelopment and in some cases scoliosis.

Aquatic Sports

The aquatic environment offers a unique medium for the human body, which has evolved for locomotion across dry land. In some cases the body is immersed in the water, within which gravitational forces are counteracted by buoyancy while air resistance is replaced by fluid resistance and drag.

Swimming and Water Polo

Swimmers with square shoulders and upright trunks who possess long clavicles and large scapulae appear to have lower levels of shoulder flexion-extension than those with sloping shoulders. If square-shouldered swimmers need high levels of flexibility, they must undergo a particularly intensive program to improve this physical capacity. Swimmers with inverted feet (pigeon toes) are admirably suited for back and front crawl or butterfly kicking, while those with everted feet (duck feet) are very well suited to breaststroke kicking. Brodecker (1952) indicated that leg hyperextension is very prevalent in swimmers, and some sports medicine physicians have suggested that this occurs because the cruciate ligaments of the knee slowly stretch from constant kicking, thus allowing more recurvatum to develop. High-level coaches and biomechanists, however, do not seem to have reached a consensus on whether or not this is an advantage to the swimmer; some state that it makes little difference in kicking while others suggest that this posture gives a greater range of anterior and posterior motion at the knee joint.

Rowing and Canoeing

Rowers and canoeists do not appear to need any inherent postural characteristics in order to gain an advantage over other competitors in their sports. It is true that many of them have a slightly rounded back, but this phenomenon is thought to be the result of intensive training with slightly hunched shoulders rather than a contributing factor to their self-selection for these two sports.

Gymnastics and Power Sports

Gymnastics and power sports are examples of sports in which natural selection based on posture, body type or both is common.

Gymnastics

Female gymnasts with lordosis and an APT are able to hyperextend their spine more easily than those who are more flat backed. If they have protruding buttocks along with these characteristics, they will possess hip extension advantages over flat-buttocked competitors and will be able to spring very effectively in the floor exercise part of their programs. However, skills should not be taught to utilize excessive lordosis, as this can place greater stress on the posterior elements of the spinal vertebrae, causing stress fractures. Muscular control of the available range of movement or any increase in flexibility is most important to avoid injury.

Weightlifting

Weightlifters do not appear to need any one major postural characteristic in order to gain a distinct advantage over their fellow competitors. From a postural perspective, they are a highly variable group.

Track, Field and Cycling

Success at the highest level in track, field and cycling has a high genetic component, contributing, for example, to the type of muscle fibre (e.g., fast- or slow-twitch); however, athletes in these sports may also have characteristic postural types.

Sprinting

Athletes with APT as well as protruding buttocks (provided they have all the other necessary physical capacities) are usually excellent sprinters (figure 7.10). This postural phenomenon is commonly found in Africans (currently in or originally from Africa) and is less common in Europeans, while Asian males very rarely

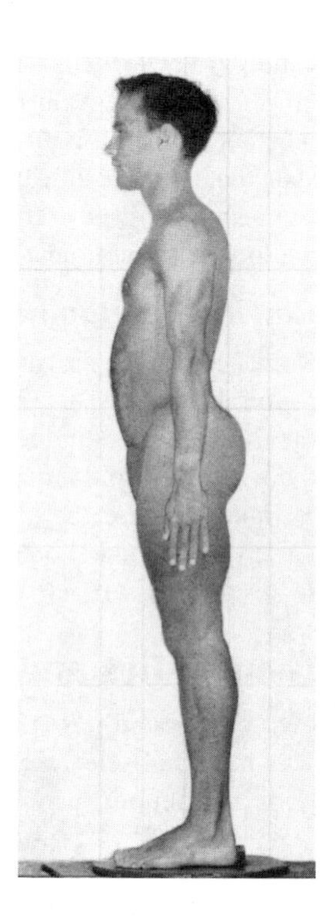

Figure 7.10 This sprinter demonstrates an anterior pelvic tilt and protruding buttocks.

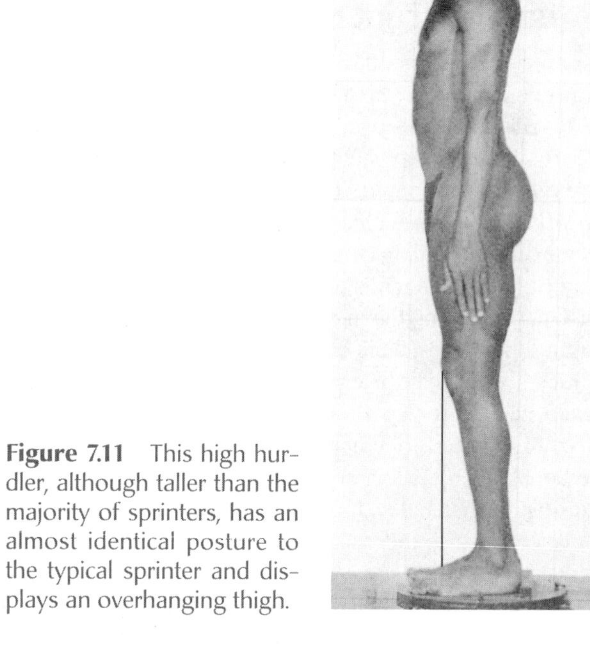

Figure 7.11 This high hurdler, although taller than the majority of sprinters, has an almost identical posture to the typical sprinter and displays an overhanging thigh.

possess it. One should note, however, that this characteristic is more common among European females and is sometimes found in Asian women. Webster (1948, 331) appears to be the first coach to have commented on it when he stated that "actual dissection of negroes has shown that they have a more forward pitch of the pelvic bones and consequently a more forward hang of the thigh". Brodecker (1952) supported this observation and linked it to protruding buttocks. He further stated that many African American athletes possess APT and that it is also "typical of the female". He also suggested that the "overhanging knee joint" *(overhanging thigh)*, where the patella is well forward of the junction of the anterior part of the ankle joint when viewed laterally (the exact anatomical position is with the patella vertically above the tarsometatarsal joint), occurs in individuals with a tilted pelvis (figure 7.11). This is a contrasting leg posture to that of swimmers, who generally have little pelvic tilt, relatively flat buttocks and almost no overhanging thigh. Brodecker (1952) suggested that individuals with APT and protruding buttocks are able to exert more force in running as the leg is extended back. This is in agreement with many sprint coaches, who believe that this posture gives the sprinter an optimal driving angle in the extension phase of the running cycle. The high hurdler, although taller, has an almost identical posture to the sprinter; figure 7.11 illustrates this phenomenon.

These postural characteristics in very fast runners are accompanied by large and powerful buttocks (gluteus maximus muscle) and thighs, particularly the hamstring muscle group (biceps femoris—long head, semitendinosus, semimembranosus), which, with the gluteus maximus, extend the thigh with a powerful "driving action".

Middle and Distance Running

Athletes in middle distance running events, particularly those in the 400 m and some in the 800 m events, have a similar buttock posture, although usually not as extreme, to that of sprinters (figure 7.12). As the races get longer, the protruding buttocks characteristic disappears, with long distance runners having reasonably flat backs and buttocks (figure 7.13).

Jumping

The group of field athletes who compete in jumping, although different to sprinters in their proportions, have almost identical postures. Figure 7.14, which shows an outstanding long jumper, illustrates this and demonstrates APT and particularly the protruding buttocks that are accompanied by the overhanging thigh of the vast majority of agility athletes.

Throwing

Elite-level throwers, like weightlifters, have no single postural characteristic that is obvious. Their performances are related more to body bulk, proportionality and explosive power.

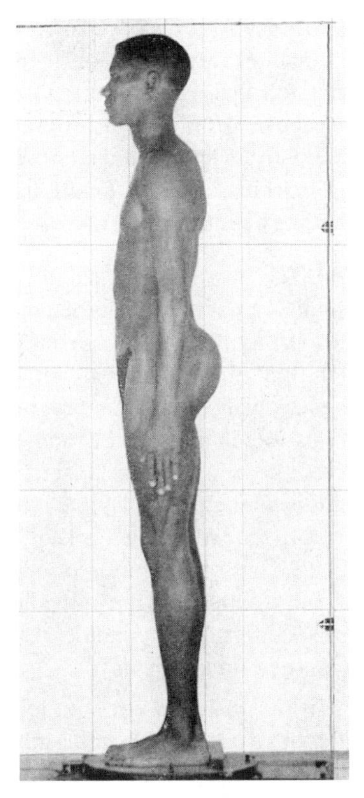

Figure 7.12 This 400 m runner has a similar buttock posture to sprinters.

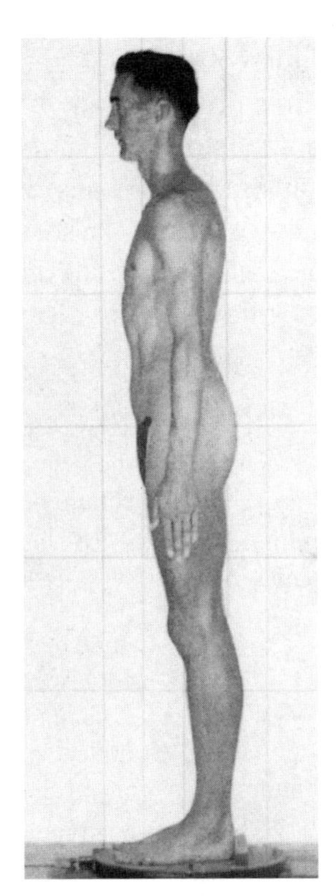

Figure 7.13 This distance runner displays the flat back and buttocks commonly observed in many of the athletes in his group.

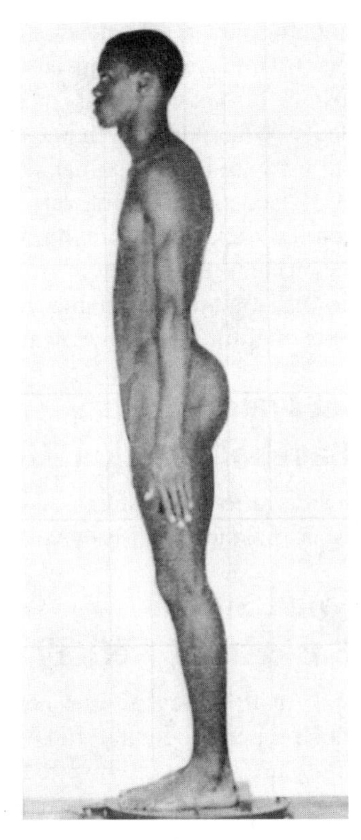

Figure 7.14 This jumper has a similar hip and buttock posture to sprinters.

Cycling

Cyclists have definite postural characteristics, but these are thought to relate more to their heavy training routines over prolonged periods of time than to their constitutional bone shape. They tend to have slightly rounded backs and possess the overhanging thigh phenomenon. It is thought that the latter posture is caused more by the heavy musculature in their thighs and buttocks, particularly in sprint cyclists, than by the natural APT and protruding buttocks displayed by the majority of sprint and agility athletes. In other words, the very heavy power training of cyclists has influenced their posture.

Mobile Field Sports (Field Hockey, Soccer, Lacrosse)

Athletes in mobile field sports are usually a varied group with no rigid postural requirements. Players in positions in which they have limited territory to cover may have a specialized posture such as inverted feet. Where a high level of speed is needed for a reasonable distance, for example in wing positions in various sports, APT and protruding buttocks can be of value, provided that the player has the necessary skill and other physical requirements to be a valuable team member. In these sports, a nonrigid or non-upright spine is an advantage because the player is in a slightly flexed position for a reasonable period during the game. It is an advantage, therefore, for these athletes to have at least moderate lumbar and thoracic curves. As with all agility sports, genu recurvatum (figure 7.4) is an inferior leg posture in comparison to the overhanging thigh posture (figure 7.11), not only because the knee joint is already hyperextended and takes a longer time to straighten than the slightly flexed knee joint, but also because the genu recurvatum posture creates an unstable knee joint for twisting and turning movements.

Contact Field Sports (Rugby Codes, Australian Football, American Football)

Players in contact field sports are similar in posture to those in mobile field sports but generally have more body bulk. Advantageous postural characteristics are a reasonable degree of spinal curvature rather than a rigid upright posture; inverted feet, for situations in which quick stepping and elusive running are needed for short distances; and APT, protruding buttocks and an overhanging thigh for players who need bursts of speed. Genu recurvatum is a poor posture for contact sport; therefore players who have this characteristic and who are also swimmers should be advised against doing very much front or back crawl kicking, which accentuates the problem, making the knee joint more unstable and less functional for agile movement.

In some sports, static and dynamic postures may have advantages or disadvantages. Certain sports may also cause adaptive postures. Assessment of the athlete can help determine optimal posture for the sport.

Set Field Sports (Baseball, Cricket, Golf)

A wide variety of postures are seen in set field sports, and no one posture seems to give any player an advantage.

Court Sports (Basketball, Netball, Volleyball)

Basketball, volleyball and netball are agility sports in which superior height, reflexive movement and jumping ability are important. Inverted feet, which promote fast steps and good balance, are desirable, as are APT and protruding buttocks, accompanied by an overhanging thigh, along with reasonable spinal curvatures in both the lower and upper back.

Martial Arts (Judo, Wrestling, Boxing)

In the grappling sports, good balance is important; therefore inverted feet could be advantageous. Having reasonably accentuated spinal curves will also give the competitor more trunk mobility, which should be an advantage.

Modifying Posture and Technique to Improve Performance

It will be quite obvious to the reader by now that there are several postures that, if not too extreme, may be advantageous to top athletic performance and that posture is another way in which self-selection can occur for various sports or events. Thus there should be no attempt to modify these characteristics in any way even though at first glance they may appear to be partially defective. Other postures, however, may be detrimental to performance, so that a strategy should be worked out to modify them.

The coach must decide whether the athlete should undergo a modification program or not. In many cases, this will not be necessary. However, modification should be considered if it is felt that a high-level athlete can benefit from such a program, particularly in sports in which winning margins are very small. The following actions can be taken.

- The first approach should be to modify any static defects that may need correction, using a series of exercises that will stretch tight muscles and strengthen those that have become slack. This corrective program should be undertaken if possible in pre- or early adolescence. In postadolescence, correction will take longer; but reasonable results can still be obtained at this time if the program is intensive enough. One must always keep in mind that static defects that are evident just prior to and in adolescence will become more extreme as the individual ages, and that physical discomfort and the incidence of injury will probably increase as time goes on.
- The second action is one that only a few enlightened coaches have used thus far but that will become more popular in the future: accentuating those postures that are known to be advantageous for various sports. This is particularly important for individuals who appear to have almost all of the other necessary physical capacities for optimal performance but lack the postural characteristics needed for highly specialized events. The following two examples will illustrate this point. If a sprinter has all the other characteristics to run very fast but does not have enough APT, then a minor postural modification may be made that will enable a slightly better "driving angle" during the thigh extension phase of the running cycle: that is, tilting the pelvis forward. To do this the trunk extensor and thigh flexor muscles must be strengthened and at the same time the abdominal and thigh extensor muscles stretched. Also, agility athletes who need speed over a short distance can slightly increase their level of tibial torsion (inverted feet or "pigeon toes") by strengthening the medial rotators of the hip joint and stretching the lateral rotators. It should be stressed again that posture modification will occur only with an intense intervention program, often over several years.
- Dynamic posture can also be changed with good skills coaching, and enlightened coaches have done this for the past half century in technique-oriented sports such as track and field, gymnastics, diving, rowing, golf and swimming. These coaches know how to incorporate various postural positions into their athletes' techniques—for example tilting the pelvis, rounding the back, tucking in the chin, squaring the shoulders or everting the feet. However, less attention has been directed to dynamic posture in team sports in which agility is an important factor in

the game and more specialization in various positions is becoming necessary. For example, the footballer who is straight-backed with little mobility in his spine has a poor posture for collision sports, in which players are running into rucks, mauls and packs, or for closed-field running, in which tacklers are able to hit the ball carrier with little notice. The round-shouldered player has a natural advantage in this situation because he is able to "tuck in" or "cover up" very quickly, whereas the straight-backed player is not in as good a natural position to take the heavy contact and can easily become injured. Well-informed coaches can develop this dynamic characteristic with intensive practice in the correct postural position, thereby making their players more physically effective and less prone to injury.

Summary

To conclude this chapter, we stress again that a large number of physical, physiological, psychological and skill factors contribute to top sport performance and that posture may play only a small role in some sports, almost none in others. It is important, however, for coaches to be aware of the situations within which posture is very important and to be able to appreciate its value in their athletes. If a particular posture appears to be an essential characteristic for a highly specialized sport, then coaches need to have enough knowledge to better select their athletes, or to at least modify athletes' postures so that they can improve their performance.

eight

Strength

Michael McGuigan, PhD; and Nicholas Ratamess, PhD

Maximum strength is a major factor influencing performance in a variety of sports. Athletes for thousands of years have used resistance training to enhance strength in an attempt to improve their sporting performance. It has been reported that athletes used handheld weights in 708 BC to improve performance in long jumping (Minetti and Ardigo 2002). Ancient Greeks used detailed periodization strategies based on four-day training cycles, so that there was only one intensive training session every four days (Sweet 1987). The use of resistance training is thousands of years old, but only during the past 30 years has it been transformed from the pursuit of a relatively small number of strength athletes into an integral part of the training routine for most athletes. While resistance training has historically been seen as a means to enhance muscular strength and size, it is currently used to increase power, speed and endurance; assist in rehabilitation and injury prevention; and aid in the maintenance of muscular function into old age.

This chapter reviews the current use of resistance training in the development of muscular strength. It provides background on the anatomical and physiological basis of strength. The chapter also outlines the latest developments in resistance training and compares various training strategies. A great deal of research has been performed on the effects of resistance training. This chapter, while relying heavily on both research findings and theory, is written from an applied perspective that can be easily understood by both the athlete and coach.

Relationship Between Strength, Power and Strength-Endurance

Strength and power, of importance in many sports, are related but distinct capacities. An athlete can be strong but not powerful, although in order to be powerful an athlete must be strong. Understanding the difference between these two capacities has important implications for the correct design of resistance training programs for athletes. Strength and endurance are more closely related to each other, in that the stronger athlete produces a given amount of force with less effort than an athlete with lower levels of strength and therefore has greater endurance.

Strength Versus Power

Muscular strength is the amount of force (or torque) a muscle group can exert against a resistance in one maximal effort. Power is the ability to produce high levels of work (the product of force and the distance through which the force acts) quickly. The physical capacities of strength and power are important qualities for many sports. Muscular power is discussed in more detail in chapter 9.

To describe the difference between strength and power, three force–time curves are depicted in figure 8.1. These theoretical curves represent the force-generating capacity of differing athletes when performing a maximal contraction. Curve 1 represents the force–time characteristics of an individual who is strong but not powerful, that is, a

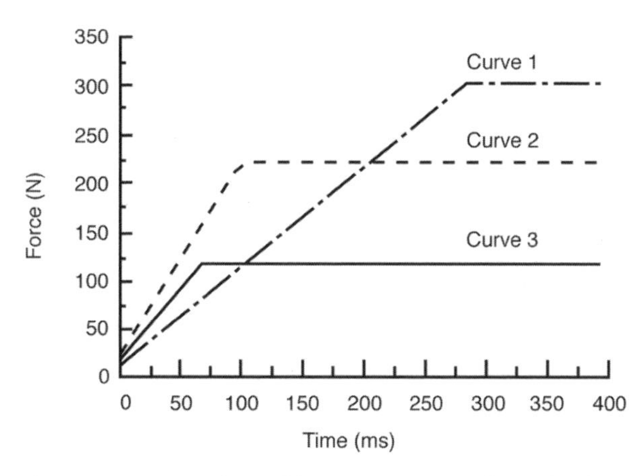

Figure 8.1 Hypothetical force–time curves for individuals of differing strength and power capabilities.

powerlifter. These individuals would be effective in performing strength movements, such as a squat or bench press lift, but less effective in a power event like a high jump or shot put, as they are unable to produce relatively high force levels quickly.

> Strength and power are related but distinct capacities that are important in many sports. Correct program design should take into account the difference between strength, power and endurance.

Curve 2 represents an individual who has less absolute strength than the athlete in curve 1; however, this competitor is able to generate a greater force over a brief period (i.e., 150 ms). Although not capable of lifting as much weight, this individual may outperform the athlete in curve 1 in a dynamic power event such as a high jump. Indeed Häkkinen and colleagues (1984) reported that elite wrestlers developed force significantly more rapidly when compared to elite powerlifters, and research by McBride and colleagues (1999) has shown that weightlifters are more powerful than sprinters and powerlifters.

Curve 3 represents an athlete who can produce the available force very rapidly but does not have a high strength level. Consequently, this individual is ineffective in both strength- and power-dominated events. Curve 3 demonstrates the fact that the level of maximum strength of an athlete places an upper limit on the performance of power events; thus power athletes must be strong (Stone et al. 2002). There is little value in having the capacity to develop force quickly if the overall level of force is particularly low. Consequently, some researchers have reported strong relationships between maximal strength and power (Rutherford et al. 1986). However, in comparing curves 1 and 2, one can see that simply being strong does not necessarily make an individual powerful. Thus, power athletes must be strong and able to generate high levels of force over relatively short periods of time.

These analyses clearly demonstrate that although strength and power are related, they are different capacities. This is one of the reasons why the strongest man or woman in the world is not necessarily the best javelin thrower or shot-putter. Regardless of technique, these people's muscles may not be capable of generating the applied force rapidly, as is required in most dynamic sporting performances. This is one of the reasons why athletes may increase strength in the gymnasium but not alter their competitive performance. The realization that strength and power are different qualities is very important in the correct formulation of resistance training for athletes, and this has often been misunderstood by coaches. Research from Häkkinen, Komi, and Alén (1985) has shown that while heavy weight training results in increases in the maximum force, the rate of force development remains unaltered. That is, the force–time curve is extended so that higher forces are generated over the long time periods (figure 8.2). Explosive resistance training, including ballistic movements and plyometrics, tends to have the opposite effect. There is a smaller change in maximal strength, but a greater increase in the rate of force development. This is observed as a shift in the force–time curve in order that greater forces are produced over shorter periods of time (figure 8.2).

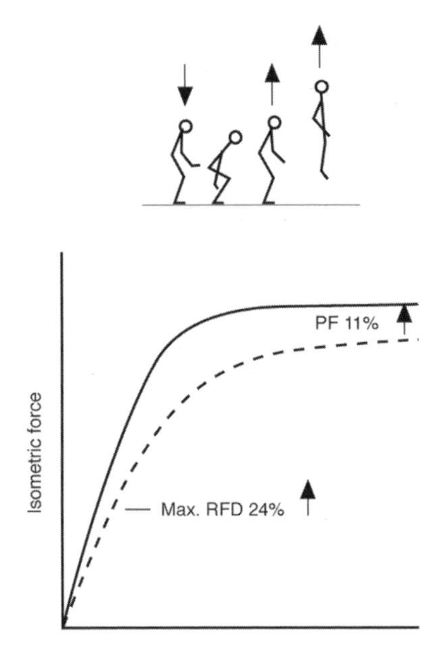

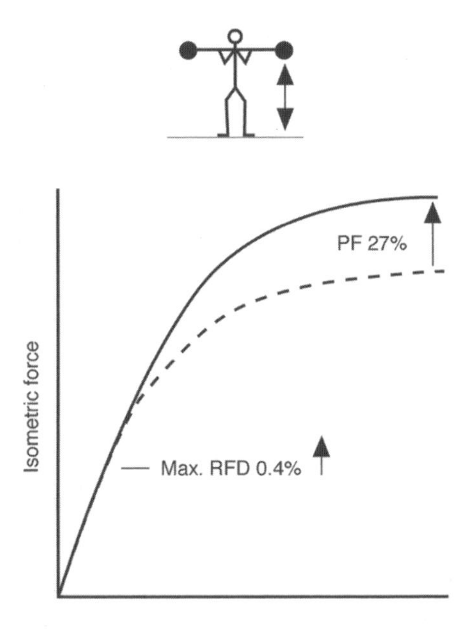

Figure 8.2 Changes in the force–time curves in response to plyometric and strength training.

Adapted from K. Häkkinen, P. Komi, and M. Alen, 1985, "Effect of explosive type strength training on isometric force- and relaxation-time, electromyographic and muscle fiber characteristics of leg extensor muscles," *Acta Physiologica Scandinavica* 125: 587-600.

Strength Versus Endurance

With all other factors being equal, a stronger muscle will have greater endurance when compared to a weaker muscle, as a smaller percentage of maximum strength is used in performing a given task, thereby delaying the onset of fatigue. At a given velocity, if a rower is required to produce a mean force against the oar of 500 N/stroke and has a maximum strength capacity of 1000 N, then the rower is working at 50% of his or her maximum during each stroke. However, if the rower's maximum strength capacity is only 750 N, the athlete will be working at 67% of maximum and may fatigue more quickly.

A similar example of the importance of strength to endurance is seen by the inverse relationship between the load used and the maximum number of repetitions (reps) that can be achieved. On the basis of data provided by McDonagh and Davies (1984), an individual with a maximal bench press of 100 kg should be capable of performing six to seven reps at a load of 75 kg (i.e., 75% of maximum). If, however, the individual's maximum is 150 kg, then the 75 kg load will represent only 50% of maximum and the person should be able to perform 12 to 13 reps. Consequently, through a 50% increase in maximal strength (100-150 kg), a 100% increase in endurance is realized (6-12 reps). The importance of strength and power to endurance can be seen in the relatively high relationships recorded by Sharp and colleagues (1982) between arm power and swimming performance over 500 m. Therefore, the physical capacities of strength and endurance are closely related, and strength-endurance athletes, such as 400 m runners, need to have relatively high levels of absolute strength.

Value of Strength in Sport

Among coaches and sport scientists there is a lack of agreement regarding how much strength is required for optimal performance in most sports (Stone et al. 2002). However, maximum strength can discriminate between athletes of different performance levels within sports such as American football (Fry and Kraemer 1991) and volleyball (Fry et al. 1991). Miyashita and Kanehisa (1979) have reported a strong relationship between sprint swimming performance and strength, while Sharp and colleagues (1982) have reported a stronger relationship between swimming performance and power.

The importance of resistance training to sport performance has been supported by studies demonstrating that resistance training enhanced some competitive performances, mostly vertical jumping ability. Newton and colleagues (1998) reported that an eight-week period of ballistic resistance training resulted in a 5.9% increase in vertical jump height in elite volleyball players. Furthermore, McBride and colleagues (2002) showed a trend toward improved 20 m sprint times with explosive resistance training, and Blazevich and colleagues (2003) showed improved sprint times with only five weeks of resistance training. However, other studies have shown that resistance training enhanced muscular strength but failed to induce changes in sporting performance such as sprinting (Fry et al. 1991). Bloomfield and colleagues (1990) observed a significant relationship between strength and throwing velocity in elite water polo players. However, when these athletes were placed on a resistance training program that resulted in significant improvements in strength, throwing velocity did not vary as a consequence of the training.

While traditional weight training results in large changes in strength among untrained subjects, and strength appears to be an important physical capacity in most sports, whether standard strength training methods can enhance sporting performance appears to depend upon the particular sport. Strength-dominated sports that involve the production of large forces over relatively long time periods (such as weightlifting) would appear to be readily improved by strength training. However, more speed-oriented activities such as throwing, hitting, punching and kicking seem less responsive to traditional strength training techniques and may require more power-based training methods such as ballistic resistance exercise.

Muscle Structure and Function

We turn now to the structure of human skeletal muscle, which is composed of a variable number of individual muscle fibres, and the theory most often used to explain how muscle develops force to pull on tendons and thus create movement, the sliding filament theory. Other considerations include the categorization of muscle fibre types, the adaptations of muscle fibres to training, the process whereby fibres are recruited as muscular tension is developed and the three types of muscular actions.

Muscular Structure

A single muscle is composed of a number of individual muscle fibres whose number varies considerably, depending on the size and function of the muscle. Each muscle fibre is between 1 and 40 mm in length and is 10 to 100 microns (μm) in diameter. Inside the muscle fibres are numerous myofibrils that range in diameter from 1 to 2 μm. These myofibrils are aligned in columns and have distinctive markings. The repetitive markings define the contractile unit of muscle, the sarcomere, which is bound on each end by a Z-line. Each myofibril is composed of numerous sarcomeres that are joined end to end (in series) at the Z-lines. Each sarcomere is approximately 2.3 mm in length and composed of a number of microscopic

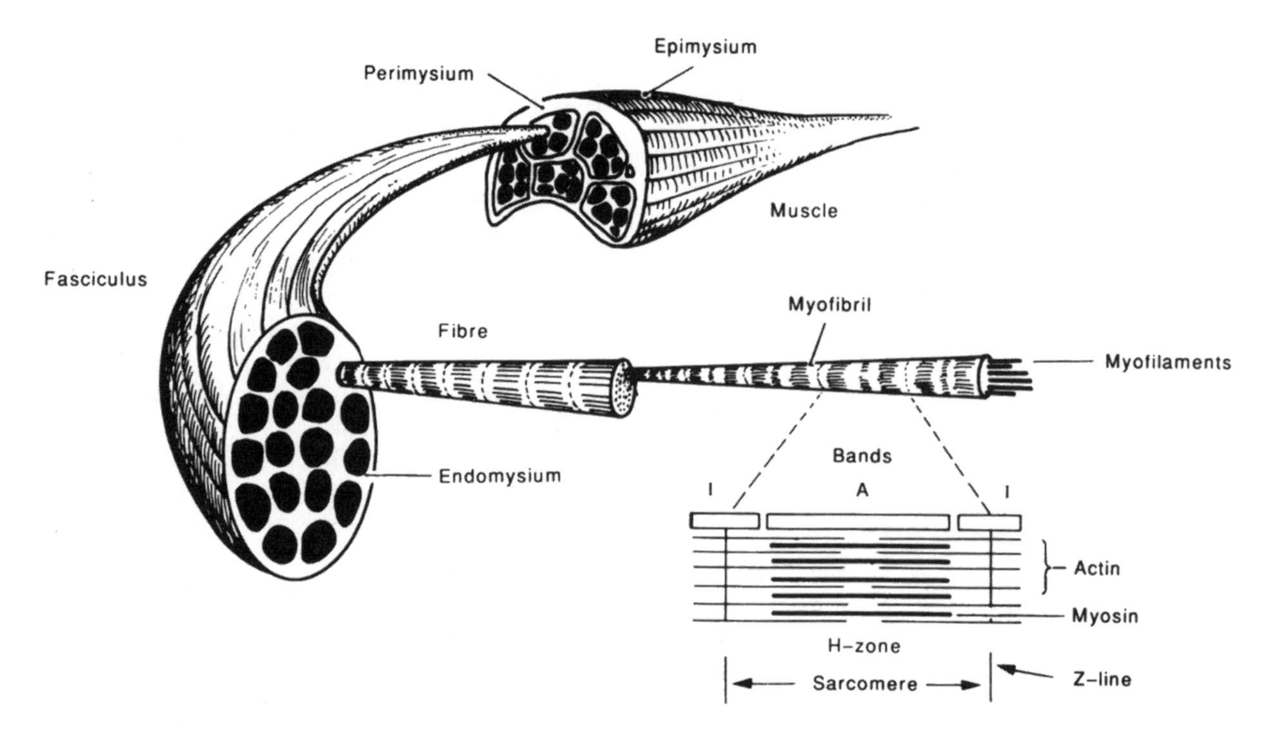

Figure 8.3 Structural design of human skeletal muscle.

Reprinted, by permission, from J. Bloomfield, P.A. Fricker, and K.D. Fitch, 1992, *Textbook of science and medicine in sport* (United Kingdom: Blackwell Scientific Publications).

myofilaments called actin and myosin. The thinner of the two small filaments is the protein actin, while the protein myosin forms the thicker filament (figure 8.3).

Sliding Filament Theory

In developing force, a muscle contracts and shortens. The precise mechanism used to shorten the sarcomere is not fully understood; but there is evidence to suggest that when stimulated, the actin and myosin filaments slide past one another. This motion is accomplished by the pulling action of cross-bridges that reach out from the myosin filaments and attach themselves to the actin filaments. After binding, the cross-bridges suddenly rotate, drawing the two protein threads past one another and reducing the length of the sarcomere by 20% to 50%. This action takes place simultaneously in millions of myofibrils, resulting in a forceful pull on the tendons (figure 8.3).

Muscle Fibre Types

Traditionally, two broad categories of muscle fibre types, slow- (ST or type I) and fast-twitch (FT or type II), have been identified based on their histochemical staining qualities as well as their physiological characteristics. More sophisticated techniques, however, have identified more categories (table 8.1). It is now generally recognized that skeletal muscle fibres do not exist in discrete forms at the subcellular level, but rather in a continuum based on the multitude of combinations of myosin heavy and light chain isoforms, polymorphic expression of protein isoforms, metabolic potential and Ca^{++} handling characteristics.

The ST fibres account for approximately 50% of normal human skeletal muscle fibres and are characterized as possessing an aerobic endurance quality. Fast-twitch fibres may be further divided into three categories based on their stained appearance, as well as their propensity for recruitment. Fast-twitch type "a" (FTa or type IIa) constitute about half the FT muscle fibres, with the remainder predominantly FT "x" (type IIx). A small number of FT fibres are classified as "ax" (FTax or type IIax). It should be noted that fibres previously classified as type IIb in humans have been renamed as type IIx according to their myosin heavy chain (MyHC) component, resembling that of the MyHCd(x) isoform of rats (Smerdu et al. 1994). The FT fibres produce more force than ST fibres, but they fatigue more rapidly (table 8.1). For this reason, ST fibres are preferentially recruited during low-intensity activities; and as the tension requirements increase, more FT fibres are activated.

In general, ST muscle fibres have good aerobic endurance compared to FT fibres. Faulkner, Claflin, and McCully (1986) reported that the peak power output of FT fibres was four times that of ST fibres. The differing

Table 8.1 Physiological Characteristics of Muscle Fibre Types

Characteristic	Type I	Type IIa	Type IIx
Fibre diameter	Small	Large	Large
Contraction speed	Slow	Fast	Fast
Force production	Low	Intermediate	High
Endurance	High	Intermediate	Low
Fatigability	Low	Intermediate	High
Capillary density	High	Intermediate	Low
Aerobic enzymes	High	Intermediate	Low
Anaerobic enzymes	Low	High	High
ATPase activity	Low	High	High
Mitochondrial density	High	Intermediate	Low
Myoglobin	High	Low	Low
Colour	Red	White	White

capabilities of these muscle fibres mean that the relative percentage of FT and ST fibres is important in elite sport performance. A high percentage of type IIa fibres is apparent in both strength-power athletes and endurance athletes (Fry et al., "Male Olympic-Style Weightlifters" 2003, "Competitive Powerlifters" 2003). Newton and colleagues (2002) showed that there were greater improvements in strength and power in both older and younger subjects with a higher proportion of type II fibres.

Muscle Fibre Adaptations to Training

The dominant mechanism sustaining long-term enhancement in muscular strength is an increase in the size of the muscle, and there is a very strong relationship between the size and the strength of muscle in males and females. Long-term development of strength is dependent on the continual increase of muscular size. Two mechanisms have been proposed that result in increases in muscular size:

- **Muscular hypertrophy:** The individual muscle fibres increase in size.
- **Muscular hyperplasia:** The individual muscle fibres split (or increase in number based on satellite cell proliferation) and as a result the number of muscle fibres increases.

Research indicates that muscle hypertrophy is the dominant mechanism underlying the increase in muscular size. Muscle hypertrophy is attributable to the following changes in the muscle:

- Increased number and size of myofibrils per muscle fibre
- Increased amounts of protein filaments, particularly the myosin filaments
- Increased size and strength of tendons, ligaments and connective tissue

Changes in muscle fibre size and transitions in FT fibres in particular are important adaptations to resistance training.

It is interesting to note that hypertrophy, in response to high-intensity resistance training, occurs in both the FT and ST fibres (Kraemer et al. 1995). The vast majority of research has demonstrated that the relative percentage of FT and ST muscle fibres is genetically determined and varies little in response to training. With endurance training, FT fibres may enhance their endurance capacity; however, they will not turn into fully functional ST muscle fibres, and the reverse situation also applies. It is generally accepted that resistance training can promote changes within the population of FT fibres (i.e., type FTx to FTa) (figure 8.4).

Figure 8.5 outlines the typical fibre type composition of elite athletes in different sports. Endurance-based athletes, such as cross-country skiers, possess a greater proportion of ST fibres compared to the untrained population, whereas power athletes such as sprinters have proportionally more FT fibres. It appears that transitions and proportions of type II fibres are critical for strength-power performance (Fry 2004).

Size Theory of Motor Unit Recruitment

A motor unit includes the alpha motor neuron and all the fibres it innervates. Individual muscle fibres do not operate independently but are innervated by motor neurons or nerves that originate in the spinal cord. Each muscle has 100 to 1000 motor units that control the firing of the individual muscle fibres. A single

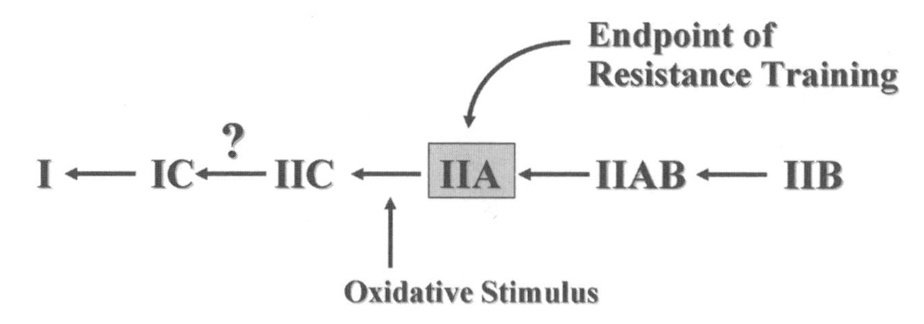

Figure 8.4 Fibre type transformations in response to resistance training. The question mark refers to the possibility of subtype shifts toward IC fibres, but this has not been confirmed in human studies. N.B. Some researchers prefer to use IIx rather than IIb.

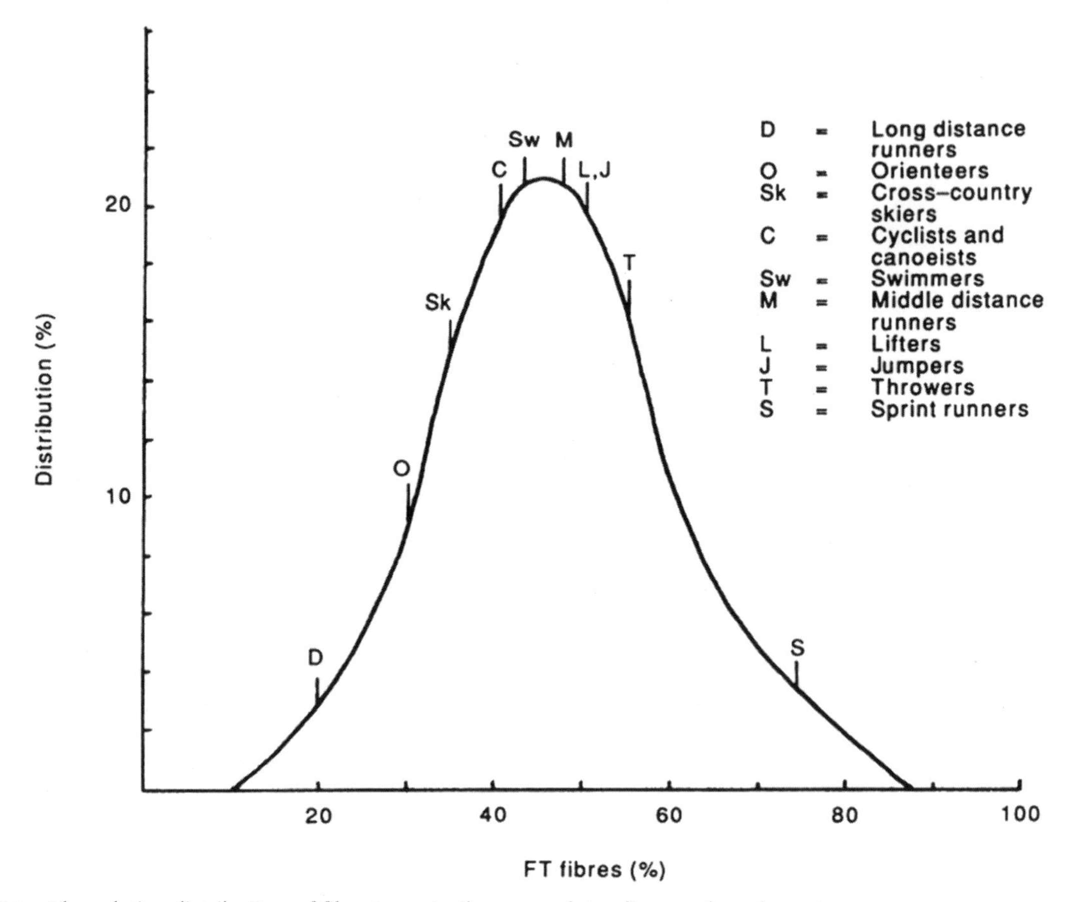

Figure 8.5 The relative distribution of fibre types in the vastus lateralis muscles of sportspersons.

Reprinted, by permission, from J. Bloomfield, P.A. Fricker, and K.D. Fitch, 1992, *Textbook of science and medicine in sport* (United Kingdom: Blackwell Scientific Publications).

motor neuron may innervate as few as three muscle fibres in small muscles that require fine precision, such as those controlling the eyes. In larger muscles such as the gluteus maximus, as many as 800 muscle fibres may be controlled by one motor neuron.

All muscle fibres within the same motor unit will consist of the same fibre type. Typically, ST units include between 10 and 180 fibres, while FT units are larger and include 300 to 800 fibres. As muscular tension is developed, motor units are recruited on the basis of their size (Henneman et al. 1974); initially the smaller, ST motor units are recruited, followed by the larger, FT motor units (figure 8.6). Recruitment starts with the lower motor units and progresses to a higher electrochemical threshold as motor units are recruited to meet force demands. Increases in tension will, at first, be produced via increase in the firing rate of the ST motor units. Greater tension is developed by the additional recruitment of progres-

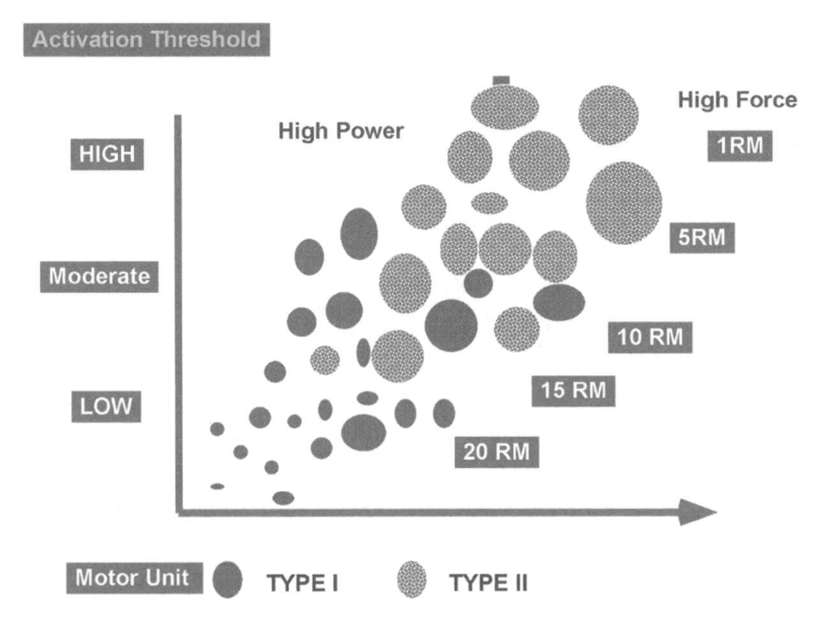

Figure 8.6 Graphic representation of the "size principle", whereby motor units that contain type I (slow-twitch) and type II (fast-twitch) fibres are organized according to some "size" factor. RM = repetition maximum.

Reprinted, by permission, from National Strength and Conditioning Association, 2008, *Essentials of strength training and conditioning*, 3rd ed., edited by T.R. Baechle and R.W. Earle. (Champaign, IL: Human Kinetics), 97.

sively larger FT motor units and through increase in their firing rate. This process continues until maximal force is achieved when all of the available motor units are recruited and fired at their maximal rate. Interestingly, preferential recruitment of FT fibres may occur during ballistic training.

Based on the size theory of motor unit recruitment, the large FT motor units, which are predominantly responsible for the production of powerful movements, will be recruited only if relatively large forces are required. This theory is often used as a rationalization for the use of heavy training loads in the development of muscular power and athletic performance (Schmidtbleicher 1988). It is suggested that training with light loads does not allow for the recruitment of FT muscle fibres, as the force requirement is simply too low. In contrast, heavy-load training requires near-maximal forces to be developed, and all motor units, fast and slow, must be used.

Muscle Actions

Once activated, a muscle will attempt to shorten and exert force on the tendon; this force is then transferred to the skeletal structures. The resistance encountered during the muscle action will determine the resulting movement. The three types of muscular actions are as follows:

- **Concentric (CON) or shortening contraction:** This results when the tension developed is greater than the resistance and the muscle shortens. Such an action is produced in the lifting (concentric) phase of resistance training exercises such as a bench press, squat and arm curl.
- **Eccentric (ECC) or lengthening contraction:** This results when the tension developed is less than the resistance and the muscle attempts to shorten but is actually lengthened. Such a muscular action is produced in the lowering (eccentric) phase of resistance training exercises.
- **Isometric (ISOM) or static contraction:** This results when the tension developed is the same as the encountered resistance and the length of the musculature remains essentially unaltered. Such a muscular action is observed when one attempts to exert a force against an immovable object such as a wall.

The action that a muscle group performs during an activity is used to define its role. There are three differing roles:

- ***Agonist* or prime mover:** This is the muscle or muscle group that causes the action (e.g., the biceps brachii during the arm curl exercise).
- ***Antagonist:*** This muscle or muscle group, located on the opposite side of the joint, acts to resist the activity (e.g., the triceps brachii during the arm curl exercise).
- ***Stabilizer* or fixator:** This muscle or muscle group action serves to stabilize the skeletal structures so that tension can be effectively developed by the prime mover. For example, the musculature about the wrist and shoulder joints will isometrically contract during the performance of an arm curl exercise to stabilize the skeletal structures so that the biceps brachii muscle group can effectively generate force.

Muscle Mechanics and Neuromuscular Considerations

The capability of a muscle to produce force depends on two relationships, the length–tension relationship and the force–velocity relationship; that is, the maximal force a muscle can exert is associated with its length and also with the rate of change in its length. After addressing these relationships, this section considers adaptations in the nervous system that result from resistance training as well as issues pertaining to muscle mechanics.

Length–Tension Relationship

Muscular tension is most effectively achieved at the normal resting length of muscle, as this enables maximum binding of actin and myosin protein filaments (figure 8.7). As muscle contracts, its filaments become overlapped and cross-bridge linkages cannot bind completely, with the result that the contraction becomes less effective. Conversely, if the muscle is extended beyond resting length, the overlap between filaments is reduced and not all of the filaments can bind. Therefore, muscular force capability is reduced. However, if a muscle is stretched beyond its normal resting length, additional tension is produced as a result of the elasticity of passive structures (e.g., connective tissue) that lie parallel to the contractile elements. The combined effect of the muscular and elastic factors is depicted in figure 8.7, which illustrates the classic length–tension curve.

Force–Velocity Relationship

All factors being equal, high force capacity of a muscle will be realized during an isometric contraction, and the muscle's capacity to exert tension will decrease as a function of increased speed of concentric movement (figure 8.8). This is due, in part, to the viscous fluid in the muscles, which resists motion in proportion to the magnitude of velocity. Furthermore, the cross-bridges break and rejoin more often at higher velocities, resulting in reduced tension, since at high contraction speeds there is not enough time to generate maximum force. As observed in figure 8.8, maximum force is realized in rapid eccentric movements. In fact, maximal eccentric force is approximately 1.3 times higher than maximal concentric force (Griffin 1987). This fact is often used to justify the performance of heavy eccentric training as an effective method for enhancing strength, as more force can be developed in the eccentric phase than in the isometric or concentric phase.

Neural Adaptations to Resistance Training

Resistance training elicits adaptations in the nervous system in addition to musculoskeletal adaptations. Training-induced neural adaptations are perhaps best illustrated by the fact that unilateral strength training has been consistently observed to produce strength increases in the untrained contralateral limb (Munn, Herbert, and Gandevia 2004). It has been suggested that early adaptations to resistance exercise are neural in nature, followed later by muscle hypertrophy. These neural adaptations are believed to occur through a number of processes, such as disinhibition, increased activation of agonists as a result of recruitment of additional motor units or increased firing rate, motor unit synchronization, effect of learning, reduced activation of the antagonist muscles or some combination of these. However, the exact mechanisms for neural responses to training are unknown (Carroll, Riek, and Carson 2001).

Observation of large strength gains without changes in muscle cross-sectional area provides the basis for adaptations of the neural system as dominant in the early stages of training (Rutherford and Jones 1986). After this early period of training, neural adaptations still appear to form an important part of the adaptation process; however, they

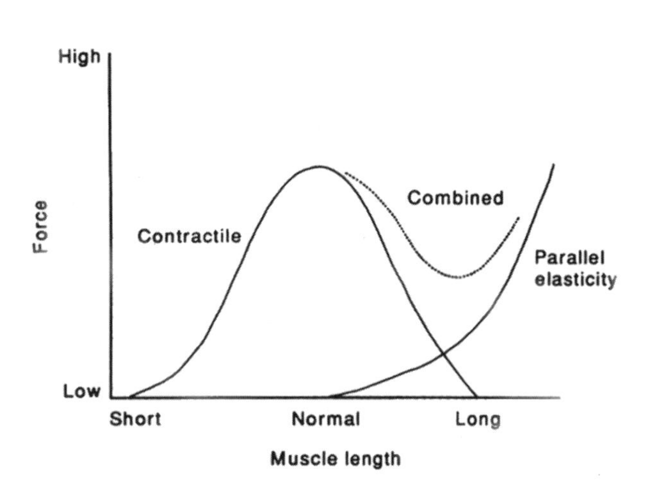

Figure 8.7 The length–tension curve for an isolated muscle.

Reprinted, by permission, from J. Bloomfield, P.A. Fricker, and K.D. Fitch, 1992, *Textbook of science and medicine in sport* (United Kingdom: Blackwell Scientific Publications).

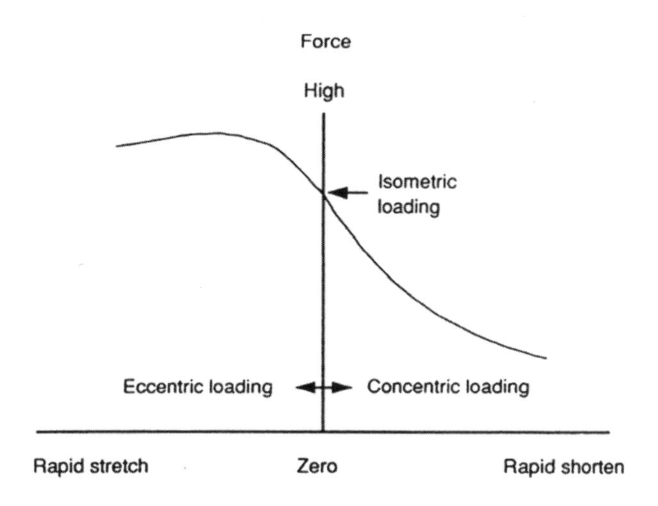

Figure 8.8 The force–velocity relationship for muscle.

Reprinted, by permission, from J. Bloomfield, P.A. Fricker, and K.D. Fitch, 1992, *Textbook of science and medicine in sport* (United Kingdom: Blackwell Scientific Publications).

become less important than morphological adaptations in the musculature itself (Aagaard 2003). However, Staron and colleagues (1994) showed that muscle adaptations occurred as soon as two to four weeks after initiation of heavy resistance training with changes in the proportion of type IIb fibres. This means that after only several workouts, significant changes in fibre distribution were taking place. This may be relevant from a strength training perspective, and increases in the content of myofibrillar protein could occur much earlier than previously thought (Phillips 2000).

> Resistance training elicits adaptations in the nervous system in addition to muscular changes. Early phase adaptations appear to be predominantly neural but muscle adaptations also appear to occur during the initial stages of resistance training.

Muscle Mechanics

The forces produced by muscles act through lever systems of the skeleton. Force generated by muscle is applied through tendons to bone, causing the bone to rotate. The rotatory action of bone depends on certain mechanical properties of the muscle and the lever system. Force production is also influenced by muscle architecture, specifically how the fibres are oriented with respect to the tendon.

Force Production

The musculoskeletal system represents a lever system in which force is generated by muscles and applied through tendons to bone, which consequently attempts to rotate about its axis (joint). This rotatory tendency, termed torque, is a product of the applied force and the perpendicular distance from the force to the axis of rotation. The torque exerted is proportional to the force of muscular contraction and the distance of the tendon insertion from the joint (axis of rotation) and determines the strength of the muscle. Further, the torque is determined by the angle of pull that the tendon has with respect to the bone. The optimal angle of pull is 90°, and the torque will reduce as this angle becomes progressively larger or smaller. The angle of pull of the tendon and the muscle's ability to produce force vary with differing positions, so it is not surprising that maximal force produced throughout a range of motion varies quite considerably, as illustrated in figure 8.7.

Muscle Architecture

The orientation of fibres with respect to the tendon is an important consideration in the ability of the musculature to produce force. If the muscle fibres are oriented longitudinally with respect to the tendon, then the muscles are capable of producing low force but shorten over a relatively long range. These are known as fusiform muscles (figure 8.9). There are relatively few fusiform muscles in the human body, and most are located in the extremities (e.g., the brachioradialis). If the muscle fibres are oriented at an angle to the tendon they are termed penniform. These muscles produce larger forces but do so

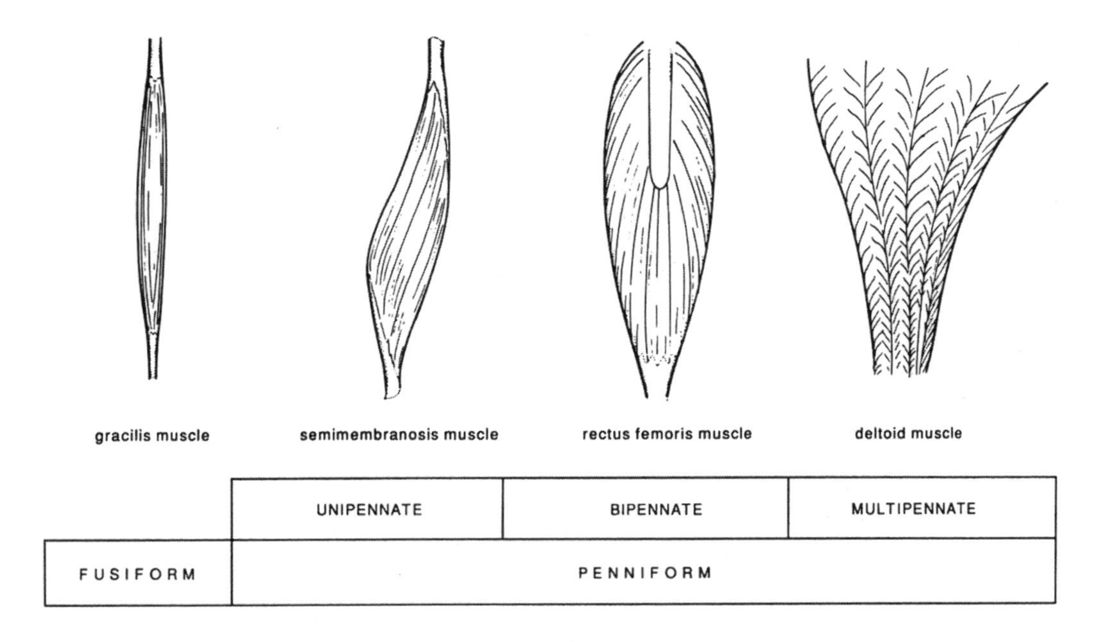

Figure 8.9 The arrangement of fibres in human skeletal muscle.

Reprinted, by permission, from J. Bloomfield, P.A. Fricker, and K.D. Fitch, 1992, *Textbook of science and medicine in sport* (United Kingdom: Blackwell Scientific Publications).

over a shorter range. They are common within the human body and account for approximately three-quarters of all skeletal muscle. There are three types:

- **Unipennate:** These muscles have muscle fibres on one side of the tendon only (e.g., semimembranosus).
- **Bipennate:** These have muscle fibres on both sides of a central tendon (e.g., rectus femoris).
- **Multipennate:** These have muscle fibres on both sides of a number of tendons (e.g., deltoid).

Factors Affecting the Development of Muscular Function

Factors that influence the development of muscular function include age, gender, genetics and psychological factors. Although muscular function tends to decline with age, an appropriately designed resistance training program will yield many health benefits in older persons; such programs are also safe and effective for young athletes. Gender affects the amount of lean body mass at full maturity and thus muscular function, though women's response to resistance training is similar to that of men and in some cases superior. Genetic factors place an upper limit on the functional capacity of muscle; nervous system arousal can optimise people's strength potential.

Age

Age has an important influence on the development of muscular function. Even though muscular function tends to deteriorate with age, this decline can be minimized with resistance training. In fact, many studies have demonstrated that inactive individuals can greatly increase muscular function with training, even into old age. It has been suggested that maximal muscular strength is achieved between the ages of 25 and 35 (Wilmore and Costill 2004), and world records (figure 8.10) show that total weight lifted then declines at a steady rate with age (Anton, Spirduso, and Tanaka 2004). It does appear that most athletic performances decline at a steady rate with age, primarily due to decrements in strength and endurance. Furthermore, Evans (2000) suggested that the decline in maximal muscular power with age has important implications for functional performance.

Resistance Training for the Older Athlete

Resistance training has been shown to provide myriad health benefits for older populations, but the key factor to successful resistance training for any population is appropriate program design. Similar resistance training routines can be used for older athletes as those adopted by their younger counterparts (American College of Sports Medicine 2002). However, one must be aware that at the commencement of the training, the individual may possess a relatively low level of strength, and correspondingly low training intensities need to be prescribed. A number of studies have demonstrated

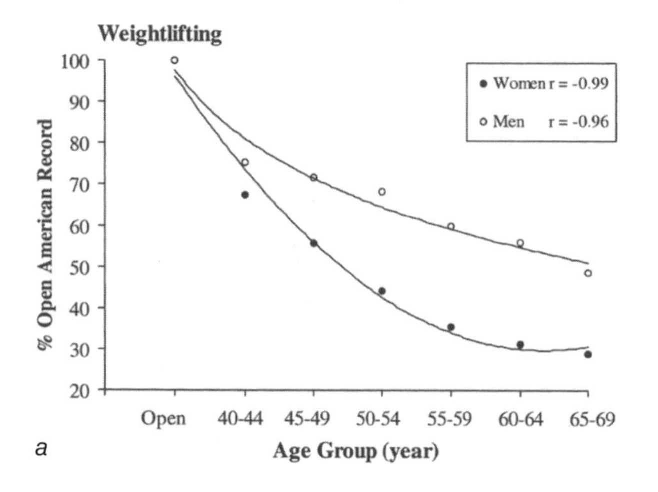

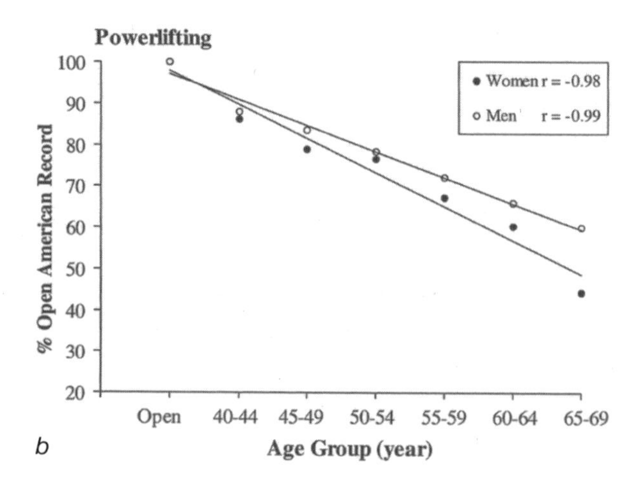

Figure 8.10 Age-related differences in weightlifting and powerlifting. The graphs show *(a)* weightlifting (an average of snatch and clean and jerk) and *(b)* powerlifting (an average of deadlift, squat and bench press) performance records for women and men.

Reprinted, by permission, from M.M. Anton, W.W. Spirduso, and H. Tanaka, 2004, "Age-related declines in anaerobic muscular performance: weightlifting and powerlifting," *Medicine and Science in Sports and Exercise* 36: 143-147.

quite substantial strength increases in previously inactive elderly subjects who commenced resistance training (Häkkinen et al. 2002).

The Younger Athlete

The competitive nature of many sports has resulted in a requirement for high levels of strength, power and endurance among children who are athletes. Consequently, many coaches want to know whether young children should be exposed to high-level resistance training and what potential hazards this may cause. Resistance training is growing in popularity due to the potential health benefits associated with this method of conditioning.

When a male reaches puberty, there is an approximate 20-fold increase in the production of testosterone, and this rapidly accelerates the development of lean body mass (see also chapter 5). In essence, the individual is on natural steroids and if properly prepared can achieve the best muscular function gains of his life. Ideally, competitors will have been exposed to weight training methods prior to reaching puberty so that they are accustomed to resistance training and the exercise techniques are well mastered. This can generally be achieved with 6 to 12 months of low-intensity training.

Weight training can also be employed by children prior to puberty. A common belief is that performance benefits will be somewhat limited, due to the low level of circulating androgens; however, recent research demonstrates that strength gains of 30% to 40% can occur in untrained children after 8 to 12 weeks of resistance training (Faigenbaum et al. 2001). With an emphasis on learning the correct technique during this period, once the athlete has reached puberty, the loads can be increased. An effort must also be made to increase the intensity of training so that the full benefits of elevated testosterone production can be realized.

Gender

Males and females differ physiologically. By the time the average female has attained full maturity, she is approximately 13 cm (5.1 in.) shorter, 15 kg (33 lb) lighter in body mass and 20 kg (44 lb) lighter in lean body mass than the average male, with considerably more adipose tissue. This large difference in lean body mass is predominantly due to a much higher production of the hormone testosterone in males—approximately 10 times higher in average men than in women. Furthermore, the larger amount of lean body mass in males results in large sex differences in muscular function.

Strength differences between genders may vary depending on the muscle group being tested. In the upper body (arms, shoulders and chest), men are considerably stronger than women, but the differences are less marked in the lower body. Wilmore (1974) reported that women have approximately 40% of the strength of men in the upper body and approximately 70% in the lower body. However, when the strength differences are expressed per unit of lean body mass, women have approximately 55% of the strength of men in the upper body and are equal to men in the lower body. The latter point highlights a very important fact with respect to gender differences in muscular function. In essence, skeletal muscle among men and women is very similar; the main difference is that males simply have a greater amount.

Resistance training has become a popular and important exercise component in training programs for women. While studies have shown that women are not as strong as men, their ability to respond to resistance training is similar, and in some cases superior. Generally, these studies have been performed for a limited duration (6-10 weeks) and have involved untrained subjects. In such instances it is believed that large performance gains are achieved mainly because of neural adaptations. Our understanding of the long-term (i.e., greater than six months) training effects on muscle hypertrophy in women is limited. In the longer term, gains in strength are mostly sustained through muscle hypertrophy; and because of the limited production of testosterone, women can increase muscle size and function with training, but not to the same extent as their male counterparts (Kraemer et al. 2004).

Genetic Factors

Strength is considered one of the most trainable physical capacities in sport. It is not unusual for maximal strength to double in response to a training intervention. Nevertheless, there are limits to the development of muscular function because of genetic traits that greatly affect this physical capacity. The relative proportion of FT and ST fibres, which is genetically determined to a large extent, is an important determinant of the functional capacity of muscle. According to Fry (2004), a larger proportion of FTa fibres may be important in the development of muscular strength.

Other factors such as tendon insertion, body type (Hart, Ward, and Mayhew 1991) and lever length are genetically determined. These limitations to maximal strength need to be recognized by athletes and coaches, as they may represent advantages for a given competitive sport. A rower with relatively long arms may be limited in the weight he or she can lift during the bench press exercise, but this limitation could represent an advantage for rowing, as the increased lever length will enhance stroke length and rowing performance.

Exciting advances during the last decade have been made in our understanding of the molecular and cellular regulation of muscle adaptation to exercise. The ACTN3 genotype (R577X) has been found to be associated with performance in Australian elite athletes (Yang et al. 2003). It is important to remember that genetic factors are just one variable among myriad factors that contribute to successful athletic performance. Muscular strength and power are more likely determined by many different genes, each explaining some of the variability that is seen in these measures.

A number of factors including age, gender, genetics and psychology appear to influence strength development.

Psychological Factors

To recruit all available motor units and fire them at their maximal rate in a relatively short period of time requires considerable mental effort, determination and aggression (Ikai and Steinhaus 1961). The nervous system may need to be substantially aroused to produce these maximal levels of strength and power. Therefore, strategies used by athletes typically include preparatory arousal, imagery and positive self-talk. The use of such psychological techniques is thought to increase the arousal level of the central nervous system, allowing for the recruitment of more motor units than normally possible. These techniques enable individuals to come closer to the true strength capacity of their muscles.

Strength Assessment

A variety of measurement strategies have been developed for the assessment of muscular strength. These have encompassed the isometric, dynamic and isokinetic modes of contraction with measurement apparatus ranging from the very simple, such as the use of a barbell to perform a 1-repetition maximum (1RM) test, to the very expensive isokinetic dynamometers. A number of field tests have been used extensively to estimate power developed by the lower limbs (see also chapter 9). Certain strength measures represent specific or independent qualities of neuromuscular performance that can be assessed and trained independently.

The coach or sport scientist might wish to accurately assess muscular strength and power for a variety of reasons (Sale 1991). Some of these are the following:

- Establishing the relevance of strength and power to a particular sport
- Determining the strengths and weaknesses of an athlete by comparing test performances with group norms—by profiling the athlete in this manner, one can use the test results for the implementation of specific resistance training regimens
- Monitoring training progress and the progress of injury rehabilitation

Although there are varying opinions, six broad strength qualities can be drawn out from this type of testing. In sports that involve repeated maximal efforts, such as swimming, a seventh quality, termed *power endurance,* can also be included. The following are the six broad strength qualities.

- **Maximum strength**—highest force capability of the neuromuscular system produced during slow eccentric, concentric or isometric contractions
- **High-load speed-strength**—highest force capability of the neuromuscular system produced during dynamic eccentric and concentric actions under a relatively heavy load (>30% of max) and performed as rapidly as possible
- **Low-load speed-strength**—highest force capability of the neuromuscular system produced during dynamic eccentric and concentric actions under a relatively light load (<30% of max) and performed as rapidly as possible
- **Rate of force development (RFD)**—the rate at which the neuromuscular system is able to develop force, measured through calculation of the slope of the force–time curve on the rise to maximum force of the action
- **Reactive strength**—ability of the neuromuscular system to tolerate a relatively high stretch load and change movement from rapid eccentric to rapid concentric
- **Skill performance**—ability of the motor control system to coordinate the muscle contraction sequences to make the greatest use of the other five strength qualities

A wide variety of methods can be used to assess these qualities. Testing must be specific, so it is important to avoid implementing tests for the sake of testing. It is also important to critically examine what tests are used and not to choose tests solely because they have been used previously, or because the equipment and expertise are available.

Maximum Strength

Maximum strength can be determined using several methods. Many prefer isometric testing because it is not confounded by issues of movement velocity and changing joint angle. Research has demonstrated the importance of isometric maximal strength in a variety of athletic populations including track and field athletes (Stone et al. 2003) and weightlifters (Stone et al. 2005). In isometric testing the athlete assumes a task-specific body position and then applies maximal effort against a fixed constraint (figure 8.11). Force transducers are used to measure the peak force output, and various measures of the rate of force development can also be determined. Maximal isometric strength tests have particularly high test-retest reliability (Viitasalo, Saukkonen, and Komi 1980), but the disadvantage of isometric testing is a lack of specificity to the dynamic movements predominant in sport.

To assess strength dynamically a 1RM test can be performed. This test determines the maximum load a person can lift for one complete rep. The 1RM test has been widely used for a variety of movements, such as the squat (figure 8.12) (Wilson, Elliott, and Wood 1992) and the bench press (figure 8.13). Prior to the performance of the test, subjects are warmed up and perform a series of submaximal lifts with progressively heavier loads. As the loads become relatively close to maximum, they are incremented by small amounts (2.5-10 kg [5.5-22 lb], or 5-10% of maximum) until the lift cannot be completed. A 3 min recovery should be interposed between lifts. With trained subjects who have a good knowledge of their maximum load, the process is relatively simple and generally requires no more than three or four lifts. Even when the procedure is used with novice subjects, the maximum is typically realized by the fourth trial.

During these tests in which maximum loads are being lifted, it is imperative that they be performed with absolute safety as a priority. It is important that spotters be used, and it is desirable to have the lifts performed in a rack, so that if the bar cannot be lifted it can be placed onto a rack or safety catches. Further, it is important that the lifts be performed in a technically correct fashion, particularly for exercises that involve the lower back, such as squats, power cleans or deadlifts. Repetition maximum strength tests, when administered using experienced subjects, have high reliability; and maximal dynamic strength is a useful predictor of performance in athletic activities (Fry et al. 1991).

Isokinetic testing is commonly used to determine muscular strength. "Isokinetic" literally means "same velocity" and refers to tests that are performed at a predetermined constant velocity. Isokinetic devices involve a dynamometer that is interfaced with a motor and serves to provide equal and opposite resistance to the force provided by the musculature once a predetermined velocity has been reached (e.g., Biodex). These devices maintain a specific velocity and record the force, or more usually the torque, applied by the athlete throughout the range of motion. A test of strength using these devices may also be performed at the constant velocity of zero to produce an isometric contraction.

Many scientists perceive isokinetic tests as the standard for assessment of the entire force–velocity curve. The predominant reason is the large degree of control over the movement in terms of the velocity of motion. The movements required by the tests are easily standardized, involving, for the most part, isolated muscular actions. For example, the popular leg extension test (figure 8.14) may be readily replicated with great reliability.

Isokinetic tests record the torque produced throughout the entire range of motion and allow for the identification of regions of strength or weakness within the range. In

Figure 8.11 Isometric midthigh pull. The exercise is being performed on a force plate for measurement of peak isometric force and with use of an adjustable power rack.

Figure 8.12 Back squat exercise. The exercise is being performed in a power rack.

Figure 8.13 Bench press exercise. The exercise is being performed with a spotter.

contrast, isotonic tests typically give a single, representative value for the entire movement, while isometric tests are specific to a single joint angle. Recent innovations, combined with the unilateral nature of many of the isokinetic machines, allow these systems to assess strength differences between alternate limbs, agonist and antagonist muscle groups and eccentric versus concentric movement phases.

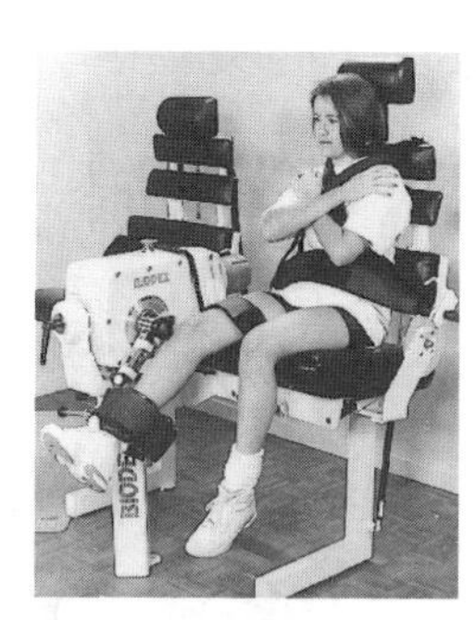

Figure 8.14 Isokinetic strength test for leg flexion-extension.

Many strength practitioners perceive that the well-controlled nature of isokinetic devices results in a lack of external validity. By controlling the movement velocity and producing isolated single-joint movements that are not particularly representative of sporting actions, these devices are considered to lack application to competitive sport. Nevertheless, sport-specific isokinetic devices are available, such as the biokinetic swim bench, and these produce data that are often highly related to athletic performance. Sharp, Troup, and Costill (1982) trained four novice subjects on the biokinetic swim bench for a period of four weeks. The training resulted in a 19% increase in power output on the swim bench and a 4% improvement in swimming velocity.

The results from research studies highlight one of the most important factors in testing, that is, the principle of specificity, which suggests that the testing of muscular function should be specific to the method of training. When one combines this idea with the principle of specificity of training, it is clear that training and testing should both be specific to the competitive performance.

Whether the results are expressed as absolute measures or relative to body weight depends on the task and the athlete, and both methods have applications. When the athlete must move body weight, then relative measures may be more important (e.g., high jump). But when momentum or total strength is the key, then absolute measures may be more instructive (e.g., rugby, football). Certainly, relative measures allow for better comparison of athletes with different body weights. Most variables can be expressed in relative terms; for example, 1RM squat can be expressed as body weight lifted, and power output during jumping can be expressed as watts per kilogram. Another, more sophisticated method is allometric modelling, in which performance in strength and power tests can be adjusted for body mass to allow for better comparison of competitors with diverse body masses (Vanderburgh and Dooman 2000).

> The key components to any training program are progressive overload, specificity, and variation. As long as these are included, any resistance training program can be effective.

General Training Principles

General training principles include progressive overload, specificity, variation, general-to-specific progression and individualization, as well as detraining and supervision. Progressive overload refers to a gradual increase in the stress imposed on the body over time. According to the specificity principle, the majority of improvements from training will be specific to the nature of the stimuli applied. Variation in one or more components of the program is necessary over time to keep the training stimulus optimal. A general-to-specific approach to progression suggests that as people advance in their training, more variation in the program is needed in order to increase the rate of progression. Detraining refers to the cessation of training or a substantial reduction in a program component. This section ends with a discussion of supervision, which is critical to progression during strength training.

Progressive Overload

Progressive overload refers to the gradual increase of the stress placed upon the body during strength training. In reality, the human body has no need to increase muscular strength unless it is continually forced to meet higher physiological demands. Progressive overload entails one or more of the following strategies.

- Resistance may be increased.
- Reps may be added to the current workload.
- Rep velocity with sub- to near-maximal loads may be increased to enhance the neuromuscular response. The intent is to lift the weight as fast as possible (i.e., *compensatory acceleration*).
- Rest intervals may be lengthened to enable greater loading.
- Training volume may be increased within reasonable limits (i.e., 2.5-5%, or periodized to accommodate heavier loading).
- Other advanced supramaximal loading training techniques may be introduced.

The use of progressive overload can overcome *accommodation* (i.e., reduced performance gain resulting from an unchanged training stimulus). There is an inverse relationship between the time used to perform a nonvaried program and the amount of performance gain. The physiological adaptations to a current training program take place within a few weeks. Proper manipulation of the acute program variables alters the training stimulus; and if the stimulus exceeds the athlete's conditioning threshold, then further improvements in muscular fitness will occur.

Specificity

All training adaptations are most specific to the stimulus applied. Although there are nonspecific improvements, the majority of improvement will be specific to the nature of the stimuli. Training adaptations are specific to the

- muscle actions involved,
- velocity of movement and rate of force development,
- range of motion,
- muscle groups trained,
- energy systems involved,
- movement pattern and
- intensity and volume of training.

Specificity becomes most evident during progression to more advanced strength training. Many studies examining untrained or moderately trained individuals have shown a substantial magnitude of transfer to the nontrained domain. Although a transfer effect is desired, trained athletes may display less potential than lesser-skilled individuals; therefore, the training program needs to target specific components of performance.

Muscle Action

Strength training with ECC, CON and ISOM muscle actions elicits substantial strength improvements. In comparison, ISOM-, ECC- and CON-only training increase peak ISOM torque and ECC and CON isokinetic strength significantly, and may increase strength in the untrained muscle action to a lesser extent. When ISOM is compared to dynamic strength training, ISOM training can increase dynamic strength (especially when multiple joint angles are trained) and dynamic training can increase ISOM muscle strength; but with the exception of a few studies that have shown similarities, research indicates that the trained muscle action will yield the greatest improvement when tested specifically. Although evidence supporting and contradicting specificity has been reported, it appears that the greatest gains in muscle strength occur via the trained muscle actions.

Velocity of Movement

Velocity specificity suggests the greatest strength increases will take place at or near the training velocity. Although specificity has been observed, there have been significant carryover effects to nontrained velocities, as well as carryover velocity effects between muscle actions. Isokinetic studies have shown strength increases specific to the training velocity with some carryover above (as great as 210°/s) and below (as great as 180°/s) the training velocity mostly at moderate and fast velocities, but not at very slow ones (i.e., 30°/s). Many of the studies showed greater carryover below the training velocity whereas few showed carryover to faster velocities. Training at moderate velocity (180-240°/s) produces the greatest strength increases across all velocities.

Strength training with dynamic muscle actions demonstrates specificity and carryover increases to other nontrained lifting velocities. However, the greatest carryover effects are seen in untrained or moderately trained individuals. Perhaps most critical to discussing velocity-specific adaptations is the intent to maximally accelerate the weight during strength training.

Range of Motion

Specificity of range of motion (ROM) can be seen in limited ROM dynamic, isokinetic or ISOM training. Dynamic limited ROM training was shown to increase strength in the trained ROM mostly, with some carryover increases throughout full ROM. Isometric training strength increases are specific to the joint angles trained (i.e., *angular specificity*) but have shown carryover to ±30° of the trained angle. The magnitude of carryover appears greatest at joint angles corresponding to greater muscle lengths. Although the largest improvements in strength occurred at the trained area of the ROM, some strength carryover was shown in nontrained areas of the ROM.

Muscle Groups Trained

Adaptations to strength training take place predominantly in the trained muscle groups. Ideally, a strength training program will target all major muscle groups. Nevertheless, some areas may be untrained if the individual fails to include exercises that stress all major muscle groups. Muscle strength and hypertrophy increases will take place only when muscle groups are specifically trained.

Energy Systems

Adaptations to strength training are specific to the metabolic demands. The interaction between volume, intensity, rep velocity and rest interval length is critical to eliciting acute metabolic responses that target different energy systems. Although all of our metabolic systems are actively engaged, one may predominate based on the training stimulus. Much of the energy demand of resistance exercise is met by the adenosine triphosphate–creatine phosphate (ATP-CP) and glycolytic metabolic pathways. Adenosine triphosphate and CP depletion during acute strength training is rapid, especially in FT fibres. Anaerobic glycolysis becomes increasingly important during intense, long-duration sets (at least 8-12 reps) and when short rest intervals are used. Bodybuilding workouts stress the glycolytic pathway to a greater extent than strength training.

Movement Patterns

Specificity of movement patterns is observed during strength training. Although a transfer of training effects will occur, and is desired when it comes to performing athletic motor skills, specificity in program design relates to movement patterns. Movement patterns examined have included those with free weights versus machines, open- versus closed-chain kinetic exercises, unilateral versus bilateral training and movement-specific training.

Free Weights Versus Machines Free weights consist of barbells, dumbbells and associated equipment (i.e., plates, collars, benches) whereas machines encompass a variety of specifically designed pieces of equipment that provide resistance within some prescribed ROM and movement pattern. Both free weights and machines are very effective for increasing muscle strength (table 8.2). It is difficult to state which favours greater strength increases. Free weight training leads to greater improvements in free weight tests, and machine training results in greater performance on machine tests. When a neutral testing device is used, strength improvements from free weights and machines appear similar. Free weight training appears more applicable to athletic skill enhancement. Multiple-joint exercises such as the squat and Olympic lifts display similar motor patterns to vertical jumping, and some studies have shown that these exercises can enhance athletic performance. Although free weights and machines have advantages and disadvantages (see table 8.2), both are substantially effective for strength training, and the coach or athlete can decide on the balance of use between the two in order to elicit specific training adaptations.

Closed- Versus Open-Chain Kinetic Exercises A closed-chain kinetic exercise is one in which the distal segments are fixed (squat, deadlift), whereas an open-chain kinetic exercise (leg extension, leg curl) enables the distal segment to freely move against a resistance. Many closed-chain exercises stress multiple joints and major muscle groups whereas many open-chain exercises stress one major muscle group or joint. Both stress each joint differently. The effects on athletic performance are less clear. Augustsson and Thomee (2000) showed moderate correlations between closed- and open-chain exercises and vertical jump performance. However, some studies have shown closed-chain exercises to be superior for enhancing vertical jump performance. Blackburn and Morrissey (1998) reported high correlations between closed-chain exercises and vertical and standing long jump performance. Augustsson and colleagues (1998) reported that a closed-chain exercise (squat) increased vertical jump by 10% whereas no difference was observed after open-chain exercises. Training studies tend to favour closed-chain exercises for the enhancement of athletic performance.

Table 8.2 Comparison of Free Weights and Machines

Advantages of free weights	Advantages of machines
Safe, but may require a longer learning period	Safe and easy to learn
Lower cost	Easier to change the resistance or load (i.e., via weight stack)
Less maintenance	Enable the performance of exercises that may be more difficult with free weights (e.g., leg extension, leg curl)
Promote the complete development of neuromuscular coordination as stabilizer muscle activity is required for motion control	Some machines provide variable resistance, thereby altering loading based on a single-joint ascending–descending strength curve
Better able to provide sufficient resistance for very strong athletes	Isokinetic machines control velocity and may be useful for rehabilitation and testing
Enable more variability for exercise performance in terms of foot–hand placement, position, or posture	Machines usually do not require the use of a spotter
Enable performance of total-body Olympic lifts that are essential for optimal power development	
Enable greater bar acceleration for strength-power training	
Enable several exercises to be performed with very few pieces of equipment	
Appear most effective for enhancing athletic performance	

Unilateral Versus Bilateral Training Exercise performance with only one limb or with two limbs simultaneously affects the neuromuscular adaptations to strength training. *Cross-education* refers to strength gained in the nontrained limb during unilateral training. Strength increases in the contralateral limb and may be specific to muscle actions trained and training velocity. In a recent meta-analysis, Munn, Herbert, and Gandevia (2004) reported that contralateral limb strength increased up to 22%, with a mean increase of nearly 8% compared to pretraining values. The strength increase is accompanied by greater electromyography (EMG) activity, suggesting that neural mechanisms underlie the strength improvement. *Bilateral deficit* refers to the phenomenon that maximal force produced by both limbs contracting bilaterally is smaller than the sum for the limbs contracting unilaterally. Unilateral training (although it increases bilateral strength) can contribute to a greater bilateral deficit whereas bilateral training reduces the bilateral deficit. Specificity is observed as unilateral strength training results in better performance of unilateral tasks than bilateral training. It appears that athletes involved in sports in which unilateral strength and power are important (e.g., pitching, boxing) and athletes with glaring weakness on the contralateral side may benefit from unilateral training. Bilateral performance may be enhanced via unilateral training. Optimal strength training may involve the inclusion of both bilateral and unilateral contractions, with the ratio of bilateral to unilateral contractions based on the needs of the sport.

Movement-Specific Training Movement-specific (e.g., *functional*) training entails the use of exercises that train movements rather than individual muscle groups. The intent is to enhance athletic performance by sport-specific strength training and to provide a link between muscular strength gained through traditional methods and functional strength. Training consists of several multiplanar movements (often in unstable environments) to enhance stabilizer muscle function, as well as a variety of exercises using equipment such as bands, medicine balls, dumbbells, stability balls and other devices.

Overweight Implements Performing sport-specific exercises against a resistance is common among athletes. Overweight implements allow the athlete to overload a sport-specific motion, thereby eliciting a resisted motor pattern similar to the motion itself. Training with overweight (and underweight) implements has been performed for many years to enhance power. It has been suggested that the implement be 5% to 20% greater than the normal load (or 5% to 20% less for underweight objects) and that the frequency ratio be 2:1 for weighted to normal-weight implements for throwing athletes. Escamilla and colleagues (2000) reported that most studies on the use of overweight and underweight implements in baseball players showed enhanced throwing velocity with regulation-size balls. Overweight implements can be used for virtually any motor skill. For example, a chute, sled or weighted vest can be used during sprinting and jumping; weighted bats can be swung and weighted bands employed during training. It is important that overweight implements be used in conjunction with normal training, as the velocity attained with heavier implements may be slower; thus normal (or underweight) implements or training is needed to prevent the athlete from losing velocity (see also chapter 10).

Variation

Training variation requires alterations in one or more program variables over time to keep the training stimulus optimal. Because the human body adapts rapidly to the stress placed on it, variation of the training stimulus is critical in order for subsequent adaptations to take place. Studies show that the systematic variation of volume and intensity is most effective for long-term progression compared to nonvaried programs. Variation, or *periodization,* became apparent for strength training via the work of Selye (1956). His theory *(General Adaptation Syndrome)* proposes that the body adapts via three phases when confronted with stress: (1) shock, (2) adaptation and (3) exhaustion or staleness. *Shock* represents the response to the initial training stimulus in which soreness and performance decrements are produced. Performance increases during the second stage, *adaptation,* in which the body accommodates to the training stimulus. Once the body has adapted, no further adaptations will take place unless the stimulus is altered. This produces the third stage, *exhaustion* or *staleness,* in which a plateau is encountered.

Historically, periodization was used extensively among athletes to achieve peak performance for competition. Currently, periodized strength and conditioning programs are common. A survey of 137 Division I strength coaches showed that ~93% used periodized models in their programs (Durell, Pujol, and Barnes 2003). The importance of periodization lies in the planned periods of intensity and volume variation, as well as active rest, which enable the athlete to attain peak performance while minimizing the risk of overtraining. As the athlete progresses to a high level of strength, it becomes increasingly difficult to maintain that strength year-round. This would entail continual heavy training, and the body cannot withstand that magnitude of voluminous stress without experiencing overtraining symptoms. The critical strategy is to peak strength performance at the right times while maintaining the rigors of practice and competition. Advanced periodized training involves a yearly cycle, or *macrocycle.*

Within a macrocycle, smaller two- to three-month cycles *(mesocycles),* as well as shorter weekly cycles *(microcycles),* are incorporated.

Resistance training program design should begin with general recommendations and proceed more specifically as one progresses.

Although there are many ways to periodize training programs, two general models have been studied. The *classic model (linear periodization)* is characterized by high initial training volume and low-to-moderate intensity. As training progresses, volume decreases and intensity increases in order to maximize strength, power or both. Each training phase is designed to emphasize a particular physiological adaptation until peak performance is achieved. Following peaking, a period of active rest or unloading is used for recovery and to prepare for the next training mesocycle. Comparisons of classic strength-power periodized models to nonperiodized models have shown the classic model to be superior for increasing maximal strength (e.g., 1RM squat, cycling power, motor performance and jumping ability), except for one showing similar performance improvements (Baker, Wilson, and Carlyon 1994).

The *undulating* or nonlinear program enables variation in intensity and volume within each 7- to 10-day cycle by rotating different protocols daily. Only one characteristic is trained in a given workout. The use of heavy (3-5RM), moderate (8-10RM) and lighter resistances (12-15RM and less) may be randomly rotated over a training sequence. This model compares favourably with the classical model and nonperiodized multiple-set programs, and was shown to be more effective for increasing 1RM strength over 12 weeks of training than the classic model (Rhea et al. 2002) but inferior to the classic model for enhancing in-season strength in freshman college football players (Hoffman et al. 2003).

Progression: A General-to-Specific Approach

The rate of progression depends on several factors including the individual's training status. The largest rates of strength improvement are seen in untrained individuals, as the "window of adaptation" is greatest during this time. Because neural adaptations predominate initially, it is difficult to differentiate the efficacy of various training programs. Several low-intensity, low-volume programs have yielded similar results to higher-intensity, higher-volume programs. However, many of these programs yielded inferior results in individuals with greater levels of strength and experience. Trained individuals show a slower rate of progression (Häkkinen et al. 1987). Therefore, it can be concluded that basic programs can be very effective initially. However, program design needs to incorporate the principles of progressive overload, specificity and variation to a greater degree in order to progress at a higher level (figure 8.15).

Individualization

Different athletes respond differently to strength training. We have shown that the same program elicited >20 kg (44 lb) increase in 1RM squat in four subjects but <3 kg (6.6 lb) increase in four others (Ratamess et al. 2003; figure 8.16). Genetics, training status, nutritional intake and the program itself play significant roles in the amount of adaptation that will take place. The coach and athlete should be aware of individual response patterns and the need for variation if the response is minimal. In light of this information, the most effective

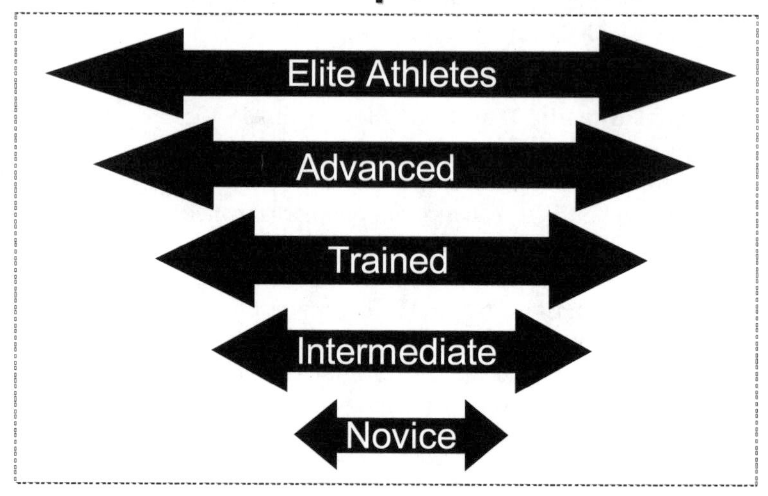

Figure 8.15 General-to-specific model of resistance training progression. This is a simplified schematic representing a theoretical continuum of the amount of variation needed in a consistent strength training program targeting progression. The narrow segment (i.e., novice) suggests that limited variation is suitable in this population, as most programs are effective at this level. It is important to begin gradually (i.e., learn proper technique, allow adequate recovery time); therefore, a general program design is recommended (i.e., building a solid strength foundation). However, as one progresses, the arrows become wider. This suggests that more variation (i.e., specific training cycles) is necessary in order to increase the rate of progression.

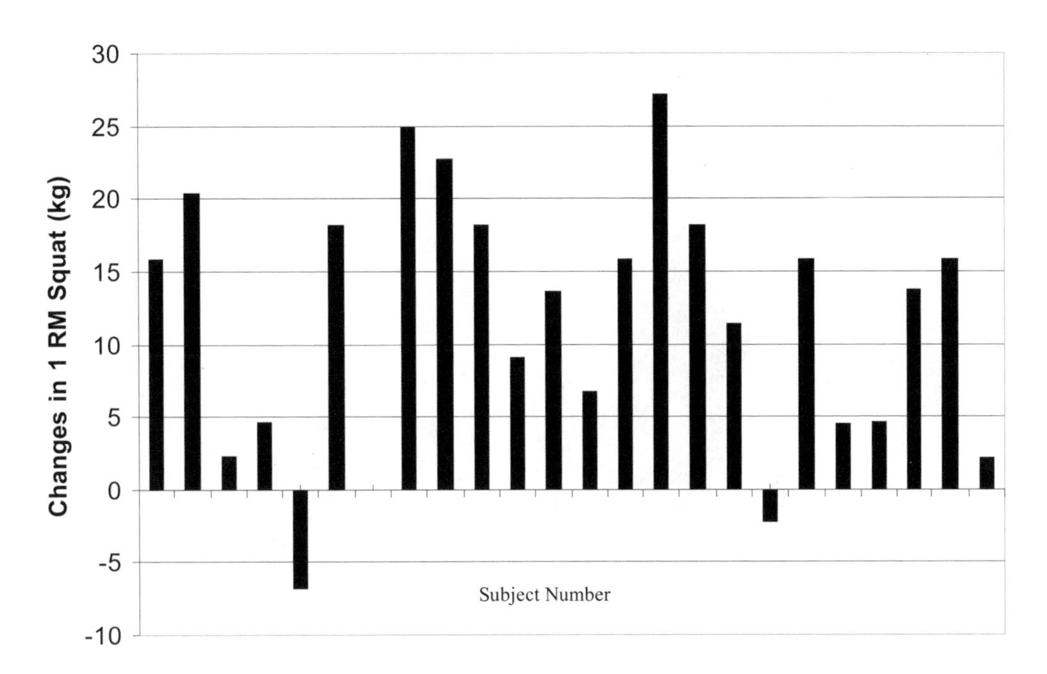

Figure 8.16 Subject variability in response to a four-week overreaching protocol. Data show the change in 1-repetition maximum squat (kg) after four weeks of overreaching.

Adapted from N.A. Ratamess et al., 2003, "The effects of amino acid supplementation on muscular performance during resistance training overreaching: evidence of an effective overreaching protocol," *Journal of Strength and Conditioning Research* 17: 250-258.

strength training programs are those designed to meet individual needs.

The magnitude of strength desired is of interest to the athlete. For example, an athlete may make great progress in one exercise and apply a "strength cap" to that exercise (maintenance of the desired level of strength) and focus on improving others. Maintenance training differs from progressive strength training in that the level of intensity and volume can be lower and still be effective. Therefore, there's no simple answer regarding how much strength is enough, but continually trying to improve it certainly bodes well for the athlete.

Detraining

Detraining refers to the cessation of strength training or a substantial reduction in frequency, volume or intensity that results in performance decrement and a loss of some of the physiological adaptations associated with strength training. Strength loss depends on the length of detraining and training status of the athlete. Decrements may occur in as little as two weeks and possibly sooner in highly trained athletes. However, in recreationally trained men, very little change may take place. Strength reductions appear related to neural mechanisms initially, with atrophy predominating as the detraining period extends. The amount of muscle strength lost rarely reduces strength to below pretraining values, indicating that strength training has a residual effect when the stimulus is removed. However, when the athlete returns to training, the rate of strength retention is high, suggesting the concept of "muscle memory".

Supervision

Supervision is critical to progression during strength training. Athletes who are supervised tend to progress at a faster rate than athletes training without assistance (Mazzetti et al. 2000; figure 8.17). Recently, Glass and Stanton (2004) have shown that novice trainees tend to self-select loads below the threshold of adaptation (i.e., 42-57% of 1RM) and do not lift heavy enough loads to increase maximal strength. Therefore, direct supervision, apart from reducing the risk of injury, ensures higher levels of progression.

Strength Training Program Design

A strength training program is made up of several variables that can be systematically altered to attain a specific goal. The critical element is that each program be individualized to the athlete when applicable. Program individualization is based on goal setting via a *needs analysis*. Here the coach addresses pertinent questions based on strength training goals. When the relevant information is compiled, program development begins. The program

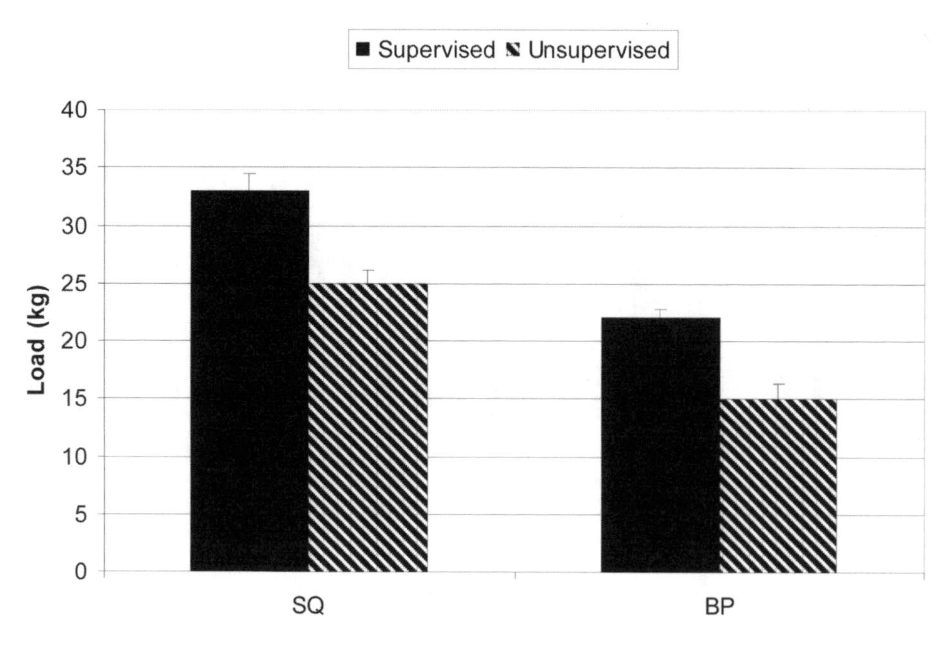

Figure 8.17 Comparison of supervised versus unsupervised strength training. Subjects used a periodized program for 12 weeks in either a supervised or an unsupervised condition. Subjects in the supervised group increased 1-repetition maximum squat (SQ) and bench press (BP) to a greater extent.

Adapted from S.A. Mazzetti et al., 2000, "The influence of direct supervision of resistance training on strength performance," *Medicine and Science in Sports and Exercise* 32: 1175-1184.

will consist of several variables that are structured in order to meet the athlete's goals, as follows:

- Muscle actions used
- Exercise selection
- Exercise order and workout structure
- Intensity
- Volume (total work)
- Set structure (for multiple-set training programs)
- Repetition velocity
- Rest interval length
- Training frequency

Consider the following when setting goals via a needs analysis:

- Are there health or injury concerns that may limit exercise selection or intensity?
- What type of equipment (e.g., free weights, machines, bands or tubing, medicine balls, functional equipment) is available?
- How many days per week will the athlete work out, and are there any time constraints that may affect workout duration?
- How many muscle groups will be trained per workout?
- What muscle groups need special attention (note that all major muscle groups are trained but that some may be given special priority based on strengths and weaknesses)?
- What are the targeted energy systems (e.g., aerobic or anaerobic)?
- What types of muscle actions are needed?
- What are the most common sites of injury for the given sport?
- Is progression or maintenance the goal?
- How will other training modalities (e.g., cardiorespiratory exercise, flexibility training) be integrated into the strength program?
- Are other goals (e.g., muscular endurance, hypertrophy, power, speed, bone growth) sought besides strength gains?

Muscle Actions

Most strength training programs include dynamic reps with CON and ECC muscle actions, whereas ISOM actions play a secondary role. The role of muscle action manipulation during strength training is minimal. Some advanced strength training programs may include different forms of ISOM training, the use of supramaximal ECC muscle actions and accommodating resistance devices in order to maximize gains in strength and hypertrophy.

Exercise Selection

Two types of exercises may be selected: single and multiple joint. *Single-joint exercises* stress one joint or major muscle group, whereas *multiple-joint exercises* stress more than one joint or major muscle group. Both types are effective for increasing strength. Single-joint exercises (e.g., leg extension) have been used to target specific muscle groups and are thought to pose less risk of injury due to the reduced level of skill required. Multiple-joint exercises (e.g., squat) involve more complex neural activation and coordination; and due to the larger muscle mass involvement (and amount of weight used), these exercises have been regarded as the most effective exercises for increasing muscular strength. These exercises, when performed by large muscle groups, also elicit the greatest acute anabolic hormonal responses involving, for example, testosterone and growth hormone. Both types of exercises are recommended for strength training, with emphasis on multiple-joint exercises for advanced training.

> The intensity of resistance exercise is a critical variable when viewing potential neuromuscular adaptations to training. Everyone has a threshold intensity that needs to be trained at or above for progression to take place. Failure to reach this critical threshold may result in limited health and performance improvements.

Exercise Order and Workout Structure

The sequencing of exercises and number of muscle groups trained during a workout affect the acute expression of muscular strength. Three workout structures include total-body workouts, upper-lower body split workouts and muscle group split routines. *Total-body workouts* involve performance of exercises stressing all major muscle groups (i.e., one or two exercises per major muscle group) and are common amongst athletes and Olympic weightlifters. *Upper-lower body split workouts* involve performance of upper body exercises during one workout and lower body exercises during another and are common amongst athletes, powerlifters and bodybuilders. *Muscle group split routines* involve performance of exercises for specific muscle groups during the same workout and are most popular among bodybuilders or for hypertrophy training. All three structures are effective for increasing strength. The major difference is the amount of specialization between workouts (i.e., three or four exercises per muscle group for split routines vs. one or two exercises per muscle group in a total-body workout) and the amount of recovery between workouts. Similar strength improvements may be seen with each.

Acute lifting performance can change depending on where an exercise is placed in sequence. Sforzo and Touey (1996) reported a 75% and 22% decline in bench press and squat performance, respectively, when these were performed later in the workout versus early. Simao and colleagues (2005) reported 28% and 8% reductions in the number of reps performed, respectively, when the bench press and lat pulldown were performed at the end of the workout versus early.

Large muscle mass exercises performed early have a stimulatory effect on small-mass exercises performed later in the workout. Hansen and colleagues (2001) compared two protocols: eight sets of 8 to 12 reps of arm exercises versus eight sets of 10RM of the leg press followed by the same arm workout. They found greater acute elevations in growth hormone and testosterone and greater ISOM strength increases following the latter training protocol. This facilitation of hormone secretion played a critical role in muscle strength enhancement. Performing these exercises early not only takes advantage of a nonfatigued situation, but also is ergogenic as it creates a favourable anabolic hormonal environment for strength and hypertrophy gains. Some recommended sequencing strategies for strength training include, but are not limited to, the following:

- Total-body workout
  - Perform large before small muscle group exercises.
  - Perform multiple-joint before single-joint exercises.
  - For Olympic lifts, perform the most to least complex, and perform these before basic strength exercises.
  - Rotate upper and lower body exercises or opposing (agonist–antagonist relationship) exercises.
- Upper-lower body split
  - Perform large before small muscle group exercises.
  - Perform multiple-joint before single-joint exercises.
  - Rotate opposing exercises (agonist–antagonist relationship).
- Split routines
  - Perform multiple-joint before single-joint exercises.
  - Perform higher-intensity before lower-intensity exercises.

Intensity

Intensity describes the amount of weight lifted during strength training and is highly dependent upon other variables such as exercise order, volume, frequency, muscle action, rep speed and rest interval length. Intensity prescription depends on individual training status. Light loads of ~45% to 50% of 1RM may increase strength in untrained individuals, as this initial phase of lifting is characterized by improved motor learning and coordination. However, greater loading is needed to increase maximal strength as one progresses from intermediate to advanced levels of training (≥80-85% of 1RM; Häkkinen, Komi, and Alén 1985; Hoffman and Kang 2003). This is important because neural adaptations are crucial to maximal strength development as they precede muscle hypertrophy. Although motor unit activity increases with fatigue, it seems that there are specific motor unit recruitment patterns during the lifting of very heavy or near-maximal loads that do not appear to be attainable with light to moderate loading. Thus, heavy loading is necessary at times, but the periodization of intensity appears more critical to optimal strength training. Zatsiorsky (1995) characterized the training practices of elite weightlifters and showed that for the competition lifts, ~8% of training encompassed loads of 60% of competition best or less, 24% was dedicated to 60% to 70% loading, 35% of time was dedicated to 70% to 80% loading, 26% of time was dedicated to 80% to 90% loading and only 7% of time was dedicated to maximal weights (>90% of competition best). Intensity prescription for the Olympic lifts varies (~70-100% of competition best) based on mesocycles; most sets are performed for less than six reps (mostly two to three reps).

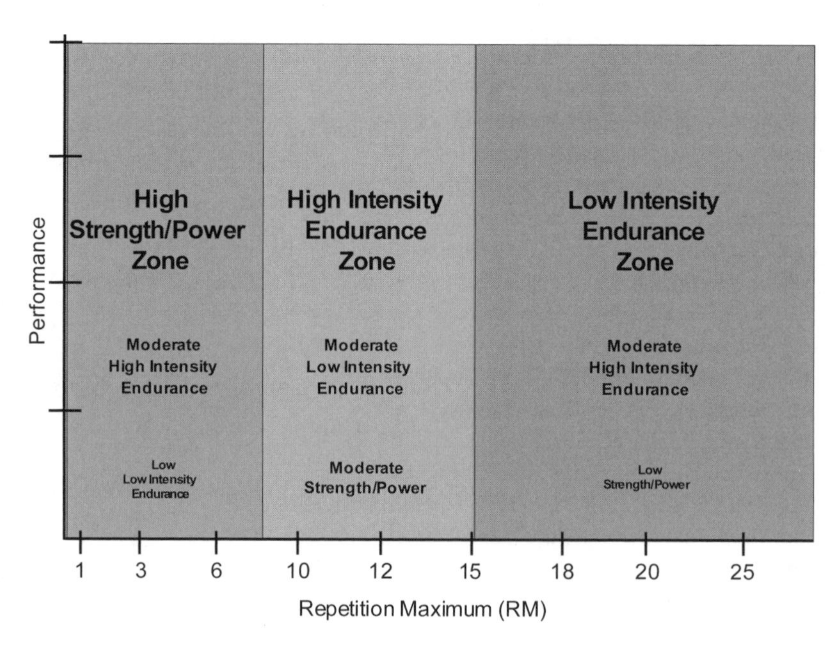

Figure 8.18 Theoretical repetition maximum (RM) continuum. In zone 1 (1-6RM), strength and power is most readily trained at this intensity, whereas high-intensity (HI) endurance may be moderately improved and low-intensity (LI) endurance is not well trained in this zone. In zone 2, HI endurance is maximized whereas moderate improvements in LI endurance and strength-power are seen. In zone 3, LI endurance is maximized whereas HI endurance is moderately increased and strength-power is very low.

Adapted from S.J. Fleck and W.J. Kraemer, 2004, *Designing resistance training programs*, 3rd ed. (Champaign, Illinois: Human Kinetics).

There is an inverse relationship between the amount of weight lifted and the number of reps performed (figure 8.18). Several studies have shown that training with loads ~80% to 85% of 1RM and beyond (e.g., 1-6RM) was most effective for increasing maximal strength (Campos et al. 2002) (figure 8.19). Although substantial strength increases were shown with loads ≈70% to 80% of 1RM (e.g., 6-12RM), it is believed that this range may not be as effective for advanced strength training as heavy loading but may be more effective for increasing hypertrophy. Lighter loads (12-15RM) target local muscular endurance. Although each "training zone" has its advantages, devoting 100% of training to one general RM zone or intensity may be counterproductive, since intensity is exercise dependent. Hoeger and colleagues (1987) showed that 80% of 1RM corresponded to a 10RM for the bench press, leg extension and lat pulldown but to only a 6RM for the leg curl, 7- to 8RM for the arm curl and 15RM for the leg press. Thus, muscle mass involvement and fibre type composition affect RM zones.

There are three basic and effective methods to increase loading:

- Increasing relative percentages
- Training within a RM zone
- Increasing absolute load

One contemporary program design issue relates to whether each set should be performed to muscular exhaustion *(training to failure)*. In this design, reps are performed with RM loading until another rep cannot be performed with proper technique or without assistance. The rationale for training to failure is to maximize motor unit activity and muscular adaptations. The challenge, however, is to assign sufficient sets while minimizing the risk of overtraining and injury. Drinkwater and associates (2005) showed a greater improvement in 6RM bench press with training to failure (9.5%) compared to the nonfailure regimen (5.0%); in contrast, others have shown no advantage for strength but have shown that training to failure was advantageous for enhancing muscular endurance. Although it remains

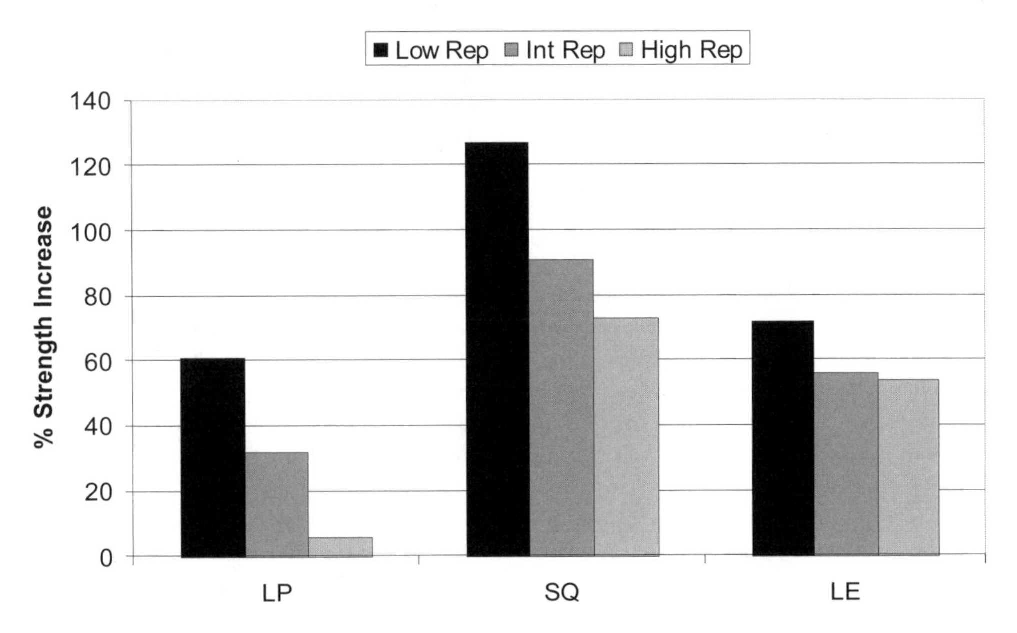

Figure 8.19 Relative increase in maximal strength from training in different RM zones. Comparison of three training programs in untrained men over eight weeks. The low-rep group trained at four sets of 3- to 5RM with 3 min rest intervals; the intermediate-rep group trained at three sets of 9- to 11RM with 2 min rest intervals; the high-rep group trained using two sets of 20- to 28RM with 1 min rest interval. These data show that the largest increases in the leg press (LP), squat (SQ) and leg extension (LE) occurred in the low-rep training group.

Data from G.E.R Campos et al., 2002, "Muscular adaptations in response to three different resistance-training regimens: specificity of repetition maximum training zones," *European Journal of Applied Physiology* 88: 50-60.

unclear how sets to failure should be best included in program design, evidence supports both philosophies for optimal strength training, and the goal of the exercise selected may be critical to the decision. Complex exercises (e.g., Olympic lifts and their variations) and ballistic exercises that require high force, velocity, power production (high quality of effort), and proper technique are best performed with minimal fatigue. Consequently, training to failure may be counterproductive. Yet there is evidence that fatigue may enhance the strength training process for less complex exercises. Strength training with multiple-joint, basic strength and single-joint exercise to failure, at least part of the time, may be beneficial. Not every set is performed to failure, but the inclusion of at least a few sets may be advantageous.

Training Volume

Training volume is estimated from the total number of sets and reps performed during a workout. One can alter volume by changing the number of exercises performed, the number of reps performed per set or the number of sets performed per exercise. There is an inverse relationship between the number of sets per exercise and the number of exercises performed in a workout. Strength training is synonymous with low-to-moderate training volume as a low-to-moderate number of reps are performed per set. Intensity and volume are inversely related such that high-intensity phases coincide with lower-volume periods. Training volumes of athletes vary considerably. It appears that the periodization of volume and intensity is the key provided that a threshold of volume is used during advanced strength training. In fact, periods of reduced volume following intense training periods (i.e., "taper") are effective for peaking maximal strength.

One facet of volume is the number of sets performed per muscle group or workout. There are few data directly comparing strength training programs of varying total sets, thus leaving numerous possibilities for the strength and conditioning professional when designing programs. In a recent meta-analysis, Peterson and colleagues (2004) reported that performance of about eight sets per muscle group produced the largest effect size for strength increases. Although this was not a comparative analysis, it appears that more than one exercise for a muscle group may be advantageous. Most studies have used two to six sets per exercise and shown substantial increases in strength in trained and untrained individuals. Typically, three to six sets per exercise are most common during strength training.

Most training volume studies have compared single- and multiple-set programs. In most studies, one set per exercise performed for 8 to 12 reps at an intentionally slow lifting velocity has been compared to both periodized and nonperiodized multiple-set programs. In novice individuals, similar results have been reported with single- and multiple-set (three sets) programs, whereas some studies have shown multiple sets to be superior. Thus, it appears that volume may not be critical during the first 6 to 12 weeks and that either type of program may be used effectively. However, periodized, multiple-set programs are superior as one progresses to intermediate and advanced stages of training. A recent study in resistance-trained postmenopausal women showed that multiple-set training resulted in a 3.5% to 5.5% strength increase whereas single-set training resulted in a 1.1% to 2.0% reduction in strength (Kemmler et al. 2004). No study has shown single-set training to be superior to multiple sets. It is important to point out that not all exercises need to be performed with the same number of sets, and that emphasis on higher or lower volume is related to program priorities as well as to the muscle or muscles trained in an exercise movement.

Set Structure for Continuous Multiple-Set Programs

Performing multiple sets enables variation within the exercise paradigm. The intensity and volume during each exercise can vary according to personal preference (figure 8.20). Advantages and disadvantages for each system shown in the figure have been identified. Ascending pyramids have been criticized because of the wide range of intensities (70-100% of 1RM) and because the athletes tend to progress to near 100% of 1RM, which could lead to overtraining. This may be true for the classical ascending pyramid scheme, but pyramid protocols have been modified to target specific intensities. Some experts suggest that ascending pyramids are most effective when a narrow range of intensity (10-20%) is used. Other methods can be used to develop a pyramid that peaks at a lower intensity, based on the goals.

Combination and undulating methods are common. Undulating models combine two or three of these basic

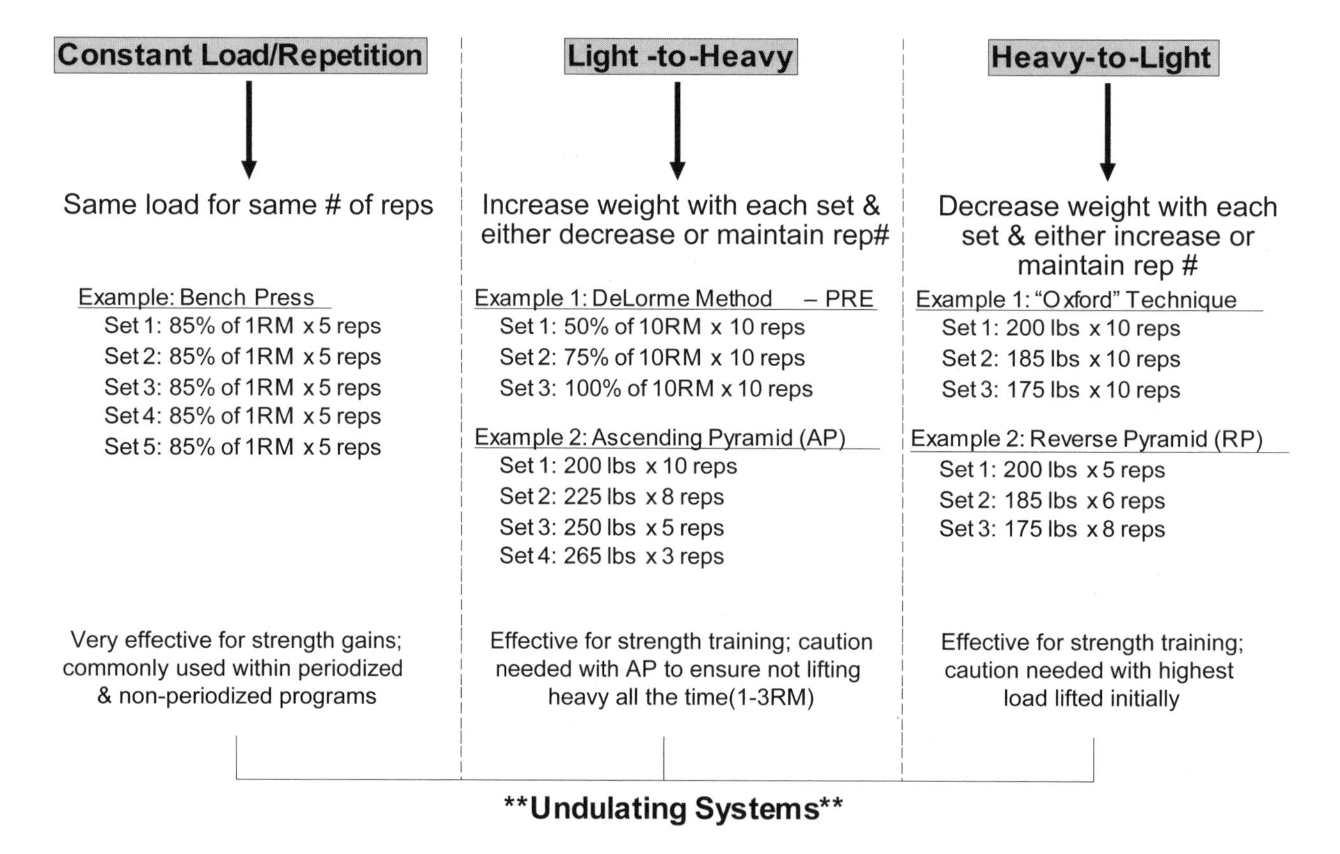

Figure 8.20 Each of these set structure systems has been shown to be effective for increasing muscular strength. Variations or combinations of these can be used (i.e., undulating systems) within one exercise paradigm. Use of each one will depend on the goal of the particular exercise.

structures and cycle between heavy and moderate loading; these models are used for multiple-joint exercises comprising several sets (less than five sets) in which the number of reps and the loading fluctuate. An example is the use of *wave loading* during peaking. One method requires the athlete to perform three reps, two reps and one rep, respectively, with 90%, 95% and 100% of 1RM in the first wave. A second wave is performed for three reps, two reps and one rep, with 1.0 kg added to each set. The athlete continues until the wave cannot be completed (Poliquin 2001). Although many structuring systems are possible, few comparative studies have been conducted on such systems.

Dynamic Repetition Velocity

The velocity of dynamic reps affects the neural, hypertrophic and metabolic responses to training. The intent to control velocity is critical during dynamic (isotonic) training. Since force is the product of mass and acceleration, reductions in force production are observed when the intent is to lift slowly. Two types of slow-velocity contractions exist—unintentional and intentional. *Unintentional* slow velocities are seen during high-intensity reps when either loading or fatigue is responsible. *Intentional* slow-velocity reps are used with submaximal loads over whose velocity the athlete has control. It was shown that CON force production was lower (e.g., 771 vs. 1167 N) for an intentionally slow velocity compared to a moderate velocity with a corresponding lower level of motor unit activation (Keogh, Wilson, and Weatherby 1999) (figure 8.21). The lighter loads that accompany intentionally slow velocities may not provide an optimal stimulus for improving 1RM strength. Moderate-to-high velocities (1-2 s CON or less; 1-2 s ECC) are more effective for enhanced muscular performance as determined by the number of reps performed, work and power output and volume.

Critical to strength training is the intent to move the weight as quickly as possible. Recent studies have shown intentionally fast velocities to be more effective for advanced training than traditionally slower velocities with similar loading. Jones and colleagues (2001) found that 1RM squat increased by 11.5% after 10 weeks of strength training with a light load (40-60% of 1RM). This technique, known as *compensatory acceleration,* requires the individual to accelerate the load maximally throughout the ROM during the CON action to maximize bar velocity. A major advantage is that this technique can be used with heavy loads, especially for multijoint exercises.

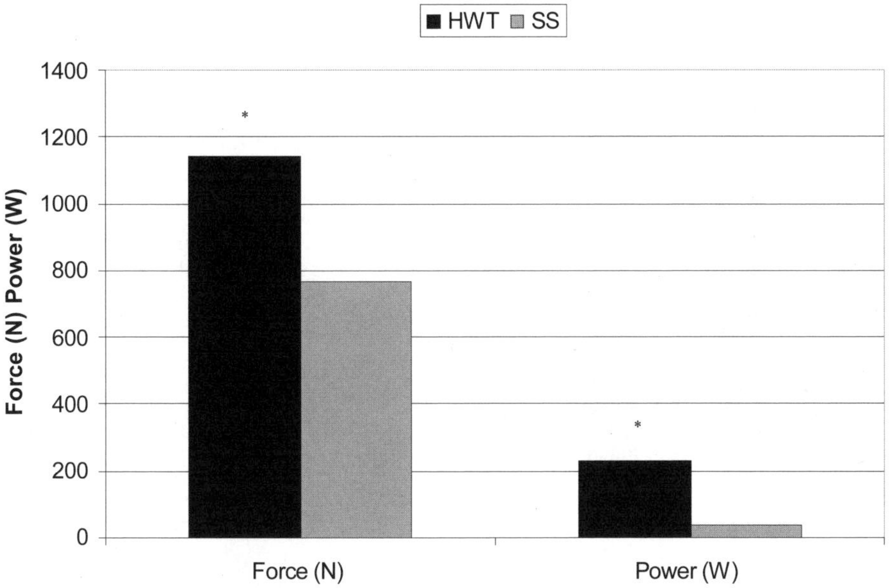

Rest Intervals

Rest interval length depends on training intensity and goals, fitness level and the targeted energy system. The amount of rest between sets and exercises affects the metabolic, hormonal and cardiovascular responses to resistance exercise, as well as performance of subsequent sets and training adaptations. Acute force production may be compromised with short (1 min) rest periods, as this may not be sufficient to replenish muscle phosphagen levels (figure 8.22). We have recently developed a continuum using 5 sets × 10 reps of the bench press with 0.5, 1, 2, 3 or 5 min rest intervals. The results in figure 8.23 show that no reductions in performance were observed with 5 min of rest over the first four sets; performance was maintained over two to three sets with 2 to 3 min of rest, respectively; and performance was reduced with each subsequent set when 30 s or 1 min of rest was used. It appears that at least 2 to 3 min of rest is most beneficial for multiple-joint exercise performance during advanced strength training.

Figure 8.21 Comparison of concentric force and power between heavy resistance training (HWT) and super-slow (SS) training. These data show that lifting with an intentionally slow velocity significantly reduces force and power production. The asterisk indicates a significant difference ($p < 0.05$) between conditions. Data were averaged over five repetitions of the bench press.

Adapted from J.W.L. Keogh, G.J. Wilson, and R.P. Weatherby, 1999, "A cross-sectional comparison of different resistance training techniques in the bench press," *Journal of Strength and Conditioning Research* 13: 247-258.

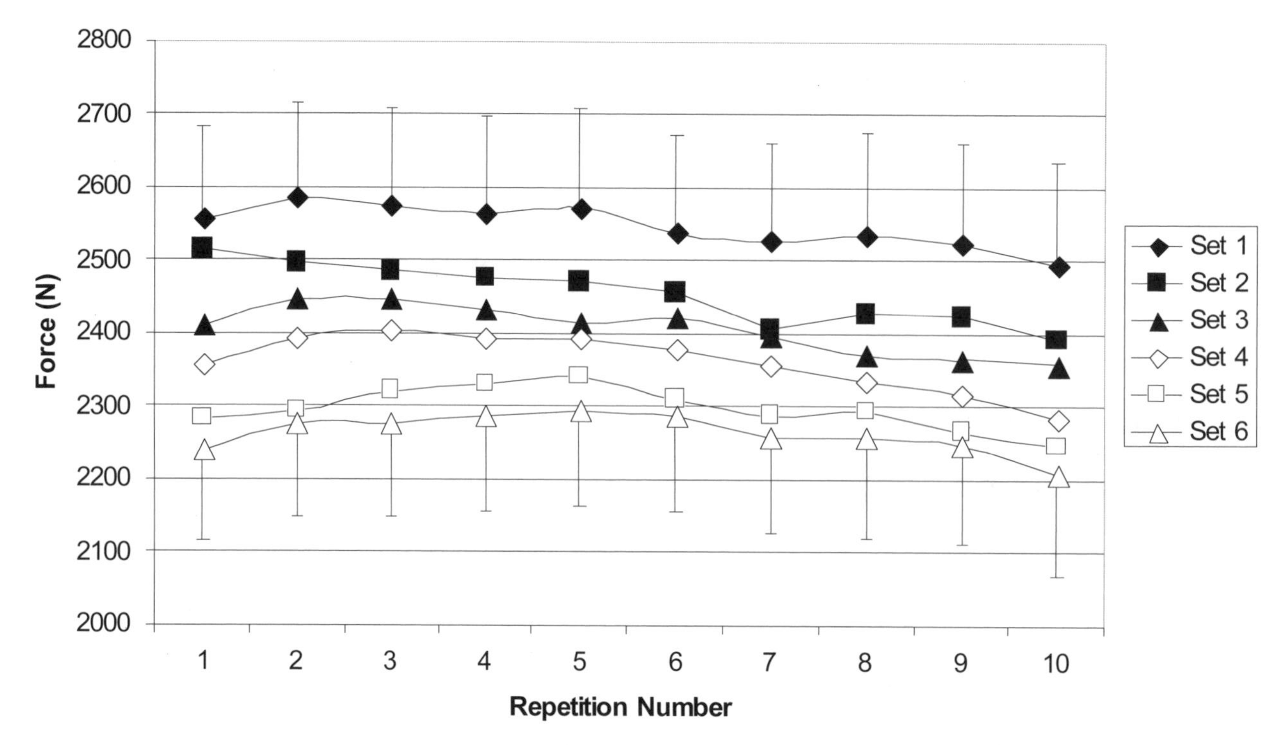

Figure 8.22 Kinetic profile of a squat protocol. Data were collected via a force plate during a squat protocol consisting of six sets of 10 reps with 75% of 1RM using 2 min rest intervals (Ratamess et al. unpublished observations). These results showed a gradual reduction in average force with each subsequent set.

From Ratamess et al. unpublished observations.

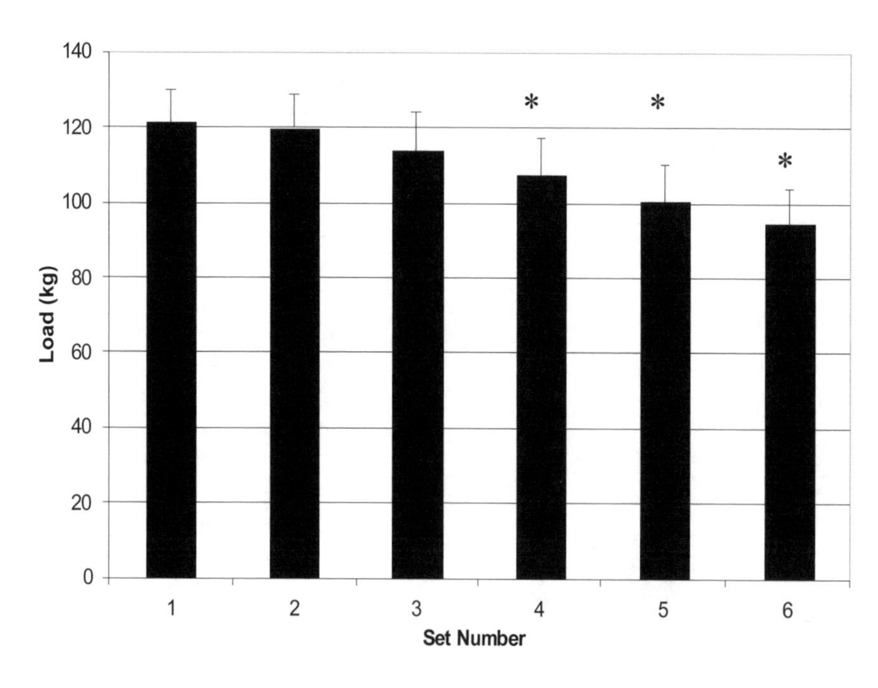

Figure 8.23 Loads lifted during a squat protocol. Data were collected during a squat protocol consisting of six sets of 10 reps with 75% of 1RM using 2 min rest intervals (Ratamess et al. unpublished observations). The loads used during sets 1, 2 and 3 did not significantly differ. However, sets 4, 5 and 6 were all lower (*$p < 0.05$) in comparison to the first three sets.

From Ratamess et al. unpublished observations.

Longitudinal studies have shown greater strength increases with long versus short rest intervals. In training for absolute strength, rest intervals of at least 3 to 5 min are recommended for multiple-joint exercises. Rest interval length varies according to the goals of the particular exercise, and not every exercise will use the same rest interval. One needs to consider fatigue effects from previous exercises in the workout when selecting rest intervals for strength training. Strength may be increased with use of short rest periods but at a slower rate (i.e., it will take a longer period of time to achieve a certain level of strength enhancement with use of short rest periods compared to long), thereby demonstrating a need to establish training goals early.

Frequency

The number of training sessions performed and the number of times certain exercises or muscle groups are trained during the week may affect strength training adaptations. Several factors are involved, including training volume and intensity, exercise selection, level of conditioning or training status (or both), recovery ability, nutritional intake and training goals. Numerous studies have shown two or three alternating days per week in untrained individuals to be very effective whereas one or two days per week are effective for maintenance. In other studies, four or five training days per week were superior to three days, two days and one day per week.

Progression in strength training does not require a change in frequency, but may coincide with alterations in other acute variables such as exercise selection, volume and intensity. Higher training frequencies may enable greater specialization via exercise selection and volume per muscle group in accordance with more specific goals, but research results vary. Advanced weightlifters and bodybuilders use high-frequency training, with four to six sessions per week, and the frequency for elite weightlifters and bodybuilders may be greater. Double-split routines (two training sessions per day with emphasis on different muscle groups) are common and may result in 8 to 12 training sessions per week. Frequencies as high as 18 sessions a week have been reported for Olympic weightlifters. The rationale for high-frequency training is that frequent, short sessions followed by periods of recovery, supplementation and food intake allow for high-intensity training. Elite powerlifters typically train four to six days a week. Only certain muscle groups are trained per workout, so that each major muscle group may be trained only two to three times per week despite the large number of workouts.

Strength Training Methods and Techniques

Besides conventional dynamic strength training, other types and techniques of training may be used for persons at advanced levels. These methods and techniques include isometric training, variable resistance training, dynamic training beyond failure, supramaximal loading, partial ROM training, combining sets and movement-specific (functional) training.

Isometric Training

Isometric, or static, training was performed for many years but has been somewhat overlooked in recent times. Types of ISOM training include contracting against an immovable object, position holds (holding a weight at a specific position), manual resistance (via a partner or opposing limb) and functional isometrics. Gains in ISOM strength are specific to the trained joint angles, the number and intensity of contractions and training duration. Although not applicable to some sports, having significant ISOM strength (i.e., grip strength) may be advantageous for sports such as wrestling, football and powerlifting and for strength competitions. Increases in ISOM strength depend on the intensity and number of contractions, frequency of training, duration of the contractions, speed of tension rise, rest between contractions and joint angle trained. Guidelines for ISOM training include the following:

- **Intensity:** Maximal effort applied (or >80% of maximal voluntary contraction)
- **Duration:** At least 3 to 6 s (up to 10 s) per rep
- **Volume:** Three to five sets per exercise (will depend on contraction duration)—the product of volume and duration should be >30 for strength training
- **Rate of tension application:** Reaching maximum tension as quickly as possible
- **Frequency:** At least three days per week (up to seven)
- **Rest intervals:** Greater than for dynamic strength training (i.e., 30 s to 1 min between sets)
- **Joint angles:** Training at multiple joint angles comprising full joint ROM, especially the area of the "sticking region"

Functional isometric (FI) training stresses a specific area of the ROM (usually the "sticking point" or the region of maximal strength). Because ISOM strength is greater than CON strength, the rationale is to provide greater force at specific areas of the ROM to maximize dynamic strength development. Functional isometric training is generally performed in a power rack (or Smith machine with stoppers) and involves a two-phase movement requiring the performance of a short, limited ROM ballistic contraction for 5 to 20 cm (2-8 in.) followed by a forceful ISOM contraction. The pins are set in two places; the first set is placed in the starting position, which allows the barbell to rest prior to initiation, and a second set is placed at the targeted area of the ROM. The athlete exerts a ballistic force to the bar and presses-pulls the weight against the second set of pins for ~2 to 6 s with multijoint exercises such as the bench press, squat and clean pull.

Functional isometric training is an effective way to increase maximal strength at the "sticking point" of an exercise.

Functional isometric training can enhance maximal dynamic strength. Jackson and colleagues (1985) found that FI increased strength by 19.4%, whereas a dynamic-only group increased strength by 11.9%. O'Shea and O'Shea (1989) reported a greater increase in 1RM squat strength with FI than with dynamic training (31.8 kg vs. 13.2 kg, respectively). Giorgi and colleagues (1998) found similar strength increases with FI (13-25%) and dynamic (11-16%) training. However, analysis of the six strongest subjects showed that FI training led to 5% to 10% greater strength. Functional isometric training may have greater potential to enhance dynamic strength in athletes who are closer to their genetic ceiling.

Variable Resistance Training

Variable resistance training machines modify the loading throughout joint ROM. These machines have an oval-shaped cam so that loading is varied through alterations in the distance between the load and pivot point. Load modification via the cam arrangement is based on human strength curves. In theory, through the matching of human strength curves, inertia may be reduced and the muscle may be stimulated maximally throughout the ROM (figure 8.24). However, not all machines can precisely match strength curves for every individual. For most single-joint movements, torque production throughout joint ROM is depicted by an ascending-descending curve. Peak torque is produced near mid-ROM and deviates as joint angles increase and decrease. The curves differ for different multiple-joint exercises (i.e., ascending curves for the bench press, squat; descending curves for upright row, bent-over row). An advantage to variable resistance training is that loading can be reduced through the "sticking region" and increased in stronger areas of the ROM. The "sticking region" is the weak point of the exercise where lifting velocity is minimal. It is visible during heavy sets or the last few reps of a set to near exhaustion. By reducing the load in this region, variable resistance training allows other areas of the ROM to be trained to a greater extent. Strength increases are common with variable resistance training and tend to be most specific to testing on variable resistance machines.

Besides machines, other variable resistance devices are used with free weight training. Bands and chains are used to modify free weight training (primarily multiple-joint exercises with ascending strength curves). For the squat, chains applied to both ends of the bar are suspended in the air (or at least greater than half of the chain will be suspended) during lockout (the athlete supports the majority of chain weight in this strong ROM area). As the athlete descends, links of the chain are supported by the floor, reducing weight as the athlete descends to the parallel position. During ascent, more weight is applied progressively as the chain links are lifted from the floor. Chains can be used in combination with free weights on the bar or used as the sole source of resistance (plus the weight of the bar). A similar effect is gained

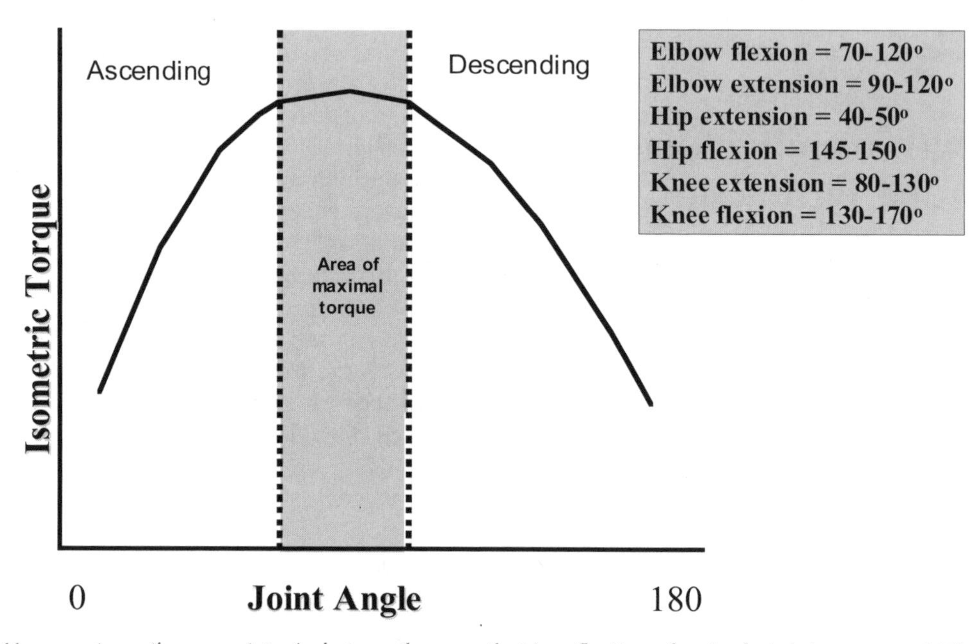

Figure 8.24 Human strength curve. A typical strength curve that is reflective of a single-joint movement. The curve is an ascending-descending curve, which is characteristic of most joints. Torque increases through the ascending region, reaches a peak at the area of maximal torque and decreases through the descending region. The shape of the curve will vary depending on the joint selected. To the right, several joint actions and their corresponding area of maximal torque production are listed. Data reported are ranges obtained from numerous investigations.

Adapted from K. Kulig, J.G. Andrews, and J.G. Hay, 1984, "Human strength curves," *Exercise and Sports Science Reviews* 12: 417-466.

with stretchable bands attached to the floor and bar. The farther the band is stretched, the more resistance is applied to the bar. Ebben and Jensen (2002) compared traditional squatting with squatting with either a chain or band (each comprised ~10% of 1RM) and found that all three produced similar muscle activation and mean and peak ground reaction forces. Further study is needed, as these modalities have been successfully used by elite powerlifters.

Training Beyond Muscular Failure

Failure occurs when the athlete cannot perform another rep with the current load without assistance. Some techniques allow the athlete to extend the set despite fatigue. These include forced reps, drop sets and partial ROM reps in a mechanically advantageous ROM.

Forced Reps

Forced reps are those completed with assistance from a spotter beyond failure to increase strength and hypertrophy. Minimal assistance is applied to allow gradual movement of the weight for about two to four reps. Forced reps are used at the end of a set when exhaustion occurs or exclusively in a set as a form of overload, although little is known about the ergogenic potential of using forced reps. Forced reps produce a greater anabolic hormone response and stress the neuromuscular system to a higher degree (Ahtiainen et al. 2003). Forced reps need to be used with caution (to avoid overtraining), as greater fatigue may ensue. The response may be more pronounced in strength-trained individuals and may be more effective as an advanced training technique.

Descending (Drop) Sets

Drop sets involve reducing the load with minimal rest and allowing the athlete to perform several more reps. Single or multiple breakdowns may be used. The rationale is that when failure occurs, there is still potential to perform more reps with less weight. Drop sets are most effective when a spotter is present to remove weights or change pins on machine weight stacks. Historically, drop sets with moderate loads were used to enhance muscle hypertrophy and endurance. However, drop sets can be used to target maximal strength. A near-maximal weight can be lifted for one to two reps, 5% of the load can be subtracted, and one or two additional reps are performed and so on until the targeted number of reps is achieved. In this manner, a six-rep set can be performed with loading greater than 6RM. Although the advantage of drop sets is unclear, it has been used successfully in the training of elite athletes.

Partial ROM Reps

Muscular failure occurs when the athlete can no longer complete the CON phase of a rep in a full ROM without assistance. However, the athlete can still perform some partial reps in the area of maximal strength. A set may be extended beyond failure if the athlete performs additional partial reps. Using partial reps to extend a set is mostly used to enhance endurance and hypertrophy. However, use of heavy loads is necessary for strength enhancement if one uses partial reps for a complete set (i.e., for every rep, not to extend a full ROM set).

Supramaximal Loading

Supramaximal loading techniques involve performing sets with greater than 100% of CON 1RM. The rationale is to greatly stimulate the nervous system (reduce neural inhibitions) with supramaximal loading to increase 1RM. Conventional loading mostly stresses the sticking region, thereby leaving other aspects of the reps trained submaximally. Targeting areas not limited to the CON "sticking region" is thought to provide a strong stimulus for adaptation. Supramaximal techniques include *heavy negatives, partial ROM reps* and *overloads.*

Heavy Negatives (Eccentrics)

Skeletal muscles have a greater ability to develop force during the ECC phase compared to the CON phase. Heavy negatives involve supramaximal ECC contractions lowered with a slow cadence (~ 3-4 s) for a series of reps. Reps during conventional sets can be enhanced with force applied to the ECC phase via a spotter *(forced negatives)*, while CON phases are performed with spotter assistance (athletes can spot themselves during some unilateral exercises). Certain machines have multiple loading capacities that enable greater loading during the ECC phase. Alternatively, the athlete can perform a machine CON phase bilaterally yet lower the weight unilaterally to supramaximally load the ECC phase. The ECC phase may be loaded with 20% to 40% more weight than the CON phase.

Lowering a heavy weight (120% of CON 1RM) during the bench press can enhance 1RM performance. It is thought that supramaximal loading leads to greater muscle hypertrophy and strength gains. Brandenburg and Docherty (2002) compared nine weeks of traditional training to ECC-accentuated training (3 × 10 reps, 75% of 1RM for CON, 110-120% of 1RM for ECC) and found similar strength increases for the preacher curl. However, ECC-accentuated training led to greater (24%) strength improvements in the elbow extensors (15%). Many strength athletes use heavy negatives to overload the body in a way not possible with traditional set schemes. Because of the greater ECC load, heavy negatives are more conducive to delayed-onset muscle soreness and possibly injury. Using this technique sparingly during maximal strength cycles (four to six weeks) may be beneficial, although more research is needed in this area.

Partial ROM Training (Area of Maximal Strength)

Dynamic partial ROM training is an advanced technique used for overload with near-maximal or supramaximal loading in the area of maximal strength (although it can be used in other areas of the ROM). The rationale is to reduce neural inhibitions by applying a supramaximal load and train a ROM that is submaximally trained with conventional loading. Training in a partial ROM where sport performance may be specific (e.g., a half squat performed to enhance vertical jump performance) has been conducted with some athletes.

Maximal force production for the bench press occurs near the lockout phase, and supramaximal loads may be lifted in this ROM as the "sticking region" is bypassed. Mookerjee and Ratamess (1999) compared partial to full ROM 1RM and 5RM bench presses and showed that partial ROM strength was ~11% to 18% greater than with full ROM lifts. Between two testing sessions, partial ROM weights lifted increased (~4.5%) and the partial ROM bench presses tended to yield higher integrated EMG than full ROM lifts, thereby suggesting greater neural activation.

The magnitude of strength enhancement and carryover to sport performance requires further study. Strength athletes such as powerlifters may benefit from enhanced strength in this area, as many compete with bench press shirts. These shirts are highly ergogenic and provide substantial assistance at the lower part of the ROM but less near full lockout. Greater strength near full lockout could translate into greater competition success. Partial ROM lifts can be incorporated into strength peaking mesocycles, perhaps sequenced following the full ROM core exercise but before subsequent assistance exercises (figure 8.25). Some common variations include partial squats, bench presses and deadlifts in a power rack using the pins as the low point (with or without pause), bench pressing off the floor or using boards placed on the chest to limit ROM and box squats using a high box or bench to limit full descent.

Overloads

Overloads entail holding a supramaximal load in order to make subsequent sets feel "lighter" or to engage potential neural inhibiting mechanisms. An athlete preparing to attempt a 1RM of 160 kg may precede this trial by first loading the bar to ~180 kg and holding the weight in the locked-out position. The intent is to potentiate the neural response to the 160 kg trial. Many powerlifters use this technique for the squat. As part of training, lifters will "walk out" with a supramaximal load and train the neuromuscular system to adapt. The efficacy of this technique is relatively unknown. However, studies show that neuromuscular potentiation does exist following high-intensity muscular contractions and can potentiate sub- or near-maximal performance within a short time period.

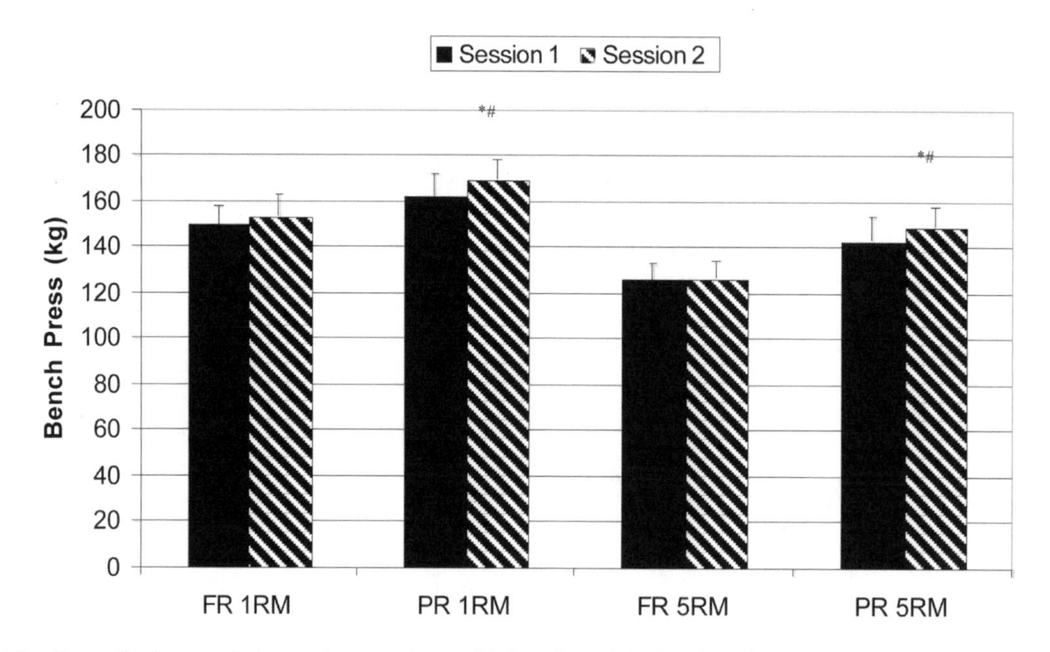

Figure 8.25 Strength changes between two sessions of full and partial ROM bench press assessments. Partial ROM 1RM and 5RM were significantly greater (*$p < 0.05$) than full ROM 1RM and 5RM, respectively. In addition, partial ROM bench press performance increased significantly (#$p < 0.05$) between session 1 and session 2. FR = full range of motion; PR = partial range of motion.

Adapted from S. Mookerjee and N.A. Ratamess, 1999, "Comparison of strength differences and joint action durations between full and partial range-of-motion bench press exercise," *Journal of Strength and Conditioning Research* 13: 76-81.

Noncontinuous Sets

Noncontinuous sets involve the inclusion of a pause-rest interval between reps within a set. The rationale is to maximize performance of each rep by minimizing fatigue. We have shown that during a continuous set of 10 reps of the squat (75% of 1RM), fatigue manifests during reps 6 through 10 (figure 8.26). All 10 reps were performed, although peak force and velocity declined.

Pause Between Reps

Inserting a rest interval in between reps reduces the accumulation of metabolites during training, which may be advantageous as each rep can be performed in a less fatigued state. Research has shown the importance of metabolite accumulation (i.e., lactate, H^+) for hypertrophy and strength gains. Light loads (coupled with restricted blood flow and increased metabolites) can produce similar hypertrophy gains to heavier loads. Because of the potential enhancing effects of metabolites, some researchers have designed studies to compare continuous sets versus sets including intraset rest intervals. Rooney, Herbert, and Balnave (1994) found that strength increased 56% with use of continuous reps compared to 41% with use of rest intervals. Lawton and colleagues (2004) reported that strength increased 9.7% with continuous reps compared to 4.9% in the noncontinuous protocol. However, Folland and colleagues (2002) reported similar increases with continuous and noncontinuous reps. These studies show that both modes are effective for increasing strength, although continuous reps are more advantageous at least in novice to moderately strength-trained individuals.

Rest-Pause Training and Variations

Rest-pause training allows more reps to be performed with maximal or near-maximal weights via short intraset rest periods. The rationale is to increase volume per set with heavy loading in order to increase motor unit activity. Short pauses enable partial recovery. For example, an athlete performs eight reps with a 3RM in the bench press of 135 kg. The athlete performs three reps without assistance, racks the weight and rests for 15 to 30 s, then proceeds to perform one or two additional reps with the same load and rests for 15 to 30 s. These cycles continue until the targeted numbers of reps are performed. Many variations exist within rest-pause training. It is suggested that 6 to 10 reps be performed for only one set with several days of recovery in between training sessions.

Few scientific data exist concerning rest-pause training, though many elite powerlifters have used this training method. One popular protocol is the *dynamic method* (Simmons 1996), whereby 8 to 10 sets of a core exercise (bench press, squat, box squat) are performed for two or three explosive reps (using compensatory acceleration) with ~60% of 1RM and with 45 to 60 s rest between sets. Although two or three reps are performed consecutively, the large set number with substantial rest illustrates a

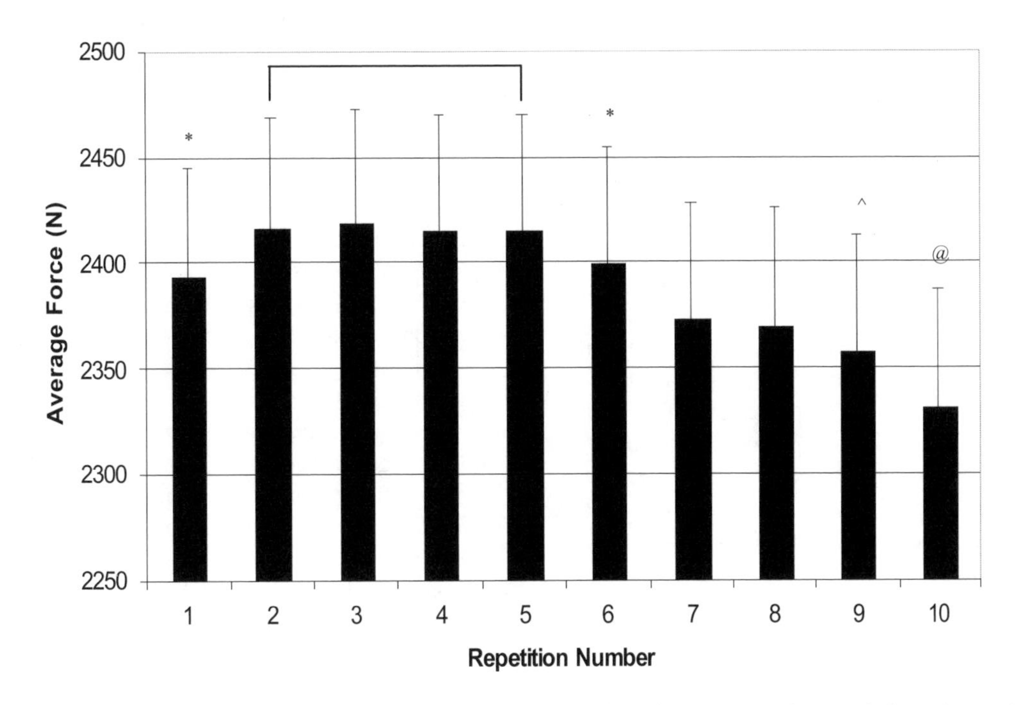

Figure 8.26 Kinetic profile of a 10-repetition set of squats. These data demonstrate that peak force is produced during repetitions 2 through 5. * indicates $p < 0.05$ compared to reps 2 through 5. ^ indicates $p < 0.05$ compared to reps 1 through 6. @ indicates $p < 0.05$ compared to reps 1 through 9. Beyond five repetitions, peak force is gradually reduced.

From Ratamess et al. unpublished observations.

variation of rest-pause training. One theoretical advantage is that multiple reps are performed. We have found that many athletes do not reach peak force or power on the first rep of a set. Rather, peak force was seen consistently on the second or third rep (figure 8.26). Multiple reps may enable the athlete to attain peak force or power prior to resting between reps. This philosophy is applied to Olympic weightlifting. Haff and colleagues (2003) studied rest-pause training (known as *cluster training*) in Olympic weightlifters and track and field athletes. They compared it to traditional methods and showed that clusters (with 30 s rest between reps) led to greater bar velocity and displacement during 1 × 5 reps of clean pulls.

> Circuit resistance training programs are effective for increasing the continuity of a workout, decreasing total workout time, and increasing aerobic capacity.

Combining Exercises

Combining exercises involves performing two or more exercises consecutively with minimal rest. Multiple actions can be combined into a single exercise to produce combination lifts. This is common with the use of Olympic lifts. A combination lift may involve cleaning the weight from the floor, front squatting it and finishing with a push press. This sequence is performed for a series of reps. Another strategy is to perform all reps for one exercise followed by consecutive performance of one or more other exercises. The following terminology has been used to denote the type of training:

- **Supersets:** consecutive performance of two exercises (either for the same muscle group or for different muscle groups)
- **Tri-sets:** consecutive performance of three exercises
- **Giant sets:** consecutive performance of four or more exercises

These combinations are used for increasing local muscular endurance and hypertrophy, especially if one is trying to reduce total workout duration. Although strength can increase, most strength enhancement is seen in lesser-trained individuals. Combination lifts are used to increase muscular endurance as the weight lifted for each exercise is less (and dependent on the weakest of the exercises). Many bodybuilders use supersets, tri-sets and giant sets for increasing muscle hypertrophy. It is common for these athletes to combine exercises stressing the same muscle groups. The largest benefit is enhanced muscle endurance and size due to the large fatigue effect. Strength athletes may use supersets to increase muscle strength. Many times the supersets involve opposing or unrelated muscle groups so adequate recovery takes place and greater loading can be used.

Circuit training programs consist of a series of exercises intended to stress most major muscle groups consecutively with minimal rest (10-30 s) between exercises. Once the series is complete, the individual may conclude the workout or repeat the circuit. Circuit training is time efficient, and because of the continuous nature of the program, produces moderate cardiovascular effects. This continuity of exercise requires that light-to-moderate loading be used for ~8 to 12 reps. Although circuits can be designed with heavier weights, it is the first few exercises that may benefit the most while the athlete is minimally fatigued. Circuit programs may not be the most effective for maximal strength training. Although strength increases have been shown in novice to intermediate-trained individuals, advanced to elite strength athletes benefit very little from the short rest intervals coupled with the moderate intensity.

Strength-Endurance Training Methods

Athletes not only need to be strong, but also need to have the ability to repeatedly develop a high level of force. Sports like wrestling, football and boxing and strength competitions require high force and power production over an extended period of time. Training of these athletes requires a periodized approach with specific strength-endurance mesocycles, especially near the beginning of the competition season (endurance may be enhanced more efficiently once the athlete completes a strength-power peaking phase). Undulating models entail simultaneous training for both strength and endurance within the same week. Rhea and colleagues (2003) compared the effects of linear (25-, 20- and 15RM loading cycled every five weeks), undulating (25-, 20- and 15RM trained each week) and reverse periodization (15-, 20- and 25RM cycled every five weeks) and reported that reverse periodization led to the highest improvement in endurance (73%), although the difference between conditions was not significant.

Submaximal and high-intensity (strength) endurance may be trained. *Submaximal endurance* refers to the ability to perform submaximal contractions for an extended period of time. *High-intensity (strength) endurance* is the ability to maintain maximal or near-maximal force for an extended period of time or to maintain a number of reps for a number of sets with heavy weight with minimal fatigue.

Traditional Strength Training

Local muscular endurance improves during strength training. Strength training can increase the maximal number of reps performed with an absolute load; however, minimal effects are observed with relative endurance. Despite adaptations to skeletal muscle, an athlete will typically be able to perform a similar number of maximal reps with a specific percent of 1RM before and after strength training. Moderate-to-low resistance training with high reps is most effective for improving absolute or submaximal muscular endurance. Training to increase local muscular endurance implies that the individual performs high reps (long-duration sets) or minimizes recovery between sets (for strength-endurance) or both. Exercises stressing large muscle groups yield a high acute metabolic response, and metabolic demand is an important factor for improving muscular endurance (i.e., increased mitochondrial and capillary number, fibre type transitions, buffering capacity).

Loading should be multidimensional. Light loads (>15-20RM) coupled with higher reps is most effective for increasing local muscular endurance. Campos and colleagues (2002) showed that training with 20- to 28RM with 1 min rest intervals increased muscular endurance the most compared to 3- to 5RM with 3 min rest intervals or 9- to 11RM with 2 min rest intervals. Moderate to heavy loading (coupled with short rest periods) is effective for increasing high-intensity endurance. It is recommended that short rest periods be used for endurance training, that is, 1 to 2 min for high-rep sets but <1 min for moderate (12-20 reps) sets.

Strength-Endurance Training

Strength-endurance training mostly targets the glycolytic energy system with intermittent, high-intensity exercise. Glycolysis predominates during moderate- to high-intensity exercise lasting 20 s to a few minutes in duration. Lactic acid (an increase in H^+) is produced at a high rate, which is a major contributor to fatigue, and favourable adaptations take place. Acidosis can be tolerated during heavy exercise and is buffered more efficiently. The athlete can substantially improve strength-endurance by training this system specifically. High-volume strength training programs (3 to 6 sets × 10-12 reps or more) of moderately high intensity (70-80% of 1RM) using short rest intervals (1-2 min) between sets are effective for increasing buffer capacity. For progression, the athlete can gradually increase the volume, increase the duration, reduce rest intervals between sets or increase the loading (with the same volume). One needs to consider the metabolic demands of the sport when designing the strength-endurance program. A preseason program for wrestling may comprise a total-body workout (perhaps in circuit fashion) with minimal rest in between sets. The athlete can be given a rest interval that mimics the break between wrestling periods. In American football the offence or defence averages ~12 to 15 series per game and 4.5 plays per series; each play averages ~5.5 s of explosive activity, and the time between plays averages 26 to 34 s (Kraemer and Gotshalk 2000). Thus, a preseason conditioning phase could be designed based on these metabolic demands.

Strength and Aerobic Exercise Compatibility

Aerobic training benefits muscle endurance capacity, allowing the anaerobic athlete to recover faster during rest intervals; it also results in less lactic acid formation and permits a more efficient metabolism of by-products. However, how much aerobic exercise a strength-power athlete should undertake has been a topic of discussion, as has the impact of concurrent strength and endurance training. Because the two lead to specific antagonistic physiological adaptations, it has been suggested that one may counteract the other. Although strength and endurance will improve through concurrent training, one or the other may be attenuated if both are trained intensely. Most studies have shown that concurrent training may not attenuate endurance performance as long as the frequency of aerobic training is maintained to accommodate strength training, although one study has shown that the addition of strength training can hinder $\dot{V}O_2max$ improvements (Glowacki et al. 2004). The concurrent effects on strength development have been variable, with some studies showing that maximal strength is compromised and others showing no incompatibility.

Key programming issues relating to potential incompatibility include whether or not aerobic and strength training are performed in the same session; the sequence, volume and intensity of each; and training status of the individuals. Researchers who have reported an incompatibility have suggested the following possible reasons:

- **Inadequate recovery between workouts:** Too high a frequency in relation to the volume or intensity of concurrent training leads to overtraining.
- **Residual fatigue:** Performing endurance training prior to strength training leads to suboptimal effort during strength training due to fatigue.
- **Altered neuromuscular recruitment patterns and adaptations:** These are based on the antagonistic neuromuscular adaptations of the two modalities (i.e., greater fast-to-slow fibre transitions and attenuated type I muscle fibre hypertrophy with concurrent training).

Several studies that have shown an incompatibility have examined such protocols as three days per week of strength training plus three days per week of endurance training on alternating days, or four to six days per week of combined high-intensity strength and aerobic training; thus overtraining could have played a role. When both were performed during the same session (yielding three days per week frequency with at least one day off in between workouts), the incompatibility was not evident (McCarthy, Pozniak, and Agre 2002). However, Sale and colleagues (1990) showed that four separate training sessions per week was better than two concurrent days for increasing leg press 1RM (25% and 13% improvement, respectively). Thus, increasing recovery between workouts may attenuate the incompatibility. Häkkinen and colleagues (2003) reported similar strength increases following 21 weeks of concurrent training and strength training only; however, the strength training–only protocol resulted in a greater rate of force development, whereas concurrent training attenuated this increase.

Explosive strength may be compromised to a greater extent than unintentionally slow-velocity maximal strength, and the sequence may play a critical role. Leveritt and Abernethy (1999) examined lifting performance 30 min after a 25 min aerobic exercise protocol and found that the number of reps performed during the squat was reduced by 13% to 36% over three sets (figure 8.27). At least lower body exercise performance could be reduced when aerobic training is performed first. In summary, it appears that strength and endurance training can be performed simultaneously provided that adequate recovery is allowed between workouts, that strength training is performed first and that the volume and intensity of aerobic exercise are not too high.

Strength Training, Injuries and Injury Prevention

Research indicates that strength training reduces injury risk. Although injuries do occur during training, these are far less common with supervised strength training in comparison to other sporting activities. Strength training is safe regardless of the training methods used. The most common site of injury during strength training is the lower back, and the type of injury mostly encountered is a muscle strain. Injuries are often attributable to improper exercise technique, muscular imbalances, improper loading, overtraining, weight room accidents and unsupervised training. Competitive lifting (weightlifting, powerlifting, bodybuilding and strength competitions) may pose a greater risk for injury than noncompetitive lifting. After studying 358 bodybuilders and 60 powerlifters, Goertzen and colleagues (1989) showed that muscle injuries were common amongst 84% of these athletes and that major shoulder or elbow injuries had occurred in 40%, with the incidence twice as high in the powerlifters. Raske and Norlin (2002) reported that weightlifters exhibited a higher frequency of low back and knee injuries whereas shoulder injuries were more prominent in powerlifters.

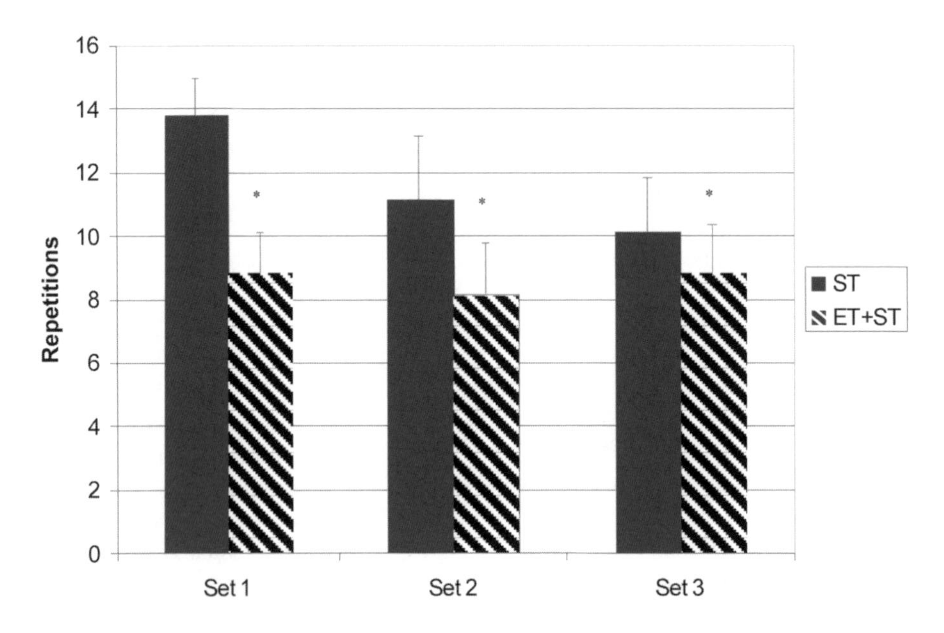

Figure 8.27 Strength training performance following aerobic exercise. These data demonstrate a significant reduction (*$p < 0.05$) in the repetitions performed for the squat exercise over three sets when lifting occurred within 30 min of completion of an aerobic exercise protocol. ST = strength training–only session; ET+ST = strength training following an aerobic workout (25 min of 5 × 5 min intervals of cycle exercise ranging from 40% to 100% of $\dot{V}O_2$ peak).

Adapted from M. Leveritt, and P.J. Abernethy, 1999, "Acute effects of high-intensity endurance exercise on subsequent resistance activity," *Journal of Strength and Conditioning Research* 13: 47-51.

Injury Prevention

Strength training can reduce the risk of injury in several ways. The enhanced muscular strength increases joint stability, which serves to dissipate high levels of joint reaction force during explosive sprints, jumps, landings and high-agility movements. Connective tissue (tendons, ligaments, bone) structures adapt to strength training. Strength training increases tendon stiffness and bone mineral density. Taken together, these data demonstrate that strength training leads to connective tissue adaptations that ultimately reduce the risk of injury.

One important aspect of strength training is to improve strength of weak muscle groups. Balance needs to be maintained between muscles on the two sides of the body and those that have antagonistic relationships. Appropriate muscle balance is critical to reducing the risk of injuries such as knee ligament sprains and strains, hamstring tears, shoulder injuries and low back pain or injury. For example, Tyler and colleagues (2001) studied adductor muscle strains in National Hockey League (NHL) players and found that players were 17 times more likely to strain an adductor muscle if its strength was ≤80% of abductor group strength.

Gender differences exist, making females more susceptible to injury (e.g., anterior cruciate ligament tears). Among other hormonal, anatomical and physiological differences, female athletes show greater strength imbalances, altered recruitment patterns, altered activation timing and greater side-to-side deficits in the lower extremity compared to male athletes. Muscle balance between the internal and external rotators of the glenohumeral joint is important for reducing shoulder injuries. Because external rotators are weaker than internal rotators, the inclusion of exercises emphasizing the external rotators is critical for throwing, striking and grappling athletes.

Hamstring injuries are common among athletes involved in explosive, anaerobic sports. The hamstring muscle group is multiarticular and is important for knee support, propulsion and velocity control during explosive motion. Balance between the hamstrings and quadriceps is critical to reducing injury risk. It has been suggested that the ratio of hamstring to quadriceps strength be at least 0.60 (hamstring strength should be >60% of quadriceps strength) and that the strength difference between dominant and nondominant legs be <10%. These standards may be greater for higher-calibre athletes. Knapik and colleagues (1991) showed that female collegiate athletes were at higher risk for lower extremity injury if there was >15% difference in knee flexor strength between right and left legs and a hamstrings-quadriceps ratio <0.75. Extensive hamstrings strength training is critical to injury prevention.

Mechanical Ergogenic Aids

Specialized strength training equipment has been developed to reduce the risk of injury (Escamilla 1988). Although these pieces of equipment provide substantial joint support and reduce injury risk, most strength athletes use them for their ergogenic potential. Lifters use straps and chalk to enhance grip during heavy lifting, especially for pulling exercises. The following are some additional types of equipment.

- **Lifting belts:** Belts augment intra-abdominal pressure during lifting by 13% to 40% and reduce compressive forces on the back by ~6% (Lander, Simonton, and Giacobbe 1990). They also enable lifters to use more weight, perform more reps and perform faster reps during trunk-hip extension exercises. These benefits depend on the type and tightness of the belt (tighter belts provide more support). Quadriceps muscle activity increases during squatting with a belt; however, trunk muscle activity is reduced. Belts should not be overused, as they limit strength development of the core. It is recommended that they be used only during performance of lifts with maximal or near-maximal weights. Furthermore, belts should be used only during a set and should be loosened or removed in between sets, as they reduce blood flow back to the heart and increase blood pressure.

- **Joint wraps:** Knee wraps directly enhance performance of exercises involving knee extension by providing a rigid support that resists knee flexion. The magnitude is dependent upon the type of wrap used, the wrapping technique, the angle of knee flexion during wrapping and tightness of the wrap. Long, thick and heavy wraps with sufficient elasticity provide the most support. Some lifters report at least 10 to 15 kg (22-33 lb) enhancement of squat lifts. Similarly, elbow wraps are effective for reducing elbow pressure and enhancing performance of extension exercises (e.g., bench press), while wrist wraps reduce pressure on the wrist by limiting joint flexion and extension and allow for greater stability during exercises using heavy weights. As with lifting belts, wraps should not be overly used.

- **Bench press shirts:** The majority of powerlifters use bench press shirts during competition and during the final stages of precompetition training, and substantial gains in the bench press are commonly reported. Although research on this is lacking, gains of 10 to 20 kg (22-44 lb) are common, with some elite lifters reporting increases of >45 kg (100 lb).

- **Suits (super suits, briefs, erector shirts):** Super suits and briefs are tight compression garments used primarily by powerlifters to enhance stability and performance of the squat and deadlift. The super suits extend from the upper quadriceps area to the upper abdomen and have straps that are placed over the shoulders. More support is gained the farther a lifter descends, and the suit serves to assist the lifter during ascent through the sticking region. Testimonials from lifters estimate a 25 kg (55 lb) increase in the squat, while some experienced lifters report a more than 70 kg (154 lb) benefit. Escamilla (1988) reported that lifters with knee wraps and suits, compared to without, lifted an average of ~13% more weight. The mean total time needed to complete a rep was almost 0.5 s less when suits and wraps were used.

Summary

It is clear that the development of muscular strength is a complex area with many acute training variables that need to be considered as part of the design of the strength component within a training program. This chapter has outlined the use of strength training in sport from a performance enhancement perspective. Current training practices and principles have been discussed; further information on specific training exercises and programs is provided in chapter 13.

nine

Power

Robert U. Newton, PhD; and William J. Kraemer, PhD

This chapter begins by defining maximal muscle power and then discussing why this aspect of neuromuscular function is so important to human performance. The underlying neural and intramuscular mechanisms that contribute high power capacity are explored, with particular attention to the effects of training. Subsequently, maximal power production and training program design are addressed, including advanced strength and conditioning techniques for developing power output.

Maximal Power Production Defined

Power can be defined as the applied force multiplied by the velocity of movement (Knuttgen and Kraemer 1987). As *work* equals the product of force and the distance moved, and velocity is the distance moved divided by the time taken, power can also be expressed as work done per unit time (Garhammer 1993). Power output for an athlete can range from 50 W produced during light cycling or jogging to around 7000 W produced during the second phase of the pull for the Olympic clean in weightlifting (Garhammer 1993). The main focus of this chapter is on the highest level of power output, that which can be produced in one or two muscular contractions. This has been termed *maximal instantaneous power* (Gollnick and Bayly 1986); however, for the purposes of this chapter the term *maximal power* is used.

The main focus of this chapter is on the highest level of power output—that which can be produced in one or two muscular contractions and termed *maximal power.*

Why Is Maximal Power Important?

Maximal power output is the main determinant of performance in activities requiring one movement sequence with the goal of producing a high velocity at release or impact (Young and Bilby 1993) as in throwing, jumping and striking activities. In addition, sudden bursts of power are required when one is rapidly changing direction or accelerating during various sports or athletic events (e.g., football, basketball, soccer, baseball and gymnastics).

As an example, the height to which an athlete jumps when rebounding in basketball is determined purely by the velocity with which he or she leaves the ground. At the bottom of the movement the body stops momentarily (figure 9.1); and as the athlete extends the trunk, hips, knees and ankles, the body is accelerated upward to a maximum takeoff velocity as the athlete leaves the ground. This takeoff velocity is determined by the force that the muscles can generate against the ground multiplied by the time during which the forces are applied, termed *impulse,* minus the impulse due to the body weight (Winter 1990).

Once the athlete has left the ground he or she can no longer apply force; and the faster the body is accelerated vertically, the shorter the time between the bottom of the movement and takeoff (see figure 9.1 = 268 ms). It is here that we can appreciate the importance of maximum muscular power. As an athlete attempts to maximize his or her power output, the time over which force can be applied to accelerate the body decreases. Therefore,

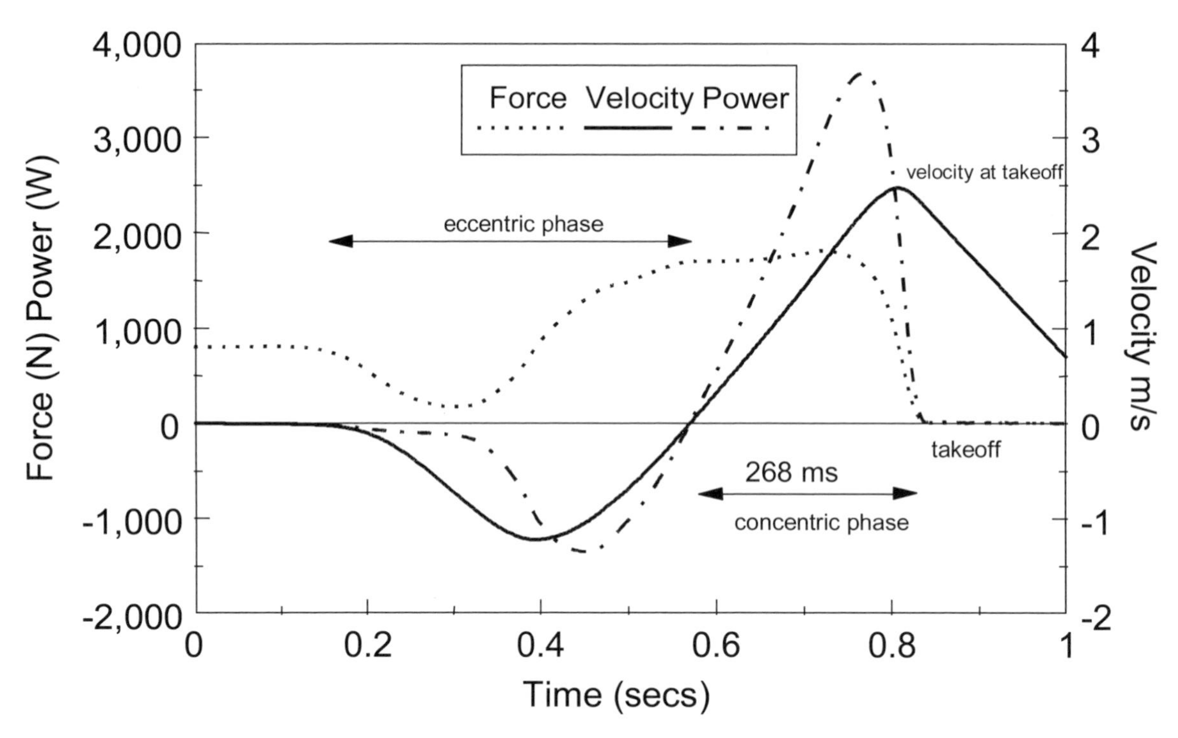

Figure 9.1 Vertical ground reaction force, centre of mass velocity and power output of the subject during a vertical jump with countermovement. Note that the concentric muscle action is only 268 ms in duration. The resulting takeoff velocity is determined by the sum of the forces (impulse) that can be produced during this short time period.

Adapted, by permission, from R.U. Newton and W.J. Kraemer, 1994, "Developing explosive muscular power: implications for a mixed methods training strategy," *Strength and Conditioning Journal* 16(5): 20-31.

three mechanical properties of the neuromuscular system become limiting factors to performance:

- The ability to develop a large amount of force in a short period of time—this has been termed the maximum rate of force development (mRFD)
- The ability of muscle to produce high force at the end of the eccentric phase and during the early concentric phase
- The ability of muscle to continue producing high force output as its velocity of shortening increases

A number of factors contribute to maximizing these three properties. Discussion of each factor will assist our understanding of the effects of different training strategies and how they may influence training efficiency.

Factors Contributing to Maximal Power Output

Maximal power output is affected by both intracellular and neural factors. Important intracellular factors are muscle cross-sectional area, muscle architecture, energy availability and muscle fibre type. Neural factors that may contribute to or limit maximal power output include, among others, increased activation of agonists and neural contribution to rate of force development.

Intracellular Factors Affecting Maximal Power Output

Maximal power is produced by a maximal rate of cross-bridge cycling at the level of actin and myosin (see chapter 8) and the total quantity of myofilaments. It is instructive then to examine the factors described in table 9.1 that may influence these two quantities.

Neural Factors Affecting Maximal Power Output

Human strength and power is determined not only by the size of the involved muscles and their contractive characteristics, but also by the ability of the nervous system to appropriately activate these muscles. To produce high power output the *agonists* must be fully activated, muscles assisting or coordinating the movement *(synergists)* must be appropriately activated, and the muscles that produce force in the opposite direction to the agonists, termed *antagonists,* must be appropriately activated (Sale 1992). Therefore, the control of movement by the nervous

Table 9.1 Intramuscular Factors Possibly Contributing to or Limiting Maximal Power Output

Cross-sectional area (CSA)	Strength and thus power are a function of the total CSA of the muscles that can be activated. Hypertrophy training to increase muscle size is an important component of power development.
Muscle architecture, strength and power	Pennate muscles adapt quite specifically to high-force versus high-velocity training. Increased pennation angle favours force production, but power training reduces pennation angle to facilitate high contraction velocity (Blazevich et al. 2003).
Energy availability	Adenosine triphosphate availability is not a limitation to maximal muscle power production, as sufficient stores are immediately available in the muscle for such short-term efforts.
Muscle fibre type	Maximum contraction velocity and power output of fast-twitch muscle fibres are approximately fourfold those of slow-twitch types (Faulkner, Claflin, and McCully 1986). Although considerable shift between fibre characteristics is possible, genetic endowment of fibre type is an important factor in maximal power capability.

system is critical to strength and power production, and this explains the large increases in strength that are apparent in the first few training sessions for athletes performing an exercise to which they are not accustomed (Sale 1992). Adaptive changes in the nervous system optimise control of the muscles involved in the exercise through a number of possible mechanisms listed in table 9.2.

Performance Qualities Contributing to Power

We turn attention next to the many factors that affect power performance. Maximal power performance responds to training using a countermovement in which muscles are first stretched and then shortened (stretch-shortening cycle). It responds more fully to specific power training than to heavy resistance training since power training involves both force and velocity, as well as a shorter deceleration phase in which muscle activation decreases. Ballistic resistance training, in which the load is propelled against gravity or thrown, overcomes the problem of deceleration. Each component contributing to maximal power production appears to have its own window of adaptation, suggesting that the training program for an athlete should use mixed methods and should target those components with the greatest potential for adaptation—that is, the components in which that athlete is weak.

Stretch-Shortening Cycle

Most powerful activities involve a countermovement during which the muscles involved are first stretched and then shortened to accelerate the body or limb. This action of the muscle is called a *stretch-shortening cycle* (SSC) (Komi 1986) and involves many complex and interacting neural and mechanical processes. A great deal of research has been directed toward the study of the SSC (Bobbert et al. 1996; Bosco and Komi 1979; Ettema, van Soest, and Huijing 1990; Gollhofer and Kyröläinen 1991) because it has been observed that performance is greater in SSC movements than it is when the activity is performed with a purely concentric action (Bosco and Komi 1979). For example, differences in jump height of 18% to 20% have been observed between squat jump (SJ) and countermovement jump (CMJ) (Bosco et al. 1982). A SJ is a purely concentric jump initiated from a crouch position. The CMJ is initiated from a standing position; the athlete performs a preparatory dip movement and then jumps upward.

Although several mechanisms have been proposed (Bobbert et al. 1996), it would appear that the difference in CMJ and SJ height is due primarily to the fact that the countermovement allows the subject to attain greater force output at the start of the upward movement. This results in greater forces exerted against the ground and subsequently an increase in impulse ($F \times t$) and thus acceleration of the whole body upward. The other proposed mechanisms appear to play a secondary role in the enhancement of performance by the SSC (Bobbert et al. 1996).

Maximal power performance has been shown to respond to training that involves the athlete's performing SSC movements with a stretch load of greater magnitude and more rapidly than he or she is accustomed to. These activities have been termed *plyometrics* and have been found, in a number of studies, to be effective for increasing jumping ability (Adams et al. 1992; Clutch et al. 1983; Schmidtbleicher, Gollhofer, and Frick 1988; Wilson et al. 1993). Plyometric training results in an increase in the overall neural stimulation of the muscle and thus force output; however, qualitative changes are also apparent (Schmidtbleicher, Gollhofer, and Frick 1988). In subjects unaccustomed to intense SSC loads, some studies have shown a reduction in electromyographic (EMG) activity

Table 9.2 Neural Factors Possibly Contributing to or Limiting Maximal Power Output

Increased activation of agonists	According to the size principle, motor units are recruited in a certain order according to size, and thus from slow- to fast-twitch types, as a voluntary contraction increases from zero to maximal (Hannerz 1974). With strength and power training the ability to recruit the higher-threshold motor units is improved.
Neural contribution to rate of force development	Power training involving rapid, ballistic movements results in increased rate of onset of motor unit activation, which produces a higher rate of force development (Häkkinen and Komi 1985).
Premovement silence	During ballistic movements from a relaxed muscle state, the neural activation to the muscle may become greatly reduced to bring all the motor units into a ready state to contract synchronously and rapidly. The importance of this mechanism in power development has been discounted by recent research (Hasson et al. 2004).
Preferential recruitment of motor units	As a result of power training, the neural system may learn to overcome the size principle of motor unit recruitment and preferentially recruit the more powerful fast-twitch motor units immediately (Grimby and Hannerz 1977).
Selective activation of agonists within a muscle group	Fast muscles (i.e., those with a relatively high proportion of fast-twitch motor units) may be preferentially activated over slow muscles when high-velocity movements are attempted (Sale 1992). This is most likely a learned ability resulting from power training.
Coordination of movement pattern and skill	Power performance is affected by the interaction between agonists, antagonists and synergists involved in the movement. Resistance must be low in order to produce a fast movement velocity, and flow of power from joint to joint in the kinetic chain must be timed optimally to maximize the end segment velocity. This is a learned phenomenon developed through appropriate skills practice; thus it is important to combine practice of the target activity with strength and power training (Bobbert and van Soest 1994).

starting 50 to 100 ms before ground contact and lasting for 100 to 200 ms (Schmidtbleicher, Gollhofer, and Frick 1988). This is attributed to a protective mechanism by the Golgi tendon organ reflex acting during sudden, intense stretch loads to reduce the tension in the tendomuscular unit during the force peak of the SSC. After periods of plyometric training the inhibitory effects are reduced, in an effect termed disinhibition, and increased SSC performance results (Schmidtbleicher, Gollhofer, and Frick 1988). Plyometric training places considerable forces on the musculoskeletal system, and it is recommended that the athlete have a preliminary strength training base prior to commencing a plyometric training program (e.g., squat 1.5 times body weight) (Chu 1992). The potential for injury is thought to be much higher for drop jumps, which should not be attempted by the beginner (Schmidtbleicher 1992).

Muscular Strength and Heavy Resistance Training

Strength, the ability of the muscle to exert a maximal force or torque at a specified or determined velocity (Knuttgen and Kraemer 1987), varies for different muscle actions such as eccentric, concentric and isometric contractions (Kraemer 1992). Often coaches and athletes associate the term strength only with the force that can be exerted during slow-speed, or even isometric, muscle actions. This is often determined using a 1-repetition maximum (1RM) test in which strength is assessed as the maximum weight the athlete can lift once throughout the complete movement. The development and assessment of 1RM strength have received a great deal of research attention, and the interested reader may refer to the relevant literature (Atha 1981; Berger 1962; Häkkinen 1989). During lifting of the maximal weight that the athlete is capable of lifting, the limiting factor is muscle strength at slow contraction velocities. Muscle strength as required in 1RM lifts, however, is needed in a limited number of athletic endeavours (e.g., powerlifting); most sports require high force output at much faster velocities of movement.

Research findings (Häkkinen and Komi, "Changes" 1985; Wilson et al. 1993; Wilson, Murphy, and Walshe 1997) and anecdotal evidence from strength and conditioning specialists indicate that if an athlete's strength at slow movement velocities increases, then power output and athletic performance will also improve. This occurs since maximum strength, even at slow velocities, is a contributing factor in maximal power.

Muscle strength as required in 1RM lifts is needed in a limited number of athletic endeavours (e.g., powerlifting); most sports require high force output at much faster velocities of movement.

When one is attempting to maximize power output, the concentric phase follows the eccentric phase and as such starts from a zero velocity. Therefore, the force produced during the later part of the eccentric phase, the changeover from lengthening to shortening (which includes a period when the muscle is contracting isometrically) and the subsequent concentric contraction is determined by the maximum strength of the agonist muscles during slow eccentric, isometric and concentric contraction. If maximal strength is increased, then higher forces can be exerted during this period, resulting in increased impulse and therefore increased acceleration.

However, as the muscles begin to achieve high velocities of shortening, strength capacity at slow movement velocity has a reduced impact on the ability of the muscle to produce high force (Duchateau and Hainaut 1984; Kaneko et al. 1983; Kanehisa and Miyashita 1983). This fact becomes increasingly important as the athlete attempts to train specifically for maximal power development. In terms of training, several studies have shown improved performance in power activities (e.g., vertical jump) following a strength training program (Adams et al. 1992; Bauer, Thayer, and Baras 1990; Clutch et al. 1983; Wilson et al. 1993). For example, one study demonstrated a 7% improvement in vertical jump following 24 weeks of intense weight training (Häkkinen and Komi, "Changes" 1985).

Despite these observed improvements, specific power training appears much more effective (Häkkinen and Komi, "Effect" 1985). In one study, subjects performed movements in which they attempted to maximize power output with relatively lighter loads and showed a 21% increase in vertical jump. These results indicated that there might be specific training adaptations to heavy resistance versus power-type training. Heavy resistance strength training using high resistance and slow velocities of concentric muscle action leads primarily to improvements in maximal strength (i.e., the high-force, low-velocity portion of the force–velocity curve; see figure 9.2), and the improvements are reduced at the higher velocities. In power training, which utilizes lighter loads and higher velocities of muscle action, there are resulting increases in force output at the higher velocities as well as an improved rate of force development (Häkkinen and Komi, "Effect" 1985).

Although velocity-specific training adaptations are observed, performance changes with training are not always consistent with this principle. The conflict results from the complex nature of powerful muscle actions and the integration of slow and fast force production requirements within the context of a complete movement. Another confounding influence in observing clear, specific training adaptations is the fact that in untrained people, a wide variety of training interventions will produce increases in strength and power. Depending on the training status of the individual, the response may not always follow the velocity-specific training principle (Komi and Häkkinen 1988). For individuals with low levels of strength, improvements throughout the force-velocity spectrum may be produced regardless of the training load or style used (Komi and Häkkinen 1988). A further avenue for improvement in untrained subjects is the relatively high rates of force development (RFD) that can be achieved with heavy weights, or even isometrically when the subject has the "intention of making fast movements" (Behm and Sale 1993). This may contribute to adaptations particularly in people who are unfamiliar with fast contractions.

It appears that training adaptations of single factors (i.e., high force, high power) occur only after a base level of strength and power training has been achieved. This notion is supported by the fact that if the athlete already has an adequate level of strength, then the increases in maximal power performance in response to traditional strength training will be poor, and more specific training interventions are required to further improve maximal

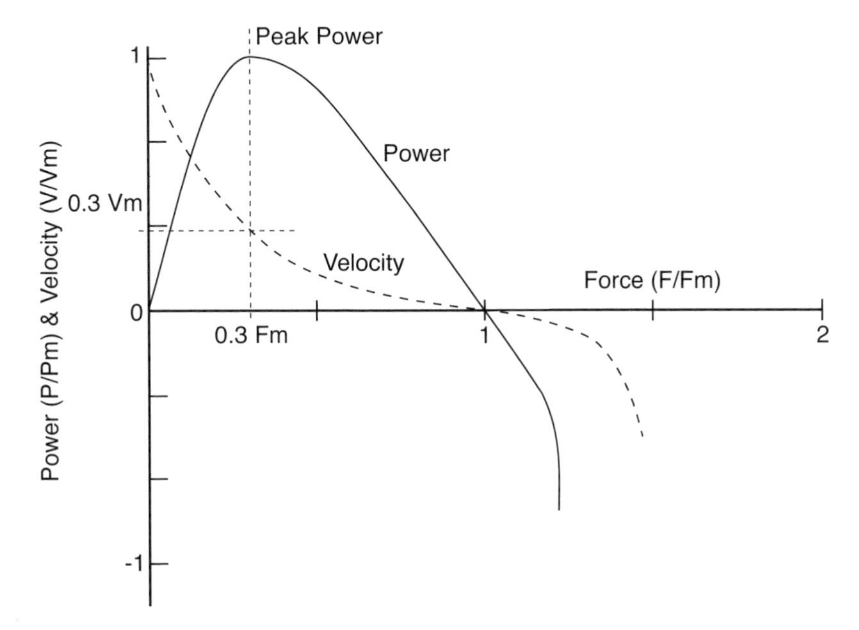

Figure 9.2 Relationship among force, velocity and power for skeletal muscle. Vm, Pm and Fm are maximum movement velocity, maximal power output and maximum isometric force output, respectively.

Adapted, by permission, from J.A. Faulkner, D.R. Claflin, and K.K. McCully, 1986, Power output of fast and slow fibers from human skeletal muscles. In *Human muscle power,* edited by N.L. Jones, N. McCartney, and A. J. McComas (Champaign, IL: Human Kinetics), 88.

power output (Häkkinen 1989). Thus, improvement of maximal power output in trained athletes may require more complex training strategies than previously thought (Wilson et al. 1993).

This contention is supported by research (Wilson, Murphy, and Walshe 1997) comparing changes in 1RM squat, vertical jump and flying 20 m sprint velocity during eight weeks of either weight training or plyometric training. Subjects were classified as "weak" or "strong" based on their pretraining 1RM squat. The results demonstrated significant negative relationships between weight training–induced improvements in sprinting, jumping and pretraining 1RM performance. The authors hypothesized that this was due to the principle of "diminished returns" whereby initial improvements in muscular function are easily attained whereas further improvements are progressively harder to achieve. Unexpectedly, the performance gains from the plyometric training were unrelated to initial strength levels.

The Need for Training Integration

Use of slow-velocity, heavy resistance training for the development of maximal power is often justified on the basis that power is equal to the product of force and velocity of the muscle action. It has often been reasoned that if the athlete increases his or her 1RM strength, then this alone is sufficient to influence power output. However, if we are to maximize improvements in power performance, then we must train both the force and velocity components. Because the movement distance is usually fixed by the athlete's joint ranges of motion, velocity is determined by the time taken to complete the movement. Therefore, if we train using methods that decrease the time over which the movement is produced, we increase the power output. Intimately linked to this concept is the mRFD.

Resistance Training and Rate of Force Development

Since time is limited during powerful muscle actions, the muscle must exert as much force as possible in a short period of time, a quality that has been termed *maximum rate of force development* (figure 9.3). This may explain to some extent why heavy resistance training is ineffective for increasing power performance. Squat training with heavy loads (70-120% of 1RM) has been shown to improve maximum isometric strength (i.e., movement velocity = zero); however, it did not improve the mRFD (Häkkinen, Komi, and Tesch 1981) and may even reduce the muscle's ability to develop force rapidly (Häkkinen 1989). On the contrary, an activity during which the athlete attempts to develop force rapidly (e.g., maximal power jump training with light loads) increases an athlete's ability to increase force output at a fast rate. Specifically, maximal power–type resistance training increases the slope of the early portion of the force–time curve (Häkkinen and Komi, "Effect" 1985) as can be observed in figure 9.3.

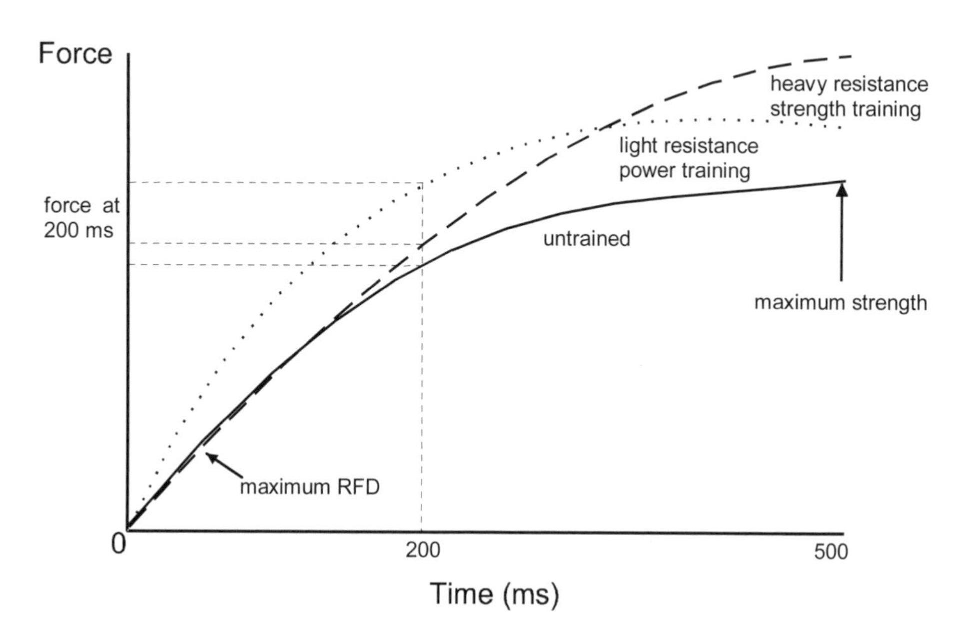

Figure 9.3 Isometric force–time curve indicating maximum strength, maximum rate of force development and force at 200 ms for untrained, heavy resistance strength–trained and light resistance power–trained subjects.

Adapted from K. Häkkinen and P.V. Komi, 1985, "Changes in electrical and mechanical behavior of leg extensor muscles during heavy resistance strength training," *Scandinavian Journal of Sports Sciences* 7(2): 55-64.

Although heavy resistance training in this study increased maximum strength and thus the highest point of the force–time curve, it did not improve power performance appreciably, especially in athletes who had already developed a strength training base (i.e., who had had more than six months of training) (Häkkinen 1989). The reason may be that the movement time during powerful activities is typically less than 300 ms (Young 1993) and most of the force increases cannot be realized over such a short period of time. The athlete does not have the time to utilize the strength gains achieved through heavy resistance training.

The Deceleration Phase and Traditional Weight Training

The results of many studies (Berger 1963; Wilson et al. 1993; Young and Bilby 1993) highlight a further problem with traditional weight training and power development. It has been observed that when one is lifting a maximal weight in a bench press, the bar is decelerating for a considerable proportion (24%) of the concentric movement (Elliott, Wilson, and Kerr 1989). The deceleration phase increases to 52% when one performs the bench press lift with a lighter resistance (e.g., 81% of 1RM) (Elliott, Wilson, and Kerr 1989). In an effort to train at a faster velocity that is more specific to sporting activity, athletes may attempt to move the bar rapidly during the lift. However, this also increases the duration of the deceleration phase, as the athlete must slow the bar to a complete stop at the end of the range.

Previous research has compared the kinematics, kinetics and neural activation of a traditional bench press movement performed with the intention of maximizing power output and the ballistic bench throw in which the barbell is projected from the hands (figure 9.4) (Newton et al. 1996). Significantly better performances were produced during the throw movement compared with the press for average velocity, peak velocity, average force, average power and peak power. The average muscle activity during the concentric phase for pectoralis major, anterior deltoid, triceps brachii and biceps brachii was higher (19%, 34%, 44% and 27%, respectively) for the throw condition. Further analysis of the velocity and force profiles revealed a deceleration phase during the press, lasting 40% of the concentric movement, that was associated with a decrease in muscle activation. It was concluded that performing traditional press movements

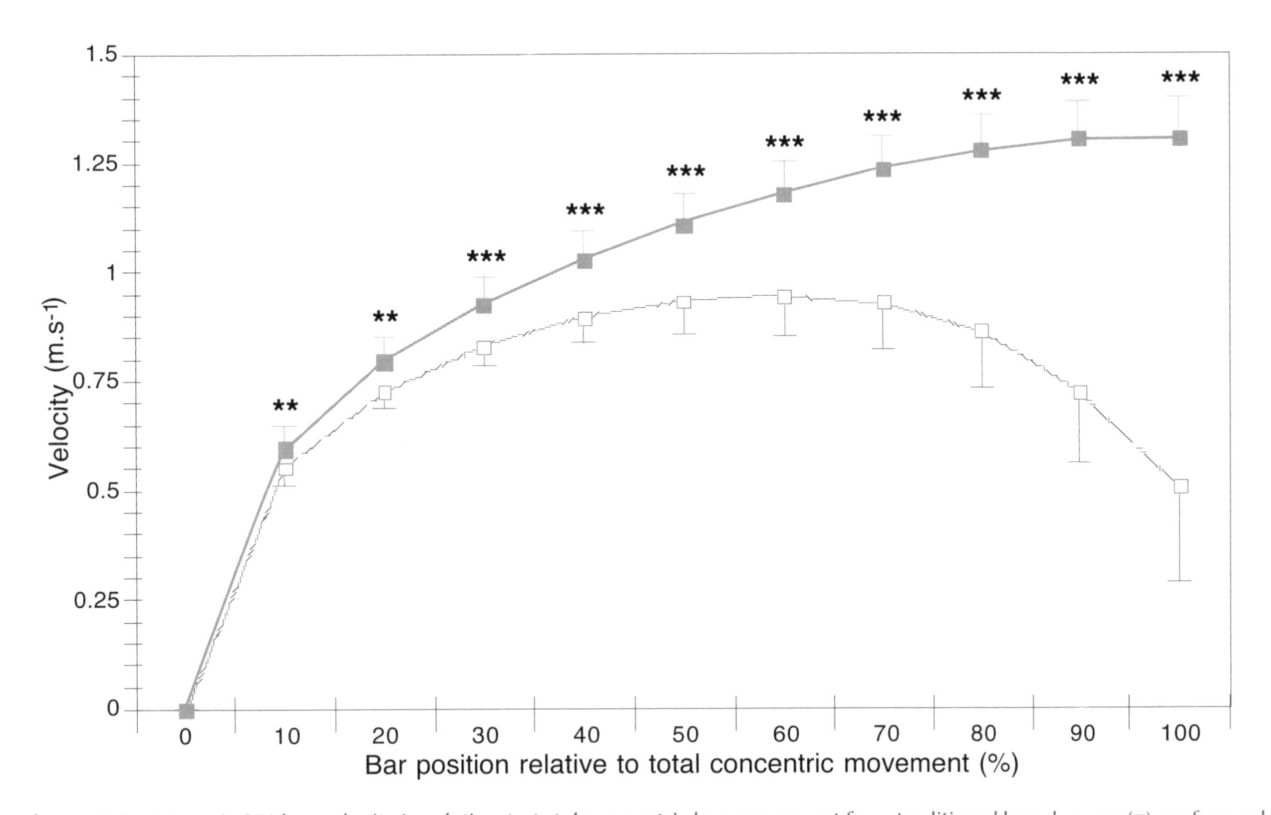

Figure 9.4 Mean (±SD) bar velocity in relation to total concentric bar movement for a traditional bench press (□) performed as rapidly as possible and a bench throw (■) (**$p < 0.01$; ***$p < 0.001$).

Adapted, by permission, from R.U. Newton et al., 1996, "Kinematics, kinetics, and muscle activation during explosive upper body movements," *Journal of Applied Biomechanics* 12: 31-43.

rapidly with light loads does not create the ideal loading conditions for the neuromuscular system with regard to maximal power production. This was especially evident in the final stages of the movement in which ballistic weight loading conditions—with the resistance accelerated throughout the movement—resulted in greater velocity of movement, force output and EMG activity. Plyometric training, weighted jump squats and the weightlifting movements avoid this problem by allowing the athlete to accelerate all the way through the movement to the point of projection of the load (i.e., takeoff in jumping, ball release in throwing, impact in striking activities).

Ballistic Resistance Training

As previously demonstrated (Newton et al. 1996), the problem of the deceleration phase can be overcome if the athlete actually throws or jumps with the weight. This type of movement is most accurately termed *ballistic resistance training.* "Ballistic" implies acceleration of high velocity and with actual projection into free space. The common English meaning of the word "ballistic" as defined in the *Macquarie Dictionary* (Delbridge and Bernard 1988) is "of or pertaining to the motion of projectiles proceeding under no power and acted on only by gravitational force and the resistance of the medium through which they pass". As projecting the load into free space such that it becomes a projectile is the essential aspect of this type of training that differentiates it from other forms, the term ballistic resistance training seems most appropriate.

Heavy Versus Light Resistance

With reference to either traditional or ballistic resistance training, there is considerable controversy over the resistance to be used for the development of maximal power performance (Wilson et al. 1993; Young 1993). If training is limited to traditional resistance techniques, then heavy (>80%) resistances are preferable because it is not possible to overload the muscle sufficiently using light resistances while stopping the weight at the top of the range of motion (Hatfield 1989). With ballistic resistance there is perhaps no optimal intensity or resistance. Both heavy (>80%) and light (<60%) resistances have application in the training of muscular power, with each affecting different components of muscle power production. If the coach or athlete has to choose a single resistance, then the resistance that produces the greatest power output (30% maximal voluntary contraction, MVC) has been shown to be optimal (Kaneko et al. 1983; Wilson et al. 1993). In reality, there is a wide selection of resistances, and greatest training adaptations will result when athletes train with resistances that span the concentric force–velocity capability (Kaneko et al. 1983).

The Velocity-Specific Training Controversy

Velocity specificity of resistance training is currently one of the most contentious issues in the field of muscle strength and power development. Studies using isokinetic testing and training methods have shown that strength increases are specific to the velocity at which one trains (Lesmes et al. 1978). If you train at a slow movement velocity, you tend to increase strength at that velocity, while strength at higher velocities (which are more common in sport) is not improved. Based on this, it has been recommended that resistance training be performed at a high speed if the purpose of the training is to increase power output.

However, evidence has been presented (Behm and Sale 1993) that it is the intention to move quickly that determines the velocity-specific response, and that heavy resistance weight training may be effective if the athlete attempts to move the resistance as quickly as possible. The authors (Behm and Sale 1993) suggested that the principal stimuli for the high-velocity training response are the repeated attempts to perform ballistic contractions and the high RFD of the ensuing contraction. The type of muscle action (isometric or concentric) appeared to be of lesser importance. Despite this, the overwhelming majority of research supports the importance of velocity-specific training adaptations (Kaneko et al. 1983; Wilson et al. 1993). It must be kept in mind that there are considerable changes within the muscle cells, as well as the architecture of the whole muscle, that have been shown to be very specific to the contraction velocities employed during the training intervention.

The authors of several studies (Kaneko et al. 1983; Moritani et al. 1987; Wilson et al. 1993) have recommended that to increase maximal power output, athletes should train at the velocity and using the resistance that maximize mechanical power output. As can be seen from figure 9.2, the concentric force and velocity capabilities of muscle are intimately linked. Maximal mechanical power is produced at a resistance of 30% of maximum isometric strength, which corresponds to a velocity of muscle shortening of approximately 30% of maximum (Faulkner, Claflin, and McCully 1986). Heavy resistance training will increase power output at low velocities and heavy resistances, while light resistance training (e.g.,

The problem of the deceleration phase can be overcome if the athlete actually throws or jumps with the weight. This type of movement is most accurately termed *ballistic resistance training.* "Ballistic" implies acceleration of high velocity and with actual projection into free space.

30% MVC) will increase power output for light resistances (Duchateau and Hainaut 1984). Furthermore, heavy resistance training tends to shift the optimal resistance for power output toward the heavier resistances (i.e., maximal power output is produced at a heavier resistance).

Therefore, we can be fairly confident that increases in power are specific to the training resistance and velocity used. This may be the rationale behind the recommendation of a 30% MVC resistance. As this is the resistance at which power is maximized, training at this resistance will necessarily produce the greatest increases in maximal power. However, the degree to which this increase in power output will transfer to athletic performance may depend on whether the mass being moved represents a similar resistance to 30% MVC. Accelerating the leg to kick a football or throwing a baseball represents a much lighter resistance than 30%.

To test the transference of heavy- versus light-load power training to athletic performance, three groups of athletes with previous resistance training experience completed eight weeks of training with heavy- versus light-load jump squats (McBride et al. 2001). The athletes performed sessions of jump squats with either 30% (n = 9) or 80% (n = 10) of their 1RM in the squat or served as controls (n = 7). An agility test, 20 m sprint and jump squats with 30%, 55% and 80% of their 1RM were performed before and after training. The 30% load training group attained significant increases in power output and peak velocity during the jump squats at all loads tested. This group also significantly increased in the 1RM with a trend toward improved 20 m sprint times. In contrast, the 80% load training group significantly increased both peak force and peak power, but only in jump squats performed at loads of 55% and 80%, and significantly increased in the 1RM, but ran significantly slower in the 20 m sprint. In terms of muscle activation, the group training with 30% load increased EMG activity but only for the jump squat tested at the 30% load. The group training with the 80% load significantly increased EMG activity, but only for jump squats performed with the 80% load. These results suggest high task specificity in terms of both load and velocity. This investigation indicates that training with light-load jump squats results in increased movement velocity capabilities and that velocity-specific changes in muscle activity may play a key role in this adaptation.

Further research is required; however, it may be prudent to continuously adjust the resistance used in training to ensure increased power output at both slow and fast movement speeds. Also, the presentation of a range of loads and resulting movement speeds may be more effective in that the changing stimulus to the neuromuscular system may elicit greater adaptations. This concept remains speculative, however, and warrants further research.

It may be prudent to continuously adjust the resistance used in training to ensure increased power output at both slow and fast movement speeds. Also, the presentation of a range of loads and resulting movement speeds may be more effective in that the changing stimulus to the neuromuscular system may elicit greater adaptations.

Impact Forces During Training for Maximal Power Performance

Although ballistic resistance training is effective for improving power performance, it does present the problem of the high forces exerted on athletes when they land from the jump or catch the falling weight (Newton and Wilson 1993). However, weight training equipment can be adapted to reduce the forces experienced during the landing phase (Newton and Wilson 1993; Humphries, Newton, and Wilson 1995).

In one study (Humphries, Newton, and Wilson 1995), 20 subjects performed a series of loaded jumps for maximal height, with and without a brake mechanism designed to reduce impact force during landing. Vertical ground reaction force data were collected for each jump condition. Peak impact force, impact impulse and maximal concentric force were then calculated. The brake served to significantly reduce peak impact force by 61% and impact impulse by 67% (figure 9.5). No significant differences were found for peak concentric force production. It was concluded that the braking mechanism significantly reduced ground impact forces without impeding concentric force production. The reduction in eccentric loading using the braking mechanism may reduce the incidence of injury associated with landings from exercises designed to increase maximal power.

Periodization, Ballistic Resistance Training and Injury

It may be advisable for ballistic resistance training to progress gradually from unloaded to loaded conditions, with the athlete having completed a prior strength training program. This is based on the recommendation in a position paper from the U.S. National Strength and Conditioning Association that prior to the performance of high-intensity plyometric training, an athlete should be capable of squatting at least 1.5 times body weight (Wilson, Murphy, and Walshe 1997). However, there is no research evidence at present to support or refute this recommendation.

It appears that there is a constant proportion of collagen and other noncontractile tissue, regardless of

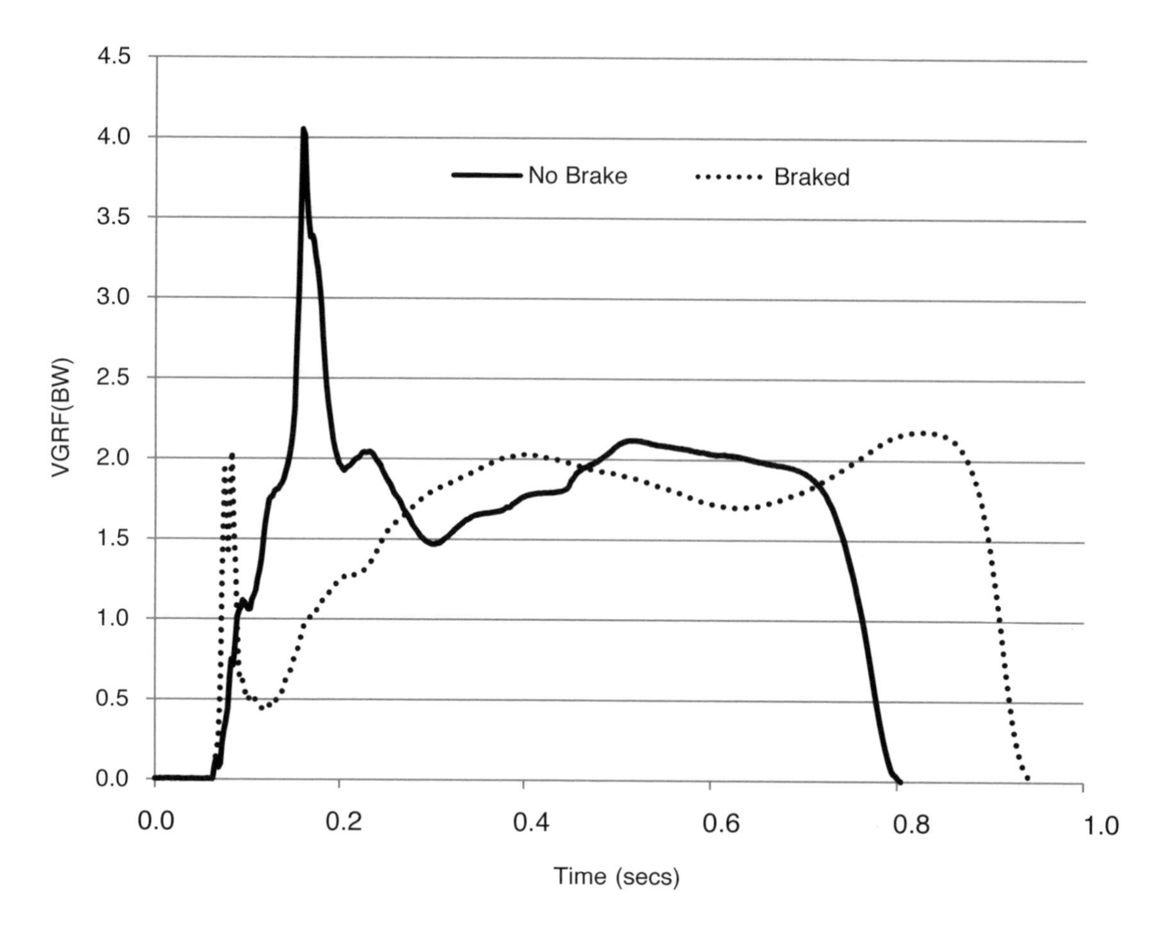

Figure 9.5 Vertrical ground reaction force expressed in body weights during a landing and subsequent vertical jump. Examples for the "non-braked" and "braked" jump conditions are shown. Initial impact spike of landing is reduced from approximately 4 times body weight to 2 times body weight in this example.

the muscle size or state of training. Thus, the absolute amount of connective tissue is considerably greater as a result of heavy resistance training compared with pretraining levels as the training-induced hypertrophy of muscle fibres is accompanied by a proportional increase in connective tissue (MacDougall 1986). Within the muscle cell, endo- and exosarcomeric cytoskeletal proteins create series and parallel connections between contractile proteins, resulting in a meshwork across which force can be transmitted in practically any direction (with respect to the fibre axis) onto the connective tissue matrix and thus to the tendons (Patel and Lieber 1997). It could be postulated that adaptations in the cytoskeleton would have a protective effect in terms of injury from subsequent plyometric or ballistic resistance training. Prudence suggests that a preparatory phase for development of basic strength levels may be necessary before progression to ballistic training techniques, as has been similarly recommended for plyometric training (Chu 1992).

The reasons for completing a prior strength training program are twofold. First, strength training with heavy loads, and thus slow movement velocity, uses controlled movements to overload the musculoskeletal system. Structural changes may result that increase the strength of these structures and thus prepare the body for the higher forces that may be exerted during ballistic resistance training. This may have a protective role in terms of injury prevention, although this idea has not been positively demonstrated in the research literature. Second, several theories of periodization (Bompa and Fox 1990; Mateyev 1972; Medvedyev 1988) state that it is desirable to increase muscle size and maximal strength prior to entering a training phase directed at improving maximal power, as the final increases in power performance will be greater. As there is a strong relationship between strength and power, it has been suggested that one cannot have a high degree of power without first being relatively strong (Wilson, Murphy, and Walshe 1997). Once again, however, due to the difficulty of executing such long-term training studies, no scientific longitudinal training research has been published to confirm this theory. However, a short-term (eight-week) training study has shown that the performance gains from a plyometric training program are unrelated to the subject's initial strength level (Wilson, Murphy, and Walshe 1997).

The Window of Adaptation

Several studies have compared the effectiveness of plyometric training, resistance training and a combination of the two. Although specific training protocols vary, in general plyometric training alone has been shown to be effective for increasing power performance in both trained and untrained individuals (Adams et al. 1992; Clutch et al. 1983; Di Brezzo, Fort, and Diana 1988; Duke and BenEliyahu 1992; Holtz, Divine, and McFarland 1988; Schmidtbleicher, Gollhofer, and Frick 1988; Wilson et al. 1993). Traditional resistance training has resulted in increases in power output in the majority of the research (Adams et al. 1992; Bauer, Thayer, and Baras 1990; O'Shea and O'Shea 1989; Williams 1991; Wilson et al. 1993; Young and Bilby 1993), with a limited number of papers indicating no change in already strength-trained subjects (Häkkinen and Komi, "Effect" 1985; Komi et al. 1982).

When resistance training is combined with plyometrics, power output can be increased (Bauer, Thayer, and Baras 1990; Blakey and Southard 1987; Clutch et al. 1983), and this is perhaps a greater stimulus to maximal power production than either weight or plyometric training alone (Adams et al. 1992). These findings highlight the multifaceted nature of power performance, with a mixed training methods approach being most effective in that it develops more components of muscle power. Furthermore, it has been demonstrated that as an athlete develops one component to a high level (e.g., strength), the potential for that component to contribute to further increases in power output diminishes (Häkkinen and Komi, "Effect" 1985). Thus, each component can be thought of as having a *window of adaptation* to the larger window of adaptation in maximal power production. This concept is summarized in figure 9.6.

The window of adaptation refers to the magnitude of potential for training adaptation. The principle of *diminishing returns* relates to the observation that weight training is less effective at enhancing athletic performance as the strength level increases (Wilson, Murphy, and Walshe 1997). In a similar manner, an athlete who has undergone a program of plyometric training will exhibit a shrinking window of adaptation to this form of stimulus. As this window shrinks, training time will be more efficiently spent on other training methods such as heavy resistance weightlifting or skills training. Furthermore, training must be targeted to increase performance in those components in which the athlete is weakest, because herein lies the largest window for

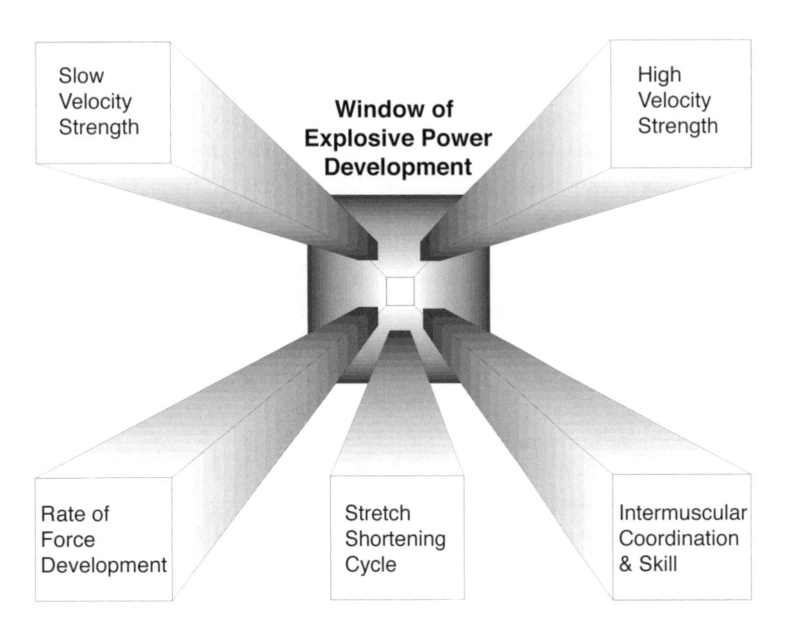

Figure 9.6 Schematic diagram representing the components contributing to maximal power production. Each component can contribute to the overall window for power adaptation. The greater the development of a single component, the smaller that component's window of adaptation and thus potential to develop muscle power. Thus, it may be more efficient to design the training program to target those components with the greatest windows of adaptation, that is, the components in which the athlete is weak (Newton and Kraemer 1994).

Adapted from R.U. Newton and W.J. Kraemer, 1994, "Developing explosive muscular power: implications for a mixed methods training strategy," *Strength and Conditioning Journal* 16(5): 20-31.

adaptation and thus the greatest potential increase in maximal power production.

The higher the level of the athlete's ability and depth of training background, the greater the challenge to produce training adaptations in maximal power activities such as vertical jump. Often novel methods need to be utilized. For example, the effects of ballistic resistance training have been examined in elite jump athletes (volleyball players) (Newton, Kraemer, and Häkkinen 1999). Sixteen male volleyball players from a National Collegiate Athletic Association Division I team participated in the study. All players had a considerable background in traditional resistance training and plyometric training. Standing vertical jump and reach (SJR) and jump and reach from a three-step approach (AJR) were measured as well as force, velocity and power output during various loaded and unloaded jumps. Following pretesting, the subjects were randomised to control or treatment groups. All subjects completed the usual preseason volleyball on-court training combined with a weight resistance program that included knee flexion and extension exercises but no squats or leg press exercises. In addition, the treatment group completed eight weeks of loaded vertical jump training twice each week using loads of 30%, 60% and 80% of their previously determined 1RM squat, while the control group completed squats and leg press exercises. During each session the treatment group completed six sets of six repetitions.

An electronic braking mechanism (Humphries, Newton, and Wilson 1995) was used to reduce the load on the subject during the down phase of each jump. Both groups were retested at the completion of the training period. The treatment group produced a significant increase in SJR and AJR of 5.9 ± 3.1% and 6.3 ± 5.1%, respectively, whereas the control group did not change. Analysis of the data from the other jump tests suggested that increased overall force output and increased RFD were the main contributors to the improved jump height. These results lend support to the effectiveness of ballistic resistance training for improving maximal power production in elite athletes. The specific nature of this form of training has produced improvement in vertical jump beyond that attained after prolonged traditional resistance and plyometric training. It would appear that the window for adaptation of neuromuscular performance specific to ballistic vertical jump training was not well developed in these athletes and thus allowed them to produce further performance improvements.

Assessment of Muscular Power

Evaluating an athlete's muscular power involves assessing neuromuscular performance qualities, diagnosing performance in order to prescribe the appropriate training program and determining key performance characteristics in order to achieve the desired performance goal. Power performance qualities are measured with equipment such as the force plate and transducer systems. Testing focuses on such power qualities as maximum strength, speed-strength, rate of force development and reactive strength. "Gold standard" tests are highly task specific and target several qualities; tests can also be individually designed to assess particular aspects of a sport. Relative and absolute measures of performance components may be more or less valuable depending on the task and the athlete. Performance diagnosis is an important aspect of power assessment that can detect not only an athlete's strengths and weaknesses but also problems with training.

Neuromuscular Performance Qualities

Certain measures represent specific or independent qualities of neuromuscular performance that contribute to maximal power, and these qualities can be assessed and trained independently. Performance diagnosis is the process of determining an athlete's level of development in each of these distinct qualities. When specific performance qualities are targeted with prescribed training, greater efficiency of training effort can be achieved, resulting in enhanced athlete performance. As we have already discussed in the context of the window of adaptation, there are strength, speed-strength, reactive strength, SSC and power endurance dimensions. Because elite athletes tend to be genetically predisposed to their sport and train to enhance their abilities, specificity of these qualities is inherent to a particular sport or athletic event. In other words, each sport or event requires a certain level of these performance qualities to underpin a competitive advantage.

Certain measures represent specific or independent qualities of neuromuscular performance that contribute to maximal power, and these qualities can be assessed and trained independently.

Performance diagnosis is the process of determining an athlete's level of development in each of these distinct qualities. When specific performance qualities are targeted with prescribed training, greater efficiency of training effort can be achieved, which in turn results in enhanced athlete performance.

Performance Diagnosis and Prescription

The implementation of performance diagnosis and prescription to enhance maximal power performance should flow according to a logical sequence (figure 9.7). The initial step requires a determination of the important qualities in the target activity (i.e., a performance needs analysis). A test battery is then established to assess these qualities in an efficient, valid and reliable manner. A training program is developed based on the performance diagnosis, which will improve performance in the target sport. The final aspect is perhaps the most important, as isolated testing has little utility. It is only with frequent, ongoing assessment that a complete profile of the athlete is compiled and manipulation of training variables can be coordinated to advance the athlete toward performance goals. The test-retest cycle (figure 9.7), with frequent adjustments to the training program, is a key feature of the performance diagnosis and prescription process. We address various aspects of this process next.

Determination of Key Performance Characteristics

The first step toward achieving the desired performance goal is to determine the key performance characteristics of the target activity. For example, if the task is to maximize takeoff velocity in the high jump, then those strength and power qualities that influence takeoff velocity need to be determined. This can be achieved through several processes, such as biomechanical evaluation, analysis of high-level athletes and pre-post testing. The best approach may be to combine all three to gain the greatest understanding of the target performance.

- Biomechanical evaluation leads to an understanding of the forces exerted by the jump leg, minimum knee angle and contact times that are observed during the high jump. Speed, range of motion and contraction type of the other body segments should also be determined.
- Analysis of high-level athletes in the sport can provide information on strength and power qualities. It can be assumed, though with caution, that an athlete who is performing well in the sport possesses the necessary levels of these qualities.
- A third method is to test athletes before and after phases involving certain training emphases. If they respond with large improvements in the targeted strength or power quality, then it could be that they are deficient in that quality and that this may require further attention. Certainly, if a component is improving rapidly, it may be prudent to maintain the emphasis until some plateau occurs. However, there are caveats to this approach. First, it is wasteful to continue to seek improvement if the quality is not of significance to the task or if the athlete has an

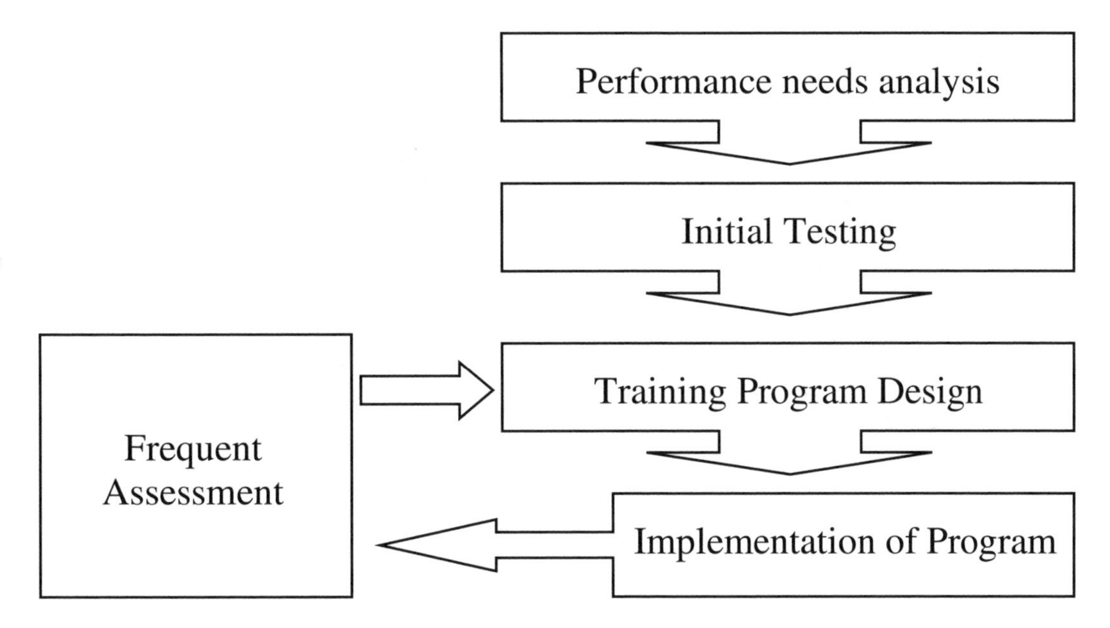

Figure 9.7 Test-retest cycle of strength diagnosis and subsequent program design results in greater training efficiency. An important aspect is ongoing assessment to quantify training progress, gauge effectiveness and detect staleness, overtraining or injury (Newton and Dugan 2002).

Adapted from R.U. Newton and E. Dugan, 2002, "Application of strength diagnosis," *Strength and Conditioning Journal* 24(5): 50-59.

adequate level of the quality such that other qualities may be more limiting. Second, it may be better, for the sake of training variety, to take note of the large response and return to the given quality at a later phase of training.

Equipment for Measuring Power Performance Qualities

Equipment commonly used to measure power performance characteristics includes the force plate, which directly measures the force an athlete produces, and the linear displacement transducer, which can provide highly accurate measurements of displacement. The data obtained with these devices can be used to derive a large variety of other performance measures.

Force Plate

One of the most versatile items for assessing power performance is the force plate (figure 9.8). This is a platform containing force transducers that directly measure the force produced by the athlete. Force needs to be measured in only one direction for most tests and so a uniaxial plate will suffice. Triaxial force plates are more versatile, as vertical and horizontal components of force output can be measured, but they are also much more expensive and often not as portable.

As the athlete pushes against a force plate during the leg press or squat, force–time data can be obtained. Alternatively, if the equipment is fixed so that no movement can occur, isometric measures of peak force and RFD can be gathered. During dynamic movements, it is possible to obtain measures of dynamic strength such as the highest force produced.

Vertical jumps performed on a force plate provide a rich array of performance data. If the athlete is isolated on the plate (i.e., does not touch any other surface), then the impulse–momentum relationship can be used to derive velocity–time and displacement–time datasets from the force–time recording. Combining force and velocity data allows instantaneous power measurements to be derived throughout the jump, and summary variables such as peak and mean power can also be calculated (figure 9.8).

Linear Displacement Transducer

Transducer systems (figure 9.9) can provide very accurate displacement–time data at sampling rates of 500 Hz or more. These data can be used to derive velocity–time, acceleration–time, force–time (if mass is known) and power–time datasets. The transducers have an extendable cable that can be attached to an athlete or an implement such as a barbell. As the person or object moves, the displacement is measured and recorded with a computer

Figure 9.8 Athlete performing loaded jump squats on a force platform. Measurement of force–time data allows calculation of a range of indicative performance measures such as peak force, average force, flight and contact times, takeoff velocity and power output.

Figure 9.9 Linear position transducer provides accurate measurement of displacement and velocity of the athlete or barbell. Commercial computer software can be used to calculate force and power through the movement as well as various summary performance variables.

system. Such systems are particularly useful for measuring performance during jumping with a barbell, upper body movements such as bench throws and the weightlifting movements. For the latter, attaching the transducer to the barbell provides information on velocity, force and power applied to the barbell during weightlifting movements (e.g., hang clean, high pull and snatch), which correspond closely with the athlete's power capacity.

Choice of the system to use comes down to budget and setting. Some movements are not amenable to measurement by position transducer or force plate. The best system is a combination of both direct force and displacement measurement, because this provides the most accurate data. Examples of key performance measures that can be derived from such systems are provided in figure 9.10.

Testing for Specific Power Qualities

Tests, like training, must be specific for the strength or power quality in question. Although opinions within the literature differ, six broad strength and power qualities can be assessed as specific and independent aspects of neuromuscular performance. In sports that involve repeated maximal efforts such as sprinting or swimming, a seventh quality termed power endurance should be included. However, this chapter focuses on maximal mechanical power outputs (i.e., performances requiring only a few consecutive maximal efforts):

- **Maximum strength**—highest force capability of the neuromuscular system produced during slow eccentric, concentric or isometric contractions
- **High-load speed-strength**—highest force capability of the neuromuscular system produced during dynamic eccentric and concentric actions under a relatively heavy load (>30% of max) and performed as rapidly as possible
- **Low-load speed-strength**—highest force capability of the neuromuscular system produced during dynamic eccentric and concentric actions under a relatively light load (<30% of max) and performed as rapidly as possible
- **Rate of force development**—the rate at which the neuromuscular system is able to develop force, measured by calculating the slope of the force–time curve on the rise to maximum force of the action
- **Reactive strength**—ability of the neuromuscular system to tolerate a relatively high stretch load and change movement from rapid eccentric to rapid concentric contraction
- **Skill performance**—ability of the motor control system to coordinate the muscle contraction sequences to make greatest use of the other five strength and power qualities such that the total movement best achieves the desired outcome

Maximum Strength

Maximum strength can be determined using several methods (see especially chapter 8). Many prefer isometric testing because it is not confounded by issues of movement velocity and changing joint angle; however, others believe that isometric testing lacks specificity to the dynamic movements predominant in sport (Murphy and Wilson, "Correlations" 1996). Dynamic tests such as the 1RM test require minimal equipment, but can increase the risk of injury if good technique is not enforced; and, in particular, the testing personnel must be well trained. Alternatively, lighter weights can be used for a 3- to 10RM test and regression formulas used to estimate 1RM strength (Mayhew et al. 1995).

Speed-Strength

One way to assess an athlete's lower body speed-strength is through the use of jump squats. These can be performed with a free weight or in a Smith machine to limit the movement to the vertical plane. For free weight testing, the jump should be performed within a power rack with bottom stops in case the athlete falls (see figure 9.11). In most cases, the athlete can perform a single jump, dipping down and then jumping upward for maximal height.

Sophisticated systems are available to measure force output and bar kinematics (displacement and velocity with time) as explained earlier in this chapter. From these systems, data such as height, velocity, force and power output can be calculated

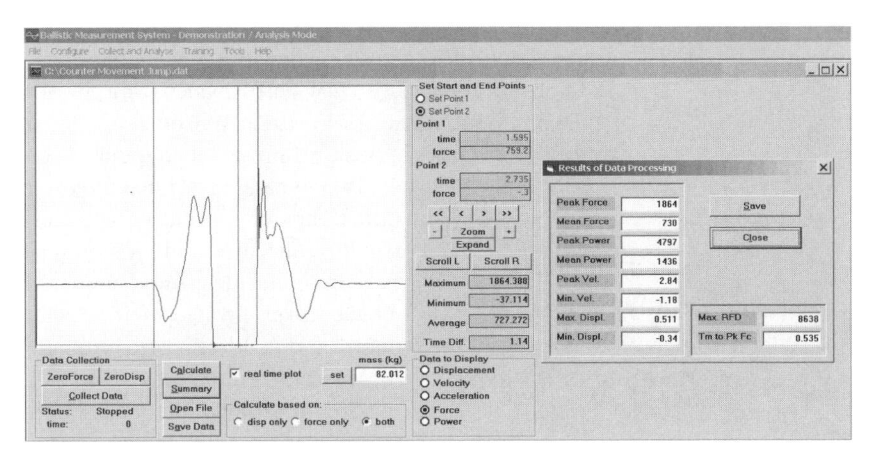

Figure 9.10 Sample data from performance analysis systems that incorporate displacement or force measurement technology or both.

Figure 9.11 Purpose-built power cage for training and assessment of upper and lower body power. Bottom stops reduce injury risk if the athlete falls or misses the barbell. A braking mechanism can be used to reduce landing impact or eccentric loading. A force plate and linear transducer are used to record performance characteristics for feedback to the athlete and performance diagnosis.

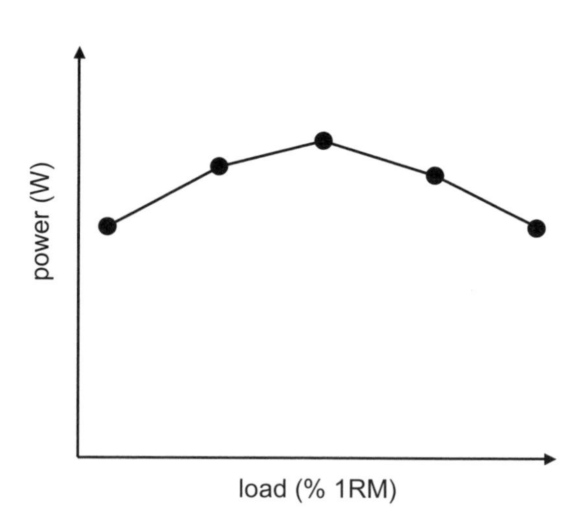

Figure 9.13 When power output is plotted against load, the optimal load is the load that elicits the highest power output.

for the jump squat. The loads used depend on the athlete and task, but a spectrum is useful so that an impression of the athlete's performance under heavy and light loads can be gained (i.e., high-load or low-load speed-strength). One scheme is to use loads of 30%, 55% and 80% of 1RM (McBride et al. 2001); another method is aimed at determining the *optimal load* for power production (Baker, Nance, and Moore, "Bench Press Throws" 2001, "Jump Squats" 2001; Wilson et al. 1993).

To determine the optimal load for the jump squat, the athlete first performs the movement with a preselected load (figure 9.12). After each trial the load is adjusted up or down; and, because of the relationship between force and velocity, the power output will change. Only one load will produce the highest power output (Wilson et al. 1993); this is termed the optimal load (figure 9.13). This may be expressed as an absolute force (or mass) or as a percentage of maximum isometric force or 1RM. Recent research has suggested that the load at which mechanical power is maximized shifts in response to training demands (Baker 2001). Although more research is required, this measure may provide a useful tool for monitoring the effects of changing emphasis in program periodization as well as for detecting overtraining, illness and staleness. For example, Baker (2001) has found that the optimal load increases during phases that emphasize strength and decreases during speed training.

Figure 9.12 Athlete performing loaded jump squat for determination of optimal load for power production.

Although more research is required, measurement of optimal load may provide a useful tool for monitoring the effects of changing emphasis in program periodization, as well as for detecting overtraining, illness and staleness.

A final system to be noted uses absolute loads, and changes in performance are measured as height, velocity or power increases. Use of the same absolute load may better reflect strength improvements in terms of power increases compared to measuring power output at a given percentage of maximum strength; a relative load. Common loads for bench throw and jump squats are 20 kg (standard Olympic barbell weight [44 lb]), 40 kg (Olympic barbell plus 2 × 10 kg plates [88 lb]) and 60 kg (Olympic barbell plus 2 × 20 kg plates [132 lb]).

Rate of Force Development

There are several options to choose from for testing RFD. One common protocol uses isometric squats, but this incorporates many of the same advantages and disadvantages as testing maximal strength with isometrics (Murphy and Wilson, "Correlations" 1996). Rate of force development can also be determined during both concentric and eccentric phases of dynamic tests, which may have greater relevance to task performance (Murphy and Wilson, "Assessment" 1996, 1997; Wilson et al. 1995); however, this has not been well researched to date.

Two dynamic tests that are often utilized are the concentric-only jump and the concentric-only jump squat. For the concentric-only jump, the athlete squats down to a self-selected depth and holds that position for 3 to 4 s, then attempts to jump for maximum height without a preparatory movement. This can be difficult to perform and may require several trials to yield accurate data.

Concentric-only jump squats are performed in a similar fashion as standard jump squats. For this variation, mechanical stops are positioned in the squat rack or Smith machine at the appropriate angle for the bottom position of the jump, and this becomes the starting position for the jump squat. Ground reaction force and bar displacement can be recorded, as well as such derived variables as jump height, power output and peak force developed. The highest force produced during a concentric-only movement has been termed maximal dynamic strength (MDS) (Young 1995)—a strength and power quality with good predictive and discriminatory capability between athletes of different levels. Heavier external loads will result in greater MDS values, and greater test specificity will be obtained if the selected load is similar to the target task. Dynamic concentric-only tests (as with isometric tests) allow calculation of several measures of the ability to rapidly develop force, such as maximum RFD or the impulse (F × t) over the initial 100 ms. The interested reader may consult the extensive discussion provided in other texts on the subject (see Zatsiorsky 1995).

Reactive Strength

The most common test for assessing an athlete's reactive strength is the depth jump. The athlete drops down from a box, lands and then jumps upward for maximum height (figure 9.14). A contact mat system or force plate can be used to record the characteristics of the performance. It has been reported that the instructions given to the athlete affect the results and that athletes should attempt maximum jump height while minimizing ground contact time (Young, Pryor, and Wilson 1995). Trials completed at increasing drop heights provide insight into how the athlete responds to increasing stretch loads. A common progression is to employ 0.30, 0.45, 0.60 and 0.75 m (0.3, 0.5, 0.65 and 0.8 yd) drop heights. Calculated variables include jump height, flight time, contact time and flight time divided by contact time (which is also termed *reactive strength index* or RSI). The "best" drop height can be determined as the one that elicits the highest RSI (figure 9.15). An athlete with a reasonable level of reactive strength should be able to produce a better jump height following a drop compared to a countermovement jump (which effectively has a drop height of zero).

Figure 9.14 Athlete performing a drop jump onto a force plate with linear transducer attached to track displacement.

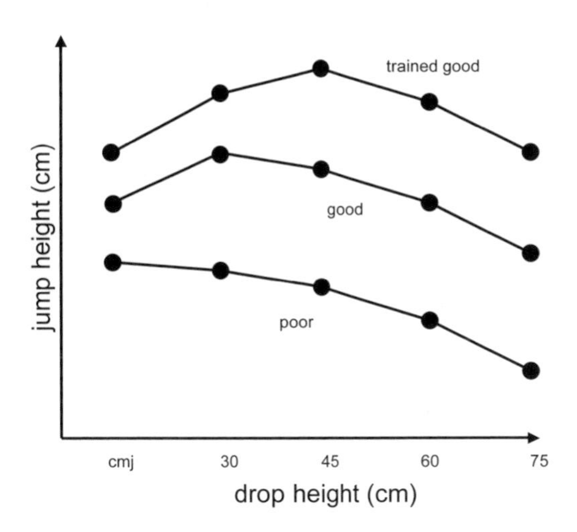

Figure 9.15 Plot of reactive strength index against drop height. The best score corresponds to the optimal drop height.

"Gold Standard" and Specific Tests of Skill Performance

It is useful to include a test that is highly task specific and that incorporates several strength and power qualities. We may refer to this as the "gold standard" for the target task. For example, in volleyball, the approach jump and reach is commonly used as a "gold standard" test (Newton, Kraemer, and Häkkinen 1999). In basketball, the athlete could use an approach run onto a contact mat and then jump for maximal height while performing a jump shot action, landing back on the contact mat. In athletics, the actual field event performance (e.g., long jump or shot put distance) can be used.

Sometimes it is necessary to design highly specific tests that assess particular aspects of a sport. For example, power output produced under fatigue or following repeated impacts (or both) is important to the sport of rugby. In this instance, an obstacle course could be developed simulating a game, with outcome measures including time to complete certain sections, as well as performance in a power test such as a vertical jump.

The countermovement jump is a basic test of vertical jump capacity (figure 9.16). The athlete dips down and then immediately jumps upward, attempting to maximize the height jumped. With the use of a force plate and position transducer, ground reaction force and displacement can be recorded (McBride et al. 1999). From these data,

Figure 9.16 The countermovement vertical jump is a very common test of leg extensor power.

a number of variables that characterize the performance can be calculated. Jump height is an obvious measure, but the power output during the jump, peak force produced and time to attain that force can also be determined.

Relative or Absolute Measures

Whether the results are expressed as absolute measures or normalized relative to body weight depends on the task and athlete. Both methods have application. When the athlete must move body weight against gravity (e.g., high jump), then relative measures may be more important. However, when momentum or total strength is key (e.g., rugby, American football), then absolute measures may be more instructive. Relative measures allow for better comparison of athletes of different body weights, and most variables can be expressed in relative terms. For example, a 1RM squat can be expressed as the number of "body weights" (BW) lifted, or power output during jumping can be expressed as watts per kilogram body mass.

What to Look For

Apart from assessing an athlete's relative strengths and weaknesses (see also chapter 2), performance diagnosis can also detect problems with the athlete, the athlete's training program or both. For example, if there is a lack of progression of a target quality, then either the athlete has reached his or her genetic potential for that attribute, or the training is inappropriate or the athlete has adapted to the current program and requires variation to stimulate further gains.

Apart from assessing an athlete's relative strengths and weaknesses, performance diagnosis can also detect problems with the athlete, the athlete's training program or both.

If a decline in a strength and power quality is detected, this may simply be a result of changing emphasis in the training. However, if it was deemed important to develop this attribute in an earlier phase, then a significant decline should be avoided. For example, if a certain level of reactive strength was attained preseason and a 15% decline in-season was seen, some modification to the training regimen may stem the loss. Furthermore, decline of a strength and power quality may indicate tiredness, overtraining, staleness or injury, so these causes should also be investigated.

Assessing Imbalances

In the interests of both injury reduction and performance enhancement, it is instructive to investigate imbalances between agonist and antagonist muscle groups as well as between left and right sides of the body. There is considerable literature (Aagaard et al. 1995, 1998) on the former, so we will confine our discussion to assessing imbalances between the left and right leg extensors. Most people exhibit some dominance resulting in differences in performance between, for example, hops performed on the left versus the right leg. Differences of more than 15% may indicate existing pain and injury, inadequate recovery from previous injury or an undesirable imbalance in muscle strength and power qualities. It is easy to assess such differences by having the athlete perform unilateral movements such as single-leg hops and comparing flight and contact times. Cutting and sidestepping tests using timing lights or contact mats to measure speed in each direction are also useful.

One should take the specific sporting movements into consideration when assessing left- to right-side imbalances. In sports that require a one-legged takeoff, for example the high jump or long jump, it may be quite normal for the dominant or takeoff leg to be stronger than the contralateral leg. This would be expected due to the nature of the event, so time spent trying to eliminate imbalances in leg power might be counterproductive.

Training Methods for the Development of Power

Methods for developing power in athletes include heavy resistance training, in accordance with the power requirements of the sport; ballistic training, which should constitute a considerable proportion of the training volume; plyometrics; and weightlifting. Strength training will translate to gains in power performance, but probably not immediately. Tapering and recovery are important aspects of a training program that should vary according to the performance demands of the sport.

To be powerful, the athlete must also be very strong in the movements requiring high power expression. This relationship becomes more direct as the power requirements of the specific sport gain greater emphasis.

Heavy Resistance Training

To be powerful, the athlete must also be very strong in the movements requiring high power expression. This relationship becomes more direct as the power requirements of the specific sport gain greater emphasis. As a general rule, the athlete should be able to generate three times the muscle tension required during the actual sporting performance. This corresponds to the observation that the neuromuscular system can generate greatest power output when it is working against a load of approximately 30% of maximal strength. For example, in a sport for which high vertical jump is crucial (e.g., volleyball or basketball), the athlete should be strong enough to squat with a load of 2BW on the barbell. That is, when athletes are jumping with BW alone, they are working around a load that is 30% of their maximum strength. In sports like triple jump, long jump and sprinting, in which the driving action is off a single leg, the leg extensor strength should be even higher. However, keep in mind that a single-leg press or squat of 3BW is not realistic as no athlete could ever achieve this. However, such a theoretical analysis does indicate the need for very high strength development in athletes who require very high maximal power. Detailed information on assessing and developing strength is provided in chapter 8.

Ballistic Resistance Training

The problems of traditional heavy resistance training and power development have already been discussed. We have suggested that ballistic training, in which the load is propelled against gravity or thrown, provides much better performance improvement than other forms of training. When athletes are training for power, ballistic training should compose a considerable proportion of the overall training volume—this reflects the *specificity principle.*

The issue of different load ranges has also been discussed, so we will now concentrate on specific training movements that will then be revisited in chapter 13. The most effective exercise for developing leg extensor power is the vertical jump. The overload can be applied using a barbell held in the back squat position or dumbbells held in the hands. Weighted vests are also effective, but with these it can be difficult to apply the required resistances when targeting 60% to 90% 1RM.

The loads used should be varied between 30% and 90% of 1RM. It may be advantageous to perform a heavy set (80% 1RM) immediately before a lighter set (30% 1RM) because of the postactivation facilitation effect. Another method is to perform light and heavy days rather than to mix loads within a session. This scheme is particularly useful for athletes in team sports when lighter days are required early in the week following weekend competition.

Ballistic training for the upper body can be performed with medicine balls or power bags. If a greater range of load is required, then movements that mimic the target sport can be performed with barbells. A Smith machine uses vertical sliders to limit movement to a single plane. In this case the bar can be loaded to the desired level and then various types of throws performed. Bench throws are particularly effective for a boxer or football player who requires high force to punch or fend off an opponent. These are performed like a normal bench press, but the bar is thrown and then caught. For the shot put, use an incline bench and adjust the angle so that the throw is in the same plane as the athlete's put. Pulling movements can also be performed ballistically, with the athlete lying on a bench and pulling the bar rapidly upward. A rubber stopper or foam dampener is required to soften the impact with the bench.

Specifically designed equipment is available for ballistic resistance training. Due to considerations of specificity, free weight bars are being used more frequently with a special "power cage" that provides bottom stops in case of a missed catch or bungled jump. The equipment may also incorporate braking systems to limit impact forces; the system shown in figure 9.11 is an example.

Plyometrics

Plyometric training involves the performance of principally lower body, but some upper body, exercises, which really emphasize the SSC. Body weight alone is usually used for lower body plyometrics. Exercises involve either bilateral or unilateral landings, and a vast array of bounding, hopping, jumping and striding movements can be incorporated (Chu 1992). For the upper body, exercises such as clap push-ups and drop push-ups are used. The goal is to provide a very rapid and powerful contact phase, with the specific goal of "tuning down" the muscle inhibition reflexes. In other words, plyometric training aims to expose the neuromuscular system to high stretch loads in an effort to desensitise the Golgi tendon organs (see chapter 12). The reader is referred to the many books dedicated to plyometric training for detailed explanations of exercises and how to incorporate these into an athlete's overall training program.

Weightlifting

The sport of weightlifting involves two lifts, the *snatch* and the *clean and jerk.* Elite athletes in this sport exhibit very high power output in vertical jump, broad jump and sprinting acceleration. It is thought that the training they perform, in combination with the genetic predisposition to being very strong and powerful, results in their extraordinary power capacity. For this reason, and also because many weightlifting coaches become strength and conditioning coaches, weightlifting training techniques have become very popular for athletes in all sports requiring power development. At this time there are few research papers that provide strong evidence for training transfer to other sport performance, but the increasing use of such methods is pervasive.

The snatch and the clean and jerk are very complex movements that require many years of training to perform well. However, there are many other preliminary techniques used as training movements for these lifts, including the hang pull, hang snatch, high pull, push press and hang clean (see also chapter 13).

Translation of Strength Gains to Power Performance

We have already discussed the concept that increasing muscle strength does not necessarily translate immediately to increased power output. The athlete must be given time to "practice" with the adapted muscle strength. This occurs somewhat automatically as part of the periodization of training, which is discussed fully in chapter 13. Translation can also occur through the performance of "complex training". Here, the athlete performs heavy resistance training and then, immediately on completing the set, attempts a very sport-specific set of exercises. For example, a set of heavy back squats could be followed by a set of vertical jumps, or a set of heavy split squats could be followed by a 40 m sprint run. The most important points to remember are that strength training may not immediately translate into increased power performance and that the athlete must be given power training exercises that are very close to the sport movements to assist in this translation.

Tapering and Recovery to Optimise Power Performance

As we have discussed, a wide range of neural and muscular factors must combine optimally to produce maximal power output. Certain training modes affect these components in a negative manner. For example, heavy resistance training alters muscle pennation architecture in the opposite direction desired for power production; so, in preparation for power- and speed-oriented events, heavy resistance training must be tapered up to four weeks prior to competition. This is not to say that training is stopped altogether, because that would result in strength decrement and thus power loss. The volume of heavy resistance training must be reduced markedly to perhaps one to three sets performed once per week leading up to competition. This strategy is quite variable depending on the performance demands of the sport. If high strength and power is required as in rugby, then heavy resistance training must continue right through the preseason and in-season periods. In fact it is desirable that "personal bests" in strength be set toward the end of the competitive season in rugby and other collision-combative sports, because this is the period when the hardest and most important games occur.

Another mode of training that can really be detrimental to power performance is endurance exercise. Long bouts of low-intensity, predominantly aerobic exercise have a considerable impact on both strength and power (see also chapter 8). The cause is probably persisting neural fatigue as well as the fact that aerobic training moves the muscle fibre and contractile protein expression away from what is optimal for strength and power. For sports that are entirely strength and power dominated, the athletes never perform aerobic exercise. However, in sports that require some combination of strength, speed, power and endurance, the periodization and tapering process must be managed very carefully. Team field sports such as Australian football, soccer and hockey are good examples of such sports. Tapering for competition requires little or no continuous endurance training and a shift toward power training and interval-type training that will have less of an effect on strength and power.

Summary

The mechanisms that contribute to maximal power production are many and relatively complex. The exquisite movements that make sport so exciting, such as the slam dunk or the blistering burst of speed in football, require very high power outputs and the optimisation of a wide range of neural, muscular, mechanical and skill components. For this reason the development of power performance has been a controversial topic in the research literature. What is known is that a "mixed methods" approach is best, in which the various factors contributing to the target performance are determined and assessed and then specific training phases are implemented with frequent follow-up testing. While maximal strength is very important, development of this component alone will decline in efficiency as the training age of the athlete increases. More sophisticated training methods incorporating heavy- and light-load ballistic training, plyometrics and even unloaded or overspeed techniques will be of benefit. Perhaps the two key concepts of this chapter are specificity, in terms of matching the target activity in velocity, range and type of movement; and variation of loading, in terms of resistance, volume and intensity, to continue to elicit gains in power performance.

ten

Speed

John Cronin, PhD; and Anthony J. Blazevich, PhD

In sport, the term "speed" can be used in many contexts, referring to the average speed to finish a marathon, the speed through the circle for a discus thrower, the maximum takeoff speed of a triple jumper or the speed of a serve or pitch. "Speed" in mechanical terms refers to the time taken to complete a linear or angular distance, which is the result of reaction speed (reaction time, RT) and movement speed.

Reaction time is the time it takes from the detection of a stimulus to the first movement. In the sprint events, the RT of elite athletes is ~130 to 140 ms (Mero, Komi, and Gregor 1992). In terms of a 10 s, 100 m run, this represents only 1.3% to 1.4% of the race and would seem relatively unimportant to train. However, sequential placing in the 100 m can differ by as little as 1 to 10 ms. Importantly, if one athlete's RT is shorter than that of a second athlete who is accelerating at an identical rate, the first athlete will always be at a more advanced state of acceleration (i.e., as they both accelerate, the athlete with the shorter RT will always be at a higher speed and therefore the distance between the athletes will increase). Thus, the distance between the athletes is more than the distance covered by the athlete with the quicker RT before the second athlete reacts. This increase in distance is important for many sports. If all other factors are equal, then small changes in RT can make a difference to a place on the podium or the outcome of a sporting contest. Given that RT is a conscious, voluntary action and not a reflex, it can be improved through signal recognition and efficient execution of movement patterns, as is evidenced by the quicker RTs of elite compared to novice sprinters. This chapter does not deal with reaction speed or RT, as this capacity fits better into the motor control and learning area; but it is important to point out that it is an essential capacity for many sports and one that coaches must be capable of improving in their athletes.

Movement speed, for the purpose of this chapter, refers to the ability of athletes to move themselves or their limbs at a very rapid rate for relatively few repetitions. Movement speed is often associated with running speed, and there is no doubt that a high level of running speed is essential for many sports. It is with this in mind that many of the examples throughout this chapter are discussed in terms of sprint running. However, it should be noted that specific limb speed or very rapid limb movements are also a part of many other sports, and many of the suggestions provided throughout can be extrapolated to such movements.

An athlete's speed capability was traditionally thought to be largely dependent on genetic factors, with only relatively small improvements resulting from training. However, considering the adaptive potential of the human body and the knowledge that small changes can have substantial influences on an athlete's ranking, sport scientists, conditioners and coaches clearly focus on training athletes to improve their speed potential. Given this information, the present chapter addresses those factors that may influence speed development. The review is for the most part based on empirical evidence, and, where possible, the practical implications of the research are discussed and training methodologies suggested.

Neural Considerations

In order for a muscle group to generate its greatest force at high shortening speeds, two factors need to be considered:

- *Intramuscular coordination,* which refers to the relationship between the excitatory and inhibitory input to the agonist muscle. This involves maximizing motor unit recruitment, firing rates of nerves innervating the muscles and synchronization of motor units as well as improving excitatory reflex feedback (e.g., muscle spindles) while reducing inhibitory reflex feedback (e.g., from Golgi tendon organs, load and pressure receptors).
- *Intermuscular coordination,* which refers to the ability of the agonist, antagonist and synergist muscles to contribute fully to the target activity and involves minimizing antagonist coactivation and maximizing synergistic contribution.

With respect to intramuscular coordination and increasing the activation of agonist muscles, it has been well established that both heavy resistance exercise and explosive or plyometric-type training have significant effects that occur after a single session and continue to improve for many weeks as long as the intensity of training remains high (figure 10.1). Agonist recruitment, as estimated through electromyogram (EMG) analysis, has also been shown to increase along with lifting performance in Olympic weightlifters during periods of high-intensity training, but to decrease when training intensity was lower (Häkkinen et al. 1987). This finding is consistent with the fact that agonist recruitment is higher when individuals attempt to move faster, regardless of the load actually moved (Aagaard et al. 2000). The increased EMG reported in such studies might partly reflect other neural-mediated adaptations such as increase in motor unit synchronization, which has been demonstrated to positively influence the muscle's rate of force development (Semmler 2002), and an increased nerve conduction velocity.

While they remain speculative, there may be other means by which agonist activation might be improved. In terms of force production, selective activation of fast-twitch (FT) motor units has been reported for ballistic (Zehr and Sale 1994) and eccentric contractions (Nardone, Romano, and Schieppati 1989; Nardone and Schieppati 1987), and one study has demonstrated that sprint athletes may be able to selectively activate FT motor units (Saplinskas, Chobotas, and Yashchaninas 1980). These researchers found that sprinters (as compared to untrained subjects and long distance runners) had the highest motor unit

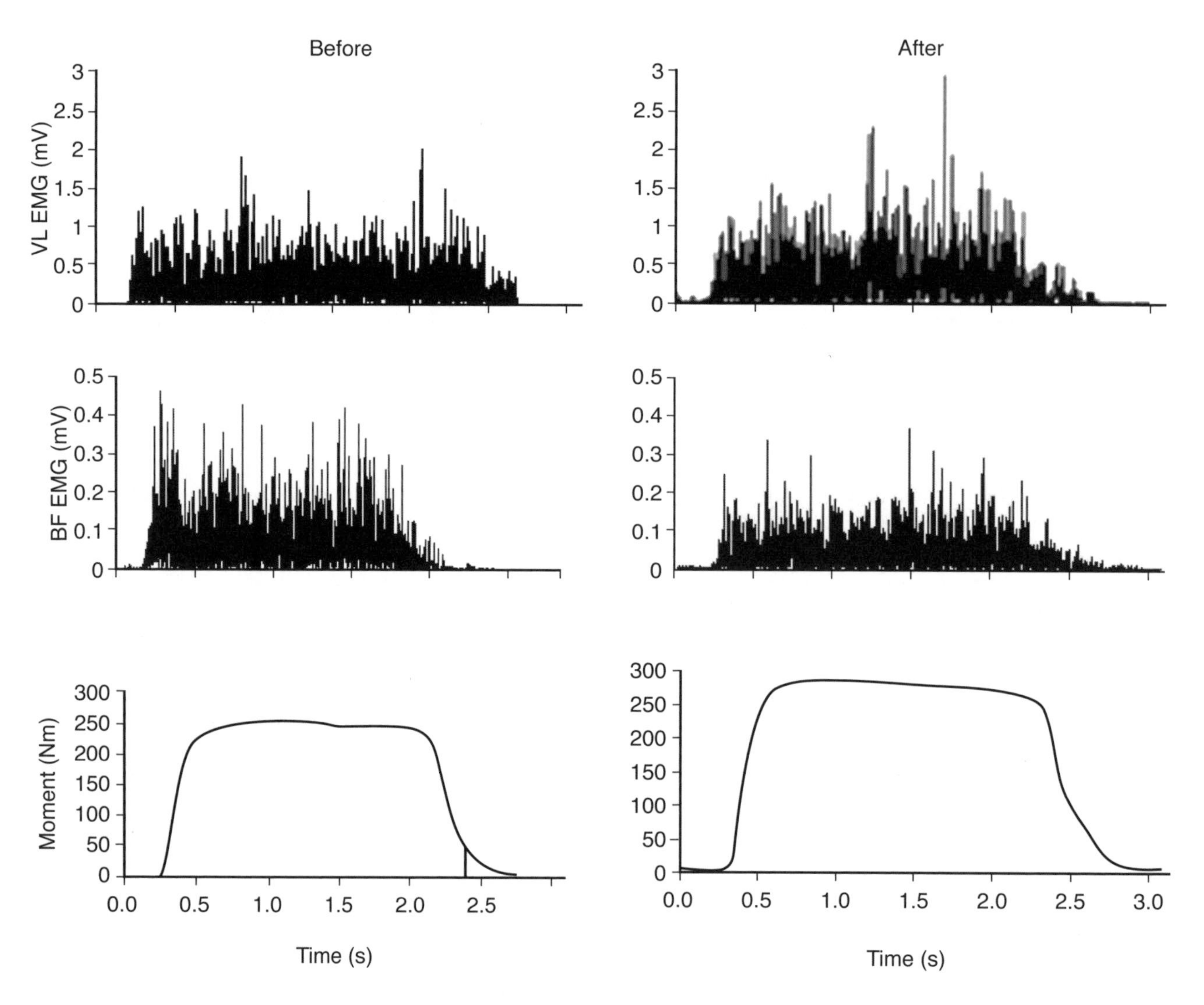

Figure 10.1 Changes in agonist (VL: vastus lateralis) and antagonist (BF: biceps femoris) electromyogram (EMG) during an isometric knee extension contraction performed with maximum rate of force development after high-speed knee extension training. An increase in VL EMG is simultaneous with a decrease in BF EMG, which largely accounts for the increase in knee extensor moment (bottom right graph).

firing rates in the tibialis anterior, although it is not known if this occurs in muscles involved in propulsion. This allows the possibility that such training might induce chronic changes in the ability to recruit FT fibres, but no research has specifically examined this contention.

> Maximise intramuscular coordination training using near maximal loading and plyometric techniques. Maximise intermuscular coordination by using movement or sport-specific and technique training. A mixture of both training methods (combination training) may offer the best option.

Also, *reflex potentiation,* the reflex increase in muscle activity elicited during maximal voluntary contractions, has been shown to increase after strength training (Sale et al. 1983); and a cross-sectional study by Upton and Radford (1975) showed reflex potentiation to be enhanced in sprinters. Whether enhanced reflex potentiation has performance advantages is debatable, but it may improve preactivation prior to foot strike and subsequently aid rapid force production and increase the stiffness of muscle (see later section, "Muscle Stiffness"). Training to increase excitatory input and decrease inhibitory input to the agonist muscle is addressed later in this chapter when muscle stiffness is discussed in more detail.

Finally, *electromyostimulation* (EMS), which involves the application of a high-voltage current to the skin overlying muscles, has been shown to cause a significant activation of FT fibres. This is particularly the case when EMS is applied to the quadriceps muscles. The greater activation of FT fibres is attributable to the current's being applied superficially to the muscle (i.e., over the skin) where the proportion of FT fibres is greater; ST fibres tend to be found in greater number deeper in the muscles. Since EMS results in excitation of the intramuscular branches of the nerves rather than the muscle fibres directly, short-term (i.e., less than four weeks) strength gains mostly result from increases in neural activation (Maffiuletti, Pensini, and Martin 2002). It was recently demonstrated that eight weeks of EMS training applied four times a week resulted in an increased isometric knee extensor force, muscle activation, muscle cross-sectional area and fascicle pennation in the vastus lateralis (Gondin et al. 2005). Shorter periods (four weeks) of EMS applied to the quadriceps and calf muscles have also been shown to improve speed performance in athletes (Maffiuletti et al. 2000), although it has often been observed that the greatest gains occurred two to four weeks after the cessation of EMS training as long as the athlete had continued his or her sport training (Brocherie et al. 2005; Malatesta et al. 2003). Therefore, short-term EMS training is beneficial for increasing muscle activation and speed performance, although longer-duration training causes gross muscular adaptations also.

Reductions in antagonist coactivation are also beneficial in most activities. There is strong evidence that coactivation is reduced after periods of high-speed movement training (figure 10.1); indeed, coactivation seems to be reduced upon learning of most movements, whether they are performed rapidly or not (Schmidt 1987). Similarly, maximizing synergistic contribution involves movement-specific training such as technique or coordination training. Certainly, then, the performance of specific task practice and similar high-speed movements will benefit speed athletes.

To summarize, with regard to maximizing intramuscular coordination, near-maximal (weightlifting loading parameters) and plyometric-explosive contractions would seem the methods of choice. Moving the load as explosively as possible should be the intention of both training methods. Lifting heavy loads may also have the added benefit of reducing inhibitory reflex feedback to the agonist muscle. Plyometric-explosive training (including technique drills) may have the added benefit of increasing reflex potentiation and allow a greater activation of the FT motor units. Eccentric strength and EMS training may also have a place in the training regimen if selective activation of FT motor units is thought desirable. In terms of maximizing intermuscular coordination, it would seem that training should simulate the movement pattern of the target activity as closely as possible (e.g., technique training, specific resisted training) without interfering with the technique itself. Given this information, it appears that some form of combination or contrast training might best achieve the desired neural adaptations for increased movement speed.

Morphological Considerations

High-speed movements are produced after the performance of two sequential actions:

- Rapid muscle shortening, creating a high-speed force whose energy is largely stored in the muscles' elastic tendons
- High-velocity recoil of the tendons after they are stretched, resulting in limb movement (figure 10.2)

Because of its velocity-dependent damping properties, tendon stores more energy when the muscle contracts faster. Therefore, in order to optimise movement speed, it is important to increase both the force and speed of muscle contraction and optimise tendon mechanical properties.

Gross Muscle Structure

Muscles that contain long fascicles (with groups of fibres that typically do not run the length of the fascicle, but interdigitate longitudinally) generate forces over longer

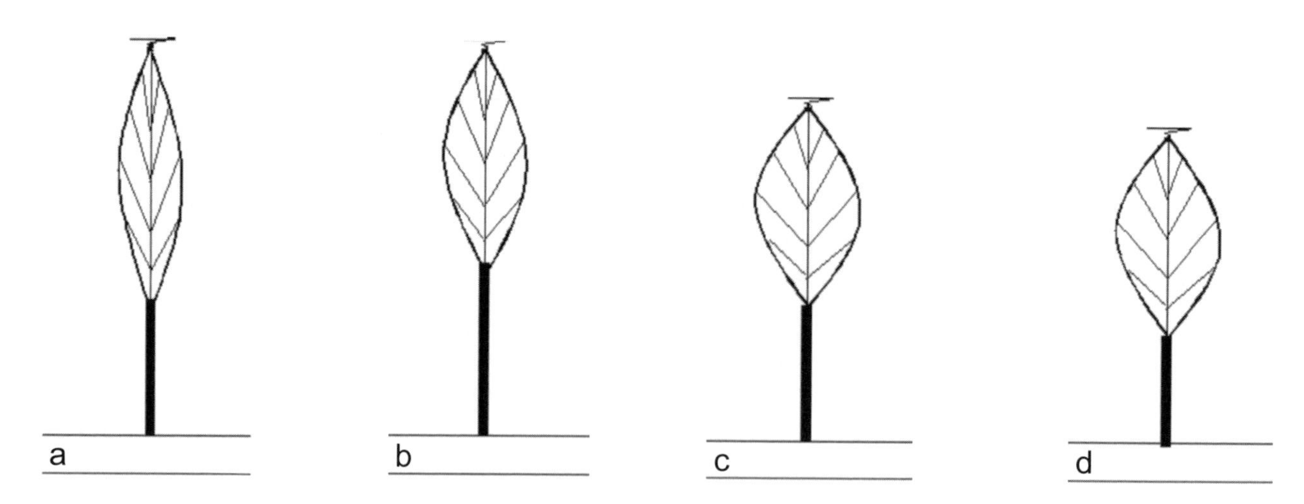

Figure 10.2 The concentric (shortening) phase of rapid movements, regardless of whether the muscle-tendon unit completes a stretch-shorten cycle, occurs as the muscle first shortens while the compliant tendon lengthens and the tendon subsequently shortens rapidly. In the final phase, the muscle is often quasi-isometric, with the majority of the shortening of the muscle-tendon unit taking place in the tendon.

ranges and at higher velocities than those with short fascicles. In contrast, muscles with fascicles that attach at a large angle to the tendon (or aponeurosis), compared to fusiform muscles (chapter 8), usually generate greater peak tension, for two reasons:

- They possess a greater physiologic cross section relative to their muscle volume, with more contractile tissue able to attach to the tendon or aponeurosis.
- The fascicles rotate as they shorten, thereby allowing the tendon to travel a distance that is greater than the length of fascicle shortening. This allows muscle fibres to shorten at lower speeds for a given tendon movement velocity and provides the opportunity for fibres to operate closer to their optimum length for a given tendon movement distance.

Due to space and mass constraints, it is rare for muscles to contain very long fascicles attaching at large angles; thus a trade-off exists such that a muscle's main architectural features suit its main purpose. On a basic level, it would be optimum for muscles that need to operate at high shortening speeds to have longer fascicles, whereas muscles that produce large forces at lower velocities or participate in stretch-shorten contractions would benefit from shorter fascicles attaching at larger angles. This indeed seems to be the case, as it has been shown that elite sprinters, when compared to elite distance runners and controls, have longer fascicle lengths and lesser pennation angles in their functionally important vastus lateralis muscles (Abe, Kumagai, and Brechue 2000).

In terms of training it is important to note that muscle architecture is highly plastic. Chronic heavy weight training (e.g., >14 weeks) has been shown to increase muscle fascicle angle in younger adults (Aagaard et al. 2001; Kawakami et al. 1995), as has eight weeks of EMS training (Gondin et al. 2005). To the best of our knowledge, the elimination of weight training simultaneously with an increase in high-speed training in athletes results in a significant increase in fascicle length (Blazevich et al. 2003). These results suggest that muscle architecture adapts specifically to a training stimulus, with low-speed, high-load training causing a shift in properties more consistent with maximum force development and high-speed training causing a shift consistent with higher speeds of muscle shortening. In both athletes (Blazevich et al. 2003) and previously untrained adults (Seyennes, de Boer, and Narici 2007; Blazevich et al. 2007) significant changes seem to take place within several weeks. In order to optimize muscle architecture for high-speed force production, an athlete should use at least some heavy weight training on muscles whose force generation requirements are based on a high force production over a small range of motion, such as the calf muscles, which often participate in stretch-shorten actions. Alternately, a reduction in heavy weight training and an increase in high-speed training would likely benefit muscles that shorten rapidly over larger length ranges, such as the vastus lateralis of sprint runners (Kumagai et al. 2000).

Fibre Type

A muscle's contractile properties are also strongly dependent on the characteristics of its fibres. Since type IIb fibres, compared to type I fibres, have higher shortening velocities (up to 10 times), have higher maximum power outputs (about two times), produce their peak power at higher loads and velocities and have a greater efficiency of force production at fast shortening, they are ideally suited to high-speed force production. Type IIa fibres are intermediate and are therefore also very useful. Several studies have shown that male and female sprinters have a high percentage of FT (type IIa and IIb) muscle fibres in their leg muscles. Thus, the use of training strategies to increase FT fibre proportion should be central to a program aimed at increasing movement speed.

It is now reasonably clear that the proportion of FT fibres (comprising types IIa and IIb) can change after prolonged (several weeks to months) physical training. While endurance training leads to a general increase in the proportion of type I fibres, albeit with an increase in shortening velocity of those fibres (Widrick et al. 1996), sprint training has often been associated with an increase in FT fibres. Generally, sprint bouts that have resulted in such shifts have lasted less than 30 s with near-complete recovery allowed before a repeat sprint. Such shifts are seen with both sprint running and cycling training, indicating that training mode might not be a significant factor. Explosive weight and plyometrics training also causes a slow-to-fast transition (Häkkinen, Komi, and Alén 1985), while prolonged heavy weight training tends to cause a shift from type IIb to IIa, with little shift between slow and fast (Abernethy et al. 1994).

Interestingly, Paddon-Jones and colleagues (2001) recently showed an increase in type IIb fibre proportion after high-speed, but not low-speed, eccentric strength training of the quadriceps. Also, while periods of inactivity have previously been associated with increases in type IIb fibre proportion with a decrease in force generation, one study showed a shift toward the faster fibre type with an increase in unloaded knee extension velocity and muscle power three months after the cessation of a heavy weight training program (Andersen et al. 2005). Thus, although more research is required to fully explain how to use muscle unloading as a "training aid" for improving unloaded limb shortening velocity, it seems that periods of high-speed eccentric training or strict rest might be of significant benefit to speed athletes aiming to improve their FT fibre proportion.

While increasing the size of slow-twitch (ST) fibres would increase limb mass without an increase in fast force production, an increase in the size of FT fibres would be of great benefit since their force generation capacity is proportional to their size. Heavy weight training increases the size of both FT and ST fibres, although the increase in FT fibres usually precedes, and is of greater magnitude than, the increase in ST fibre size. The increase in fibre size usually occurs after at least a few weeks of training, but seems to be nearly complete within three to four months (Häkkinen, Alén, and Komi 1985); certainly increases are very limited in highly trained bodybuilders (Alway et al. 1992). Sprint and explosive weight and plyometrics training are potent stimulants of FT fibre hypertrophy. Since ST hypertrophy is relatively minor in response to this training, and thus the increase in fast force production would occur with proportionally less increase in total muscle mass than with long periods of heavy weight training, explosive type training should form the basis of programs aiming to increase the size and contractile strength of FT fibres.

Tendon Properties

Tendon recoil is largely responsible for the high movement velocities attained in skills such as throwing and jumping. Therefore, optimization of tendon mechanical properties is an important goal of training. Stiffer tendons are stretched, and therefore able to store elastic energy, only when subjected to heavy loads, although they recoil with greater force and at higher speeds than more compliant tendons when exposed to such loads. Stiff tendons are therefore ideal for storing and releasing energy in movements of small amplitude, at high velocities or under high loads. For example, the Achilles tendon of a sprint runner would be optimized at a higher stiffness than for a marathon runner, as distance running involves lower forces and longer ground contact times. More compliant tendons are ideal for movements of larger amplitudes, at slower velocities or under lighter loads; an example is the Achilles tendon of a spiker in volleyball. It is currently very difficult to know the optimum stiffness for various tendons in the body, and it is also very difficult to test tendon stiffness. So performance testing is important for monitoring the effects of training on tendon stiffness and performance.

While the effects of physical training on tendon stiffness have only recently begun to be investigated in humans, it has been shown that resistance training can increase tendon stiffness (Kubo, Kanehisa, and Fukunaga 2002). It is not known how different forms of strength training affect the tendon; however, a decrease in physical activity (e.g., bed rest) is known to reduce tendon stiffness (Kubo et al. 2004), although muscle force production is also compromised after prolonged rest. While acute (10 min) stretching was shown to reduce tendon stiffness (Kubo et al., "Influence of static stretching" 2001), three weeks of stretching had no long-term effect (Kubo, Kanehisa, and Fukunaga 2002); it is possible that longer periods of stretching are required before tendon stiffness is affected. Indeed, Wilson, Elliott, and Wood (1992) showed that eight weeks of stretching training decreased whole muscle-tendon unit (MTU) stiffness mostly under heavy loads, a condition in which muscle stiffness is almost maximal and tendon stiffness (series elastic component [SEC] stiffness) is the main determinant of whole MTU stiffness. This is strong evidence for a decrease in tendon stiffness, or that of another series elastic element, after prolonged training.

No data are available describing the effects of explosive or plyometric training on human tendons. In summary, it seems that tendon stiffness can be altered with appropriate physical training, but more research is needed before specific training plans can be developed, particularly for athletes who need to reduce the stiffness of their tendons while continuing to train hard.

Muscle Stiffness

Muscle stiffness can be defined as the resistance of a muscle to an increase in length under an applied load. The relationship between muscle stiffness and running speed is assumed to be strong, as many authors have speculated that a stiff MTU will enhance the rapid transmission of force (Aura and Komi 1987; Komi 1986). That is, in events such as sprinting that require maximal force production in a very short time (~100 ms), a stiffer MTU, and hence muscle, is thought advantageous (Mero, Komi, and Gregor 1992). Importantly, since tendons recoil faster than muscles can shorten, and energy storage during muscle-tendon lengthening is greater in the most compliant part of the system, an increase in muscle stiffness would be ideal to increase energy storage in the tendon and thus improve recoil speed of the muscle-tendon system in stretch-shorten activities (figure 10.3). Thus, to a point, increases in muscle stiffness would be ideal for the performance of rapid movements, particularly stretch-shorten actions.

In terms of stiffness regulation, any increase in muscle activation should also increase muscle stiffness. As discussed previously, maximizing motor unit recruitment, firing rates and synchronization, as well as improving excitatory reflex feedback (e.g., muscle spindles) and reducing inhibitory reflex feedback (e.g., Golgi tendon organs), will all affect muscle stiffness. Increasing the cross-sectional area of muscle through hypertrophy training will also influence stiffness by resulting in a greater number of available cross-bridges for activation. Alterations in intramuscular connective tissues and changes in structural proteins within the sarcomeres also improve force transmission through the muscle. Of greatest importance for this last point is the possible role of *titin* in muscle stiffness. It has been shown recently that force enhancement after rapid stretch of a muscle fibre is greatest in fibres that exhibit the highest passive stiffness (Herzog 2005), of which titin properties are a major determinant.

One can use a force–length (stiffness) curve to understand some of the factors that may influence the slope of this curve (figure 10.4). The reader needs to be mindful that very little research has addressed some of these areas, particularly with respect to excitatory and inhibitory reflexes affecting muscle stiffness. For a given length change (#1-2), an increase in force output occurs due to the contribution of the contractile component, which includes contribution from passive (series [SEC] and parallel [PEC] elastic components) and active com-

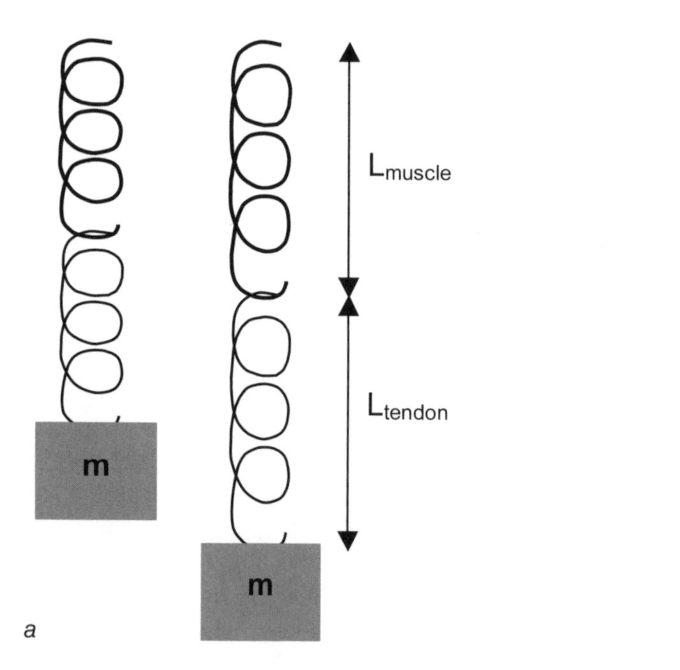

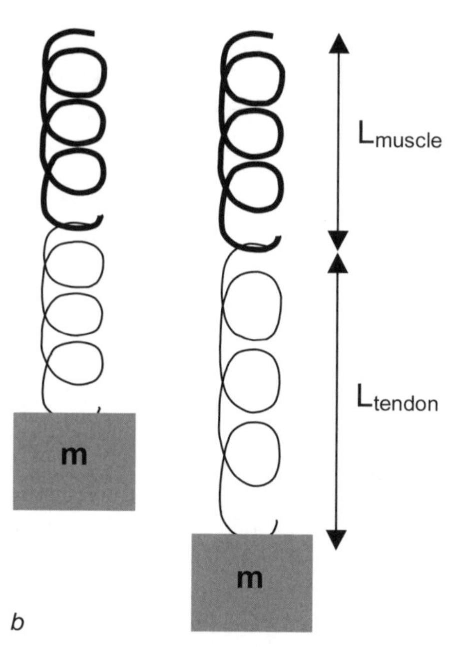

Figure 10.3 During the muscle-tendon lengthening phase of a stretch-shorten cycle, stretch occurs in both the muscle (L_{muscle}) and tendon (L_{tendon}), which act like springs. When muscle stiffness is increased, more of the stretch occurs in the tendon (*b* compared to *a*). Since the recoil speed of tendon is greater than the shortening speed of muscle, the shortening speed of the whole muscle-tendon unit is increased.

ponents (cross-bridges). Additional force results from this length change (#3) due to the length feedback component (muscle spindles) through the stretch reflex. At the same time the force feedback component (e.g., Golgi tendon organs) inhibits agonist muscle activation and decreases the force output (#4), resulting in the final force (F_{final}) for the length change (L_{final}). This information leads to the following proposals:

> Prolonged heavy resistance training can change fascicle angle, cause a shift in type IIb to IIa fibres and possibly desensitize the inhibitory influence of the Golgi tendon organs. High-speed training, eccentric training, or both can affect fascicle length, FT fibre proportion, and improve sensitivity of the muscle spindles.

- Strength training will increase the contribution of the contractile component, leading to a greater number of cross-bridges, increased connective tissue, increased or stronger tendinous tissues and greater activation of existing motor units (Klinge et al. 1997; Wood, Singer, and Cresswell 1986).
- Plyometric and technique training will probably affect the length-feedback component by improving preactivation and increasing the sensitivity of the muscle spindles, resulting in increased stiffness (Mero, Komi, and Gregor 1992).
- Heavy strength training will decrease the magnitude of inhibitory reflexes that normally affect performance in high-load movements.
- Training involving eccentric loading, such as traditional resistance training and some forms of plyometric training, increases muscle stiffness by possibly influencing titin properties.

Endocrine Considerations

Adaptations within the nervous and musculotendinous systems are contingent on appropriate hormone release, since repair and regeneration are hormone dependent. Testosterone both affects and is affected by training. Various types of training also influence peptide and thyroid hormones, as well as catabolic hormones.

Testosterone

Testosterone, in addition to its well-known positive effects on muscle protein synthesis, increases neurotransmitter release, nerve regeneration and cell body and dendrite size in the nervous system (Brooks et al. 1998; Marron et al. 2005; Nagaya and Herrera 1995). There is no clear evidence that it affects FT fibres preferentially or promotes fibre type transitions, although maintenance of FT fibre proportion has been demonstrated to be testosterone dependent in some animal muscles (Lyons, Kelly, and Rubinstein 1986). Acute increases in testosterone appear for a few hours after intense, large muscle group exercise in both men and women, with heavy resistance training being a strong stimulus (Kraemer and Ratamess 2005). This response seems to be greater after chronic weight training in men, although the response in women is less clear. There is also an increase in androgen receptor content in muscles (Bamman et al. 2001; Kadi et al. 2000),

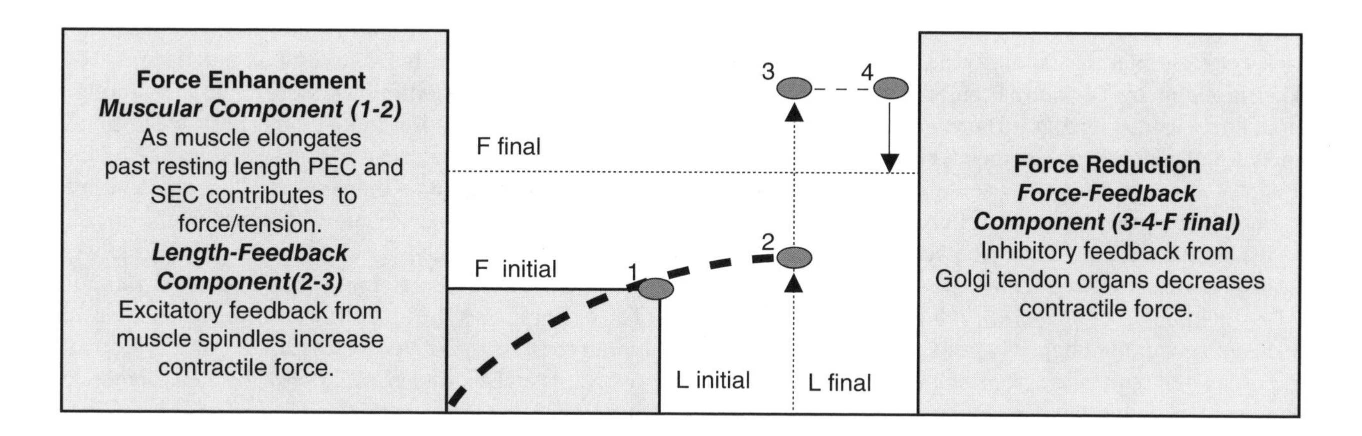

Figure 10.4 Hypothetical factors contributing to the slope of the stiffness curve.

Adapted from P. Komi, 1986, "Training of muscle strength and power: interaction of neuromotoric, hypertrophic, and mechanical factors," *International Journal of Sports Medicine* 7:10-15.

which enhances testosterone's efficiency. Increases in acute hormone release and androgen receptor number occur over periods of a few weeks to months, so training phases that target them should last at least a few weeks.

Peptide, Thyroid and Catabolic Hormones

Other hormones are also affected by training. Growth hormone (GH), which has modest effects on protein accretion and fat utilization, is significantly elevated by intense anaerobic exercise bouts including repetitive-sprint and resistance training. Such exercise also causes a delayed increase in the GH-mediated insulin-like growth factor (IGF) response and an increase in IGF binding proteins, both of which have a strong, positive effect on muscle anabolism (Le Roith et al. 2001). Nonetheless, intense training also causes the release of cortisol, which, while promoting fat utilization, also triggers muscle breakdown. Cortisol secretion is greater when exercise bouts are prolonged and intense, although this response is somewhat blunted within a few weeks of the beginning of a training program in young men (Kraemer et al. 1999). Importantly for athletes, cortisol's catabolic effects are more profound in FT fibres (Kraemer and Ratamess 2005), so minimizing cortisol release by reducing the number of long, high-intensity training sessions could be very important in order to optimise hormone release for high-speed movement performance.

Thyroid hormones have a significant impact on the metabolic rate of all cells, as well as being associated with slow-to-fast fibre type transitions in animals (Caiozzo, Herrick, and Baldwin 1992; Canepari et al. 1998). Liothyronine (T_3)-receptor binding has also been shown to increase androgen receptor production, which is further increased when testosterone levels are raised (Cardone et al. 2000). However, while it is understood that short periods (e.g., one week; Pakarinen, Häkkinen, and Alén 1991) of high-intensity training reduce thyroid hormone concentrations and that normal, chronic exercise has little or no effect on resting thyroid hormone levels, we do not know whether other exercise modes elicit positive effects. It is also not known whether acute increases in thyroid hormones can influence muscle fibre types in humans.

In summary, it is not clear whether training programs should be planned such that testosterone release is optimised, particularly for women who have lower levels of testosterone, although increases in testosterone are important for optimal neuromuscular adaptation. Current opinion is that targeted training toward improved testosterone and GH release, while minimizing catabolic hormones such as cortisol, is probably appropriate for athletes requiring high-speed movement performance.

Biochemical Considerations

Among the various forms of training, sprint-type training results in the greatest increases in stores of the immediate energy sources, adenosine triphosphate and creatine phosphate, mainly through effects on FT fibres. Chronic sprint-type training also leads to various adaptations in enzyme activity, muscle buffering and potassium regulation, thus improving the rate of muscle work during fast repetitive movements.

Energy Stores

Myocellular energy for small numbers of rapid movements comes predominantly from immediately available stores of adenosine triphosphate (ATP) and creatine phosphate (PCr). Even after 10 to 12 s of sprint running, 50% of ATP production comes from immediate sources, with the remainder from anaerobic glycolysis (Newsholme, Blomstarnd, and Ekblom 1992). While resting levels of ATP are similar amongst fibre types, FT fibres have greater levels of PCr and considerably more free creatine and glycogen. Moreover, the rate of ATP production and consumption is approximately twice as rapid in FT fibres. Sprint-type training seems to increase ATP consumption by increasing FT fibre activation (Jacobs et al. 1987). Fast-twitch fibres, therefore, also respond by increasing their ATP, PCr and glycogen stores more after sprint-type training than ST fibres. Other forms of training, including heavy resistance and longer anaerobic training, also result in increases in energy stores (although this is not always the case after resistance training; see Abernethy et al. 1994); however, one can best improve immediate energy stores through sprint-type training.

Enzymatic Activities

The rate at which ATP is utilized (hydrolyzed) is dependent on myosin adenosine triphosphatase (ATPase) activity, which is much higher in FT fibres than in ST fibres. The greater ATP utilization allows more rapid cross-bridge cycling and faster muscle shortening. Fast-twitch fibres also have accelerated calcium release and uptake by the sarcoplasmic reticulum compared to ST fibres. Chronic sprint-type training (i.e., bouts <30 s) results in a greater anaerobic enzyme activity, with a smaller (or nil) improvement after resistance training lasting several months (Tesch, Komi, and Häkkinen 1987). Other adaptations to sprint-type training include increases in lactate dehydrogenase activity (Esbjörnsson et al. 1996), increases in muscle buffering (Weston et al. 1997) and increased potassium regulation (via greater sodium-potassium ATPase concentration; McKenna et al. 1993), which would improve repeated high-speed movement performance. Thus, sprint training, more than resistance or endurance training, improves the rate of muscle work during fast repetitive movements.

> Increase testosterone and GH release via heavy resistance training whilst minimizing cortisol release by reducing the number of long, high-intensity training sessions. Use high-speed training to best improve short-term energy stores and anaerobic enzyme activity.

Biomechanical and Anthropometric Considerations

The ability to accelerate a body is directly proportional to the magnitude of the applied force (F) and inversely proportional to the mass (m) of the body to be accelerated ($a = F/m$). Resistance training techniques (see chapters 8 and 13) that maximize the force generation per unit of body mass are therefore best suited to improving acceleration. Application of the impulse–momentum relationship ($F \cdot t = m \cdot v$) further refines our understanding of the strength qualities that might be important in maximizing an athlete's speed potential. Given that an athlete's mass, or that of the limbs or an implement to be projected, changes little over the course of a race or event, velocity of movement will be directly proportional to the force applied and the time over which it acts (see figure 10.5).

During the maximum-velocity phase of a sprint run, for example, elite sprinters' foot contact times are approximately 100 ms, necessitating the application of large forces in a relatively short time. Therefore, training

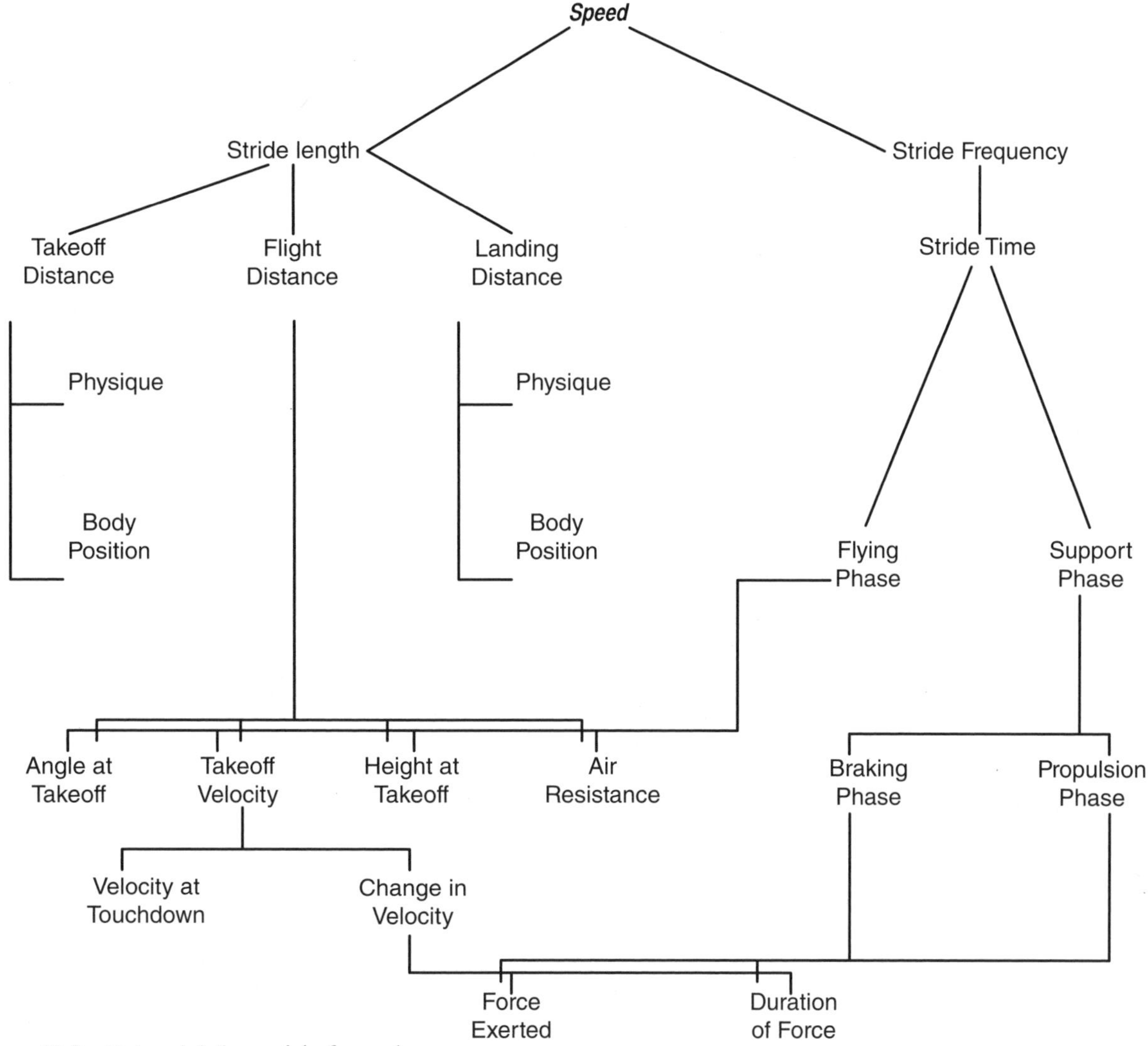

Figure 10.5 Deterministic model of speed.

Adapted from J. Hay, 1993, *The biomechanics of sports techniques* (Engelwood Cliffs, NJ: Prentice-Hall, Inc.).

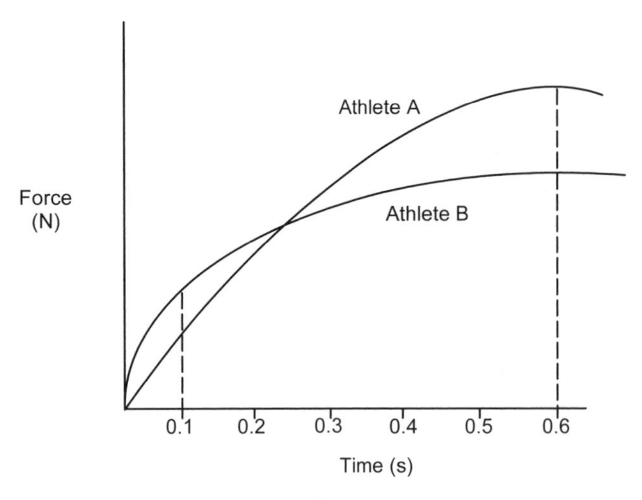

Figure 10.6 Force–time graph of two athletes with different impulse, rate of force development and maximal force characteristics.

methods that allow the development of force–time profiles similar to that of Athlete B in figure 10.6, whose initial rate of force development and force and total impulse (area under the curve) at 100 ms are superior to those of Athlete A even though athlete A has a greater maximal force capability, are important to incorporate into a program designed to improve sprint running speed.

Given that the linear speed of an athlete or an athlete's limb is a function of the angular velocity of the joints (which is dependent upon angular acceleration [α]), athletes can increase movement velocity by increasing joint torque (τ) or reducing limb inertia (I), since $\alpha = \tau/I$. They increase joint torque by improving the force generation capacity of the MTUs that act across the joint, as described earlier in the chapter. They can reduce inertia either by reducing limb mass or by locating the mass closer to the joint centre. With respect to limb mass, reductions can be achieved via decreasing fat (and water) mass or reducing the size of less essential ST muscle fibres (while maintaining FT fibre size); limb mass is also reduced in athletes with shorter limbs.

It is reasonably well established that individual muscles, as well as groups of synergist muscles, adapt in a regionally specific manner depending on the type of exercise performed. For example, well-trained sprinters have more of their mass located in the upper thigh compared to nonathletic controls (Kumagai et al. 2000), while heavy resistance training causes an increase in mass in the lower thigh. Nonetheless, in terms of resisted strength training for sprint runners, it would also be counterproductive to develop large calf muscles (thereby increasing rotational inertia) but advantageous to have most of the muscle mass distributed close to the axis of rotation (e.g., hip joint). Exponents of all sports need to be mindful of developing unnecessary muscle mass that may increase rotational inertia and decrease limb movement speeds.

Ensure lean muscle mass is maximised, impulse is maximal and the limbs are arranged as close to the joint centre as possible for high velocity movement.

Assessing Speed

Comprehensive test batteries have been designed to evaluate those factors that are thought to contribute to speed and assist in program development, such as body composition, stride length, flexibility, leg power and hamstring:quadriceps ratios. Inclusion of many of these tests is based on the relationship between sprint running and test performances, which does not allow one to establish cause and effect or offer insight into the influence of training on these factors. Whilst the use of test batteries is intuitively appealing, the practitioner needs to be critically discerning in the selection of the tests and confident that changes in a test score will also affect speed. It is obvious that a great deal of longitudinal research is needed in this area, so rather than focusing on assessment of the factors proposed to affect speed, this section concentrates on speed assessment per se and associated methodological issues.

Timing Lights

For measuring the speed of an athlete, a common method is to use timing lights connected to an electronic timer. A timing light consists of a light source (transmitter) and an optical pickup (receiver). The light sources typically used are visible, infrared, standard and monochromatic laser light. The timing light systems can vary from single- to dual- and three-beam reflector units. Dual-beam systems are probably ideal, as single-beam systems can be triggered early by a swinging arm and may produce measurement error of up to 80 ms. However, three or more beams provide accuracy improvements that are smaller than the resolution of most systems (~0.01 s), so there is little need for such setups.

Although the use of timing lights is widespread, precise details regarding their use is generally limited, so comparison between studies is problematic (see table 10.1). The following points are important to consider before testing.

- Sprint times using timing lights will be unaffected by variations in RT; consequently, the elapsed time is a good measure of "true sprint performance"; this may be a problem if the initial push-off for the sprint is an important component of sprint performance for a particular sport.
- Timing lights do not measure speed instantaneously but rather record only the average speed between gates. To obtain more relevant information during an acceleration phase, several gates may be positioned close together.
- Two light heights (shoulder and hip height) are commonly used; however, faster times are generally recorded using the lower heights (10 m [11 yd]: 3.4%; 20 m [22 yd]: 2.9%), possibly because the legs cross the beams before the torso.
- The distance to the first timing light will also have a substantial influence on sprint times; for example, a 30 cm (12 in.) compared to a 50 cm (20 in.) start distance will influence 10 m and 20 m times by ~6%.
- Using a split rather than parallel foot stance at the start will improve 10 m (~11%) and 20 m (~6%) times by allowing the body's centre of mass to travel a greater distance before breaking the gate and hence attain a higher velocity.
- Sprint times will vary depending on whether movement of the body is allowed prior to the start.
- During testing, three or four trials are generally sufficient. Unless the measurement of absolute speed is necessary, it is recommended that the average of the best two trials (rather than the best trial) be recorded, as this will improve the reliability of the measure and allow training-related changes to be established with greater certainty.
- Measures of first-step quickness (0-5 m), acceleration (0-10 or 20 m) and maximal speed (0-30 or 40 m) appear the most commonly reported parameters (see table 10.1). It has been reported that the 5 m and 10 m measure similar sprint qualities (R^2 = 84%); however, there is less common variance (52-60%) between the 5 to 10 m measures and the maximal speed measure, possibly indicating a need to assess and develop acceleration and maximal speed independently (Baker and Nance 1999; Cronin and Hansen 2005).

Table 10.1 Reported Sprint Times (Mean ± SD) for 5 m, 10 m and 30 m Distances

Author	Subjects	Start type	5 m (s)	10 m (s)	30 m (s)
Baker and Nance (1999)	9 national rugby league backs	Standing		1.71 ± 0.06	
	10 national rugby league forwards	Standing		1.75 ± 0.09	
	9 city-based rugby league backs	Standing		1.74 ± 0.05	
	10 city-based rugby league forwards	Standing		1.75 ± 0.08	
Cronin and Hansen (2005)	26 professional rugby league players	Standing 0.5 m behind timing light	0.95 ± 0.05	1.65 ± 0.06	3.99 ± 0.23
Dowson et al. (1998)	8 rugby players	Standing 1.0 m behind timing light	1.00 ± 0.06		
	8 track sprinters	Standing 1.0 m behind timing light	0.97 ± 0.09		
	8 sportsmen	Standing 1.0 m behind timing light	1.03 ± 0.06		
Ferro et al. (2001)	1999 World Championship finalists	Block start (males)		1.76 ± 0.03	3.70 ± 0.03
		Block start (females)		1.85 ± 0.02	3.97 ± 0.03
Nesser et al. (1996)	20 sportsmen	Standing		2.09 ± 0.13	
Parry et al. (2003)	22 male and female university sprinters	Block start	1.45 ± 0.01	2.18 ± 0.00	4.60 ± 0.00
Young et al. (1995)	20 elite junior track and field athletes	Block start	1.44 ± 0.07	2.19 ± 0.11	4.58 ± 0.28

- One may obtain greater diagnostic information by setting the timing lights at equal intervals (e.g., 10-20-30 m or 20-40-60 m) and expressing the times as a percentage of the maximum distance.

Video Analysis

It is also possible to use high-speed video analysis to measure movement speed, as well as the speed of balls or implements, by making use of the known frame rate of cameras (see chapter 16). There are no data specifically comparing the accuracy and reliability of video methods with those of timing lights or radar measurements. However, this technique is the easiest way to measure the velocity of implements or balls that cannot be tracked with timing systems or radar and is therefore very useful to athletes and coaches for monitoring performance.

Radar Measurements

Radar allows accurate measurement of the speed of an athlete or implement. Since rapid, multiple measurements can be made with some systems, the development of speed curves is possible. The accuracy of radar measurements is high (± 1 km.h^{-1}) as long as the angle between the radar gun and object is small. Ideally, radar guns should be placed either directly behind or in front of the moving object. Radar guns work by emitting a microwave radio frequency signal and calculating the Doppler shift (i.e., the degree to which the frequency of waves changes when they reflect off a moving object), which is proportional to object speed. The accuracy of the measurement decreases at a rate proportional to the cosine of the angle between the radar gun and the moving object. For example, if a ball's speed was measured as 10 m/s by a gun placed immediately behind release point of the ball, a second gun placed at a 20° angle would record the speed as 9.4 m.s^{-1} ($10 \times \cos 20$). A significant drawback of radar devices is that moving objects such as a racquet or club head can be difficult to track unless their trajectory is known before measurement. However, this is also a drawback of both timing lights and 2-D video methods. Another drawback is that the time taken to complete complex movement tasks (e.g., a running course) cannot be determined, as it can be with the use of timing lights or video.

> Timing lights are the most widely used equipment to measure running velocity but factors such as starting stance, distance behind the first light, first step strategy and timing light height must be taken into consideration for data to be reliable and compared to other populations.

Developing Speed

Suggestions for the training of speed have been made throughout this chapter. Therefore, the aim of this section is to direct the reader to other training components that are believed to be useful for speed development. Of course, as part of any speed development program, one should adhere to the following points.

- The athlete should have a good strength and technical foundation to build upon. Implicit in this is having adequate strength for joint stabilization and maintenance of sporting technique.
- Training needs to adhere to the principles of training as discussed in chapter 8.
- Careful monitoring of performance and indicators of overreaching, such as chronic pain, should be ongoing.

Olympic Lifting

While Olympic lifts are commonly used by athletes to improve speed and power for sport, presently, there are no conclusive data describing their benefits in this regard. Nonetheless, the Olympic lifts and derivatives form the basis of many training programmes as exercises are performed with maximum velocity and incorporate a large musculature. This would allow an extensive recruitment of fibres (Aagaard et al. 2000) and provide a strong stimulus for strength and power gains. Back, hip, knee and ankle extensors are highly involved in these exercises, so athletes requiring the development of speed or power in these areas would benefit. Also, the brevity of the time of lifting is unlikely to result in stiffened tendons and therefore elastic energy storage should not be compromised. Finally, the lack of eccentric loading on the muscle would minimize muscle damage and allow fast recovery of muscle force for subsequent training bouts.

Olympic lifts are beneficial to speed and power development for these reasons. Nonetheless, the effects of Olympic lifts differ from those of traditional weight training in a number of ways. With Olympic lifts, anabolic hormone release is small (Kraemer et al. 1992; Passelergue, Robert, and Lac 1995); the bilateral nature of the movement is specific to few sports; muscle forces are generally lower than during slow-speed, high-load movements (according to the force–velocity relationship of muscle) so there is less stimulus for protein synthesis (Fry 2004); and there is a lack of eccentric loading, which has been shown to be vital for the development of maximal contractile capacity (Dudley et al. 1991). Thus Olympic lifts have inherent limitations and do not provide a complete stimulus for strength, power and speed development. Training plans might incorporate these lifts during periods in which concentric muscle power needs to be developed (see chapter 13), when large muscle groups need to be trained in short periods of time or when it is necessary for muscle damage to be minimal. This is often the case in a competitive season, although this training should be periodized with other forms of training when a competitive season lasts more than a few weeks.

> Many components must be taken into consideration for the development of high-velocity movement (e.g., neural, biochemical, morphological, biomechanical, etc.) In terms of resistance training many techniques and loading patterns are available for speed development. Training for the most part should consist of high-speed movement training, technique development and additional high-load or explosive resistance training (or both) involving an eccentric component performed with relatively long rest periods.

Resisted Speed Training

Resisted speed training involves an athlete's moving with added load on the limbs or body or utilizing external forms of resistance (e.g., hills or stairs). While these techniques can be used in many sports, their application is common in sprint running. The adjustments in sprinting technique made during towing and uphill running seem to replicate the acceleration phase of a sprint run more closely than other resisted techniques such as the use of weighted vests. Both towing and uphill running require an increased trunk lean, stance duration and need for horizontal force production during the propulsive phase of stance, although hill sprinting also increases the need for vertical propulsion.

Conversely, weighted vests result in an increased vertical load during foot strike, thus increasing the braking forces and overloading MTUs to better effect. Since the weight of the vest is held centrally on the body, there is minimal influence on balance during the running action; however, the weight of the vest would normally cause the body to over-rotate during the acceleration phase.

The optimum load to be used during resisted sprinting has not been determined; but loads of 5% to 15% body mass appear to be the most commonly reported in the literature, and a case could be made for the use of greater resistance when training the acceleration phase compared to the maximum-velocity phase. Nonetheless, the coach or athlete should be aware of the risk that too much load may result in technique adjustments that could compromise the athlete's sprinting technique. There is also a case for the use of loads that equal or exceed body weight in the specific strengthening phase of a sprint athlete (see figure 10.7).

Figure 10.7 Specific strength training using a weighted sled.

These exercises are cyclic and are therefore much more specific than those employed in the gym environment.

Assisted Speed Training

It is thought that acceleration and maximum velocity are relatively separate and specific qualities (Delecluse et al. 1995; Young et al. 2001) and should be developed using different techniques. Interpretation of figure 10.8 would support such a contention. As athletes attain higher velocities, the increase in velocity is due to a greater contribution of stride rate rather than stride length. This would suggest that, given good stride length kinematics, greater emphasis should be placed on training for a higher stride rate to improve running speed.

Assisted speed training utilizes several strategies such as downhill sprinting (~2° slope) and towing of the athlete by machines or by a fellow sprinter using surgical tubing. Assisted training is thought to enhance maximal speed more than acceleration, as it has a greater effect on stride rate. For example, Mero and Komi (1986) found that an increase in maximum speed with towing (108.4%) was the result of increased stride rate (6.9%) and stride length (1.5%). This was mainly explained by a smaller descent of the centre of gravity during the eccentric phase, which was attributed to higher neural activation and subsequent stiffness. It would seem that assisted training offers a greater eccentric loading stimulus (i.e., increased speed of leg acceleration and deceleration), which could have interesting training applications and adaptations.

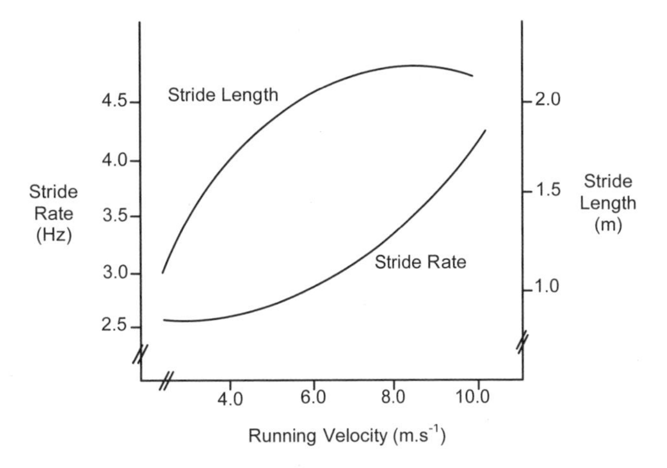

Figure 10.8 Changes in stride rate and stride length as a function of running velocity.

The reader needs to be mindful that most evidence relating to resisted and assisted sprinting is anecdotal; very few randomized controlled studies have validated the most desirable mode of training, the mechanisms behind improvements in acceleration or speed, the possible negative effect on technique and the optimal training loads. Answers to these questions are necessary if resisted and assisted sprinting is to be utilized effectively by coaches and athletes.

Summary

The training regimen for athletes who perform high-speed movements mainly consists of high-speed movement training, technique development and additional high-load or explosive resistance training (or both) involving an eccentric component performed with relatively long rest periods. These training techniques improve muscle recruitment patterns, optimise tendon properties and muscle-tendon stiffness, increase appropriate muscular development and ultimately improve high-speed force production. Although researchers have investigated the effects of these high-intensity exercises on various mechanical and neurophysiological properties, the majority of the research is cross-sectional; and little is known about the changes in these properties with long-term training programs, especially for sprint athletes. A great deal of research is needed in this area, especially longitudinal studies.

eleven

Flexibility

Patria Hume, PhD; and Duncan Reid, MHSc (Hons)

In this chapter, which covers many areas, we first present a definition of stretching and then address some of the anatomical and physiological components that are targeted for stretching. The chapter also describes various types of stretching procedures and techniques, reviews the evidence for the mechanisms of stretching, examines the effect of stretching on performance and the risk of injury, considers the factors that can influence the effectiveness of stretching and describes the techniques and equipment used to measure improvements in flexibility.

Definitions of Stretching and the Resulting Flexibility-Extensibility

Flexibility can be defined as the range of movement (ROM) in a joint or in several joints. More flexibility equates to increased ROM. Flexibility is developed by stretching of the soft tissue, primarily around a joint to increase muscle-tendon unit length (Taylor et al. 1990), and may help athletes improve their overall performance. Several literature reviews have considered flexibility as the outcome of stretching exercise (Gajdosik 2001; Harvey, Herbert, and Crosbie 2002; Shellock and Prentice 1985); however, stretching itself has not yet been well defined. Magnusson and colleagues ("Mechanical and Physical Responses" 1996, p. 77) stated that "stretching has been characterized in biomechanical terms in which the muscle-tendon unit is considered to respond viscoelasticly during the stretching manoeuvre". However, this definition of stretching refers more to a biomechanical result of stretching than to the action of stretching. In a recent review (Weerapong, Hume, and Kolt 2005), stretching was defined as movement applied by an external and/or internal force in order to increase muscle flexibility and/or joint ROM.

> Stretching is defined as movement applied by an external and/or internal force in order to increase muscle flexibility and/or joint ROM.

The terms flexibility and extensibility are used interchangeably in this chapter; either term is appropriate when one is considering the effects of stretching on joint ROM. Extensibility has been defined as the ability of a muscle to be stretched or extended (Marieb 2001). Extensibility of soft tissues is related to the resistance of tissue as it lengthens. The ability of a muscle to lengthen can also be measured in terms of its "stiffness" (or its reciprocal compliance) to movement (Magnusson 1998). Stiffness can be defined as the ratio of the change in force to the change in length of the tissue (McNair and Stanley 1996).

Anatomy and Physiology Components Targeted for Stretching

When examining the ROM in any athlete, it is important for the coach to understand the anatomical and physiological limitations that are placed on the individual. Age, gender and environmental conditions during a training session can affect the level of flexibility.

Anatomical Structures

Connective tissue (soft tissue) is widespread in the body and plays an important role in determining an athlete's ROM. Soft tissue binds together various structures and consists of both fibrous connective tissue

Acknowledgment Dr. Pornratshanee Weerapong provided information on the mechanisms of stretching and the effects of stretching on performance and injury prevention.

(collagen) and elastic connective tissue. Some joints in the body have slightly more elastic tissue, and this is one of the factors that determine their ROM. Soft tissues are elastic by nature and are designed to return to their normal length after being stretched.

When a muscle is stretched under tension during an eccentric action, the myofilaments slide apart, and elastic energy is stored in the cross-bridge linkages between the actin and myosin filaments.

Fascia is a band of fibrous-like tissue that binds many structures in the body. The deep fascia that envelops the muscle is known as the epimysium; within the muscle can be found the perimysium, endomysium and the sarcolemma, all of which bind various components of the muscle. This tissue has limited stretch and quickly resists movement.

The contractile nature of skeletal muscle means that it is not able to lengthen of its own accord but can be stretched externally. During stretching, the myofilaments can slide farther apart, and an elongation of the muscle can occur. When a muscle is stretched under tension during an eccentric action, the myofilaments slide apart, and elastic energy is stored in the cross-bridge linkages between the actin and myosin filaments. The greater the muscular tension, the more actin and myosin filaments are linked and thus more elastic energy will be stored.

Generally tendons join muscle to bone and are normally cordlike, but they can also be flat or ribbon-shaped (known as aponeuroses). Tendons consist of closely packed collagenous bundles that have a longitudinal striation. Their structure ensures that they have little stretch, and this quality enables them to transfer a muscular contraction directly to the bone to which they are attached. Nevertheless, tendons do possess some elasticity, particularly the longer ones such as the Achilles tendon, and are major storage sites for elastic energy.

Ligaments are strong bands of connective tissue joining bone; their main function is to support a joint. Ligaments consist of bundles of collagenous fibres that run parallel to one another and have a structure similar to that of tendons, but they are usually flatter in shape. A ligament's level of extensibility is similar to that of tendons, in order to allow some limited movement at the joint, but ligaments are strong enough to bind bone to bone, resisting moderate trauma or overstretching.

Physiological Components

Sense organs or proprioceptors (see figure 11.1) are involved in all movements in which precision is required and are found in the muscles, tendons and joints. During stretching, two types of sense organs come into play—the muscle spindles and the Golgi tendon organs (GTOs). These organs pick up changes in the muscle's length, or its velocity or force, and transmit information about these changes via electrical signals to the central nervous system, where they are processed and where the appropriate response is made.

Muscle spindles are the primary stretch receptors in the muscle that respond to changes in length and rate of stretch. The fusiform-shaped spindles known as intrafusal fibres (figure 11.1) run parallel to the muscle fibre. They should not be confused with extrafusal fibres, which are the contractile units of the muscle itself.

Golgi tendon organ sensory receptors, located in the tendon close to the musculotendinous junction, respond to force or tension in the muscle. Golgi tendon organs are basically inhibitory; and because they have a higher threshold than the muscle spindle, they come into play only after the muscle is vigorously stretched.

If an individual stretches a muscle or group of muscles with a ballistic motion, the muscle spindles come into play and the stretch reflex *(myotatic reflex)* is initiated. The magnitude of this reflex is dependent upon the amount and rate of stretching, so that dynamic ballistic stretches invoke the maximal stretch reflex response. When this

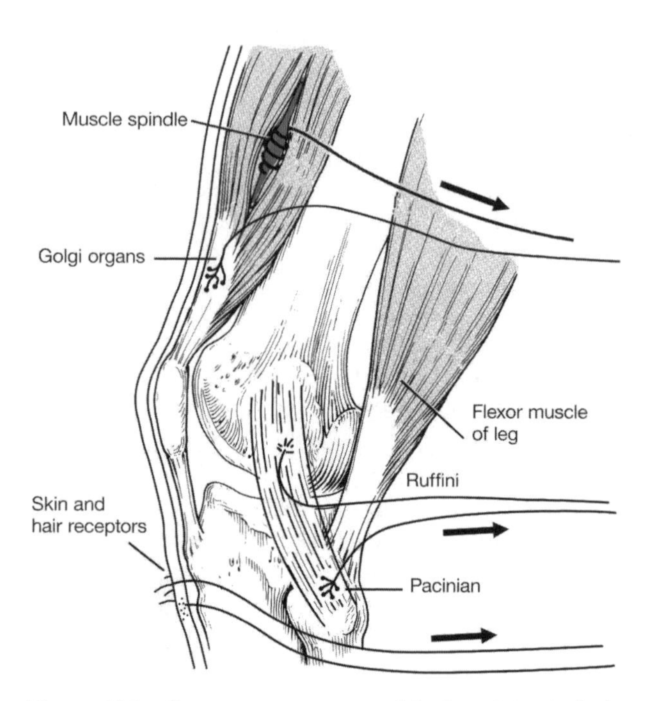

Figure 11.1 Sensory organs around the knee joint, including muscle spindles and Golgi tendon organs.

reflex fires, the muscle that is close to being overstretched suddenly contracts, reducing the extension of the limb. In this way the stretch reflex serves as a mechanism to protect the limb from being overstretched.

The inverse stretch reflex *(inverse myotatic reflex)*, also known as autogenic inhibition, occurs when a contraction or stretch exceeds a critical level. This causes an immediate reflex action that inhibits any further muscular contraction or stretching, and the tension is quickly reduced. This reduction of tension acts as a protective mechanism that prevents the muscles from injury and is made possible only by the inhibitory impulses of the GTO, which override the excitatory impulses from the muscle spindles.

Stretching techniques can be classified as either passive (static or partner stretch) or active (ballistic, dynamic or proprioceptive neuromuscular facilitation).

Types of Stretching Procedure and Technique

Liebesman and Cafarelli (1994) have defined three types of stretching procedure: acute, periodic and chronic (table 11.1). These procedures can also be further subdivided on the basis of descriptions of the stretching technique. Shellock and Prentice (1985) indicated that four different techniques are commonly used for sport activities: static, ballistic, dynamic and proprioceptive neuromuscular facilitation (PNF) (table 11.2). In testing with gymnasts, Hume and colleagues (2004) have also classified stretching technique and the

Table 11.1 Classification of Stretching Procedures

Techniques	Definition
Acute stretching	Brief, completed in one session and lasting only a few minutes
Periodic stretching	A training program in which subjects repeat a particular stretch several times a week for several weeks
Chronic stretching	A continuous stretch, generated by immobilization through casting, usually over several weeks

Table 11.2 Summary of the Advantages and Disadvantages of Stretching Techniques

Stretching techniques	Definition	Advantages	Disadvantages
PASSIVE			
Static	Passive movement of a muscle to maximum range of motion (ROM), then held for an extended period (Shellock and Prentice 1985)	Increased ROM (Halbertsma et al. 1996); simple technique	Reduced muscle strength (Cornwell et al. 2002; Fowles et al. 2000); may cause injury (Smith et al. 1993)
Partner	Passive movement of a muscle by an external force such as a partner or a piece of equipment to maximum ROM, which is then held for an extended period	Increased ROM; simple technique	Requires an external force such as a partner or equipment
ACTIVE			
Dynamic	Slow or fast movement of a joint as a result of antagonist muscle contraction throughout the ROM (Shellock and Prentice 1985)	Unknown	Unknown
Ballistic	Repetitive bouncing movements at the end of joint ROM (Shellock and Prentice 1985)	Increased ROM (Shellock and Prentice 1985)	Reduced muscle strength (Nelson and Kokkonen 2001); may cause injury (Smith et al. 1993)
Proprioceptive neuromuscular facilitation (PNF)	Reflex activation and inhibition of agonist and antagonist muscles (Burke et al. 2000)	Increased ROM (Spernoga et al. 2001)	Reduced jump height (Church et al. 2001); requires experience and practice (Smith 1994)

related ROM measurement techniques as either passive (static or partner stretch) or active (ballistic, dynamic or PNF). Table 11.2 provides a brief summary of the advantages and disadvantages of the various stretching techniques.

- **Passive stretching:** Passive stretching requires no muscular force in the muscle to be stretched. Two types of passive stretching are commonly used—static stretching and partner stretching.
- **Static stretching:** Static stretching involves holding a static position for typically 5 to 60 s after the limb has already been stretched in a slow manner (Roberts and Wilson 1999; Sady, Wortman, and Blanke 1982). Slow stretching allows muscle relaxation to occur as a result of the firing of the GTOs if the stretch is performed over a reasonable time.
- **Partner stretching:** When using partner stretching the athlete stays relaxed and makes no active contribution to the stretch. The external force produced by a partner can also be created with equipment. The partner method is well known, but the use of equipment for stretching has developed only with the variable resistance strength training machines, which can place the muscle on passive stretch with a slight change in posture near the end of each repetition.
- **Active stretching:** Active stretching requires the athlete to produce muscular force to take the joint to the end ROM (as with dynamic or ballistic stretching techniques), or to maximally contract the muscle group to be lengthened against resistance (as with PNF stretching).
- **Dynamic stretching:** Dynamic stretching involves slow or fast movement of a joint as a result of antagonist muscle contraction throughout the ROM (Shellock and Prentice 1985). The movements are usually sport-specific actions.
- **Ballistic stretching:** Ballistic stretching involves repetitive lengthening of the muscles in a cyclic manner. Ballistic stretching is considered a less beneficial type of stretching with a greater risk of injury than other types of stretching, possibly as a result of the reflex reaction of the type Ia motor neurons and the added resistance of the muscle to the cyclic changes in length (Smith 1994). However, there is little experimental evidence to support the conjecture that ballistic stretching is less beneficial and has a greater risk of injury than other types of stretching.
- **Proprioceptive neuromuscular facilitation stretching:** PNF stretching is a complicated stretching technique with a combination of shortening contraction and passive stretching. This form of stretching utilizes the concepts of reflex activation and inhibition. Proprioceptive neuromuscular facilitation usually involves taking the limb to its passive terminal position and then asking the athlete to maximally contract the muscle group to be lengthened against resistance for 3 to 6 s; this is followed by a period of relaxation during which the muscle is stretched further into the range (Osternig et al. 1990; Sady, Wortman, and Blanke 1982). Three derivations of this technique are most common: contract-relax (CR), hold-relax (HR) and agonist contract-relax (CRAC). An advantage appears to be that there is a slight gain in strength at the same time the ROM is being increased. However, PNF stretching might be harmful, as it has been found to increase blood pressure (Cornelius, Jensen, and Odell 1995) and electromyogram (EMG) activity (Osternig et al. 1987) during the contraction phases. Performing PNF technique requires some experience, and a partner is needed to help with the stretching.

Mechanisms of Stretching

The proposed mechanisms associated with changes in muscle tissue following stretching regimens include biomechanical and neurophysiological phenomena.

Biomechanical Mechanisms

Muscle-tendon units can be lengthened in two ways: by muscle contraction and by passive stretching. When muscle contracts, the contractile elements are shortened while the entire length of the muscle-tendon unit is fixed. A compensatory lengthening occurs at the passive elements of tissues (tendon, perimysium, epimysium and endomysium). When muscle is lengthening, the muscle fibres and connective tissues are elongated because of the application of external force (Taylor, Brooks, and Ryan 1997). Stretching increases muscle-tendon unit length by affecting the biomechanical properties of muscle (ROM and viscoelastic properties of the muscle-tendon unit).

The mechanical effects of acute stretching of muscle also influence the connective tissue within the muscle. Contributors to muscle tendon elasticity include the series elastic components (SEC) and the parallel elastic components (PEC) (Magnusson 1998). Perimysium, which surrounds bundles of muscle fibres, is the principal component of the PEC and functions to distribute stress evenly through the muscle and prevent overstretching. Endomysium, which surrounds individual muscle fibres, is the principal component of the SEC and functions to transfer force from the contractile components to the tendon and bone (Marieb 2001). As muscle generally contains more perimysial than endomysial connective tissue, if intramuscular connective tissue has a role in determining passive muscle elasticity, it is the contribution of the perimysium that should primarily be assessed (Purslow 1989).

Viscoelastic Properties of the Muscle-Tendon Unit

Muscle, a viscoelastic structure, exhibits several mechanical phenomena when an external load is applied. When tissues are held at a constant length, the force at that length gradually declines; this response is termed the stress relaxation response (McHugh et al. 1992). When tissues are held at a constant force, the tissue deformation continues until the tissue approaches a new length; this phenomenon is termed "creep" (Taylor et al. 1990). Creep might be another explanation for the immediate increased ROM after static stretching (Gajdosik, Giuliani, and Bohannon 1990). The musculotendinous unit also produces a variation in the load–deformation relationship between loading and unloading curves. The area between the loading and unloading curves, termed hysteresis, represents the energy loss as heat due to internal damping (Kubo, Kanehisa, and Fukunaga 2002).

> Viscoelastic properties of the muscle-tendon unit include stress relaxation (when tissue is held at a constant length and the force gradually declines) and creep (when tissue is held at a constant force and the tissue deforms to a new length).

Passive stiffness refers to the resistance of the muscle-tendon unit when external forces are applied. The slope of the force and deformation curve at any ROM is defined as passive stiffness (Gleim and McHugh 1997; Magnusson 1998). Passive torque, which occurs during passive movement, is resistance from stable cross-links between actin and myosin, noncontractile proteins of the endosarcomeric and exosarcomeric cytoskeletons (SEC) and connective tissues surrounding muscles (PEC) (Gajdosik, Giuliani, and Bohannon 1990). It appears that perimysium produces major resistance (Magnusson et al., "A Mechanism" 1996).

Active stiffness, defined as the ability to transiently deform the contracted muscle (Gleim and McHugh 1997), can be measured by the damped oscillation technique (Wilson, Murphy, and Pryor 1994). The oscillation of the contracted muscle after application of external force results from the viscoelasticity of muscle and the level of muscle activation (McNair and Stanley 1996). Passive and active stiffness provide more information on muscle-tendon unit behaviour during movement than ROM only.

In the animal model, musculotendon tissues behave viscoelastically, and under tensile loading and holding phases of stretch they exhibit properties of stress relaxation, creep and hysteresis (Taylor et al. 1990). Similar findings have been demonstrated with human skeletal muscle (Magnusson et al. 1995; McHugh et al. 1998). Although the tissue in these studies responded mechanically, these changes have been shown to be short-lived, with the effects returning to baseline levels within 45 to 60 min after the applied stretch (Magnusson 1998). Changes in ROM that occur with repeated static stretches to a specific force load reach a plateau after the third to fifth repetition, implying that the tissue reaches an optimal length quite quickly and that much less ROM is gained with further repetitions (Magnusson et al., "Mechanical and Physical Responses" 1996; McNair, Dombroski, and Stanley 2001).

Range of Motion

Numerous researchers have used ROM as a measure of change in flexibility (Henricson et al. 1984; Wiktorsson-Moller et al. 1983; Zito et al. 1997). The exact physiological mechanism of stretching resulting in increased ROM still remains unclear (Muir, Chesworth, and Vandervoort 1999), as most research has failed to show changes in muscle properties such as passive stiffness (Klinge et al. 1997; Magnusson 1998; McHugh et al. 1992, 1998) or active stiffness (McNair and Stanley 1996). However, there are two main areas of thought with respect to the mechanism associated with alteration in ROM: the addition of sarcomeres in series and stretch tolerance. Gajdosik (1991) demonstrated that straight-leg raise (as a measure of maximal hamstring length) could be influenced with a periodic stretching regimen over a three-week period. An increase in straight-leg raise and an increase in passive resistance of the hamstring muscle following the stretching intervention were observed. On the basis of animal studies demonstrating that when muscle was immobilized in a lengthened position a greater number of sarcomeres (SEC) were laid down, Gajdosik suggested that the increase in hamstring length may be due to the same phenomenon. Gajdosik also thought that the possible increase in the number of sarcomeres laid down may explain the increase in the passive length of the muscles.

An alternative explanation is that adaptations to stretching may be due to an amplified "stretch tolerance" rather than to changes in the tissue properties (Klinge et al. 1997; Magnusson et al. 1995). The stretch tolerance of a muscle is determined as the maximum limit of stretch or elongation that a subject will tolerate during a passive stretching procedure (Halbertsma, van Bolhuis, and Goeken 1996). A number of studies have used this subjective painful feeling of tension (stretch) to determine the limit of joint ROM during a passive stretch (Magnusson et al., "Viscoelastic Stress" 1996). Magnusson and colleagues ("Mechanical and Physical Responses" 1996) examined changes in peak torque, joint ROM, angular velocity and hamstring EMG using a Kincom dynamometer. Following a periodic stretching regimen to the hamstrings twice a day for three weeks, a 10° increase in knee extension ROM was achieved with a concomitant increase in torque about the knee joint. However, no changes in muscle stiffness or EMG activity in the muscle were observed. This led the researchers to conclude that changes in the ROM in the muscle were a result of increased stretch tolerance rather than extensibility of the tissue.

Neurological Mechanisms

From a neurophysiological perspective, an aim of stretching is to decrease the stretch reflex activity. Inhibition of the stretch reflex would reduce muscular resistance and improve joint ROM. However, studies by Moore and Hutton (1980) and Osternig and colleagues (1990) have demonstrated elevated EMG responses with acute stretching techniques, particularly PNF, which seems paradoxical and hence casts some doubt on the neurological mechanism.

Biomechanical responses of muscle-tendon units during stretching, on the other hand, seem independent of reflex activity as indicated by the lack of muscle activity (EMG) responses during passive stretching (Magnusson et al., "A Mechanism" 1996; Mohr et al. 1998). However, stretching techniques may affect neural responses by reducing neural sensitivity. The majority of research on the effects of stretching on neurological mechanisms has dealt with changes in the Hoffman reflex response (H-reflex)—the electrical analog of the stretch reflex but without the effects of gamma motoneurons and muscle spindle discharge. Electrical stimulation of a mixed peripheral nerve (both sensory and motor axons) will evoke the H-reflex. The activation of the motor axons directly induces the M-wave (from the point of stimulation to the neuromuscular junction) prior to evoking the H-reflex (from Ia afferents arising from annulospiral endings on the muscle spindle) via a monosynaptic connection to the alpha motoneurons. A decrease in the H-reflex has been reported during (Vujnovich and Dawson 1994) and after stretching (Avela, Kyrolainen, and Komi 1999; Gollhofer et al. 1998; Moore and Kukulka 1991; Rodenburg et al. 1994).

Mechanisms of Each Stretching Technique

Even though each technique of stretching is expected to increase muscle and joint flexibility, different stretching techniques produce increases in flexibility by different mechanisms.

Static and Partner Passive Stretching

Static stretching is the most widely used technique by athletes due to its simplicity. Partner passive stretching is also common. Static stretching has been found to affect both mechanical (see table 11.3) and neurological properties of the muscle-tendon unit, resulting in increased joint ROM.

Factors Affecting Static Stretching The duration, frequency and number of repetitions of static stretches have been shown to affect the increase in ROM following stretching regimens.

- **Duration.** Bandy and Irion (1994) randomly assigned 57 uninjured subjects to one of four groups: 15 s stretching, 30 s stretching, 60 s stretching and a control group. Subjects in the stretching groups performed hamstring stretches five times a week for six weeks. Although all three durations of stretch demonstrated an improvement in hamstring flexibility, only the 30 s and 60 s groups demonstrated a statistically significant gain in knee extension ROM (12.5° and 10.9°, respectively, with the 15 s group gaining 3.8° and the control group 0.3°). The researchers concluded that 30 s was as effective as 60 s in increasing the ROM of the hamstring group.
- **Frequency.** Bandy, Irion, and Briggler (1997) investigated the different frequencies of stretching by comparing one and three repetitions of stretch to the hamstrings over 30 s and 60 s duration. In this study, 93 uninjured subjects were randomly allocated to four intervention groups and a control group. Although all stretching groups demonstrated a statistically significant increase in knee extension ROM compared to the control, there was no difference in the magnitude of the gains in ROM between the groups. As a consequence the researchers concluded that one repetition held for 30 s was as effective as three repetitions in improving hamstring ROM.
- **Time to effect a change.** Gribble and colleagues (1999) monitored the weekly changes in straight-leg raise and active knee extension ROM with static and PNF stretching over a six-week period. Subjects in this study were allocated to one of three groups: control, static stretch and hold-relax (PNF) stretch. Subjects in the static group stretched the hamstrings using four 30 s holds four times a week. The hold-relax group undertook a stretching regimen that had a total stretching duration similar to that of the static group, and the control group did not stretch. A total increase of 33.1° was observed in the static group, of 35.2° in the hold-relax group and of 8.9° in the control group. By weeks 4 and 5, no further significant changes in ROM were observed, which indicated a plateau effect in the improvement of ROM. Willy and colleagues (2001) investigated the effect of cessation and resumption of static hamstring muscle stretching on joint ROM in 18 college students. After six weeks they increased knee extension ROM by 10°. The students then stopped stretching for four weeks, and values for knee extension ROM returned to baseline.

Table 11.3 The Effects of Static Stretching on Muscle Properties

References	Trial design	Sample	Interventions	Outcome measures	Main results
ROM AND PASSIVE STIFFNESS					
McHugh et al. (1992)	PPT	9 men, 6 women (hamstrings)	Static stretch (45 s hold) 1. At onset of EMG 2. 5° below onset of EMG (negligible EMG activity)	1. Peak torque 2. ROM 3. EMG	S: ↓ torque S: ↑ ROM
Magnusson et al. (1995)	PPT	10 men (hamstrings)	1. Static stretch (90 s hold, 30 s rest) 5 times (stretches 1-5) 2. Repeated static stretch 1 time (stretch 6)	1. Peak torque 2. ROM 3. EMG	S: ↓ stress relaxation S: ↑ ROM
Halbertsma et al. (1996)	RCT	10 men and 6 women with short hamstrings	1. Static stretch (30 s hold, 30 s rest) for 10 min (n = 10) 2. Control—rest (n = 6)	1. Peak torque 2. ROM 3. Passive stiffness	S: ↑ ROM
Magnusson et al., "A Mechanism" (1996)	PPT	7 women (one leg stretch, one leg control) (hamstrings)	Static stretch (45 s hold, 15-30 s rest, 5 times), twice daily, 20 consecutive days	1. Stress relaxation 2. Energy 3. EMG 4. ROM	S: ↑ ROM
Magnusson et al., "Viscoelastic Stress" (1996)	CCT	8 neurologically intact and 6 spinal cord–injured volunteers (hamstrings)	Static stretch (90 s hold)	1. Stress relaxation 2. Passive torque 3. EMG	NS
Magnusson et al., "Mechanical and Physical Responses" (1996)	PPT	13 men (hamstrings)	5 static stretches (90 s hold, 30 s rest) repeated 1 h later	1. Stiffness 2. Energy 3. Passive torque	S: ↓ energy, stiffness and peak torque
Klinge et al. (1997)	CCT	12 men in experimental group, 10 men in control group	4 × 45 s static stretches	1. ROM 2. Passive stiffness	NS
McHugh et al. (1998)	CCT	8 men, 8 women (hamstrings)	SLR stretch	1. Peak torque 2. ROM 3. EMG	S: ↑ ROM
Magnusson (1998)	CCT	12 men (hamstrings)	1. 90 s static stretches 2. Continuous movements 10 times at 20°/s	1. ROM 2. Passive stiffness	S: ↑ ROM
Muir et al. (1999)	RCT	10 men (one leg stretching, one leg control) (hamstrings)	1. Static stretching (30 s hold + 10 s rest) 4 times 2. Control—rest	1. Peak torque 2. Centre range (of hysteric loop)	NS

(continued)

Table 11.3 *(continued)*

References	Trial design	Sample	Interventions	Outcome measures	Main results
McNair et al. (2000)	CBT	15 men, 8 women (plantarflexors)	Static stretching 1. 1 × 60 s hold 2. 2 × 30 s hold 3. 4 × 15 s hold 4. Continuous passive movement for 60 s	1. Passive stiff ness 2. Peak torque	*Continuous movement* S: ↓ passive stiffness *Hold condition* S: ↓ peak tension
Magnusson et al. (2000)	PPT	20 men	3 static stretches (45 s hold, 30 s rest) repeated 1 h later	1. Stiffness 2. Energy 3. Passive torque	S: ↓ stress relaxation
Kubo et al. (2001)	CBT	7 men (plantarflexors)	Passive stretching to 35° dorsiflexion at 5°.s^{-1} for 10 min	1. Tendon stiff ness 2. Tendon hyster esis 3. MVC	S: ↓ tendon stiffness (10%), tendon hysteresis (34%)
Kubo et al. (2002)	CBT	8 men (plantarflexors)	Passive stretching to 35° dorsiflexion at 5°.s^{-1} for 5 min	1. Tendon stiff ness 2. Tendon hyster esis	S: ↓ tendon stiffness (8%), tendon hysteresis (29%)
Reid and McNair (2004)	RCT	43 adolescents (hamstrings)	Static stretching (30 s hold), 3 reps once a day, 5 days/week over 6 weeks	Passive force, angle and stiffness	S: ↑ ROM S: ↓ stiffness in terminal ROM
ROM AND ACTIVE STIFFNESS					
Wilson et al. (1992)	CCT	16 male weightlifters (n = 9 in experimental, n = 7 in control group)	Flexibility training (6-9 reps) of upper extremities, 10-15 min/session, twice a week for 8 weeks	1. Rebound bench press (RBP) 2. Purely concentric bench press	S: ↑ ROM (13%) S: ↑ RBP (5.4%) S: ↓ SEC stiffness (7.2%)
McNair and Stanley (1996)	CCT	12 men, 12 women (plantarflexors)	1. Static stretch (30 s hold + 30 s rest × 5) 2. Jogging (60% MHR) 3. Combined 2 + 1 Random order, each intervention for 10 min	1. ROM 2. Active stiffness	*Jogging group* S: ↓ active stiffness *All groups* S: ↑ ROM
Cornwell and Nelson (1997)	PPT	10 men (plantarflexors)	Passive stretching	1. ROM 2. Active stiffness	S: ↑ ROM
Hunter and Marshall (2002)	CCT	15 men, 15 women (n = 15 in experimental and control groups) (plantarflexors)	10 × 30 s static stretches	Active stiffness	NS
Cornwell et al. (2002)	PPT	10 men (plantarflexors)	Passive stretching (30 s, 6 times)	1. Active muscle stiffness 2. EMG 3. Jump height	S: ↓ jump height (7.4%) S: ↓ active stiffness (2.8%)

CCT = controlled clinical trial; RCT = randomised controlled trial; PPT = pre- and posttest trial; CBT, counterbalance trial; ROM = range of motion; EMG = electromyography; SEC = series elastic components; S = significant; NS = nonsignificant; SLR = straight leg raise; MVC = maximum voluntary contraction

The stretching regimen resumed again, and the 10° increase in ROM was restored after a further six weeks of stretching. In summary, the optimal prescription for stretching to make appropriate changes in ROM would appear to be 30 s holds, for one to three reps once per day, over a four- to six-week period. If the stretching regimen is not maintained, the benefits are lost within four weeks.

Effect of Static Stretching on Dynamic Flexibility

Although static stretching is effective in increasing static flexibility as measured by ROM, this type of stretching does not affect dynamic flexibility as measured by passive stiffness (Magnusson 1998) or active stiffness (McNair and Stanley 1996), but does affect viscoelastic properties by reducing stress relaxation (Halbertsma, van Bolhuis, and Goeken 1996; Magnusson et al. 1995; Muir, Chesworth, and Vandervoort 1999). The reduction of stress relaxation is an acute adaptation of the PEC to lower the imposed load across the myotendinous junction, where injury usually occurs (Magnusson et al., "Viscoelastic Stress" 1996). Cyclic stretching, or passive continuous motion, has been demonstrated to be effective for decreasing passive muscle stiffness (McNair, Dombroski, and Stanley 2001). It is believed that a less stiff muscle absorbs greater energy when forces are applied to it. As well, less muscle stiffness might be beneficial in reducing the severity of muscle soreness, as research has shown a positive relationship between passive stiffness and the severity of muscle soreness (McHugh et al. 1999).

Dynamic Stretching

In an extensive review of warm-up and stretching, Shellock and Prentice (1985, p. 272) stated that "dynamic stretching is important in athletic performance because it is essential for an extremity to be capable of moving through a nonrestricted ROM". Bandy, Irion, and Briggler (1998) investigated the effects of static stretching versus dynamic ROM stretching on hamstring flexibility in 58 normal subjects (41 men and 17 women) allocated to a static stretch group, a dynamic stretch group or a control group. Those in the static stretch group stretched the hamstrings 30 s five times a week for six weeks. The dynamic group dynamically stretched the hamstrings by actively extending the knee for 5 s, then held the stretch and the limit of knee extension for 5 s, then relaxed for 5 s. This sequence was repeated six times to equal a total of 30 s stretching. The increase in ROM was greater for the static (11.42°) compared to the dynamic group (4.27°). The results of this study demonstrated that while both static and dynamic stretching can increase ROM, static stretching is more effective at improving total ROM.

Ballistic Stretching

Ballistic stretching is thought to increase flexibility through a neurological mechanism. Researchers have reported an increase in ROM (Vujnovich and Dawson 1994; Worrell, Smith, and Winegardner 1994), decrease in EMG (Wiemann and Hahn 1997) and decreased H-reflex (Vujnovich and Dawson 1994) with ballistic stretching.

Ballistic stretching may be more harmful than other stretching techniques, as during ballistic stretching the muscle is stretched at a fast rate and rebounded back repetitively, resulting in greater tension and more absorbed energy within the muscle-tendon unit. The muscle, which is released immediately after application of a high force, does not have enough time to reduce tension (stress relaxation) or increase length (creep) (Taylor et al. 1990). Surprisingly, scientific evidence did not support the suggestion that ballistic stretching is more harmful than static stretching (Smith et al. 1993). Ballistic stretching (60 bounces per minute, 17 stretches per set for three sets) resulted in less severity of muscle soreness than static stretches (of the same intensity and duration, but held for 60 s) in college-age male volunteers.

Proprioceptive Neuromuscular Facilitation Stretching

Several PNF techniques have been used to increase flexibility, including slow-reversal-hold, contract-relax and hold-relax techniques. These techniques include the combination of alternating contraction and relaxation of both agonist and antagonist muscles (Shellock and Prentice 1985). The contractility property of muscles provides flexibility in the PNF technique on the basis of the viscoelastic characteristics of muscle and neuromuscular facilitation (Burke, Culligan, and Holt 2000). The contraction results in lengthening of the noncontractile elements (perimysium, endomysium, tendon) of muscle and consequently causes a relaxation of the muscle-tendon unit and decreased passive tension in the muscle (Taylor, Brooks, and Ryan 1997). The contraction also stimulates the muscle sensory receptors within the muscle (muscle spindle with a negative stretch reflex and GTO), which helps to relax the tensed muscle; the muscle-tendon unit becomes more relaxed after the contraction.

Some PNF techniques, such as slow-reversal-hold, require agonist muscle to contract in order to relax antagonist muscle (Shellock and Prentice 1985). *Reciprocal inhibition* occurs when the excitability signal from agonist muscle is transmitted by one set of neurons in the spinal cord to elicit muscle contraction and then an inhibitory signal is transmitted through a separate set of neurons to inhibit the antagonist muscle (Burke, Culligan, and Holt 2000). Reciprocal inhibition helps all antagonistic pairs of muscle to perform smooth movement. When antagonist muscle is inhibited, muscle is more easily stretched to the opposite direction.

Isometric contraction is commonly performed prior to passive stretching in the PNF technique. Post-isometric contraction caused a brief decrease in H-reflex response (83% by 1 s and 10% by 10 s) (Moore and Kukulka 1991). The depressed H-reflex was observed regardless of the intensity of isometric contraction (Enoka, Hutton, and Eldred 1980) and the velocities and amplitude of stretch (Gollhofer et al. 1998). A decrease in H-reflex after an isometric contraction could be a result of presynaptic inhibition. The suppression of reflex activity was brief (<10 s), indicating that passive stretching should be performed immediately after a preisometric contraction in order to yield maximal efficiency of stretching.

Proprioceptive neuromuscular facilitation stretching can result in greater improvement of ROM compared with static stretching (Magnusson et al., "Viscoelastic Stress" 1996; Sady, Wortman, and Blanke 1982). Prentice (1983) reported that after 10 weeks of stretching of the hamstrings (see table 11.4), a static stretch group had significantly increased hamstring ROM by 8° and the PNF group by 12°.

Sullivan, Dejulia, and Worrell (1992) compared PNF and static stretching with the pelvis posteriorly rotated and then anteriorly rotated across a two-week stretching program (see table 11.4). The PNF and static stretching groups showed a similar increase in ROM. The position of the pelvis was the most significant variable, with greater increases in ROM seen in the anterior pelvic tilt group. The authors commented that anteriorly rotating the pelvis would place greater tension on the hamstring and make the stretch more effective. A summary of the effects of PNF stretching of hamstrings on biomechanics and neuromuscular activity is presented in table 11.4.

Does Stretching Help Improve Sport Performance or Reduce Risk of Injury?

Stretching to improve the extensibility of muscle-tendon units has been viewed as an essential component of physical fitness, a method of improving the efficiency of movement (Godges, MacRae, and Engelke 1993) and muscle performance (Shellock and Prentice 1985; Worrell, Smith, and Winegardner 1994), reducing muscle strain and injury (Gleim and McHugh 1997; Hartig and Henderson 1999) and improving posture (White and Sahrmann 1994). However, a recent review (Weerapong, Hume, and Kolt 2005) has indicated that the benefits of stretching for performance and injury prevention may be minimal.

The Effect of Stretching on Athletic Performance

Stretching has become an integral part of the modern training program, in a similar way to strength and power, speed and mental skills training. Stretching is expected to increase flexibility and consequently enhance sport performance (Gleim and McHugh 1997). The effects of stretching on several performance parameters have been investigated, including muscle strength, power and endurance, as well as the efficiency of exercise such as running economy. However, researchers have reported detrimental effects of 3% to 12% on performance for static, ballistic and PNF stretching (see table 11.5).

Effects of Stretching on Muscle Strength, Power and Endurance

Acutely, static stretching, ballistic stretching and PNF can reduce muscle strength as determined by maximum lifting capacity (Church et al. 2001; Kokkonen, Nelson, and Cornwell 1998; Nelson and Kokkonen 2001) and isometric contraction force (Fowles, Sale, and MacDougall 2000). Nelson and colleagues ("Inhibition" 2001) reported a decrease in muscle strength for slow velocities of movement after static stretching. The negative acute effect of stretching on performance is probably explained by the change in neuromuscular transmission or biomechanical properties of muscle or both. Several studies of the effect of stretching on performance have demonstrated a reduction of performance associated with a decrease in H-reflex amplitude (Avela, Kyrolainen, and Komi 1999; Vujnovich and Dawson 1994).

Fowles, Sale, and MacDougall (2000) assessed strength performance after prolonged stretch by measuring force, EMG activity and passive stiffness. There was a maximal loss of strength immediately after stretch (28%) and strength loss persisted for more than 1 h after stretch (9%). Muscle activation and EMG activity were significantly depressed after stretching but were recovered by 15 min. Passive stiffness was recovered quickly after stretching, at 15 min, but not fully recovered within 1 h. The results imply that the impaired muscle activation was responsible for strength loss after prolonged stretching in the early phase while impaired contractile force was responsible for strength loss throughout the entire period. This implication is consistent with the results of several studies indicating that the neuromuscular inhibition mechanism is likely responsible for a decrease in performance after acute stretching (Behm, Button, and Butt 2001; Cornwell, Nelson, and Sidaway 2002).

Behm, Button, and Butt (2001) investigated the effects of static stretching on voluntary and evoked force and EMG activity of quadriceps. Maximal voluntary and

Table 11.4 The Effects of PNF Stretching of Hamstrings on Biomechanics and Neuromuscular Activity

References	Trial design	Samples	Interventions	Outcome measures	Main results
Sady et al. (1982)	CCT	1. Control (n = 10) 2. Static (n = 10) 3. Ballistic (n = 11) 4. PNF (n = 12)	1. Static: 3 × 6 s 2. Ballistic: repeated movements, 20 times 3. PNF: 3 × 6 s, 3 days/week for 6 weeks	ROM	*PNF and control group* S: ↑ ROM
Toft et al. (1989)	PPT	10 men	Contract-relax (8 s maximum contraction, 2 s relax, 8 s static stretch), 6 times	Stress relaxation	NS
Magnusson et al., "A Mechanism" (1996)	CCT	7 women (one leg stretch, one leg control) (hamstrings)	Static stretch (45 s hold, 15-30 s rest, 5 times), twice daily, 20 consecutive days	1. Stress relaxation 2. Energy 3. EMG 4. ROM	S: ↑ ROM
Magnusson et al., "Viscoelastic Stress" (1996)	CCT	8 neurologically intact and 6 spinal cord–injured volunteers (hamstrings)	Static stretch (90 s hold)	1. Stress relaxation 2. Passive torque 3. EMG	NS
Prentice (1983)	RCT	46 students (male and female); static stretch group and PNF group	Static stretch (10 s), 3 reps, 3 days/week over 10 weeks PNF slow reversal hold push phase 10 s, relax phase 10 s, 3 reps, sets as for static	ROM	S: ↑ ROM greater in PNF group
Sullivan et al. (1992)	RCT	20 subjects 1. Ant pelvic tilt (10) 2. Post pelvic tilt (10) Static stretch or PNF	Static stretch, 30 s hold, 1 rep in ant tilt, 1 rep in post tilt PNF CRC, 5 s contract, 5 s relax quads and hamstrings Both groups once a day, 4 times a week for 2 weeks	ROM Pelvic position	Pelvic position (ant) significant difference over post tilt S: ↑ ROM but static stretch and PNF same increase

PNF = proprioceptive neuromuscular facilitation; CCT = controlled clinical trial; RCT = randomised controlled trial; PPT = pre- and posttest trial; ROM = range of motion; EMG = electromyography; CRC = contract-relax-contract; S = significant; NS = nonsignificant.

Table 11.5 The Effects of Stretching on Performance

References	Trial design	Samples	Interventions	Outcome measures	Main results
STATIC STRETCHING					
Kokkonen et al. (1998)	CBT	15 men, 15 women (hamstrings)	20 min stretching (5 stretches, 3 times assisted, 3 times unassisted, 15 s hold, 15 s rest)	1. Sit and reach score 2. Maximum strength (1RM)	S: ↑ ROM (16%) S: ↓ strength (73%)
Fowles et al. (2000)	CBT	8 men, 4 women	13 maximal stretches × 135 s (= 2 min 15 s hold stretch, 5 s rest, then next stretch) static stretches, total 30 min	1. MVC 2. Twitch interpolation with EMG 3. Twitch characteristics at pre-, immediately post-, and 5, 15, 30, 45 and 60 min post-stretching	S: ↓ MVC (28%, 21%, 13%, 12%, 10% and 9% [by data collection time]) S: ↓ motor unit activation and EMG after treatment but recovered by 15 min
Knudson et al. (2001)	CBT	10 men, 10 women (quadriceps, hamstrings, plantarflexors)	3 × 15 s static stretches	1. Peak velocities 2. Duration of concentric phase 3. Duration of eccentric phase 4. Smallest knee angle 5. Jump height	NS
Nelson et al. "Inhibition" (2001)	PPT	10 men, 5 women	1 active and 3 passive stretches for 15 min	Peak torque at 1.05, 1.57, 2.62, 3.67 and 4.71 rad/s	S: ↓ strength at 1.05 rad/s (7.2%) and 1.57 rad/s (4.5%)
Behm et al. (2001)	CBT	12 men	Quadriceps stretching (45 s hold, 15 s rest for 5 sets)	1. MVC 2. EMG 3. Evoked torque 4. Tetanic torque	S: ↓ MVC (12%), muscle inactivation (2.8%), EMG (20%), evoked force (11.7%)
Cornwell et al. (2002)	CCT	10 men	3 × 30 s static stretches	1. Active muscle stiffness 2. EMG 3. Jump height	S: ↓ jump height (7.4%) S: ↓ active stiffness (2.8%)
Young and Behm (2003)	CBT	13 men, 4 women (quadriceps and plantarflexors)	2 × 30 s for each muscle	1. Concentric force 2. Concentric jump height 3. Concentric rate of force developed 4. Drop jump height	S: ↓ concentric force (4%)
Laur et al. (2003)	CBT	16 men, 16 women (hamstrings)	3 × 20 s static stretches	Perceived exertion	S: ↑ perceived exertion

BALLISTIC STRETCHING					
Nelson and Kokkonen (2001)	CBT	11 male and 11 female college students	Ballistic stretch: 15 bobs up and down once per min	1. Sit and reach score 2. Maximum strength (1RM)	S: ↑ ROM (7.5%) S: ↓ strength (7.3%)
DYNAMIC STRETCHING					
Weerapong et al. (2006)	RCT	12 male university students	1. Warm-up 2. Dynamic stretching 3. Massage	1. Leg stiffness 2. Jump and sprint performance	Stretching: S: ↑ jump and sprint performance vs. warm-up Massage: Small ↓ jump and sprint performance vs. warm-up
PNF STRETCHING (SHORT TERM)					
Wiktorsson-Moller et al. (1983)	CBT	8 healthy males	PNF: isometric contraction 4-6 s, relax 2 s, passive stretching 8 s	1. ROMs of lower extremities 2. Hamstrings and quadriceps strength	S:↑ ROM of ankle dorsiflexion and plantarflexion, hip flexion, extension, abduction, knee flexion
Church et al. (2001)	CBT	40 women	1. Static stretching 2. PNF	1. Vertical jump 2. ROM	PNF: S: ↓ jump height (3%)
PNF STRETCHING (LONG TERM)					
Wilson et al. (1992)	CCT	16 male weightlifters (n = 9 in experimental group, n = 7 in control group)	Flexibility training (6-9 reps) of upper extremities, 10-15 min/session, twice a week for 8 weeks	1. Rebound bench press (RBP) 2. Purely concentric bench press	S: ↑ ROM (3%) S: ↑ RBP (5.4%) S: ↓ SEC stiffness (7.2%)
Worrell et al. (1994)	CCT	19 participants with short hamstrings (one leg static, one leg PNF)	Static: 15 s hold, 15 s rest PNF: 5 s isometric, 5 s rest 4 reps/day, 5 days/week, 3 weeks	1. ROM 2. Con and ecc strength	S: ↑ strength Ecc: 60° and $120°.s^{-1}$ Con: $120°.s^{-1}$
Handel et al. (1997)	CCT	16 men (one leg stretch, one leg control)	CR for 8 weeks (isometric at 70% MVC, 1-2 s rest, 10-15 s passive stretching) Follow-up at 0, 4, 8 weeks	Knee flexion and extension Con: 240°, 180°, 120°, $60°.s^{-1}$ Ecc: 60° and $120°.s^{-1}$	S: ↑ torque Extension: ecc at 120° and $60°.s^{-1}$ Flexion: all velocities
Hunter and Marshall (2002)	RCT	60 participants (15 per group)	Static stretching (3 × 20 s) and PNF (submaximal contraction 10 s)	1. Drop jump 2. Countermovement jump	NS

CCT = controlled clinical trial; RCT = randomised controlled trial; PPT = pre- and posttest trial; CBT, counterbalance trial; ROM = range of motion; EMG = electromyography; SEC = series elastic components; S = significant; NS = nonsignificant; MVC = maximum voluntary contraction; con = concentric contraction; ecc = eccentric contraction; RM = repetition maximum; PNF = proprioceptive neuromuscular facilitation; CR = contract-relax.

evoked contraction decreased similarly by 12%, and muscle activation and EMG activity decreased 2.8% and 20%, respectively. Similarly, Cornwell, Nelson, and Sidaway (2002) reported that static stretching of gastrocsoleus (180 s) reduced jump height by 7.4% but that active stiffness was reduced by only 2.8%. Other researchers reported that static stretching reduced jumping performance (knee bend) by 3% (Knudson et al. 2001; Young and Behm 2003). A reduction in jumping performance was consistent with a reduction of EMG activity, but there were no changes in biomechanical variables (vertical velocity, knee angle, duration of concentric and eccentric phases). The detrimental effects of acute stretching exercise on muscular endurance were shown by Laur and colleagues (2003) when the application of acute stretching reduced the maximal number of repetitions performed with a submaximal load and also produced higher perceived exertion scores. Although the magnitude of reduction was small, it was statistically significant.

Studies on the effects of long-term stretching have demonstrated a positive effect of stretching on strength performance (Handel et al. 1997; Wilson, Elliott, and Wood 1992). Three weeks of flexibility training in both PNF and static stretching groups increased peak torque of hamstrings eccentrically (at 60°/s and 120°/s) and concentrically (at 120°/s only) (Worrell, Smith, and Winegardner 1994). Proprioceptive neuromuscular facilitation training (contract-relax technique) for eight weeks increased maximum torque of knee flexors and extensors (Handel et al. 1997). The increase in muscle strength of knee flexors was significant at all velocities, perhaps because the contraction phase of the PNF stretching technique may have the same effect as isometric muscle training. Knee flexors, which are used in normal activity less than knee extensors, showed a greater increase in muscle strength. Along more functional lines, Wilson, Elliott, and Wood (1992) reported that eight weeks of static flexibility training increased rebound bench press performance by 5.4%, in accordance with a decrease in active muscle stiffness by 7.2%.

Flexibility-induced performance enhancement might result from increased musculotendinous compliance that facilitates the use of energy strain in stretch-shorten cycle activities. In contrast with acute stretching, the combination of static and PNF training for 10 weeks, as reported by Hunter and Marshall (2002), did not have a detrimental effect on countermovement jump and drop jump but helped to increase knee joint ROM. Flexibility training of at least three weeks seems beneficial to some performance factors as indicated by increased ROM and muscle strength.

Effects of Stretching on the Efficiency of Movement

Flexibility is thought to play an important role in the efficiency of movement (Gajdosik, Giuliani, and Bohannon 1990) by enabling the use of elastic potential energy in muscle (Gleim and McHugh 1997). A compliant muscle-tendon unit needs more contractile force to transmit to the joint than a less compliant (more stiff) muscle-tendon unit, which causes a greater delay in external force generation. A stiffer muscle would provide a more efficient transmission of contractile force production, but this contradicts the aim of stretching, which is to increase muscle-tendon unit compliance.

Craib and colleagues (1996) reported that less flexible runners had a reduced aerobic demand during running (better running economy). However, the study was cross-sectional, did not control the training program of the runners and did not consider other factors that might have influenced running economy such as kinematic, anthropometric, physiological and cellular variables. In contrast, flexibility training of the hip flexors (three weeks) (Godges, MacRae, and Engleke 1993) and lower leg muscles (quadriceps, hamstrings and gastrosoleus, 10 weeks) (Nelson et al., "Chronic stretching" 2001) resulted in increased ROM but had no effect on running economy.

The Effect of Stretching on Injury Risk

There is evidence from some studies that people with reduced flexibility are more at risk of developing injuries (Witvrouw, Danneels, and Asselmann 2003); however, the relationship between stretching and injury prevention is obscure (Witvrouw et al. 2004). Pope, Herbert, and Kirwan (1998) examined 1093 army recruits and noted that subjects with poor flexibility were 2.5 times more likely to be injured than those with average flexibility. However, a recent review of the effects of stretching on the incidence of injury pointed to inconclusive results (Weldon and Hill 2003). A possible explanation is that exercise-related injury is a multifactorial, complex phenomenon involving physiological, psychological and environmental factors. The majority of research in this area has been retrospective and does not establish a clear relationship between flexibility and injury (Dubravcic-Simunjak et al. 2003; Gleim and McHugh 1997).

The type of muscle work undertaken by athletes may determine whether or not they need to perform stretching routines. Sports involving explosive-type skills (such as gymnastics, soccer and rugby) with many and maximal stretch-shorten cycles, requiring the muscle to be compliant enough to store energy and release high amounts of elastic energy, may necessitate stretching procedures as a prophylactic measure to prevent injury (Witvrouw et al. 2004). Sports that entail no or low stretch-shorten cycles such as cycling and jogging, in which most of the effort is converted directly into external work, may not require stretching exercises to improve compliance and reduce injury. An increase in ROM (or static flexibility) resulting from stretching may not be necessary for a majority of sports such as running and swimming that do not require extreme ROMs (Witvrouw et al. 2004). Dynamic flexibility might be more important because it represents the resistance of the muscle-tendon unit during movement. However, there is no research on the relationship between dynamic flexibility and rate of injury.

> Age, gender and environmental factors can influence flexibility across the life span.

In one prospective study, van Mechelen and colleagues (1993) provided a standardized program of stretching exercises to runners and assessed the number of injuries after 16 weeks. There was no reduction in injury incidence per 1000 h of running between the group that performed the experimental standardized program of stretching exercise and the control subjects, who received no stretching information. In a study of army recruits (Pope et al. 2000), a preexercise stretching program of 11 to 12 weeks did not reduce the risk of exercise-related injury. Fitness and age (Pope et al. 2000) and the early detection of symptoms of overuse injuries (van Mechelen et al. 1993) were more important factors related to injury than the stretching exercise.

In contrast to the general belief that stretching helps reduce the risk of muscle damage from unaccustomed eccentric exercise, research has shown that both prolonged static and ballistic stretching (60 s for two stretches for each muscle) induced muscle damage (Smith et al. 1993). No research has shown any benefit from stretching with respect to the severity of muscle damage or muscle soreness either before or after exercise (Herbert and Gabriel 2002). A reason for the lack of evidence for the benefits of stretching may be that earlier research addressed only static stretching. A decrease in passive muscle stiffness may be the key to reducing the severity of muscle damage, and static stretching in these studies (High, Howley, and Franks 1989; Johansson et al. 1999; Lund et al. 1998) was not held long enough to induce a decrease in passive stiffness (most studies used stretches held less than 90 s). McNair, Dombroski, and Stanley (2001) reported that dynamic stretching did reduce muscle stiffness, while McHugh and colleagues (1999) showed that passive stiffness was related to the severity of muscle damage as measured by strength loss, pain, muscle tenderness and creatine kinase activity. Any stretching technique that can reduce passive stiffness might help to reduce the severity of muscle damage.

Other Factors Influencing the Effectiveness of Stretching and the Resulting Flexibility

A number of factors may influence flexibility across the life span; these include age, gender and environmental factors.

Childhood and Adolescence

There has been debate in the literature regarding the influence of growth on the flexibility of children and adolescents. Early studies by Corbin and Noble (1980) suggested that flexibility increased in a child until adolescence, when there appeared to be a plateau effect that was followed by a steady decrease in mobility as the individual aged. Phillips (1955) did not support this finding, stating that elementary school–aged children become less flexible as they grow, reaching a low point between 10 and 12 years of age. However, Feldman and colleagues (1999) demonstrated that growth during the adolescent growth spurt did not lead to decreases in flexibility.

It has been suggested that the reduction in flexibility of adolescents during the growth spurt leads to pain, particularly in the lower extremity. However, a prospective cohort study (Shrier et al. 2001) addressed whether reductions in flexibility in 500 adolescents were associated with the development of lower limb pain. Along with other variables such as physical and occupational activities, flexibility of the quadriceps, hamstrings and flexibility in the sit and reach test were measured at baseline, 3 months, 6 months and 12 months. High growth rates and poor flexibility were not risk factors for the development of lower limb pain.

Problems occur during the many minor growth spurts in a child's life when the long bones grow rapidly and increase the tension of the musculotendinous unit (Leard 1984). This causes tightness around the joints and often places a considerable amount of stress at the epiphysial attachments. In some cases this causes an avulsion fracture to occur (i.e., the tearing away of a bony landmark to which a muscle is attached) when too much stress is placed on the apophysis (a projection of a bone to which a muscle is attached) by a forceful contraction of a muscle or group of muscles. The most common avulsion fractures in children involve the forearm flexors attached to the medial epicondyle of the humerus, sartorius muscle attached to the anterior superior iliac spine, rectus femoris muscle attached to the anterior inferior iliac spine, iliopsoas muscle attached to the lesser trochanter of the femur, abdominal muscles attached to the iliac crest, hamstring muscles attached to the ischial tuberosity, patellar tendon attached to the tibial tuberosity and the Achilles tendon attached to the calcaneus.

Micheli and Fehlandt (1992) reported 724 cases of apophysitis or tendinitis in 445 children seen at the Sports Medicine Division of the Boston Children's Hospital between 1980 and 1990. Growth played a pivotal role in the development of these injuries. The mainstay of treatment was conservative management including rest and correction of strength and flexibility imbalances.

Middle and Old Age

As the individual ages, muscles, tendons and connective tissue shorten and calcification of some cartilage occurs, with a resultant loss in the ROM. Loss of range of motion can be minimized with a well-planned stretching and strength training program provided that the individual does not overstress the musculoskeletal system.

Veteran or senior athletes must understand that as they age their muscles shorten and connective tissue becomes stiffer and tighter. Cartilage steadily calcifies and becomes thinner because of wear and tear, and as a result cannot absorb the pressure it could tolerate when the individual was considerably younger. All this leads to a steady reduction of mobility; but if regular stretching is continued that is not overly stressful, reasonable levels of flexibility can be maintained. With respect to maintaining and improving these levels of flexibility, Feland, Myer, and Merrill (2001) have demonstrated that longer durations (90 s) of hold result in greater improvement in ROM, particularly in elderly populations compared with younger populations involved in similar stretching studies.

Gender

Phillips (1955) found that elementary school–aged girls were more flexible than boys of a similar age. From adolescence onward, females appear to be more flexible, with smaller bones and less musculature than males. These findings were validated by Cornbleet and Woosley (1996), who reported that sit and reach test results were less in 199 boys than in 211 girls aged between 5 and 12 years; this was consistent with results from other studies investigating levels of flexibility using the sit and reach test. Shepherd, Berridge, and Montelpare (1990) also reported that women were more flexible than men in the 45- to 75-year-old group, indicating that these characteristics are consistent for each gender over the life span.

Environmental Conditions: Temperature

Local heating of muscle tissue has been shown to improve flexibility by promoting tissue relaxation and compliance. According to DeVries (1986), flexibility was improved by 20% by the local warming of a joint to 45° C and was decreased by between 10% and 20% by cooling to 18° C. Funk and colleagues (2001) reported that heat applied to the hamstrings of 30 male undergraduate student athletes prior to a static stretch procedure caused a significant increase in ROM over that with static stretching alone.

Warm-up has also been considered to be important prior to stretching, but this advice appears to be anecdotally based. Studies on the relationship of warm-up to improvements in flexibility have shown that static stretching alone is as effective as static stretching following a warm-up (Wiktorsson-Moller et al. 1983; Williford et al. 1986). Gillette and colleagues (1991) assessed 18- to 35-year-old males (n = 20) for core body temperature after a 20 min submaximal treadmill test followed by a 2 min stretch procedure to the hamstrings. Raising the core body temperature by 1° C did not improve flexibility in the hamstrings any more than static stretching alone.

How Do We Measure Improvement in Flexibility?

Flexibility or joint mobility is influenced by factors other than age, gender and temperature, such as body type and psychological stress. A test of flexibility should have the following characteristics:

- The test must be specific to a single joint (with the exception of the vertebral column).
- The scoring system must be independent of the size or proportionality of the athlete.
- Either a standardized warm-up or no warm-up must be permitted.
- The posture adopted by the subject must not hinder or enhance the attainment of maximum joint mobility.
- The role of the tester in guiding the movement must be clearly defined.

With these characteristics in mind, many of the "field tests" of flexibility cannot be judged as valid. For example, the sit and reach test has been used by many as a "general" test of hamstring and lower trunk mobility. However, because it tests multiple joints, the sit and reach is not valid. Individuals with varying upper and lower limb lengths, differences in hamstring extensibility or variations in the mobility of the lumbar vertebrae may all obtain the same score on this test.

The procedures for a valid test must include a clear statement regarding the extent to which the tester may influence the movement. Since the amount of force provided by an assistant cannot always be controlled (though a portable force transducer may measure the amount of force a tester applies), it is often necessary that the body segments be moved only by the subject to reach the extreme points in the ROM. Any assistance by the tester can invalidate the score and could cause soft tissue injury to the subject. Passive flexibility testing usually uses the athlete's report of tightness or pain to establish the end ROM, or trained therapists often use an "end feel" to determine the end position for a test. However, the reliability of these methods for determining the end ROM is usually less than for static tests (Hume et al. 2004).

Testing Environment

Laboratory testing allows controlled conditions that can enable test-retest comparisons. However, a common criticism of lab tests is that they are not sport specific enough. Field tests often involve a small amount of subjectivity in the rating procedure but are useful for coaches in the field. The assessment can be carried out quickly, and a rating scale is used in order to give the subject a specific numerical rating for each joint.

The temperature of muscles resulting either from environmental conditions or from warm-up can affect their flexibility (Wiktorsson-Moller et al. 1983; Williford et al. 1986). Warm-up should be standardized if there are to be repeated assessments of improvements in flexibility from stretch training.

What Joints and Stretching Technique Should Be Used for Measuring Flexibility?

In general, the joint ROM that is important for performance in the sport should be identified, and then the joint-specific motion should be measured. Whether a passive or an active test should be conducted will also depend on how ROM is achieved in the sport and on the ease of measurement. For example, passive gastrocnemius extensibility or flexibility of the ankle joint in dorsiflexion is often tested in a weight-bearing static position due to ease of measurement. However, active measurement during a movement, such as that completed by Soper, Reid, and Hume (2004) for rowers, may be more useful.

The reliability and validity of flexibility measurement techniques are important. Usually the intratester and intertester reliability range from 1° to 4° depending on the measurement technique (goniometer vs. video digitisation); therefore any changes in ROM must take into account the magnitude of the possible measurement errors when the data are interpreted for the coach or athlete.

Static Flexibility Testing

Static flexibility testing is a method of measurement carried out with athletes in a nondynamic situation, meaning that they are not performing a sport skill but rather are having their ROM assessed using specific flexibility testing positions. Scores on the tests of static flexibility are independent of the dimensions of an athlete's body segments. The positions must be attained only as a result of muscular contraction by the subject. That is, the tester must not force or assist the subject in any way except to keep the limb in the correct movement plane or to prevent other movements of the body from affecting the test score. An example of a static flexibility test for the ankle is shown in figure 11.3, where the athlete is in a static weight-bearing ankle dorsiflexion position.

Passive Flexibility Testing

Passive flexibility testing involves the examiner's moving the athlete to an end ROM position while the athlete relaxes the muscles. The end ROM is usually determined as the point at which the athlete reports tightness or pain or the examiner produces an "end feel". An example of a passive flexibility test is the passive straight-leg raise, which requires the athlete to lie supine with the nontested limb at times strapped to a plinth. The examiner then passively flexes the hip with the knee in extension until the athlete complains of discomfort or tightness in the hamstring region. This determines the "end point" of the test. The passive straight-leg raise test has been used as a measure of hamstring muscle length and also as a measure of neural mobility. There is evidence that this test may be unreliable as a measure of hamstring muscle length because of the possible contribution to the movement of posterior rotation of the pelvis during the test (Bohannon, Gajdosik, and LeVeau 1985).

The passive knee extension test uses the same procedures as the active knee extension test (described later) except that the examiner extends the knee passively until firm resistance is felt or the subject complains of discomfort or tightness in the hamstring region. This is determined as the end point of the test. Gajdosik (1991) utilized a passive knee extension test with 24 healthy subjects in a side-lying position. Supportive padding ensured minimal movement of the pelvis and thigh during the execution of test. A camera recorded the maximum knee angle and a handheld dynamometer recorded the maximal resistance to passive stretch during the test. Test-retest reliability for the maximum passive knee angle was high (an intraclass coefficient of 0.90, $r = 0.91$).

Passive flexibility can also be measured using a dynamometer. Details are provided in a later section, along with figure 11.6, which shows an example of a passive knee extension test for hamstring flexibility using a Kincom dynamometer.

Active Flexibility Testing

Active flexibility testing requires athletes to use their own muscle force to reach an end ROM position. Active end ROM is usually less than passive end ROM. An example of active flexibility testing is the active knee extension test (Gajdosik and Lusin 1983), which requires the athlete to lie supine with a strap across the anterior iliac spines and another across the thigh of the nontesting limb. The athlete actively maintains the hip of the limb to be tested at 90°. A cross-bar apparatus is sometimes placed against the anterior portion of the thigh to ensure that the hip remains at 90°. The athlete then actively extends the knee until myoclonus is observed. Gajdosik and Lusin (1983) described myoclonus as an alternating contraction and relaxation of the quadriceps and hamstring muscles at the point where knee extension becomes limited by hamstring muscle length. At that point, the subject flexes the knee slightly until the myoclonus stops, and this determines the end point of the test. The test-retest reliability for left and right extremities was shown to be high (r = .99).

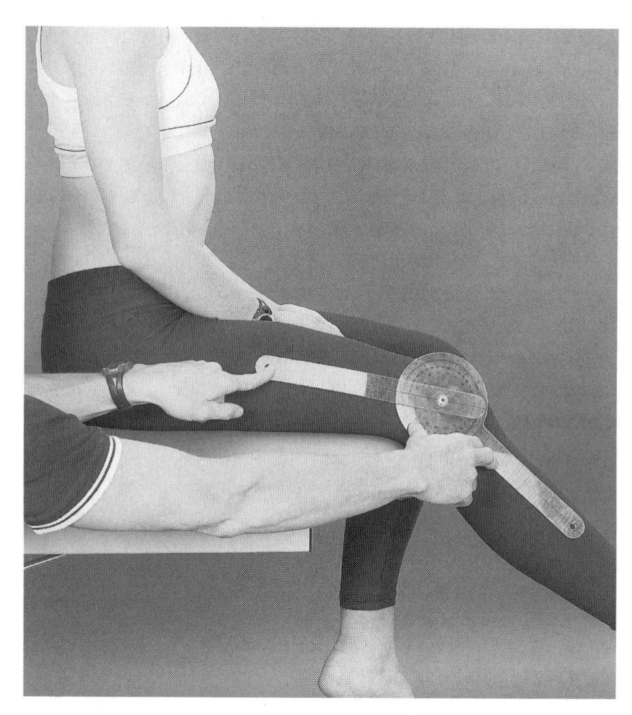

Figure 11.2 An example of use of a goniometer to measure active static knee extension angle to assess hamstring flexibility.

Reliability and Validity of Tools Used for Measuring ROM

A number of devices have been employed to determine the ROM of a body segment. These include the goniometer and arthrodial protractor. Both instruments perform essentially the same function as the Leighton flexometer; and, while they are not as adaptable to the variety of movements that may be assessed using the flexometer, they do represent a less expensive alternative. High-speed cinematography or videography to provide a 2-D or 3-D reconstruction of athletic performance is used extensively in the field of biomechanics in technique analysis. These tools may also be used to measure functional flexibility during activity. Two-dimensional video analysis with digitisation systems is now frequently used; but as yet the use of 3-D video analysis is not common, mainly because of the time and cost involved. Most ROM is measured in one plane only (i.e., 2-D using a single-axis electrogoniometer, or a sagittal plane video image), whereas new technologies such as 3-D videography and 3-D electrogoniometry can provide ROM measures in multiple dimensions.

Goniometer

The simplest device for measuring static flexibility is the goniometer (see figure 11.2), which is a protractor with two attached movable arms. Numerous researchers have used this tool for measurement of ROM (Halbertsma, Ludwig, and Goeken 1994; Ross 1999; Sullivan, Dejulia, and Worrell 1992; Webright, Randolph, and Perrin 1997; Willy et al. 2001). The goniometer measures the angle between two body segments at the extreme ends of the ROM. The tester must be very careful to locate the axis of the bones that form the joint and must be aware that the soft tissue around the joint can influence the accuracy of the measurement. Goniometers have been shown to have acceptable intra- and intertester reliability (Bandy, Irion, and Briggler 1997); but testing with a manual goniometer is really applicable only to static ROM measures, as a static hold at end range is required for the tester to be able to read off the angle on the device. Figure 11.2 shows a goniometer being used to measure hamstring flexibility via knee extension angle.

Electrogoniometer

A more sophisticated goniometer known as the electrogoniometer or "elgon" (see figure 11.3) incorporates a potentiometer at the axis of the two measurement arms. Changes in the joint angle are recorded as voltage fluctuations, providing a real-time analog display of joint motion. Elgons may be used to provide measurements of static as well as functional flexibility. Recent advances

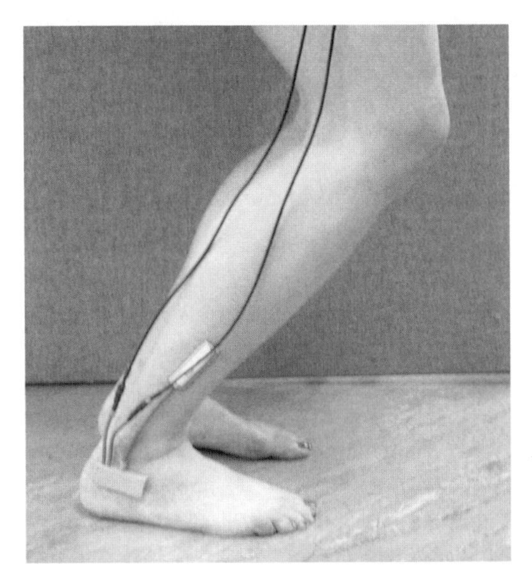

Figure 11.3 An electrogoniometer used to measure range of motion at the ankle joint.

in elgon technology have permitted the recording of 3-D movements without the previous encumbrance to normal athletic performance. The reliability of the Penny and Giles electrogoniometer has been established by Soper, Reid, and Hume (2004) for active and passive ankle dorsiflexion in a group of 10 subjects (three males and seven females). The mean changes in consecutive day-to-day measurement were –0.1 ± 2.1 to –1.5 ± 3.9. The highest standard error of the mean for these tests repeated one day apart was 2.8° (95% confidence interval [CI] = 1.9-5.1°). The test-retest correlation coefficients between day 1 and day 2 were all greater than 0.90 (95% CI = 0.62-0.99). The results indicated high levels of reliability in the measurement using the electrogoniometer.

Leighton Flexometer

The Leighton flexometer is a device strapped to the subject's limb in a standard position. The flexometer is set to zero using a gravity device; the athlete then moves the limb to end ROM, and the flexometer needle, which has moved to indicate the degrees of motion, is then clicked to a stop position and the ROM read off the flexometer dial. Figure 11.4 shows a flexometer being used to measure ROM for arm abduction.

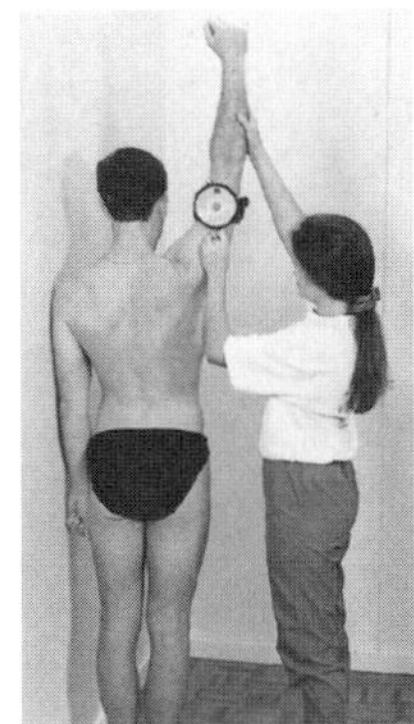

Figure 11.4 An example of use of a Leighton flexometer to measure arm abduction.

Digital Camera and Digitisation Software

With digital cameras, high-quality ROM photos can be taken that can then be analysed using computerized angle measurement systems such as siliconCOACH. Active movements, rather than just static positions, can be measured by video. For example, the extreme ROMs attained by gymnasts can be measured using grabbed video frames and siliconCOACH (see figure 11.5). Gribble, Hertel, and Denegar (2005) reported high reliability and reasonable validity for joint angles in a sagittal plane during static and slow dynamic tasks measured using SMART software (ECI Software, Inc., Boston) for digitisation. They reported intrarater ICC reliability values of 0.6 to 0.92 for the static task and 0.76 to 0.89 for the dynamic task. Measurement errors were reported to be 0.29° and 1.07°, respectively.

Figure 11.5 An example of use of a grabbed video frame and siliconCOACH to digitise hip angle for dynamic hip flexibility testing of a gymnast.

Dynamometer

A dynamometer is typically used for strength assessment, but it is also used to measure passive and active ROM (figure 11.6). With respect to the Kincom dynamometer as a reliable measure of hamstring muscle length, Magnusson and colleagues (1995) utilized a test-retest protocol on 10 male volunteers to assess reliability of measuring the final static stretch position of the hamstring muscle group with the passive knee extension test. A correlation coefficient of r = 0.99 was obtained with this study, indicating high levels of reliability.

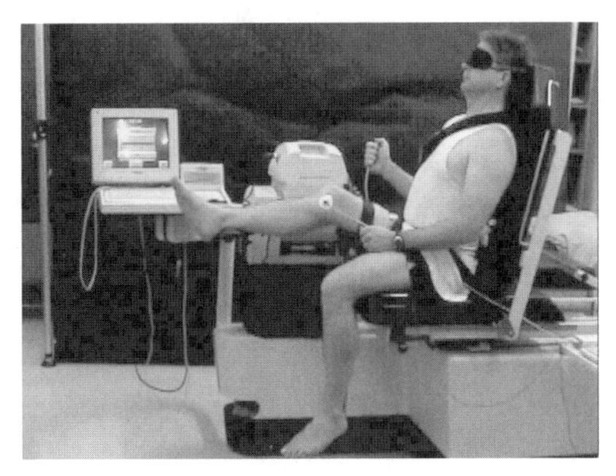

Figure 11.6 An example of passive hamstring flexibility testing on a Kincom dynamometer.

Summary

Flexibility training can increase the elasticity of the musculotendinous unit and in so doing increase the utilization of elastic energy in the stretch-shorten cycle movement.

The effects on ROM due to stretching of muscle-tendon units are thought to be both neural and mechanical, but there is debate as to which of these mechanisms apply following periodic stretching programs. There is increasing support for the idea that these changes are related to stretch tolerance rather than alterations in the mechanical properties of the muscle. The most common duration for applying the stretch is 30 s. Between one and three reps of the stretch are required to create a measurable change. The number of times per week and the number of weeks over which an athlete should stretch remain variable, but the most common frequency is five times per week for six weeks.

Common clinical practice suggests that preexercise stretching can enhance performance and prevent injuries by increasing flexibility. However, current scientific research does not support this notion. Rather, the acute effects of stretching can have detrimental effects on performance parameters such as muscle strength and jumping performance. The ideal flexibility differs for the performance of each sport activity. Compliant muscle might be beneficial to eccentric contraction, while stiffer muscle might be more suitable for concentric and isometric contractions.

Static stretching and ballistic stretching seem to have inconclusive effects on the incidence of injury and no effects on the severity of muscle damage. In order to clarify the effects of stretching, further research is recommended to

- provide information on the relationship of dynamic flexibility, performance and rate of injury;
- examine the effects of several stretching techniques, such as ballistic, PNF and dynamic stretching, on dynamic flexibility and neuromuscular sensitivity;
- compare the effects of several stretching techniques, such as static, ballistic, PNF and dynamic stretching, on different types of performance, the severity of muscle soreness, running economy and rate of injury;
- study the effects of acute stretching after long-term flexibility training; and
- provide more information on the appropriate flexibility level to enhance performance and reduce the risk of injury.

twelve

Balance and Agility

David Lloyd, PhD; Timothy R. Ackland, PhD; and Jodie Cochrane, PhD

Together with capacities such as speed, power, strength and flexibility, the coordination of muscle actions by the central nervous system plays a vital role in successful athletic performance. The ability to accurately coordinate the timing and contraction strength of skeletal muscles is essential in the related capacities of balance and agility. While both are modified by the physical structure of an athlete and may be affected by technique, balance and agility rely heavily on the development of neuromuscular control. According to Tittel (1988), this is particularly important for technical acrobatic sports such as gymnastics, rhythmic gymnastics, diving and figure skating, as well as for such activities as swimming, wrestling, fencing, boxing and ball games. In addition to neuromuscular control, agility also depends on a player's cognitive ability to read the game situation in team sports, such as football, hockey, basketball and netball.

Factors Affecting Balance

In all activities, whether stationary or mobile, balance plays an important role. Some activities require static balance whereas many sports require athletes to maintain stability during highly dynamic tasks. Static balance refers to balance in the situation in which a set position must be maintained for a period of time, as in target sports like archery and shooting; here the maintenance of a balanced and stable posture is essential for attaining accuracy. In other sports such as gymnastics and diving, stable static positions, held without excessive movement, demonstrate the athlete's strength and coordination.

Static balance can take on many forms; the simplest is quiet stance, which would be representative of the balance required in some aspects of gymnastics and diving. In other sports, however, static balance has to be maintained while a force is applied. For example, in archery there is a reaction force due to bow movement and the shooting of the arrow; in shooting, a reaction force is due to the reaction (or recoil) from shooting of the bullet. These various situations impose different neuromuscular requirements for the maintenance of static balance.

Dynamic balance is needed in sports that require stability while the athlete is in motion. Dynamic balance and the development of this capacity are essential in highly mobile sports in which the athlete must quickly react to changing circumstances. For this reason, dynamic balance and agility are closely related.

Another side to dynamic balance that is of ever-increasing concern is injury in sport. During the performance of sporting manoeuvres, athletes need to maintain control and balance of the upper body, or control stability of specific joints such as the ankle, knee and shoulder, or both. Upper body posture and motion directly influence the loads experienced by the lower limb joints; thus stability of the upper body affects the risk of lower limb joint injury. In addition, when a lower limb joint is loaded, it needs to be stabilized so that the internal structures such as ligaments and cartilage are protected from injury.

One of the main findings to come from balance and stability research is that neuromuscular and biomechanical mechanisms used for static and dynamic balance have some commonalities; however, there are also some profound differences.

Whole-Body Mechanical Considerations

High levels of balance in a sporting activity are dependent on the area of the base of support, the position of the centre of gravity and the mass (body weight) of the performer. A wide, but comfortable, positioning of the feet aids the static balance in archery and pistol shooting, while the execution of a stable handstand in diving or gymnastics requires the hands to be placed about shoulder-width apart. Thus, the area of the base of support is maximised within the ability of the performer to control his or her posture. Obviously, an exaggerated base of support caused by positioning of the feet or hands too far apart can restrict the performance of subsequent movements and can be aesthetically displeasing. In addition, an exaggerated base of support may place the limbs in postures that require greater joint torques to be generated from the joints that may not be capable of providing this support. For example, when one performs a handstand, straight elbows can more easily support vertical loads than can flexed elbows.

Another way to maintain static balance in the face of an externally applied load is to keep the body's of centre of gravity low and within the base of support. When a player adopts this posture, it is difficult for an opponent to move him or her from a set position. This balanced and stable position is sought in wrestling, for example; the low centre of gravity and the wide base of support make it difficult for an opponent to shift the line of gravity outside this base. If the opponent achieves this, then balance is lost and the supporting limbs must move to avoid a fall. These principles are used to good effect in other combative and contact sports. One can disrupt an opponent's balance by pushing or pulling above the centre of gravity and thereby moving it outside the base of support, or below the centre of gravity to take away the base of support, as in a trip in wrestling or a tackle in rugby.

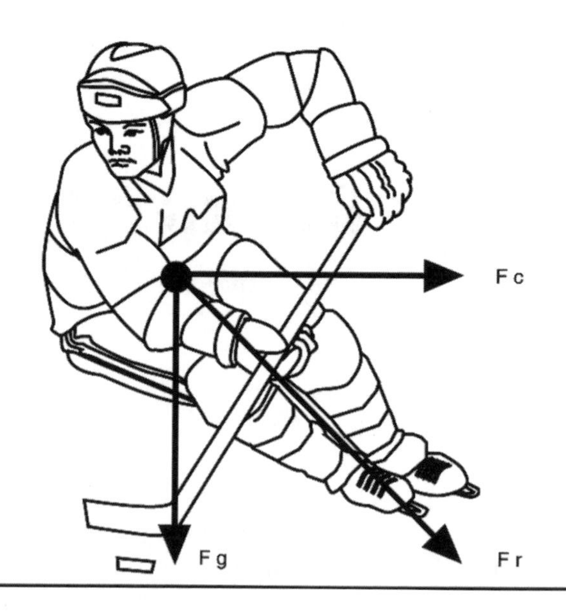

Figure 12.1 Dynamic balance in ice hockey. The resultant force (Fr) is the vector sum of the force of gravity (Fg) and the centrifugal force (Fc). Fr passes through the instantaneous base of support provided by the area bounded by the blades of the player's skates.

Using a similar logic, offensive players and ball carriers in contact football should modify their running technique when in close proximity to opponents. By shortening their stride and increasing stride rate, players are able to maintain a more "compact" and balanced gait, which keeps the centre of gravity closer to the base of support and consequently reduces the effectiveness of a tackle. Training for these players should include drills to reinforce the rapid short steps required in this situation. Furthermore, a player's inertia, or resistance to motion, is determined by his or her body mass. In contact sports, players of greater mass are more difficult to move; we see evidence for this in the body mass of linemen in American football (see chapter 3).

However, in many movements, including walking, the line of gravity projects outside the base of support yet the athlete may still be balanced. This dynamic balance is achieved due to the addition of some external force or load. For example, the ice hockey player shown in figure 12.1 would appear to be unbalanced and would fall if this were a static situation. However, the dynamic nature of the movement imposes a centrifugal force on the body. When added to the effect of gravity, the resultant force acts through the base of support to provide a posture in dynamic equilibrium. Nevertheless, dynamic balance such as this still requires the swinging leg to be placed down as the player moves across the ground to prevent falling over. Indeed, in contact football, the defensive player may stop progress of the offensive player by "taking out" or disturbing the pathway of the swinging leg using an "ankle tap". Thus one attains dynamic balance by ensuring that the dynamic, preplanned movement pattern is allowed to keep going. Many spinal reflex responses actually assist the movement's continuation rather than resisting any external perturbation to the movement (Hasan 2005).

Impacts or perturbations can be applied to the dynamically balanced player by an opponent, but also by the player him- or herself (e.g., when changing direction to avoid an opposing player, or when passing or kicking a ball). The most common self-generated large impact is the initial foot contact with the ground during landing, running or sidestepping. These impacts are very large and require considerable stabilization. It is little wonder, then, that poor control of these impacts is the cause of many lower limb joint injuries. Therefore, adjustments to the athlete's posture are needed throughout the execution of these skills as external forces vary in magnitude and direction. The rapid response required to maintain dynamic balance and stability is reliant upon a well-developed and coordinated musculoskeletal system.

Adjustments to an athlete's posture are needed throughout the execution of dynamic skills as external forces vary in magnitude and direction. The rapid response required to maintain dynamic balance and stability is reliant upon a well-developed and coordinated musculoskeletal system.

Neuromuscular Considerations

Maintenance of balance and stability relies on the control of joint posture and motion, which in turn results from constraints provided by muscular and nonmuscular tissue as well as from the nervous system's control of muscle activity, and resulting muscle force. The ability to control balance and stability is one that develops in stages during childhood as the individual learns to integrate inputs from visual and other sensory sources.

Maintenance of Balance and Stability

Maintenance of balance and stability depends on the control of joint posture and motion. Joint posture, in turn, depends on the passive constraints provided by ligaments and other nonmuscular tissue. Moreover, joint posture and motion rely heavily on the nervous system's voluntary and reflexive control of the skeletal muscles' intrinsic mechanical properties and force production capability. Voluntary control, centred in the brain, is based on the ability of the person to predict or anticipate the demands that are going to be experienced during the subsequent movement, and on conscious or unconscious selection of the muscle activation patterns appropriate to the task. Voluntary activation patterns can use the neural circuits in the spinal cord to coordinate muscle activation and reflexes can directly modulate the voluntary activation patterns. These may be coordinated by the spinal cord's neural circuits, but can also be automated responses via pathways that go through the brain.

Voluntary and reflexive control of muscle activity relies on the flow of information from sensory systems within the body and the correct interpretation of these signals so that appropriate movements or responses are initiated by the brain or peripheral nervous system or both. The *visual, vestibular* and *somatosensory* systems are involved in the control of balance and stability. *Visual* receptors (the eyes) provide information on a person's spatial orientation and positions relative to his or her environment, as well as objects and other people in the field of view. *Vestibular* apparatus provides a perception of the head's movement and orientation relative to gravity through the otolith and semicircular canal structures of the inner ear. *Somatosensory* receptors provide information regarding the relative location of one body part to another and the position of the body in space, an awareness of the body's movements and knowledge of the loads experienced by the person (figure 12.2). These receptors include joint position receptors (Ruffini endings, Golgi receptors and Pacinian corpuscles) as well as the muscle length and tension receptors (muscle spindles and Golgi tendon organs). Receptors in the skin also contribute to the sense of movement and load, especially those in the plantar surface of the foot. These three sensory systems provide information for a continuous update of movements resulting from voluntary activation patterns and contribute to reflexive or reactive actions.

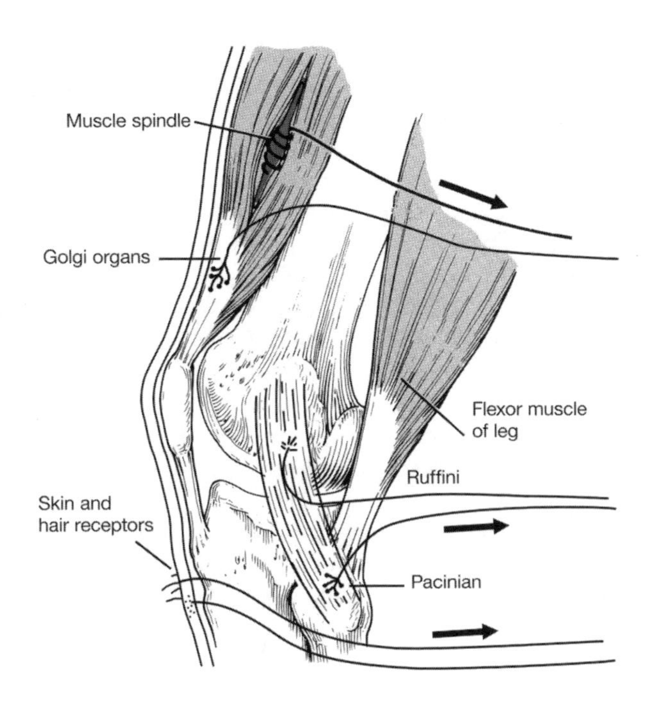

Figure 12.2 Somatosensory receptors.

Certain centres in the brain have the ability to use and integrate this sensory information differently depending on the task being performed or the demands of the situation. These neural centres can also change the action of sensory stimuli depending on the movement and joint torques that need to be generated. The adaptive use of sensory information is based on continuous update of the expected sensory information and actual information. The ability to filter and adjust the use of the sensory information is based on learning, thus emphasizing the role of training to incorporate balance and stability tasks.

All sensory information involves delays between the actual stimulating event and the neural signal and muscle activation response, with ramifications for the speed of response and therefore the effectiveness of the response. This has a large role to play in sport, especially when dynamic balance and stability have to be maintained in movements that occur at high speed, as in football, hockey, basketball and netball. Reactive strategies to correct perturbed balance or joint stability have been shown to be less effective than anticipatory activation patterns (Bennett, Gorassini, and Prochazka 1994; Hasan 2005) that are initiated before the perturbation occurs. Anticipation primes muscles for rapid and appropriate mechanical responses.

Mechanically, muscles do not just generate force; they also produce the intrinsic mechanical properties of stiffness and dampening. These mechanical properties are akin to the action of springs and shock absorbers

Like muscle force, a muscle's stiffness and dampening are greater at higher levels of activation. So, as activation increases, the muscles provide higher levels of stiffness and dampening, resulting in tighter control over joint posture and motion. These properties enable muscles to provide mechanical responses without delay.

in a motor car's suspension, which control how well your car can absorb bumps and negotiate corners. This is similar to the action of our muscles in controlling joint posture and motion. Like muscle force, a muscle's stiffness and dampening are greater at higher levels of activation. So, as activation increases, the muscles provide higher levels of stiffness and dampening, resulting in tighter control over joint posture and motion. Since stiffness and dampening are intrinsic to the activated muscle, these properties enable muscles to provide mechanical responses without delay. However, since muscle activation is subject to reflex modulation, the stiffness and dampening functions can be altered via these pathways.

By generating forces, our muscles move adjacent body parts about their common joint. Generally there two muscle actions at a joint: those of agonists and those of antagonists. An agonist muscle action occurs in the same direction as that of the movement, whereas an antagonist muscle resists the motion. Additionally, there are two general muscle activation patterns or programs that cause joint movement. The first is *reciprocating* agonist and then antagonist activation, and the second is concurrent activation of agonist and antagonist muscles that span a joint, called *cocontraction*. The brain can select either the reciprocating pattern or the cocontraction pattern, but usually we observe a combination of the two depending on the task or movement requirements.

Reciprocating and cocontraction patterns of muscle activation have different control properties. Reciprocating patterns are used in general reciprocating movement like walking and running, where, for example, the agonist starts a limb swinging and the antagonist is then activated to slow it down. Cocontraction, on the other hand, tends to "stiffen" a joint and comes into play when one is stabilizing to maintain constant joint positions and when trying to obtain tight control over points at the end of the segmental chain, such as the hand in shooting or archery. Cocontraction uses the intrinsic muscle properties of stiffness and dampening to maintain joint posture. Thus at higher levels of cocontraction, joint stiffness and dampening are greater, thereby realizing tighter position control and greater resistance to external forces.

People manifest cocontraction when they are inexperienced in performing a task or are unsure about the impacts or perturbations they may experience. Consequently, they try to stiffen their joints to tightly control their movement. As one learns to perform a task, however, cocontraction decreases, and activation patterns are used that permit the performance of smooth actions. This demonstrates the "downside" associated with higher levels of cocontraction movement flexibility and smoothness is compromised (i.e., when the antagonist's action resists the agonist muscles during performance of movements). However, use of agonist and antagonist cocontraction is sometimes appropriate, as when joint stability is required during such impacts as initial foot contact upon landing, running or sidestepping in football, basketball or netball. Here, the lower limb joints need to be stabilized so that they do not "give way" too much on impact, thereby disturbing upper body balance. But even in these situations, lower limbs that are too stiff may cause greater impact forces to be applied to the body, which in turn may make whole-body balance difficult to maintain. Many athletes must learn, via practice, well-coordinated use of muscle activation patterns that protect the body from injury and yet allow for competitive performances. Landing and sidestepping drills, for example, can be incorporated into players' training programs to develop this capacity.

Development of Balance and Stability

From recent research, we know that clear phases exist in the development of static balance among children. Shumway-Cook and Woolacott (1995) stated that visual dominance in balance control appears to recur at transitional points in human development. Infants require a strong visual dominance when learning to sit, crawl, stand and walk.

The ages from 4 to 6 years, however, represent a transitional period when stability often declines temporarily. It is postulated that the child is attempting to integrate various sensory information from the sources mentioned earlier, not just the visual cues. In the process of learning to integrate these data, conflict may exist between sensory inputs, and balance can suffer. Gradually, however, fine-tuning of the balance control mechanisms leads to adult-like control in which the postural responses occur without a great lag in time.

From 7 to 10 years and beyond, the child becomes more reliant upon somatosensory and vestibular feedback in balance control. Removal of visual stimuli does not drastically impair balance for these older children, and they may practice certain skills with the eyes closed to improve the somatosensory "feel" of the performance. Children in these age groups exhibit mature control responses, whereby the amplitude of postural adjustments during maintenance of balance diminishes and the levels of muscular activity become less variable.

As the ability to integrate all sensory information improves and the reliance on visual cues for balance control diminishes, the visual system can be freed to concentrate on other tasks such as monitoring opponents and movements of a ball, implement or target. Furthermore, since the somatosensory information is dealt with at a subconscious level, the brain is free to process information related to developing strategies and tactics.

In addition, reduced reliance on visual information for maintaining balance allows the athlete to perform maximally in poor visual environments, for example under dim or stroboscopic light conditions in artistic pursuits or during twists, spins and somersaults in gymnastics and diving.

Biomechanical Considerations

A well-developed control of posture and stability requires the coordination of joint posture and motion. Different joints, and therefore different muscles, are used to maintain posture and balance depending on the task demands. In quiet standing, a situation in which perturbations are small and are applied only in anterior-posterior directions, the *ankle joint strategies* are used to maintain balance. However, when the base of support is small or unstable, and perturbations are large or are applied in medial-lateral directions, the *hip joint strategies* are used to maintain balance (Shumway-Cook and Woolacott 1995; Winter 1995). This is important to know since it helps us define which muscle groups should be targeted in training. For static balance tasks that are performed on the balls of the feet, as in diving and gymnastics, the base of support is very small and therefore the ankles can play only a small role. So maintenance of balance in these situations is effected through the use of hip joint strategies. The notion of hip joint strategies does not mean that only the hip joint muscles are activated, since an extended leg posture still relies on coordination with knee and ankle muscles. However, movement of the head, arms and trunk (HAT), which compose about two-thirds of total body mass, is more effectively controlled at the hip joint (Winter 1995).

In dynamic balance tasks such walking and running, Winter (1995) has shown that the hip joint is the main site for maintaining control of balance, particularly of the HAT. Disturbances to the large mass of the HAT segments are due to impact forces at initial foot contact after the swing phase and during the leg drive in push-off during the stance phase. Winter (1995) has also shown that this balance is superimposed on, and tightly coordinated with, maintenance of leg extension by the ankle and knee joints.

Assessing Balance

Tests of static balance have been developed for use with athletes in sports like archery, golf and diving, as well as for sports in which the athlete must maintain standing balance when subjected to external forces. Tests of dynamic balance assess the ability of the neuromuscular system to process information from various receptors and then coordinate the appropriate motor responses.

Tests of Static Balance

A number of static balance tests have been used in the field of exercise and sport science to assess the maturity of the neuromuscular system of children. These tests cannot be applied with any great effect to adolescent or adult athletes because they do not tax the mature neuromuscular system.

More sophisticated tests and test apparatus are required to assess the athlete's ability to maintain balance in a static situation. This ability is vitally important for many performers in sports such as archery, pistol and rifle shooting, golf and diving. Static balance may be analysed using video cameras or motion analysis systems (see chapters 14 and 15) to detect postural adjustments and changes in the position of the body's centre of mass. Alternatively, centre of pressure movements can be recorded on a force platform, where accelerations of the centre of mass are detected as a change in the centre of pressure within the base of support (see figure 14.13). Postural adjustments made by the subject to counter these movements of the centre of pressure may be gross or subtle, hence the need for a visual record of the test. Stability of the hand in archery and in pistol and rifle shooting is due to both balance of the torso and relative balance of the arms with respect to the torso. Instruments such accelerometers mounted on the hand or laser pointers projecting onto a screen can be used to record and display the motion of the hand. These can also serve as a method of measuring stability and of training the person to maintain balance and stability.

A more objective assessment of standing balance where the body is subject to external forces requires the use of a generic balance task, as with the balance platform or stabilometer. The stabilometer platform, on which a subject stands, can rotate and move horizontally in either medial-lateral or anterior-posterior directions. The movable platform is instrumented to record rotational deviations from the horizontal position or linear movements. Changes in centre of pressure can also be assessed using a force plate embedded in the platform. Platform rotations or deviations of the centre of pressure (or both) are input into a computer for analysis. The test is usually conducted over a period of 15 to 30 s; subjects adopt either a "jockey" position with the feet parallel, shoulder-width apart and in a single transverse plane, or a "surfing position" with one foot forward with respect to the other. The total area under the displacement–time curve, together with the proportional deviations to the left and right, is generally used to score this test.

Tests of Dynamic Balance

Dynamic balance refers to the ability to execute movements and maintain balance while in motion. A successful test of this capacity needs to assess the ability of the neuromuscular system to assimilate information from visual, auditory, vestibular and kinaesthetic receptors and then to coordinate a series of motor responses. At a subjective level of analysis, one may do this by recording movement using video or some other form of motion capture and analysing the performance of athletes under competitive conditions. These tests tend to be sport or task specific; the reader is referred to chapters 14 through 16 for further details about image capture methodology and analysis options.

Dynamic balance refers to the ability to execute movements and maintain balance while in motion. A successful test of this capacity needs to assess the ability of the neuromuscular system to assimilate information from visual, auditory, vestibular and kinaesthetic receptors and then to coordinate a series of motor responses.

Improving Balance for Sport

Balance training methods are usually specific to each sport; however, a number of general programs have been reported in the literature. These programs involved a variety of exercises, equipment and progressions as outlined later in this chapter. Participants may begin with simple stability exercises that involve standing on one foot and progress to standing on one foot with the eyes closed (removing the visual input) and then with eyes closed and head extended (removing the visual input and confusing the vestibular apparatus).

More proficient athletes can then progress by performing these exercises on unstable surfaces (i.e., foam rubber surfaces, mini-trampoline). Still more advanced exercises can then be introduced, including balances performed with two support legs, then one support leg on tilt boards, wobble boards and Dura Discs (inflated rubber discs). Such exercises as squats, lunges and stepping onto and sidestepping off such apparatus can also be introduced to increase strength and muscle activation in more dynamic situations.

Swiss ball exercises (including hamstring and quadriceps exercises) can be implemented for very experienced participants. Here, the subject is slowly trained to balance on the Swiss ball, firstly on the knees and then on the feet with a frame support. The training program progresses such that subjects can stand, balance and perform squats on the Swiss ball without assistance.

Protection From Joint Injury

In this section we address only injuries that are due to failure of joint stabilization, that is, failure of muscles or ligaments to control joint posture. These injuries occur as a result of excessive loading that causes large movement between the two articulating bones of the joint, placing great strain and stress on the stabilizing structures such as ligaments, joint capsule and menisci. The most common joint injuries in sport are to the ligaments, particularly in the ankle and knee. Common ankle injuries are to the outer lateral ligaments (e.g., the anterior talofibular ligament) due to excessive ankle inversion. Common knee injuries are to the medial collateral ligament (MCL) from abduction and external rotation of the tibia relative to femur (Markolf, Mensch, and Amstutz 1976); the posterior cruciate ligament from posterior draw of the tibia relative to femur; and the anterior cruciate ligament (ACL) from anterior draw, abduction and internal rotation of the tibia relative to femur (Markolf et al. 1995). The ACL and MCL injuries are particularly debilitating, requiring knee joint reconstruction in many cases (i.e., repair or replacement of the ligaments or both). Research suggests that females are at a much greater risk of suffering an ACL injury than males; however, the exact mechanism has not been determined (Hewett et al. 1999).

A ligament will fail only when the applied load exceeds its strength. So how do the ligaments become highly loaded? It is the external load applied to the joint that ultimately causes these injuries, which can be categorized as either *contact* or *noncontact* injuries. Large external loading of the joint occurs in contact injuries when the athlete makes impact with another player (e.g., during a tackle in soccer) or an object (e.g., with a tree in downhill snow skiing). On the other hand, noncontact injuries occur without any contact with another player or object. However, all noncontact injuries occur when the player is in contact with the ground while performing a sporting manoeuvre. Large external loads are applied to the lower joints by impact or inertial loading, resulting from the accelerations and decelerations of the player's body mass during these manoeuvres. Examples of high-risk sporting tasks include sidestepping and sudden landings as regularly performed in all the football codes, basketball, European handball and netball, just to name a few. It is interesting to note that noncontact injuries compose a large proportion of all joint injuries. For example, noncontact ACL injuries account for over 50% of all ACL injuries and sometimes up 75%, depending on the sporting code.

The damaging external loads at the knee resulting from these manoeuvres are generally not the main flexion and extension joint motions or loads (Besier et al., "External

Loading" 2001; McLean et al. 2004), but rather the movement directions involving limited joint motion. These damaging loads are anterior or posterior draw and the out-of-plane loads in abduction and adduction, as well as internal and external rotation directions (figure 12.3) (see also Markolf et al. 1995). The out-of-plane loads can reach very high magnitudes during sporting manoeuvres (Besier et al., "External Loading" 2001, "Anticipatory Effects" 2001; McLean, Huang, and Van Den Bogert 2005) and need to be supported by internal structures of the joint. Ligaments are assumed to be the main joint stabilizers; however, there is increasing awareness that the muscles probably support most of the damaging external loads. Thus it is the failure of muscles to provide adequate support that causes the ligaments to be highly loaded and leads to damage (Lloyd 2001; Lloyd, Buchanan, and Besier 2005).

As previously discussed, cocontraction of agonist and antagonist muscles provides joint stability. In doing so, the muscle cocontraction supports the out-of-plane forces, thereby lowering the loads that have to be sustained by the ligaments (Lloyd and Buchanan 2001; Lloyd, Buchanan, and Besier 2005). Because the manoeuvres that increase ligament injury risk happen very rapidly in sport, muscle cocontraction patterns are used in anticipation of the loads to be experienced during ground contact (Besier, Lloyd, and Ackland 2003). And since the muscular stabilization of the joint relies on the intrinsic muscle properties of stiffness and dampening, with appropriate preactivation there is no delay between application of the joint loads and the muscular support provided (Lloyd 2001).

Ligamento-muscular reflexes exist whereby a ligament stretch will induce a reflexive activation of muscles to resist the stretch (Johansson 1991). There is considerable delay between the stretch and the muscular response, which in fast-moving sporting situations would make the reflex's protective action ineffective. However, stretch of the ligaments also enhances the sensory information

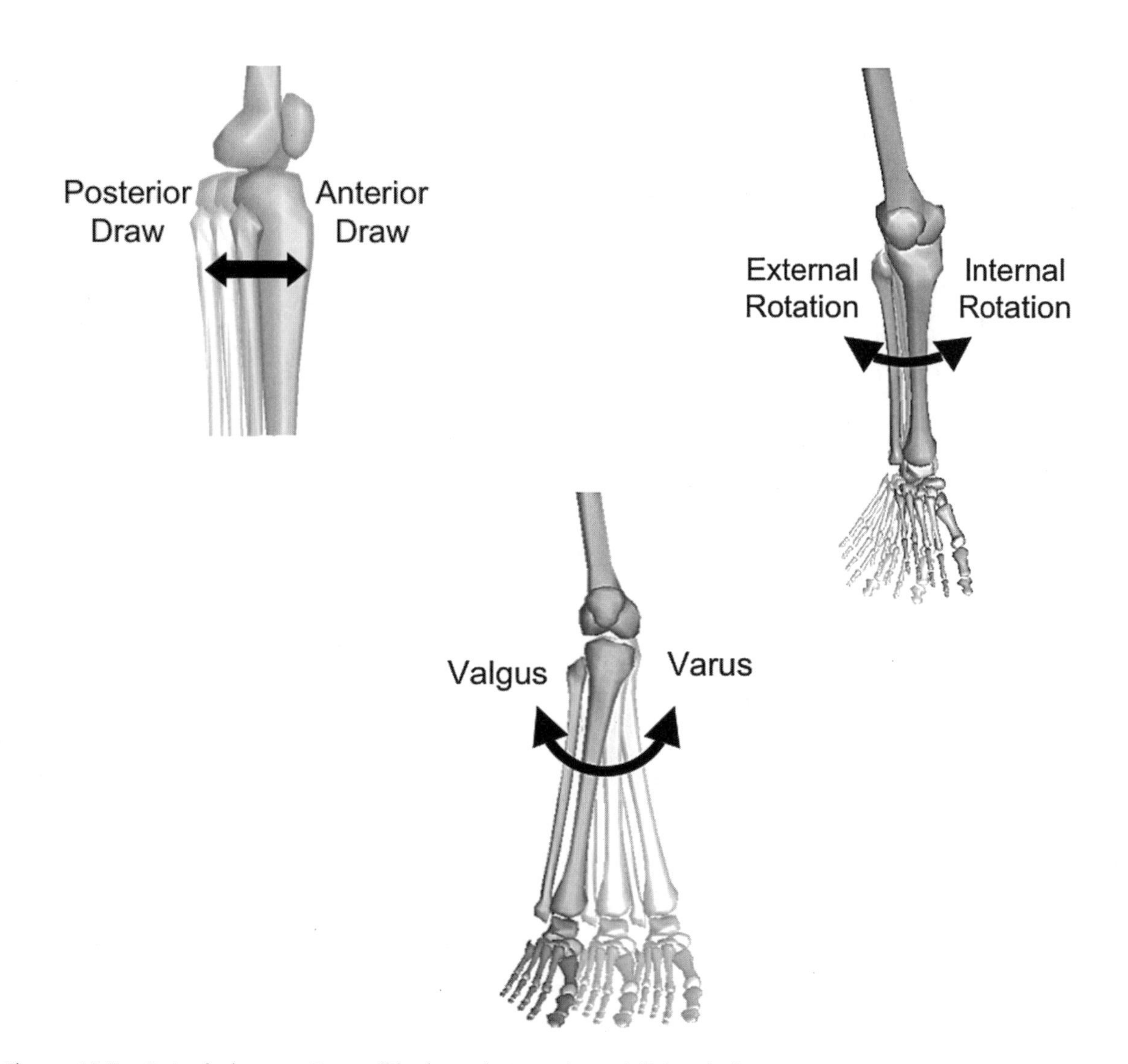

Figure 12.3 Out-of-plane motions of the knee that require stabilizing during movement.

flow from the muscles to the brain. Johansson (1991) has proposed that we use this information and learn how to prevent ligament injury by altering the voluntary muscle activation patterns to reduce ligament loading during sporting tasks (Lloyd 2001).

Reducing risk of ligament injury relies on anticipating the loads to be experienced when executing a manoeuvre. Therefore, people need specific training in order to learn how to anticipate the load and use appropriate muscle activation and movement patterns to reduce ligament damage. The best training to reduce the risk of ligament injuries appears to be the aforementioned balance training methods (Caraffa et al. 1996; Myklebust et al. 2003) with additional drills that include plyometric landing and bounding (Hewett et al. 1999) and sidestepping and landing tasks (Lloyd 2001). Players should initially perform sidesteps and landings when they have plenty of time to prepare for the task. Greater difficulty should be introduced when players become more proficient, such that the tasks are executed with little warning, or with defenders to replicate the game situation (Besier et al., "Anticipatory Effects" 2001; McLean, Lippert, and Van Den Bogert 2004). In addition, it has been suggested that the sporting tasks be performed with greater knee flexion, as extended knee joint postures can place greater loads on the ACL. When performing sidesteps, the player should avoid placing the pivot or step foot out too wide and should turn their trunk toward the new change in direction while keeping their trunk upright. Maintaining these postures during a sidestep will reduce the loads on the knee and may reduce the risk of ACL injury (Dempsey et al. 2007).

Factors Affecting Agility

Agility is a difficult capacity to define, since it incorporates elements of movement speed as well as the ability to coordinate changes in direction and modification of the normal locomotion posture. Draper and Lancaster (1985) suggest that there are several types of agility:

- Whole-body changes of direction in the horizontal plane (as in sidestepping, cutting and faking)
- Whole-body changes of direction in the vertical plane (as in jumping and leaping)
- Rapid movement of body parts, for example to control implements in sports such as fencing, tennis, squash and field hockey

Agility is an important capacity for many elite performers to develop. Agile movements of the body can be used to great effect to free an offensive player from the opposition. In cases in which the player needs to change direction quickly in sidestepping and cutting, it is important to move the body's centre of mass away from the base of support very quickly. Therefore, agility appears quite different to dynamic balance; however, it is important to remember that the latter relies on the capacity to have subsequent movements planned (such as placement of the foot for the next contact with the ground) to ensure the continued maintenance of balance.

Dintiman and Ward (1988) also recognized the importance of agility for faking and cutting manoeuvres in football and basketball. The fake is used to neutralize defenders by slowing their movement, breaking their concentration and placing doubt in their minds. Faking can cause the defender to move in a direction away from the player's intended path or draw a defender close so that a cut or side step (figure 12.4) can be effectively executed.

Mastery of all fakes and cuts as described by Dintiman and Ward (1988) requires great agility so that the player can modify the pattern of locomotion while running at high speed. Good faking skills not only improve the performance of offensive players in contact sports, but also tend to reduce the risk of injury resulting from a hard tackle.

Figure 12.4 A rugby league player using a side step.

Highly developed agility to quickly change direction is not strongly dependent on acceleration and ultimate sprinting speed but on the ability to coordinate movement and to "read the game", reacting to visual cues of the opposition to provide quick and appropriate responses to either avoid the defending or challenge the opponent(s) (Sheppard et al. 2006). If a player is able to better anticipate the play in the game, they will have more time to plan and execute a change of direction manoeuvre, such as a side step. More time to plan a side step will lower the joint loading (Besier et al., "Anticipatory Effects" 2001) and provide better tuned muscle activation patterns to stabilise the joint (Besier, Lloyd, and Ackland 2003), which may reduce the risk of joint injury. So agility and risk of injury may be linked together.

Mastery of all fakes and cuts as described by Dintiman and Ward (1988) requires great agility so that the player can modify the pattern of locomotion while running at high speed. Good faking skills not only improve the performance of offensive players in contact sports but also tend to reduce the risk of injury resulting from a hard tackle.

Agility may also be required in situations that involve no defensive player. For example, in sports such as tennis, ice hockey and badminton that entail the striking of a ball, puck or shuttle, the ability to quickly change direction and place oneself into a good position to execute a stroke is of paramount importance to the resulting speed and accuracy of the shot. The execution of a double play in baseball, or a catch or run-out in cricket, often requires great agility in order that the player can change direction quickly or move the limbs rapidly to intercept a ball.

Assessing Agility

Many field tests have been developed in order to evaluate agility. Most are specifically adapted to suit the needs of a particular sport, while others have a more general application. These tests typically involve short sprints and changes in direction, as well as the requirement that the subject accelerate rapidly from point to point on the course. Often equipment that is used in a particular sport will be incorporated into these tests.

All the agility tests presented next do not include methods to discriminate a player's ability to react to situations in a game where they have to quickly change direction. The established tests only examine a player's ability to change direction quickly. The ability to react and quickly change direction in a game situation relies on cognitive ability, and new tests incorporate features that depend on the player's cognitive skill (Sheppard et al. 2006). These tests are called reactive agility tests. However, since these tests are very new, they are yet to be accepted for normal athletic testing and normative data do not currently exist. For this reason, the reader is directed to the literature (Sheppard et al. 2006).

General Tests of Agility

Several general agility tests have been evaluated and reported in the literature (Draper and Lancaster 1985), but almost all lack standardization of protocols, established validity or published normative data for various populations. However, the Illinois agility run, described next, has been used extensively in the physical education field, and normative data are available. In addition, even though developed for cricket, the 505 test has been routinely used in different sports (Draper and Lancaster 1985).

The Illinois Agility Run

This test is a modified version of that described by Getchell (1985). A nonslip ground or floor surface at least 15 m (16 yd) long and 6.5 m (7 yd) wide is required for the test. A rectangle measuring 9.14 × 3.65 m (10 × 4 yd) must be clearly marked and divided in half along its length by four cones positioned at equidistant intervals. The centres of the first and fourth cones must be positioned directly over the midpoint of the 3.65 m sides (figure 12.5).

Administration The subject starts in a front-lying position with the vertex of the head on the starting line, forearms flexed at the elbows and hands just outside the shoulders. Two recorders with stopwatches are positioned at the start-finish line. On the command "go" they start their watches, and the subject rises from the lying position, runs as quickly as possible to the end line, stops as one foot touches or crosses it, then turns and

Figure 12.5 The Illinois agility run.

sprints back to the starting mark. After turning again, the subject then weaves in and out around the centre cones toward the end line and back to the starting line. At this point the subject again sprints to the end line, turns as one foot touches or crosses it and sprints back through the finishing line.

The watches are stopped as the subject's trunk crosses the finishing line; the two times are averaged, and the resulting score, to the nearest 0.1 s, is noted. Each subject should be permitted one practice run at a slow pace and then two trials, with the lower elapsed time of these trials eventually recorded. Grounds for disqualification of a trial score are as follows:

- Straying outside the side boundaries of the rectangle
- Any failure to touch or cross a line at either end of the rectangle
- Touching a cone, even if accidentally
- Failure to follow the prescribed course

Normative Data The comparative data for men and women displayed in table 12.1 are adapted from those reported by Getchell (1985).

Table 12.1 Comparative Norms for the Illinois Agility Run With Respect to Ratings of Excellent (1) to Poor (7)

Category	Men (s)	Women (s)
1	<16.1	<17.1
2	16.7	19.0
3	17.3	20.4
4	17.9	21.7
5	18.5	23.1
6	19.1	24.4
7	>19.3	>24.9

The 505 Test

A nonslip ground or floor surface at least 15 m (16 yd) long and 4 m (5 yd) wide is required for the test. A 15 m, 5 m and turn line must be clearly marked (figure 12.6).

Administration The subject starts at the 15 m line and sprints towards the turn line. The subject then negotiates a 180° pivot at the turn line and sprints back towards the start line. A tester stands at the 5 m line and starts timing using a stopwatch when the subject's trunk first crosses the 5 m line and stops timing as the subject's trunk crosses the 5 m line on the return run. Alternatively timing gates can be set up at the 5 m line to measure the elapsed time. The subject should be encouraged to turn with pivot foot as close as possible to the turn line so they can attain the fastest time. Each subject should be permitted one practice run at a slow pace and then three trials, with the lowest elapsed time of these trials eventually recorded to nearest 0.1 s. The test can be performed with subject turning off their preferred foot or off both left and right feet. Grounds for disqualification of a trial score are as follows:

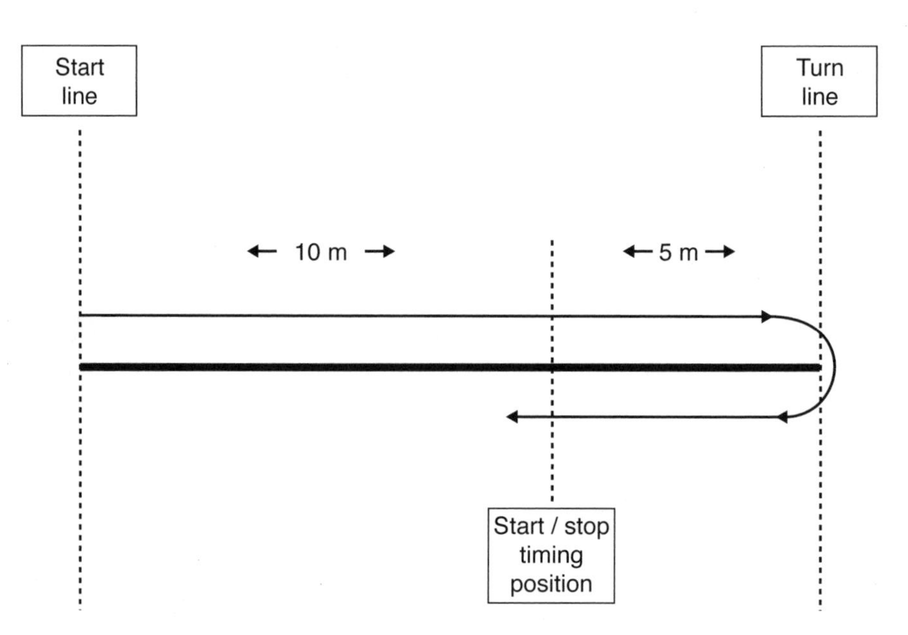

Figure 12.6 The 505 test.

- Any failure to touch or cross the 5 m line
- Any failure to touch or cross the turn line
- Failure to follow the prescribed course

Normative Data Several sources of normative data exist, including data for rugby league players (Gabbett 2006, 2007; Gabbett, Kelly, and Sheppard 2008), and basketball, field hockey, netball and tennis players (Draper and Lancaster 2000).

Specific Tests of Agility

Agility tests have been developed that are specific to basketball, baseball and softball, tennis and rugby. The zigzag drill and the ladder drill for basketball players test the time elapsed as the athlete sprints to cones on the basketball key or to specific positions on the court, respectively. In the baseball and softball agility test, athletes are timed as they sprint to a ball, pick it up and throw it to a catcher. Tennis-specific tests include the modified SEMO drill and the Quinn test, which measures dynamic balance and response time as well as agility. The rugby agility run focuses on the ability to accelerate and change direction.

Basketball Zigzag Agility Drill

The zigzag agility drill for basketball players uses marking cones (30 cm [12 in.] high) placed at the four corners of the basketball key as shown in figure 12.7. Two recorders with stopwatches measure the elapsed time as the athlete sprints from the baseline cone to the three other cones and then back to the baseline cone.

Administration

- The player begins at cone 1, with the hand in contact with the cone and feet behind the baseline.
- On a whistle or "go" command, the player sprints to cones 2, 3 and 4, then back to cone 1.
- Each cone must be touched during the circuit.
- The average elapsed time from "go" until the subject returns to touch cone 1 is noted and the better of two trials recorded.

Normative Data Comparative data for Australian age-group and senior players were reported by Draper, Minikin, and Telford (1991) and are presented in table 12.2.

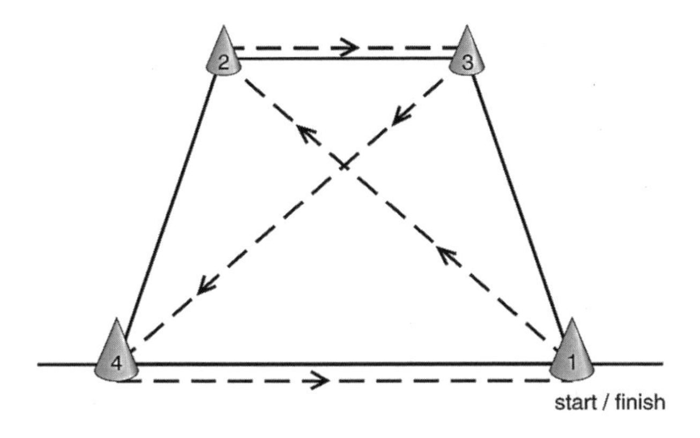

Figure 12.7 Sport-specific agility test—the basketball zigzag drill.

Table 12.2 Normative Data for Australian Age-Group and Senior Basketball Players on the Zigzag Agility Run

Team	Mean (s)	Range (s)
FEMALE		
National senior team	6.6	5.9-7.5
National junior team	6.8	6.2-7.8
17 years	6.9	6.1-7.7
16 years	7.1	6.3-7.9
15 years	7.1	6.7-7.7
14 years	7.3	6.4-7.8
MALE		
17 years	6.4	5.8-7.0
16 years	6.4	5.9-7.2
15 years	6.5	6.0-7.3
14 years	6.7	6.1-7.4

Ladder Drill or Suicides

The basketball ladder drill is more demanding than the zigzag drill in that it tests not only agility, but also other capacities. The length of the court from baseline to baseline should be 28 m (30.6 yd) (figure 12.8). Two recorders with stopwatches measure the time elapsed as the athlete, from a standing start posture at one end of the court, runs to each of four positions on the opposite end and then back to the starting line. The better of the scores for two attempts is recorded.

Administration

- The player begins at position A using a standing start posture with the leading foot behind the baseline.
- On the command "go", the player runs to and touches the opposite baseline B with one foot, turns and runs back to the starting line.
- Immediately, the player turns and runs to C and back to A, then D and back to A, then E and back to A.
- As the player's foot crosses baseline A for the last time, the elapsed time is read to the nearest 0.1 s.
- The average time is noted.
- After exactly 2 min rest, the player is permitted a second attempt, and the better score is recorded.
- It should be noted that this is also a test of anaerobic capacity, resistance to fatigue and running speed, rather than merely one of agility.

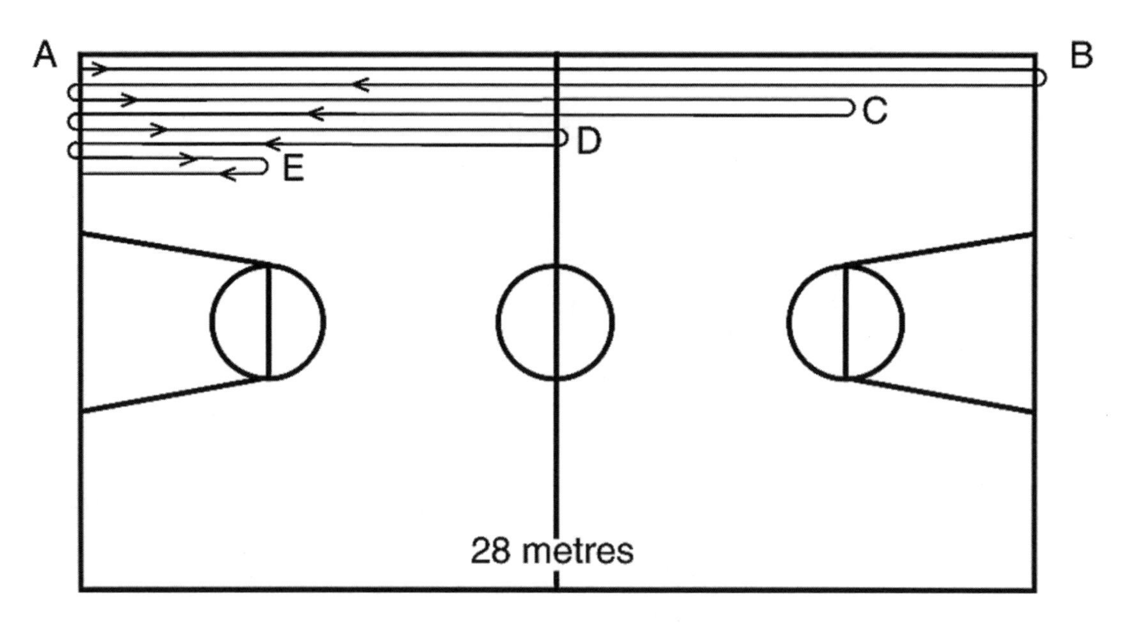

Figure 12.8 Sport-specific agility test–the basketball ladder drill.

Normative Data Comparative data for Australian age-group and senior players were reported by Draper, Minikin, and Telford (1991) and are presented in table 12.3.

Baseball and Softball Agility Test

To test the agility of baseball and softball players, a course is prepared on a level grass field as shown in figure 12.9; a catcher is located within a circle 1.5 m (1.6 yd) in diameter, and a baseball or softball is placed on the ground at the indicated position. Two recorders with stopwatches are needed.

Table 12.3 Normative Data for Australian Age-Group and Senior Basketball Players on the Ladder Drill

Team	Mean (s)	Range (s)
FEMALE		
National senior team	30.1	28.6-33.1
National junior team	30.8	29.3-34.5
17 years	32.0	29.2-36.4
16 years	31.7	28.7-35.7
15 years	31.6	29.3-33.5
14 years	32.6	30.6-34.1
MALE		
17 years	28.1	26.7-29.5
16 years	28.2	26.7-30.1
15 years	28.5	27.5-29.5
14 years	29.0	28.5-29.4

Administration

- On the command "go" the player sprints around cones A and B to the ball.
- On reaching the ball, the player picks it up and throws to the catcher.
- A score is recorded only if the ball reaches the mitt without the catcher's having to leave the circle.
- The average elapsed time from "go" until the ball reaches the catcher's mitt is noted.
- The lower score of two trials is recorded.
- It should be noted that test scores are also influenced by throwing ability.

Normative Data Normative data were reported by Draper, Minikin, and Telford (1991) for Australian junior (under 18 years) and senior baseball players. The mean and range of scores for the junior players were 6.53 and 5.50 to 7.23 s, respectively, while for senior players they were 6.13 and 5.75 to 6.55 s, respectively.

Modified SEMO Agility Test

The modified SEMO agility test was designed for tennis players to determine their agility on court. A smooth surface with a 20- × 20-foot area (6.1 × 6.1 m) is required for this test, as well as space to run around the perimeter of this square. Four cones, A, B, C and D, are placed in the four corners of the area.

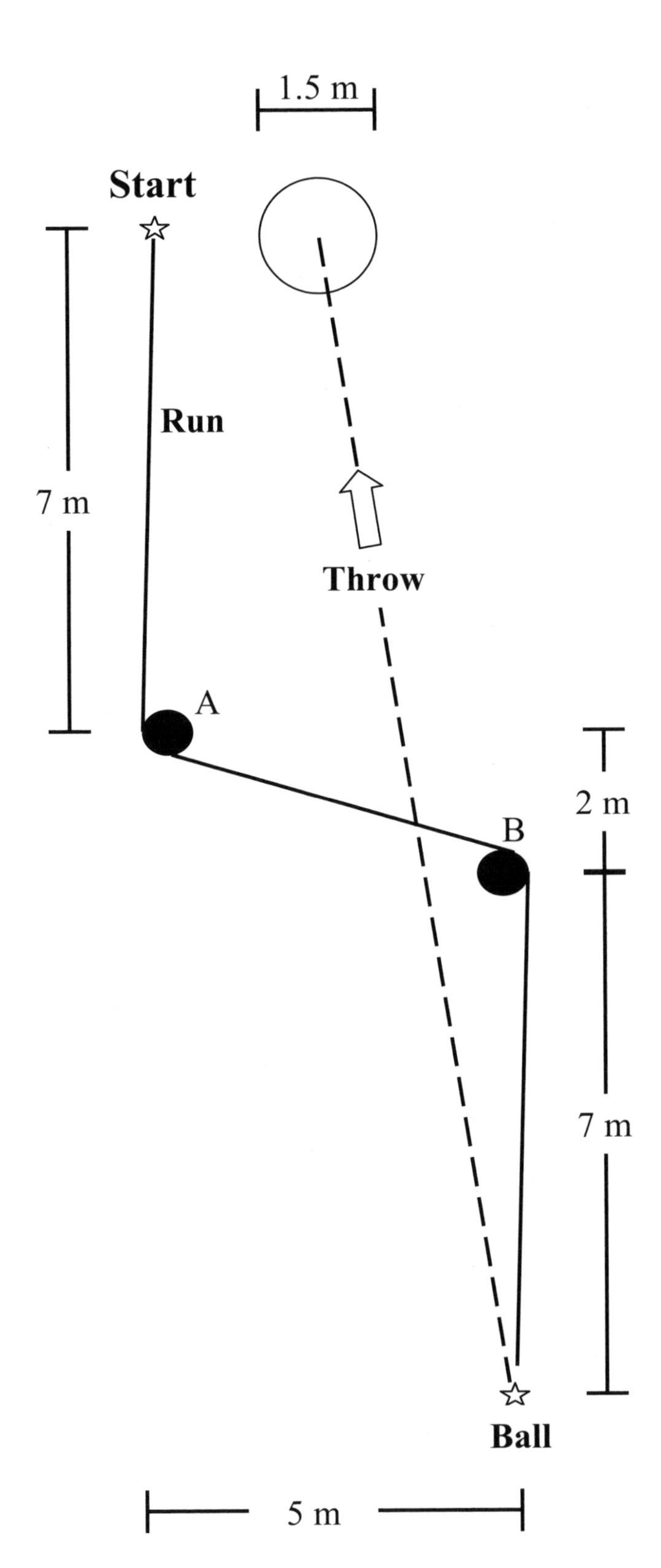

Figure 12.9 Sport-specific agility test—softball and baseball.

Administration

- The subject begins facing cone A with his or her back to the course (figure 12.10).
- On the command "go", the subject sidesteps from A to B, passing outside B.
- From there the subject backpedals to and around D, then sprints forward to and around A.
- The subject backpedals from A to C, then sprints forward to B.
- From B, the subject sidesteps to the finish line at A.
- One practice circuit should be permitted, and the better score from two trials is recorded. A recovery time of 2 min should be enforced between trials.

Normative Data Average scores for male and female players are 12.8 s and 14.2 s, respectively, according to Groppel and colleagues (1989).

The Quinn Test

The Quinn test is designed to measure dynamic balance and response time of tennis players, as well as their agility. It should be administered on half a tennis court as shown in figure 12.11.

Administration The subject begins by standing at the intersection of the service line and centre line without a racquet. While holding a tennis ball, the tester takes up a position on the centre line, approximately 2 m behind the net. On the command "go", the tester points to one of the eight positions on court using the ball and begins the stopwatch. The subject then sprints to that position, touches the spot with one foot, then sprints back to the start position.

Immediately, the tester points to another position selected at random, whereupon the subject must run to touch the spot and return to the start position. This continues until all eight positions have been visited, and the subject returns to the start position for the final time. The elapsed times for two trials are noted, and the better score is recorded. It is important to stress that the subject must watch the tennis ball, which is held by the tester at all times. This requirement is designed to help simulate the game situation. A 3 min recovery time should be interposed between trials.

Normative Data Average scores for male and female players on the Quinn test are 29.4 s and 34.5 s, respectively, according to Groppel and colleagues (1989).

The Rugby Agility Run

The rugby agility run is designed to measure the ability of a player to accelerate and change direction (A.R. Morton, personal communication 1993).

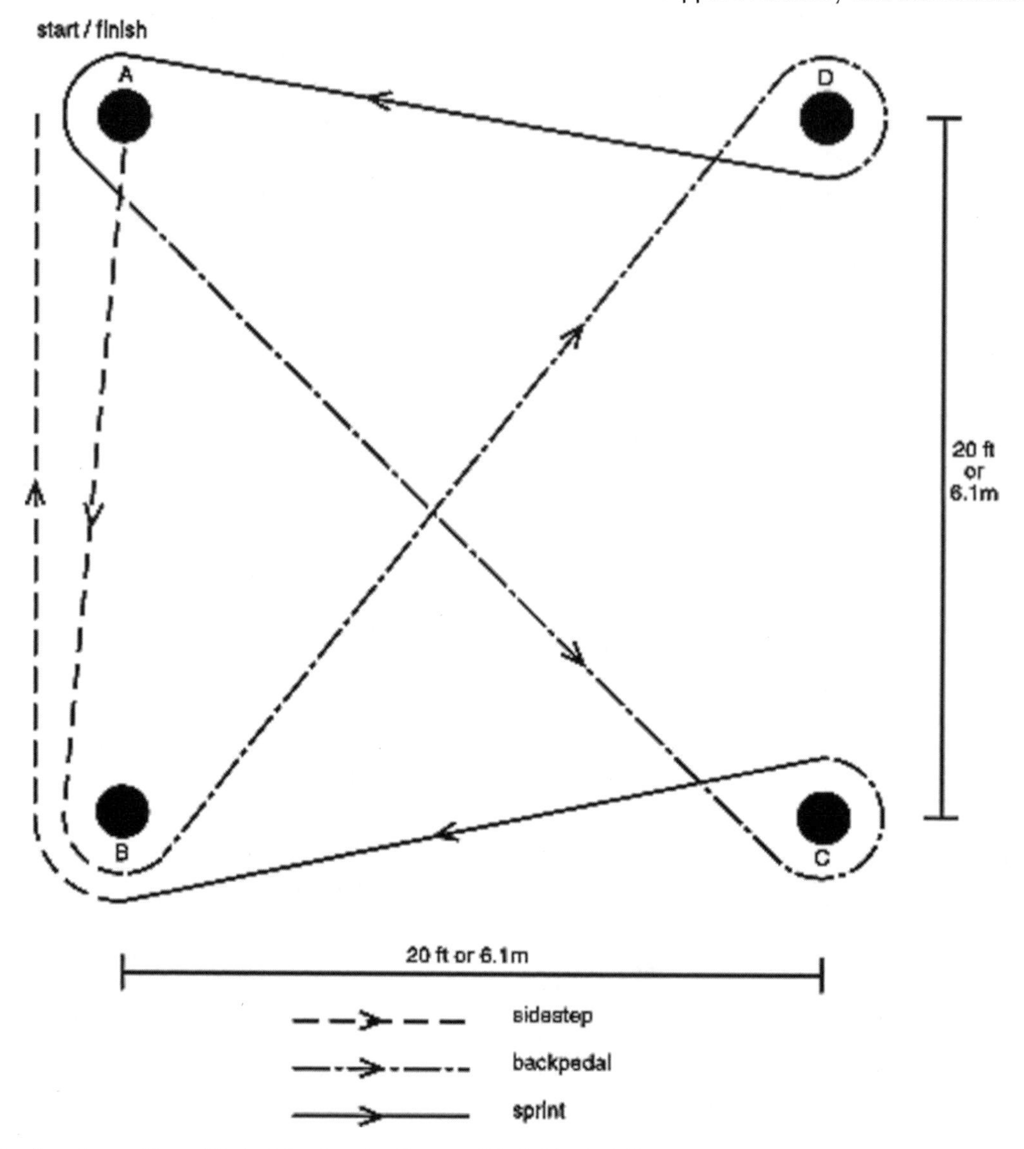

Figure 12.10 Sport-specific agility test for tennis—the modified SEMO drill.

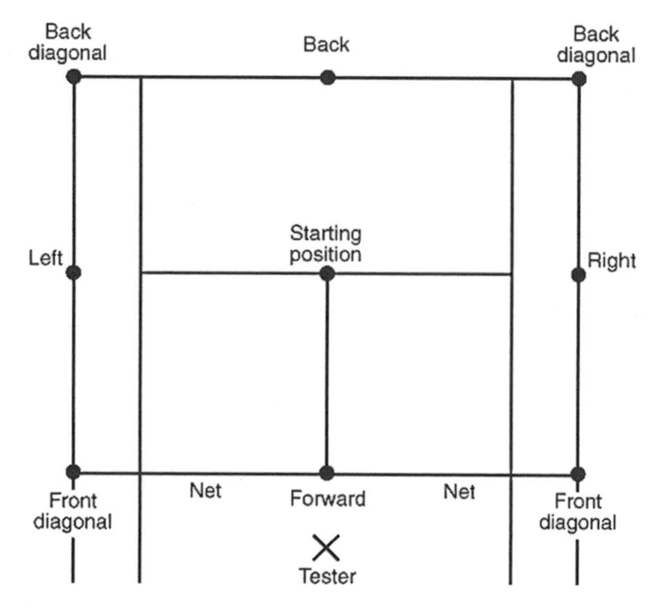

Figure 12.11 Sport-specific agility test for tennis—the Quinn test.

Administration The course is set out in figure 12.12. Cones are used to mark positions A through G and the finishing line. Tall markers, which are constructed of plastic tubing and are approximately 1500 × 30 × 30 mm (60 × 1 × 1 in.) in size to simulate opponents, are located at positions H through L.

The tester should stand at the finishing line to ensure the most accurate timing. Subjects begin by lying on their back with the head on the starting line at cone A. On the command to "go", the subject gets up and sprints around cone B, cuts back to cone C and performs a shoulder roll at cone D. Quickly accelerating from the roll to cone E, the subject then picks up a ball, passes around cones F and G and accelerates toward the tall markers. The subject weaves around the tall markers H through L and then sprints to the finishing line to score a try. Two trials are permitted, and the lower time is recorded to the nearest 0.1 s.

Normative Data The data in table 12.4 were collected by A. Morton, G. Treble, and B. Hopley and

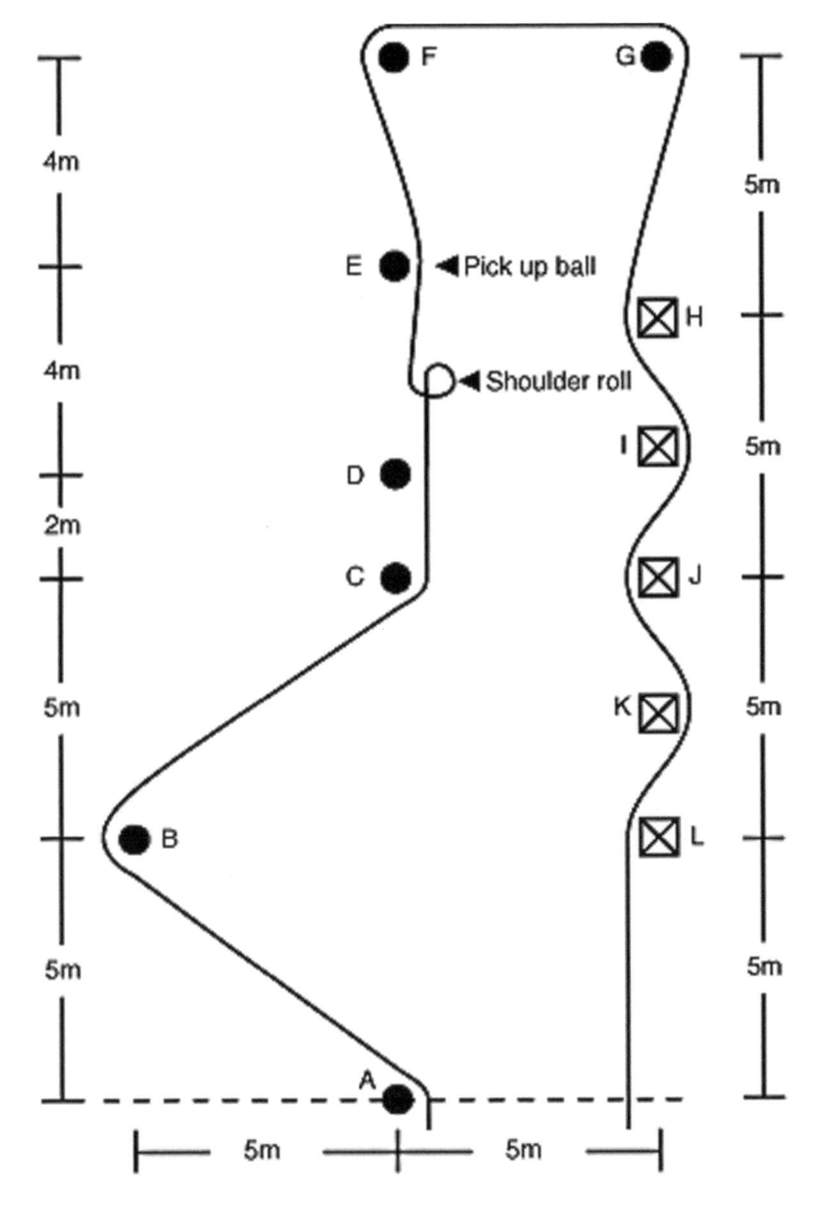

Figure 12.12 Sport-specific agility test for rugby.

Table 12.4 Normative Data for Rugby Players on the Rugby Agility Run

Group	Mean agility run time (s)
A	15.78
B	14.41
C	14.50
D	14.68

are unpublished (University of Western Australia), but represent a compilation of scores from Western Australian rugby union state representatives as well as selected players from the Sydney (New South Wales) first grade clubs. These results are reported for players in various positions, and the differences in mean scores reflect their respective roles. The tight five group (Group A) includes the props, hooker and second rowers, while Group B consists of the lock and breakaways. Group C is composed of scrum half and fly half, whereas Group D includes the centres, the wingers and the fullback.

Improving Agility for Sport

Overall, it would appear that agility is a capacity that is highly specific to each particular task. Although the ability to generate force quickly is important, as is the ability to accelerate or run fast (or both), these characteristics tend not to be related to agility skills. Agility is the ability to coordinate the movement of body segments so as to swerve or sidestep in a quick, smooth and balanced fashion.

The development of strategies to improve agility should be specific to the game or event situation. The past decade has seen a concerted effort to improve agility in the mobile court and field sports, with much more time being spent on set drills in practice. When these have been thoroughly learned, they are then integrated into the various moves that are constantly practiced in high-level sport. The days in which the brilliant performer is the only player on the team able to perform cuts, swerves and side steps are long gone, because modern training now incorporates these skills for sports in which such moves are needed.

Some tests for evaluating agility as described in this chapter can also be used to improve this capacity. While many of the tests have been criticized as being poor tests of agility, their specificity alone makes them appropriate drills for use in the development of this capacity for a particular sport. But because these tests use static markers and turning positions, their effectiveness for developing agility among contact sport players is limited. It is often more appropriate to design one-on-one drills so that a player may develop agility and respond to the defensive efforts of a mobile opponent. This will also teach the player to visually cue off various movements of their opponent so that the player has more time to respond and execute a change of direction manoeuvre, such as a sidestep. When such drills are performed repeatedly in a relatively safe, predictable environment, they will also help the player learn how to reduce the risk of lower limb joint injury.

Summary

As with many of the other physical capacities, balance and agility may be modified and can also be monitored using standardized tests. However, intensive practice is needed in sports in which players change direction fast, or in various games requiring evasive action to avoid contact, so that these manoeuvres can be performed in a flowing and balanced way and with a low risk of joint injury.

thirteen

Modifying Physical Capacities

Timothy R. Ackland, PhD; Deborah A. Kerr, PhD; and
Robert U. Newton, PhD

Our aim here is to bring together key information that has been presented in the chapters composing this section on physical capacities in order to provide specific training guidelines for athletes in a variety of sports. This is something of an instructional chapter, with specific training program recommendations that incorporate strategies for modifying body composition, strength, power, flexibility, speed and agility whenever they are considered important for competition success. However, the reader should not think that these strategies are the only definitive training programs or routines available. Indeed, coaches must exercise good judgment to modify training elements to suit their individual athletes' needs. This is the essence of good coaching, a point that is stressed in chapter 1 of this book.

Fundamental information related to resistance training and stretching exercises is provided in the first part of this chapter, followed by recommended training programs for specific sports within the nine generic sport groups that have been referred to throughout this book. In each subsection, discussion related to the existence of any self-selection parameters (those factors that give a competitive advantage but are difficult to modify) is followed by recommended programs that address the modifiable parameters—body composition, strength and power, flexibility, speed and agility.

Resistance Training Guidelines

General guidelines for all resistance training programs are outlined in this section, with topics related to rest periods, periodization strategies, number of training sessions, exercise order and the importance of warming up before training.

Rest Periods

Between repeated sets and exercises during a resistance training routine, an athlete will rest. During this rest period the energy stores of adenosine triphosphate (ATP) and creatine phosphate (CP) within the worked muscles are partially replenished and any accumulated products, such as lactic acid, partially dissipate. If the previous exercise was performed to exhaustion, it will take approximately 3 to 5 min for the energy stores within the muscle to be completely replenished. Removal of lactic acid is subject to widely varying clearance times, depending on factors such as the amount of lactic acid initially present and whether the rest period was active or inactive. Nevertheless, the more repetitions performed and the shorter the rest period, both relate directly to lactic acid accumulation. There are also differing hormonal responses with rest periods of varying length (see chapter 8).

Strength and Power Training

After the rest period, if the energy systems have not been fully replenished or the removal of accumulated products has not been completed, the subsequent work bout will be negatively affected so that a lower level of intensity and volume will result. This occurs because a lack of energy supplies or the presence of products of anaerobic energy production (or both) inhibits the ability of some motor units to fire effectively, thereby reducing the intensity of the overall contraction. Since a high intensity of muscular tension is important for the development of strength and power, it is universally recommended that rest periods of at least 3 min be imposed between repeated maximal work bouts. Put simply, maximal strength or power development will not occur if the muscles are fatigued, thus inhibiting recruitment of muscle fibres.

Often athletes are under time pressure to perform resistance training and consequently rush through workouts with insufficient rest intervals. It must be remembered that to develop muscular strength, the most important factor is the intensity of the muscular contraction, not the total number of sets or repetitions performed. Insufficient rest will result in suboptimal strength gains. If time becomes a major problem, athletes would be much better served by reducing the number of sets or exercises and ensuring that the remainder are performed at maximal intensity.

Alternatively, shorter rest intervals can be adopted providing that the athlete performs alternating upper body and lower body exercises. For example, a set of bench presses can be carried out, followed by 60 s rest, then a set of calf raises, followed by 60 s rest, and then the bench press can be performed again. This rotation is continued until the prescribed number of sets for each exercise is completed. With use of this strategy, the upper body musculature can recover whilst the lower body is exercising and vice versa.

Strength-Endurance Training

The rest periods adopted during strength-endurance training differ depending on the type of training used to develop this muscular function (see chapter 8). If the high-repetition training method is adopted, athletes are attempting to increase their tolerance to the accumulation of lactic acid and induce other local muscle adaptations. In this instance relatively short rest periods of 30 to 60 s are appropriate, for they will result in the highest accumulation of lactic acid, facilitating the desired adaptations. Conversely, if the heavy weight or variable load training methods are adopted, one of the objectives of the training is the enhancement of muscular strength. Consequently, the exercise must be performed in a fully recovered state so that relatively high training intensities can be realized; therefore, long rest periods of 3 to 5 min will be required.

The Training Year: General Periodization Strategies

The concept of periodization was introduced in chapter 8, and several periodization training strategies were presented. This section outlines how to periodize resistance training throughout a normal competitive year. The structure of the year will depend upon the dominant physical capacity that is to be maximized (i.e., strength, power or strength-endurance). Coaches must also consider the nature of the competitive season, that is, whether it involves several major meets spaced throughout a year, as in athletics or swimming, or involves a prolonged period during which games are played on a weekly basis, as with most team sports.

Periodization for Major Meets

If the sport entails several major competitions spaced throughout a year, then a number of periodization strategies are used depending on the dominant physical capacity involved in the sport.

Strength For those sports that predominantly require muscular strength, such as weight- or powerlifting, the athlete will perform a series of successive periodized strength programs. For example, if an athlete has three major competitions with 12 to 15 weeks spaced between each, the first competition will be preceded by a 12-week training period comprising a periodized routine, such as a classic or undulating program (see chapter 8). The training cycle should be organized so that a one- to two-week tapering period is imposed prior to the competition, and a one-week rest period should follow before the program is again started. This procedure is repeated throughout the competitive season. The athlete's goal, as the season progresses, is to use greater loads in each training cycle and competition.

Power If the dominant physical capacity of the sport is muscular power (e.g., with throwers, sprinters or jumpers), then the organization of the training year is somewhat different to that just described. The muscular functions of strength and power are closely related, and thus a high level of strength is a prerequisite for powerful performance. Consequently, the training year will commence with a two- to three-month period of strength training, similar to that for a strength athlete. As the competition approaches, training is progressively modified to incorporate an increased content of maximal power and plyometric training and a corresponding reduction in pure strength work. High-intensity strength training should be performed throughout the year, but the volume of this training is greatly reduced prior to the competition. Strength work is performed up to one to two weeks before the competition to maintain strength levels and continue the disinhibition of the neural system. Even within the week leading up to the competition, heavy strength training sessions should be completed, but with very low volume of only one to three sets of near-maximal lifts. The goal is to maintain or even raise neural excitability but not incur any physiological fatigue or muscle damage, which would decrease strength and power for the competition.

The resistance training strategy adopted after competition will depend upon the time period to the next major meet. If there is a relatively long period of four to six months, the athlete may perform another cycle of pure strength work prior to switching over to power-oriented training. However, if successive competitions are within two to three months, then a combined strength and power routine should be commenced, with progressively greater power content and correspondingly lower strength content as the subsequent competition approaches.

Strength-Endurance Similar to power, strength-endurance has a close relationship to pure strength, so that strong muscles will have greater endurance when working against a given absolute load. Therefore strength-endurance athletes, such as middle distance swimmers and runners, require high levels of strength and should perform a two- to three-month strength training cycle at the commencement of their training year. Several months prior to the competition, the resistance training routine should be modified to incorporate variable load training (see chapter 8). As the competition becomes progressively closer, variable load training should be modified so that a greater volume, at reduced loads, is performed. However, some high-intensity strength sessions should be maintained throughout the lead-up to the competition.

Periodization for Team Sports

Team sports generally involve one playing season per year. Therefore, the periodization model should encompass the off-season, preseason and competition phases.

Off-Season A typical season for a team sport, such as football or rugby, consists of a one- to two-month *off-season* recovery period, a three- to four-month *preseason* period and a six-month *competition* season. The off-season period is often used to rehabilitate injuries accumulated during the preceding year and involves participation in "active rest"; that is, athletes simply try to keep up a reasonable level of aerobic fitness and strength without performing overly stressful exercise. The goal is to avoid losing too much condition that will have to be regained once the preseason phase begins. During this period, athletes with specific weak points (e.g., muscular imbalances) can undergo remedial resistance training. This is also a valuable time to increase flexibility, particularly about joints where the athlete may have poor range of motion.

Preseason The *preseason* period is used to enhance the physical capacities of the players so that by the time the competitive season commences they are in peak physical condition. Training performed during the preseason for team sport athletes should be structured in much the same manner as just specified for major athletic meets. Thus during the preseason, a power athlete, such as an offensive running back in American football, would perform the same routine as a sprinter or jumper. Similarly, a strength-endurance athlete, such as a ruck rover in Australian football, would perform a similar preseason resistance training routine to a middle distance runner.

Competition During the *competition* season, so much time is spent competing, recovering from the competition and performing specific skill work that there is little time available for resistance training. However, it is very important that some resistance training be performed to maintain the physical capacities developed during the preseason. Generally, at least one intense resistance training session per muscle group per week is required to maintain the developed muscular functions. The maintenance training routine should be one that stresses the dominant physical capacity involved in the sport.

Number of Training Sessions

The number of resistance training sessions performed in a week is generally determined by the available time and the relative importance of resistance training to the sport. For example, many elite swimmers perform 10 to 12 swimming sessions per week, and consequently the time available for resistance training is limited. For the majority of athletes, two or three resistance training sessions per week are adequate. Each session should consist of exercises that train the muscle groups relevant to the sport; sessions are spaced so that at least one day of rest is imposed between workouts.

For athletes involved in sports that are heavily dependent on the physical capacities of strength and power (e.g., throwers, lifters and defensive linemen), more resistance training sessions are generally required. These athletes perform split routines, in which only a portion of the body is trained in any one session. Generally they use four to six training sessions, with each muscle group trained two to three times per week. Again, at least 48 h recovery is required between exercise sessions of the same muscle group. Some elite strength athletes, such as weightlifters, perform multiple daily resistance training sessions. They can achieve high intensities by separating the workouts into smaller units. While this is often not possible due to time demands, it is more effective in terms of strength and power gain if the resistance training program is split into two sessions for the day.

Exercise Order

Little research has been performed on the effect of the order of exercises on the training gains achieved. Nevertheless, the following training recommendations are suggested based on this limited research and the need to maximize training intensity and hence minimize fatigue.

Regardless of the training routine adopted, the musculature tends to fatigue during the workout. Thus, the exercises performed first will tend to involve the greatest training intensities. Consequently the most important exercises should be performed first. For example, a shot-putter should begin the strength training session with the inclined bench press or the squat exercise rather than the chin-up or biceps curl, as the former exercises are more important for the competitive activity. Alternatively, if an athlete has a specific weakness such as a muscle imbalance, then this muscle group should be exercised first as it is the most important muscle group to develop. Although athletes frequently perform their favourite or best exercise first, such as the bench press, often such exercises are not the most important for their competitive performance.

If athletes are attempting to develop strength or power, they should perform exercises in an order that provides the maximum amount of rest between similar exercises. For example, a freestyle sprint swimmer requires a high degree of power in the pectoralis major, deltoids, latissimus dorsi, quadriceps, triceps brachii and biceps brachii muscle groups. After performance of pectoral exercises, the deltoid and triceps muscle groups are often fatigued, and therefore exercises for these muscle groups should be spaced as far apart as is practicable. Similarly, exercises for the latissimus dorsi and biceps brachii muscles should be spaced apart to minimize the residual fatigue. Table 13.1 outlines two exercise orders for a freestyle swimmer, one designed to minimize residual fatigue and the other structured to maximize fatigue. The exercise order on the left of table 13.1 spaces exercises that use similar muscle groups as far away from each other as possible. This serves to reduce the fatigue associated with training similar muscle groups in the same session and allows the subsequent muscle groups to be exercised with greater intensity. The training order outlined on the right of table 13.1 does not provide for any recovery time between the performance of similarly acting muscle groups; hence these muscle groups will be fatigued and will experience a reduced training intensity.

Table 13.1 Differing Exercise Orders That Minimize or Maximize Residual Muscular Fatigue

Minimize residual fatigue	Maximize residual fatigue
Latissimus dorsi	Latissimus dorsi
Deltoids	Biceps brachii
Quadriceps	Pectorals
Pectorals	Deltoids
Biceps brachii	Triceps brachii
Triceps brachii	Quadriceps

If athletes are attempting to increase strength-endurance by improving their tolerance to the accumulation of lactic acid, then the exercises should be ordered to maximize the residual fatigue. Thus, exercises that act in similar ways should be performed straight after one another, as shown in the right-hand column of table 13.1.

Warm-Up

As with any other strenuous physical activity, it is important to warm the body prior to resistance training, essentially increasing the temperature of the muscles to be exercised. This should involve several minutes of aerobic exercise, such as riding a stationary bicycle, followed by light exercises for the muscle groups to be trained (e.g., light dumbbell presses to warm up the pectoral muscle group). Finally, the athlete should perform stretching exercises and, prior to the use of heavy loads, several light preparatory sets of the exercise. However, coaches should note the suggestion from recent research that static stretching during warm-up decreases muscle force production. Therefore one should take care when implementing warm-up procedures that include static stretching prior to events requiring maximal force production or high rates of force development.

The warm-up serves to facilitate performance by preparing the neuromuscular system for maximal exertion and also reduces the risk of musculoskeletal injuries. Warming up before more vigorous exercise results in less lactic acid accumulation and more force and power capacity during training or competition. It is important to realize that each new body part trained in a workout may need to be warmed. For example, if training the chest and the legs, one should not assume that the chest training will serve to warm up the legs. After the completion of the chest exercises, the legs will need to be warmed prior to commencement of leg-specific exercises.

RESISTANCE TRAINING EXERCISES

This section details 53 resistance training exercises for strength and power development that form a pool from which sport-specific resistance training programs are generated in the next section of the chapter. Unless otherwise specified, the strength training exercises should be executed so that the eccentric phase (generally the downward part) of the lift is performed under control, taking approximately 1.5 to 2.0 s to complete, while the concentric portion is executed as rapidly as possible. When using relatively heavy loads, most exercises should be performed with at least one spotter standing by in case the lift cannot be completed. Though most of the exercises presented here require barbells, dumbbells or specialized weight stack equipment, alternate forms of equipment may be substituted for different applications. These alternatives include exercises using the body weight or elastic bands to provide the resistance (see also chapter 8).

Strength Training Exercises

This section describes 41 commonly used resistance exercises. With the emphasis on improving strength, a photographic image of each exercise is provided to aid in the understanding of the technique.

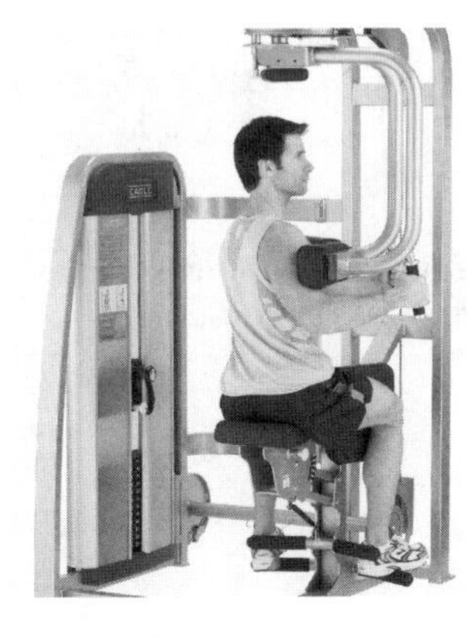

1 **Abdominal rotation**—external and internal oblique muscles

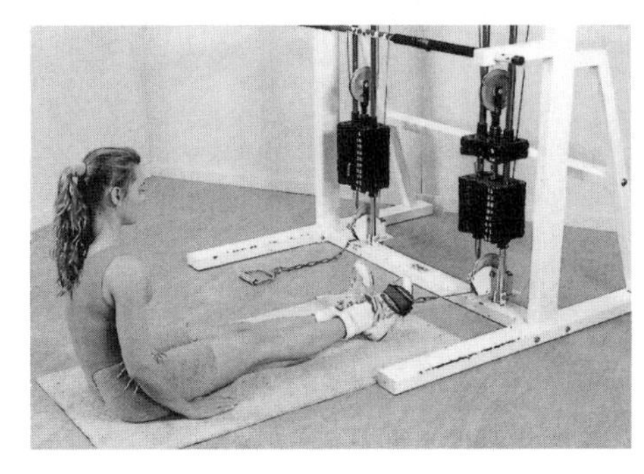

2 **Ankle curl**—tibialis anterior and other muscles of the pretibial group

3 **Trunk extension**—erector spinae, gluteal and hamstring muscle groups

4 **Barbell curl**—flexors of the forearm, notably biceps brachii, brachialis and brachioradialis

5 **Bench press**—pectoralis major, deltoids and triceps brachii

6 **Bent-over rowing and bench pull**—trunk extensor muscles, with erector spinae as stabilizers and the latissimus dorsi and forearm flexors as prime movers

7 **Close-grip bench press**—similar to 5, but with increased contribution from triceps brachii and reduced contribution from the pectoral group

8 **Close-grip pulldown**—latissimus dorsi and forearm flexor muscles

9 **Deadlift**—erector spinae, quadriceps, hamstrings and gluteal groups, as well as upper trunk and forearm musculature

10 **Dip**—pectoral, triceps brachii and deltoid musculature

11 **Internal-external rotator raise**—rotator muscles of the arm

12 **Front pulldown**—pectoralis major (initially), then latissimus dorsi and teres major muscles (one- and two-handed)

13 **Hammer curl**—forearm flexor muscles, but with emphasis on brachioradialis and brachialis

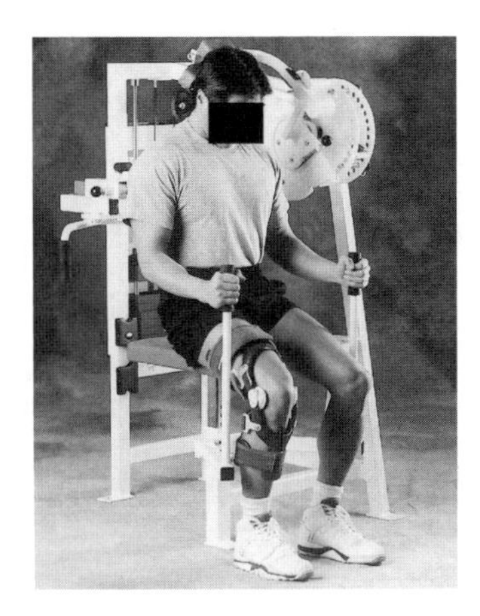

14 **Head curl**—trapezius muscle

15 **High pull**—erector spinae, quadriceps, hamstrings, gluteal and trunk stabilizing musculature

16 **Inclined bench press**—emphasizes anterior deltoid and upper pectoralis major muscle fibres

17 **Inclined sit-up**—abdominal and thigh flexor muscle groups

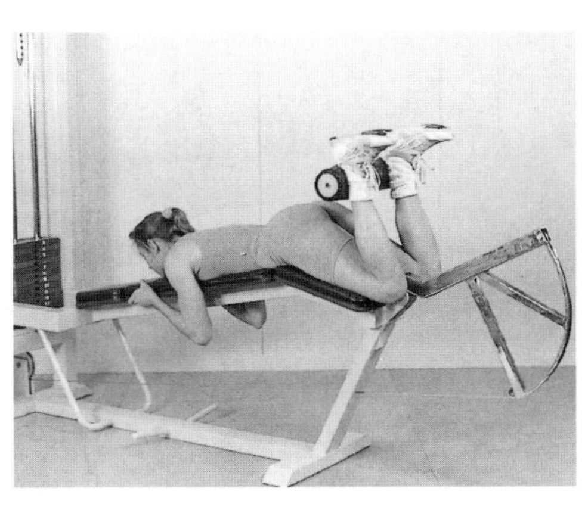

18 **Leg curl**—hamstring group and gastrocnemius

19 **Leg press**—leg and thigh extension musculature

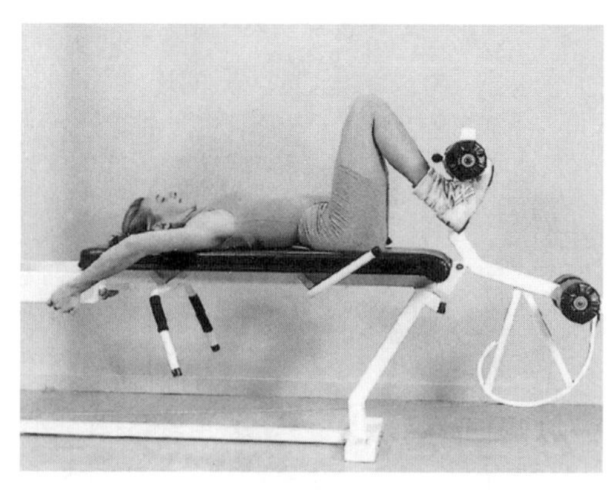

20 **Lying leg raise**—abdominal and thigh flexor muscles

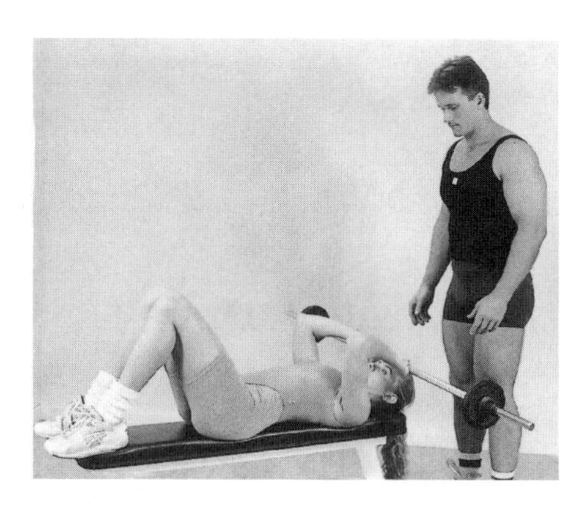

21 **Lying triceps extension**—triceps brachii muscle

22 **Pullover**—pectoralis major (initially), then latissimus dorsi and teres major muscles

23 **Reverse curl**—preferentially strengthens the brachialis muscle

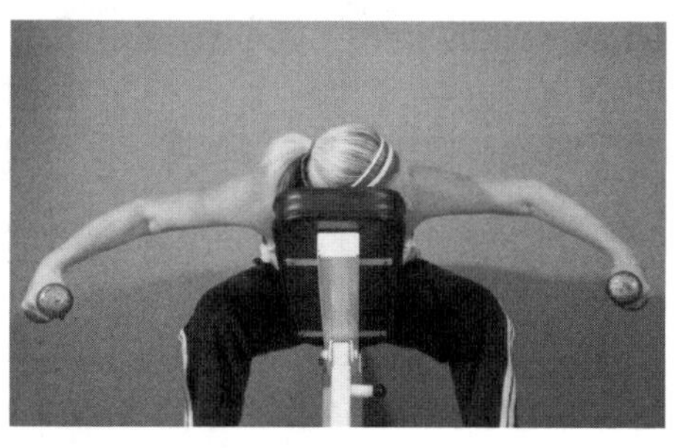

24 **Reverse pec fly**—trapezius, rhomboids, latissimus dorsi and the teres major muscles

25 **Running dumbbell**—specific exercise for muscles used in the running action

26 **Seated rowing**—rhomboids, trapezius and erector spinae muscles of the trunk, plus arm extensors and forearm flexors

27 **Shrug**—trapezius muscles

28 **Side bend**—external and internal obliques, quadratus lumborum and the erector spinae muscles

29 **Squat**—quadriceps, hamstrings and gluteal muscle groups, as well as the erector spinae muscles

30 **Standing calf raise**—gastrocnemius and soleus muscles

31 **Standing push press**—deltoid, triceps brachii and trapezius musculature as well as the quadriceps group

32 **Stiff-legged deadlift**—hamstring and gluteal groups and erector spinae group

33 **Swim bench pull**—simulate upper body arm actions used in swimming

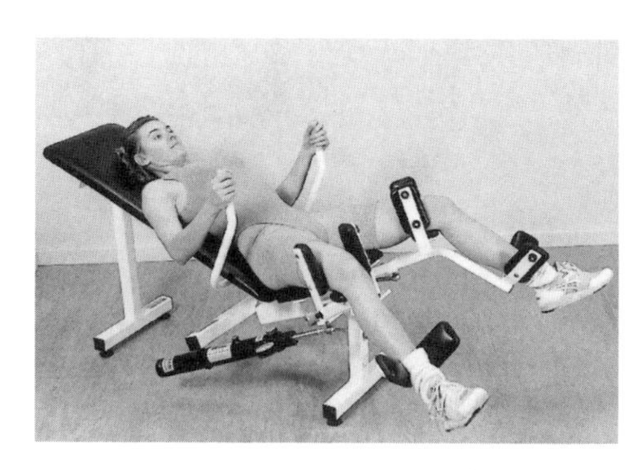

34 **Thigh abduction**—gluteus medius and minimus

35 **Thigh adduction**—pectineus, gracilis and the adductor longus, brevis and magnus muscles

36 **Thigh extension**—hamstring group, adductor magnus and gluteus maximus muscles

37 **Thigh flexion**—the iliopsoas group and the rectus femoris muscle

38 **Triceps kickback**—triceps brachii muscle

39 **Upright row**—deltoid and trapezius muscles

40 **Wide-grip chin**—latissimus dorsi and teres major and the forearm flexor muscles

41 **Wrist curl**—hand and finger flexor muscles

Power Training Exercises

A considerable volume of research has been conducted over the past decade that has revolutionized techniques for the development of maximal power. The expression and development of maximal power (see chapter 9) require performance of quite explosive or ballistic movements against load as a training stimulus. When combining this with heavy resistance training to enhance maximal strength, the athlete develops the key qualities for maximal power output. These qualities include high force capability during the transition from eccentric to concentric phases of the stretch-shortening movement, maximum rate of force development, ability to produce high force while the muscle is contracting rapidly and well-developed technique to ensure optimal flow of power through the kinetic chain. All of these aspects contribute to power production, and a quite detailed discussion of this topic is presented by Newton and Kraemer (1994).

The key training technique for power development is performance of maximal efforts against loads spanning the concentric strength range (i.e., from light loads of 15-30% 1-repetition maximum [1RM] to heavy loads of 75-80% 1RM). However, a convincing series of research papers (e.g., McBride et al. 2002) suggests that if one load is to be used, it should be the load that optimises power output (i.e., results in greatest power output and is usually around 30% of the athlete's 1RM). The most important characteristic of this training is that the athlete should jump with the load or throw the load to avoid any deceleration phase (see Newton et al. 1996); for this reason it is termed *ballistic power training*. This of course raises issues of safety; therefore specialized equipment has been developed that incorporates either specialized power racks or Smith machines, which have safety stops to catch the barbell if the athlete makes an error (e.g., www.fittech.com.au/products/bmsmbu.htm). Another issue with ballistic power training is the high impacts that the athlete experiences when landing from a loaded jump or catching a barbell; however, various braking devices have been developed that allow control of the eccentric phase either manually or by computer (Humphries, Newton, and Wilson 1994). Measurement of performance and real-time feedback to the athlete via computer are now available (www.innervations.com/ballistic/bms.htm) to further enhance the quality of ballistic power training.

Unless otherwise specified, the following exercises should be performed on the Ballistic Resistance System (BRS) (Fitness Technology, Adelaide, Australia) or a similar system using a load that maximizes the power output of the exercise (Newton et al. 1997). One determines this by modifying the load used on several trials and recording the power output on the computer system. The load that maximizes power output should be the one used in training (see chapter 9 for explanation). If the impact force is high during performance of the maximal power exercises, then the brake device of the BRS should be engaged to reduce the peak force. Auditory feedback is provided by the computer system through use of the threshold function. For performance of maximal power exercises, the bottom safety stops of the BRS should be set so that if the bar is incorrectly caught, or if an incorrect landing is performed, the bar will contact the safety stops before striking the athlete's body. All ballistic power exercises should be performed as explosively as possible, with minimum contact time.

If the athlete does not have access to a BRS, a low-friction Smith machine may be used. However, several spotters must be employed to support the bar in case it is incorrectly caught by the lifter, as this type of machine may not have bottom stops to act as safety catches. Further, a Smith machine does not provide feedback or information on optimal load, nor does it have a braking device. The performance of maximal power exercises on a Smith machine is not as effective or as safe as with use of a BRS but nevertheless is currently the next-best option.

The weightlifting exercises (snatch, clean and jerk) are also used extensively for the development of maximal power as they allow the athlete to accelerate throughout the movement and produce high power output. There are also many variations or phases of these lifts that can be used, such as power clean, hang clean, hang snatch, high pull and hang pull. A detailed discussion of these training methodologies is outside the scope of this book, and the interested reader is referred to specialized texts on the topic. However, in general, these movements, while very effective for developing weightlifting performance, do not have the same specificity as ballistic power training; thus we direct our attention to discussion of these exercises.

MAXIMAL POWER TRAINING EXERCISES

This next group of exercises are performed ballistically. For each exercise a photographic image is provided to aid in the understanding of the technique.

42 Ballistic bench press

43 Ballistic inclined bench press

44 Ballistic leg extension

45 Ballistic lying triceps extension

46 Ballistic split squats

47 Ballistic squats

48 Ballistic standing push press

PLYOMETRIC EXERCISES

The following five plyometric exercises are most commonly employed by power athletes; however, many others have been developed to suit specific sport applications.

49 Bounding

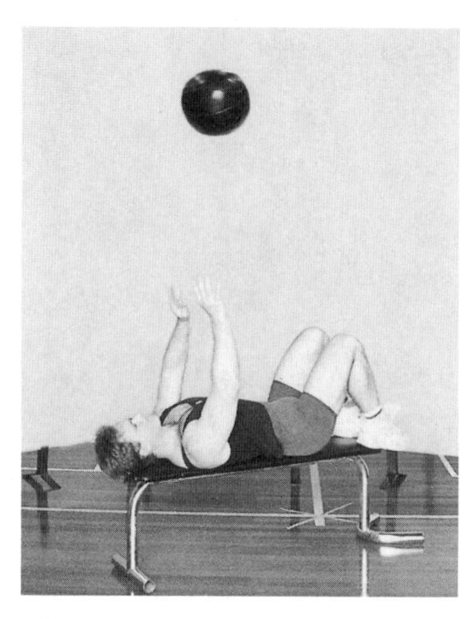

50 Chest throw

51 Depth jump

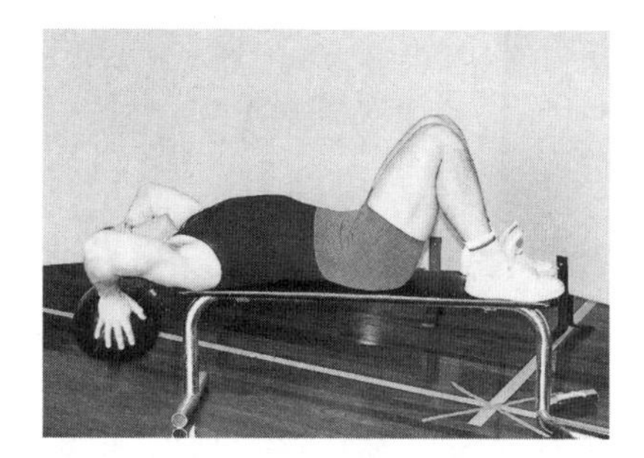

52 Overhead throw

53 Plyometric push-up

STRETCHING EXERCISES

This section details 38 stretching exercises that form a pool from which sport-specific flexibility training programs are generated later in this chapter. In chapter 11, static, passive, proprioceptive neuromuscular facilitation (PNF) and dynamic or ballistic stretching techniques were outlined. Given that most athletes use static stretching to increase range of motion, this chapter focuses on the slow static stretching techniques with additional PNF examples. For more information on other forms of stretching, readers are referred to the SportSmart Web site (www.acc.co.nz/injury-prevention/sport-safety/acc-sportsmart/index.htm).

Specific Guidelines for Static Stretching

In static stretching, the muscle group is held on stretch for a period of 20 to 30 s, but can be held for 60 s if necessary. This depends, however, on whether one can utilize the "small relaxations" that occur during the stretching exercise.

- The primary stretch is held for approximately 10 s, after which a secondary stretch is performed.
- The entire stretch (i.e., both primary and secondary) usually takes ~20 to 30 s to execute when the technique has been well developed.
- It is very important to hold the primary stretch until the inverse myotatic reflex occurs and a slight relaxation is felt.
- The primary and secondary stretches must be done very slowly with no pressure or jerkiness. With any pressure or jerkiness, there will be a rebound effect from the stretch reflex and some loss of control during the stretch.
- At the completion of the stretch, the muscle group should be released slowly and under control.
- Approximately four sets of each exercise should be done during each workout, although this may be too many in the early stages of a flexibility training program. The number can be increased to eight sets later in the program.
- It is important to keep agonists and antagonists in balance by stretching each group during a workout. The athlete should also keep the two sides of the body in balance, unless there is a logical reason not to do so.
- There is currently some disagreement as to the proper order of the stretching routine. Traditionally, exercises have been alternated over the various regions of the body. More recently, some sport scientists and coaches have suggested that the athlete concentrate on one part of the body using several exercises, then move to another part and so on. There is currently no scientific evidence available to support one method over the other.
- Flexibility training should be carried out each day in either one or two sessions depending on the sport. There should be at least one day off each week.
- It may be of value to use equipment or a partner for support purposes with some exercises. Others can be done quite effectively without such support.
- Most elite coaches now recommend static stretching at the end of a training session to help improve the range of motion, whereas sport-specific dynamic stretching is more commonly used in the warm-up.

General Guidelines for PNF Stretching

In PNF stretching, increased flexibility is gained through use of a 6 s isometric contraction of the muscles to be stretched, followed by a concentric contraction of the opposite muscle group, with light pressure being applied by the partner. The concentric contraction should last approximately 6 to 10 s. Each exercise should be repeated three or four times during a training session, and up to two training sessions per day can be performed. At least one day each week should be used for rest and should involve no flexibility training.

Specific Exercise Routines

This section describes 38 commonly used stretching exercises, including both individual and partner stretches.

INDIVIDUAL EXERCISES

Photographic images are provided for this group of individual stretches to aid in the reader's understanding of the technique.

54 Neck stretch

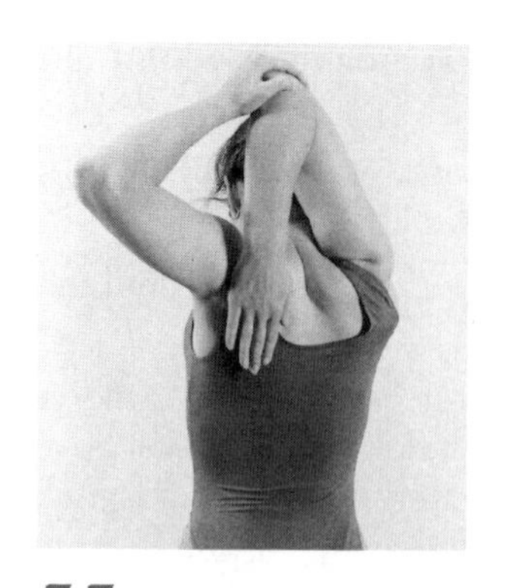

55 Triceps stretch

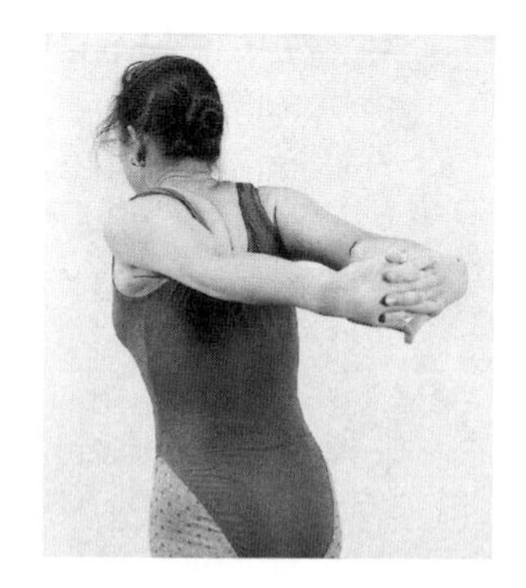

56 Backward stretch

57 Overhead stretch

58 Circular shoulder stretch

59 Hand-forearm stretch

60 Abdominal and hip stretch

61 Back roll stretch

62 Upper back stretch

63 Lateral trunk stretch

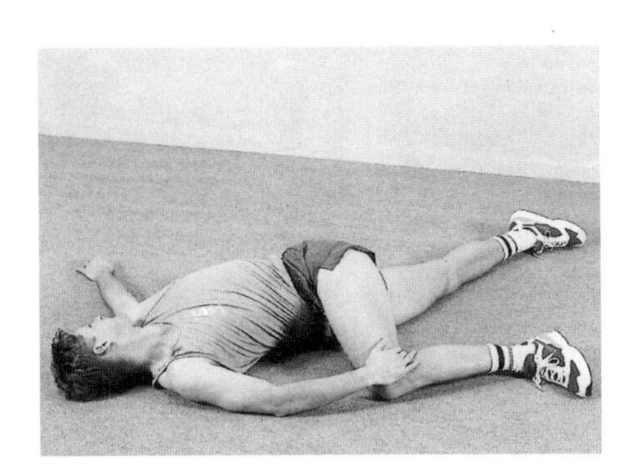

64 Lower back and hip stretch

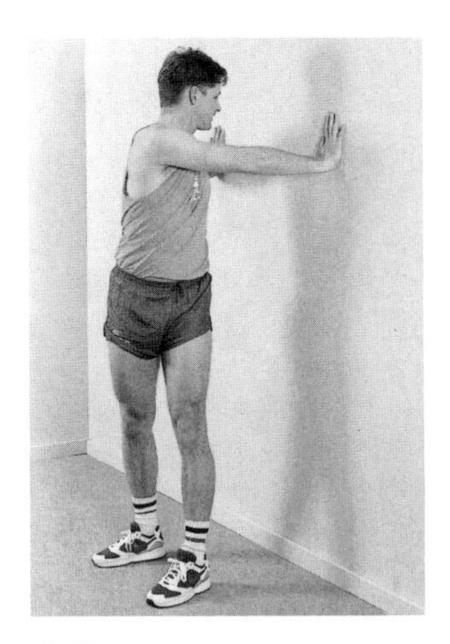

65 Rotational trunk stretch

66 Groin stretch 1

67 Groin stretch 2

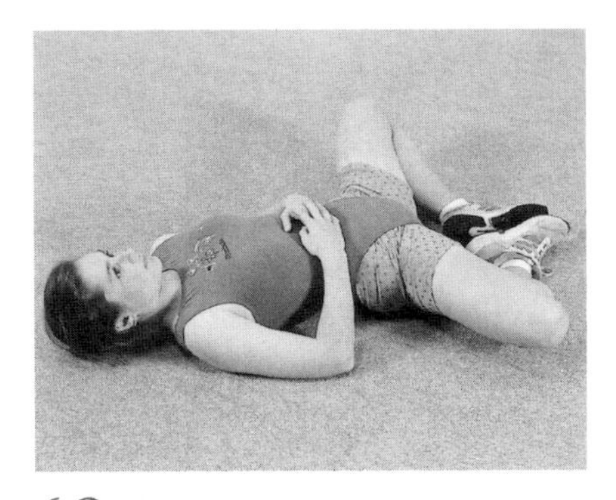

68 Groin stretch 3

69 Side split stretch

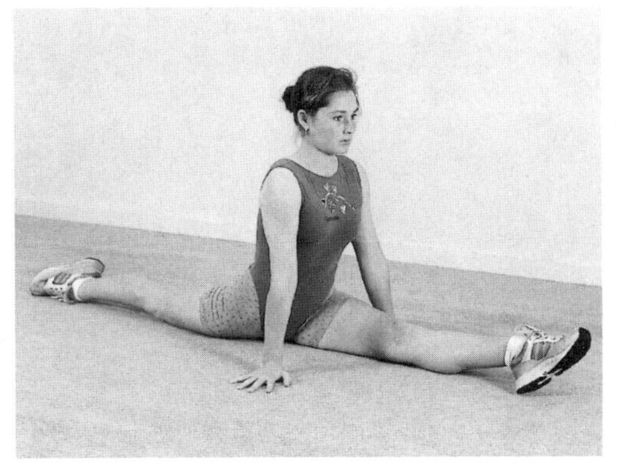

70 Split stretch

71 Hamstring stretch 1

72 Hamstring stretch 2

73 Hamstring stretch 3

74 Groin-trunk stretch

75 Quadriceps stretch 1

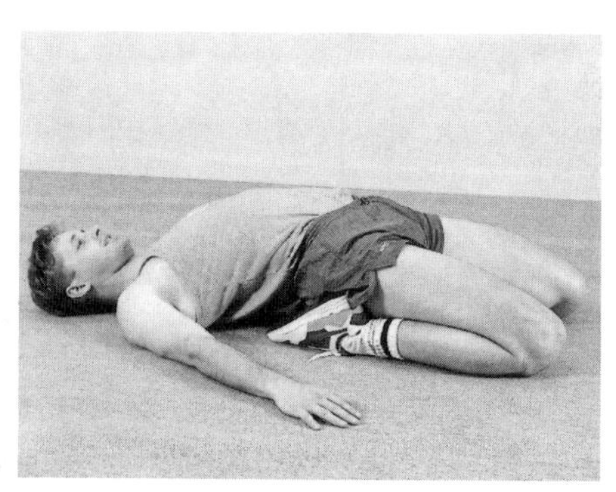

76 Quadriceps-ankle stretch

77 Quadriceps-groin stretch

78 Quadriceps stretch 2

79 Calf stretch

80 Achilles tendon–ankle stretch

81 Ankle stretch

PARTNER EXERCISES (PNF)

These 10 partner stretches can be performed statically or using the PNF principles (a = athlete, p = partner).

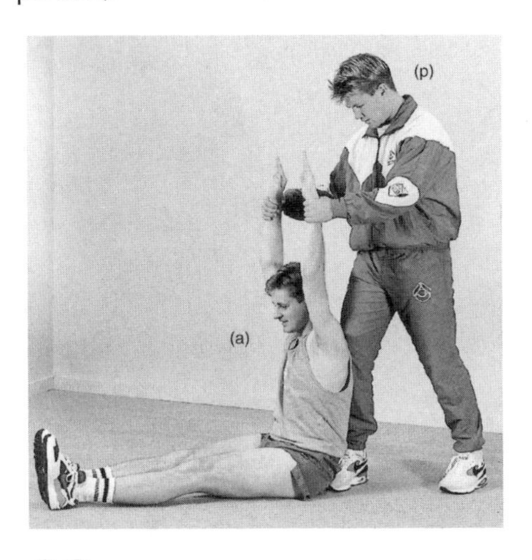

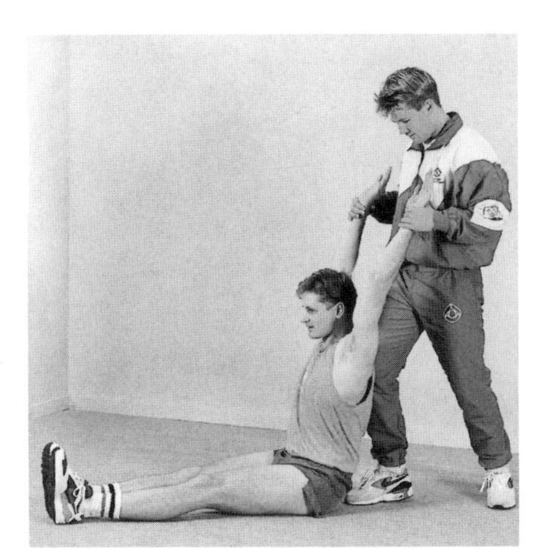

82 Arm extensor stretch

83 Arm adductor stretch

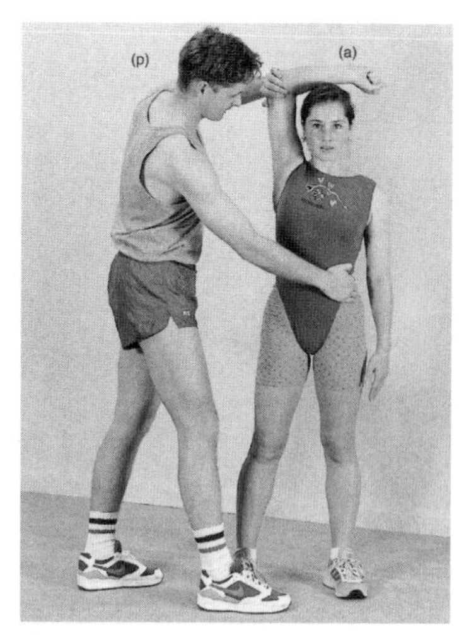

84 Trunk lateral flexor stretch

85 Split pike stretch

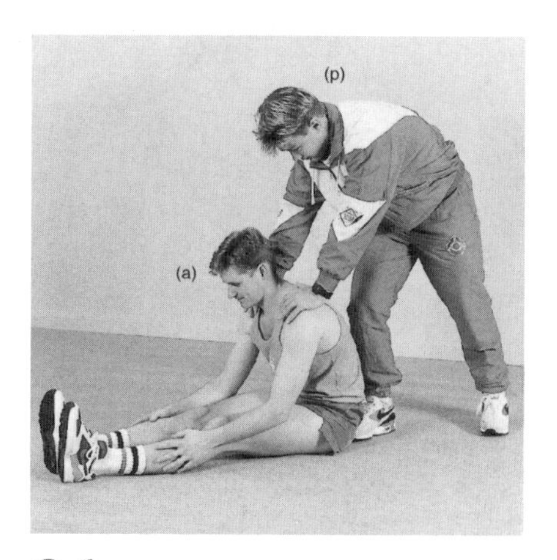

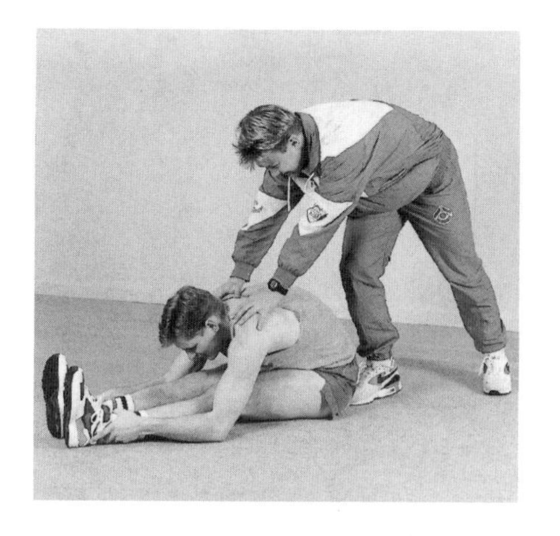

86 Trunk extensor stretch

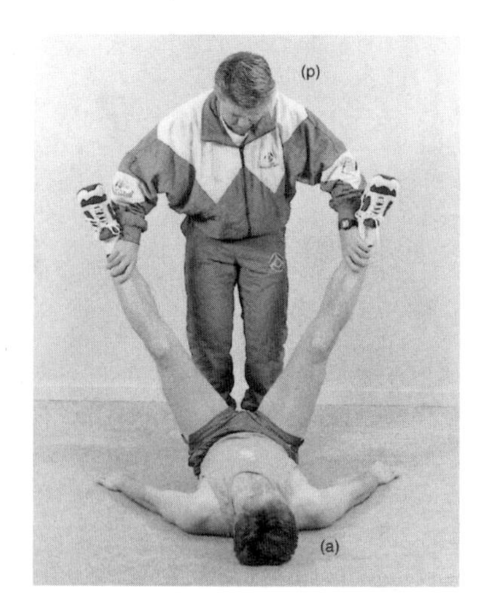

87 Thigh adductor stretch

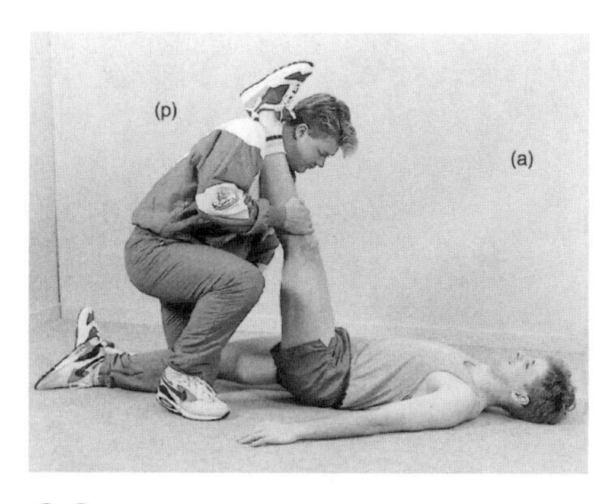

88 Thigh extensor stretch 1

89 Thigh extensor stretch 2

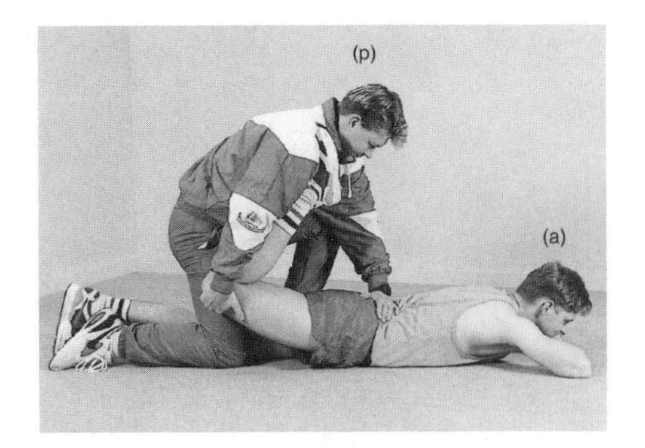

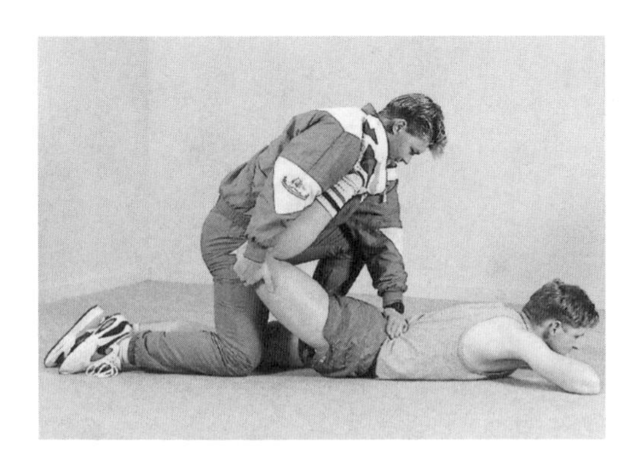

90 Thigh flexor stretch 1

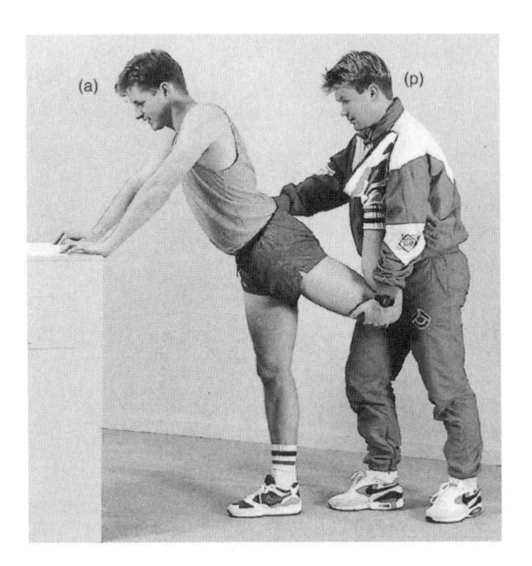

91 Thigh flexor stretch 2

Training Programs for Specific Sports

There is no "best" training regimen that can be used by all athletes, at all times and under all circumstances. However, there are exercises that are more suited to some sports than to others, as identified in this section. Sport-specific resistance training routines and stretching exercises are outlined, together with other important considerations related to size, proportionality, composition, speed, agility and other physical capacities. We emphasize, though, that these suggested routines will not necessarily be optimal for all individuals and that the information should be used only as a starting point. The coach or athlete may then modify these programs to fit the specific requirements of the sport, the phase of the year and individual characteristics of the athlete, as well as to provide variety in training.

It is assumed that before commencing the specific exercises outlined in this section, athletes will have experienced at least six months of general strength training. An example of such a basic, introductory strength training routine is outlined in table 13.2. Where highly specialized strength, power or flexibility exercises are required for elite-level performers in certain sports, the reader should consult specialist texts that describe such exercises.

Table 13.2 General Introductory Strength Training Routine

Monday	Wednesday	Friday
Bench press	High pull	Deadlift
Standing push press	Inclined bench press	Leg curl
Lying triceps extension	Close-grip bench press	Standing calf raise
Squat	Upright row	Inclined sit-up
Bent-over rowing	Trunk extension	Bench press
Inclined sit-up	Hammer curls	Standing push press
Barbell curl	Abdominal rotations	Lying triceps extension
Leg curl	Leg press	Wide-grip chin
Standing calf raise	Stiff-legged deadlift	Barbell curl

After a one-month gradual introductory phase, three or four sets of 8 to 12 repetitions should be performed for each exercise.

RACQUET SPORTS: TENNIS, BADMINTON, SQUASH, RACQUETBALL

Success in the various racquet sports depends on many capabilities, including high aerobic capacity, speed, agility and flexibility. In addition, because these are ballistic sports, the athletes must be able to perform powerful and dynamic upper and lower body movements. These athletes should also undergo training for total-body speed, inclusive of acceleration and running speed, as well as for specific limb speed and agility.

Limitations—Self-Selection Parameters

There are no morphology-related limiting factors that prevent players from reaching elite levels in the racquet sports. Tall, linear physiques provide some advantage for tennis players on certain playing surfaces, but the smaller, more agile player may be more successful on the slower clay court.

Modifiable Parameters

The following text describes at length the modifiable parameters to consider in racquet sports.

BODY COMPOSITION

Given the aerobic requirements of these sports, combined with the need for speed and agility, racquet sport players must be lean with minimal excess body fat, though not to the extreme levels observed for endurance athletes. Unless the player has an obvious problem in this regard, no universal dietary intervention is warranted for this sport.

STRENGTH AND POWER

Racquet sports are ballistic in character, requiring powerful and dynamic upper and lower body movements. In addition to concentrating on the prime mover muscles for stroke production and court coverage, exercises promoting core stability of the trunk plus scapula stabilization are essential in these sports. In addition, the resistance training exercises most suited to these sports include the following:

- Squat (exercise 29)
- Ballistic squat (47)
- Ballistic split squat (46)
- Ballistic lying triceps extension (45)
- Ballistic bench press (42)
- Overhead throw (52)
- Wrist curl (41)
- Hammer curl (13)
- Internal-external rotator raise (11)
- High pull (15)
- Trunk extension (3)
- Abdominal rotation (1)
- Thigh adduction (35)
- Thigh abduction (34)
- Leg curl (18)
- Thigh extension (36)
- Standing calf raise (30)
- Upright row (39)
- Pullover (22)
- Ankle curl (2)

Furthermore, specific arm actions can be simulated with use of a cable pulley system (figure 13.1) or elastic bands for internal-external rotation of the arm. An example of a preseason resistance training routine for racquet sports is outlined in table 13.3. Due to the unilateral nature of racquet sports, it is advisable that where possible, resistance training be performed independently on both sides (i.e., using dumbbells) to reduce muscular imbalances.

Figure 13.1 With use of a cable pulley system, specific arm actions can be simulated.

FLEXIBILITY

Racquet sports require high levels of flexibility in order for athletes to place themselves into positions from which they can hit the ball more powerfully so as to generate high segment rotational speeds when executing the shot. They will also be able to produce a greater contractile force when the stroke begins with muscles in a prestretched position.

Table 13.3 Preseason Resistance Training Routine for Racquet Sport Athletes

Monday	Wednesday	Friday
Ballistic squat	Forehand stroke	Ballistic squat
Ballistic lying triceps extension	Ballistic bench press	Backhand stroke
High pull	Ballistic split squat	Ballistic lying triceps extension
Thigh adduction	Overhead throw	High pull
Thigh extension	Thigh abduction	Thigh adduction
Standing calf raise	Pullover	Thigh extension
Abdominal rotation	Inclined sit-up	Abdominal rotation
Internal-external rotator raise	Trunk extension	Standing calf raise
Wrist curl	Hammer curl	Ankle curl
Stiff-legged deadlift		Upright row

- **Shoulder girdle:** Arm flexion-extension and internal-external rotation flexibility are essential capacities of racquet sport players. The following exercises should be carried out to achieve these capacities:
    - Triceps stretch (exercise 55)
    - Circular shoulder stretch (58)
    - Arm extensor stretch (82)
    - Arm adductor stretch (83)
- **Trunk:** Trunk flexion-extension, lateral flexion and rotation are important movements that are carried out by all racquet sport players. The following exercises should be performed to increase flexibility in these movements:
    - Abdominal and hip stretch (60)
    - Upper back stretch (62)
    - Lower back and hip stretch (64)
    - Rotational trunk stretch (65)
    - Trunk lateral flexor stretch (84)
    - Trunk extensor stretch (86)
- **Pelvic girdle and legs:** Any agility athlete must have the thigh, leg and foot flexors and extensors well stretched. The following exercises should be performed to achieve this:
    - Hamstring stretch 1 (71)
    - Hamstring stretch 2 (72)
    - Quadriceps stretch 1 (75)
    - Quadriceps-ankle stretch (76)
    - Quadriceps stretch 2 (78)
    - Calf stretch (79)
    - Achilles tendon–ankle stretch (80)
    - Thigh extensor stretch 1 (88)
    - Thigh flexor stretch 1 (90)

Furthermore, to assist on-court agility (and to prevent injury), stretching exercises for hip abduction-adduction and internal-external rotation muscles are also required, such as lower back and hip stretch (64), groin stretch 1 (66), groin stretch 2 (67) and groin stretch 3 (68).

SPEED AND AGILITY

Athletes in games such as tennis, badminton and squash or racquetball require training for total-body speed, specific limb speed and agility (chapter 12). Total-body speed, which includes acceleration and running speed, is essential in a racquet game. Running speed training that includes strength, power and flexibility activities, combined with speed-resisted and speed-assisted training, should be performed (chapter 10).

Specific limb speed can be improved with strength and power training as well as speed-assisted and speed-resisted training, but it is skill training that will provide the greatest hitting velocities in these games. The coach must constantly look for the technique weaknesses that are hindering the development of power in the various strokes, and the combination of a fully coordinated high-velocity stroke and movement speed training will greatly improve this area of the player's game.

AQUATIC SPORTS

All the aquatic sports are dominated by the need for high levels of power and strength-endurance, especially in the upper body musculature, which is responsible for developing much of the power for body, ball or craft propulsion. The relative importance of these functions is dependent upon the specific duration of the event. The mobility needs of swimmers and water polo players are almost identical, while rowers and canoeists need similar exercises for their shoulder girdles, trunks, pelvis and thighs. Unlike athletes in the majority of aquatic sports, rowers also need very high leg extensor strength and power.

Limitations—Self-Selection Parameters

Several morphological limitations exist for aquatic athletes. Mazza and colleagues (1994) reported that World Championship performance ranking for sprint and middle distance swimmers was positively influenced by tall stature and absolute limb lengths, as well as the size of hands and feet. Absolute size is also important for open competition rowers and water polo players, especially among the key position players—centre forwards, centre backs and goalkeepers. While not excessively so, open rowers and water polo players are among the tallest and most robust of the sport groups (Norton and Olds 1996). Conversely, sprint and slalom paddlers are of average size; however, elite performers generally possess a uniquely high brachial index (see chapter 6).

Modifiable Parameters

The following text describes at length the modifiable parameters to consider in aquatic sports.

BODY COMPOSITION

Most aquatic athletes endeavour to minimize body fat in order to reduce the drag of their body or craft through the water and maximize lean tissue especially in those segments of the body that are responsible for generating propulsion. Since the body is supported in a fluid medium, however, higher levels of adiposity than are observed for other elite athletes can generally be tolerated; and this does not appear to cause the same level of performance decrement as might be the case among track runners, for example. This is especially the case for open-water long distance swimmers, who, according to Ackland (1999), have much higher levels of adiposity compared to swimmers in all other events as an aid to thermal insulation and buoyancy.

Some aquatic sports, like rowing and canoeing, require athletes to compete within weight restriction rules, so athletes may be required to make a certain weight category or perhaps to "bulk up" in order to maximize their performance in open competition. Specific guidance for these strategies is offered in chapter 5.

STRENGTH AND POWER

Swimming Exercises most suited to swimming include the following:

- Swim bench pull (exercise 33)
- front or back pulldown (12)
- triceps kickback (38)
- close-grip pulldown (7)
- pullover (22)
- reverse pec fly (24)
- squat (29)
- ballistic leg extension (44)
- leg press (19)
- ballistic squat (47)
- wide-grip chin-up (40)
- ballistic inclined bench press (43)
- ballistic bench press (42)
- bench press (5)
- bench pulls (6)
- depth jump (51)
- lying leg raise (20)
- dip (10)
- trunk extension (3)
- internal-external rotator raise (11)
- wrist curl (41)
- thigh extension (36)

Table 13.4 Preseason Resistance Training Routine for Front Crawl Swimmers

Monday	Wednesday	Friday
Swim bench (slow speed)	Swim bench (moderate speed)	Swim bench (fast speed)
Ballistic leg extension	Depth jump	Ballistic leg extension
Ballistic inclined bench press	Close-grip pulldown	Ballistic bench press
Ballistic squat	Leg press	Front pulls
Front pulldown	Bench press	Lying leg raise
Lying leg raise	Bench pull	Triceps kickback
Triceps kickback	Pullover	Bench pull
Chins	Wrist curl	Internal-external rotator raise
Thigh extension		Trunk extension

An example of a preseason training routine for a front crawl or butterfly swimmer is outlined in table 13.4.

Water Polo The sport of water polo requires a high level of swimming performance; thus the exercises required for this sport are essentially similar to those outlined for swimming. In addition, water polo players require exercises that enhance the overhead throwing action, such as ballistic lying triceps extension (exercise 45) and overhead throw (52), and lower body exercises that enhance the ability of the players to rise out of the water, such as ballistic split squat (46), thigh adduction (35) and thigh abduction (34).

Rowing The resistance training exercises most suited to rowing include:

- seated rows (exercise 26)
- high pull (15)
- squat (29)
- ballistic squat (concentric movement only, 47)
- deadlift (9)
- barbell curl (4)
- wrist curl (41)
- hammer curl (13)
- stiff-legged deadlift (32)
- leg press (19)
- upright row (39)

Rowers should also perform bench press (5), leg curl (18) and inclined sit-up (17) to help balance the development of their musculature.

Canoeing Canoeing and particularly kayaking require similar muscular actions to rowing with the exception that the lower body actions are different and the upper body action is unilateral. Consequently, exercises such as leg press (exercise 19), leg curl (18) and ballistic squat (47) are of lesser importance; and additional exercises such as abdominal rotations (1), bench pull (6) and inclined bench press (16) should be performed instead. Furthermore, the arm action of a kayaker can be simulated by the performance of a front pulldown exercise (12) with a pulley system that allows each arm to be exercised alternately.

FLEXIBILITY

Swimmers depend on high levels of flexibility in the shoulder girdle for freestyle and backstroke. This enables them to keep their bodies in a straight line rather than "break" at the hips or roll, which increases frontal resistance and is the major retarding factor for the great majority of swimmers. Butterfly swimmers should have even more flexibility at the shoulder than freestyle and backstroke swimmers, as they need to be almost hypermobile (figure 13.2). This characteristic enables them to stay very flat in the water, thereby avoiding increased frontal resistance created by excessive rise and fall. If a butterfly swimmer lacks the amount of flexibility needed to perform at the elite level, then a side breathing technique can be used.

Figure 13.2 Butterfly swimmers need to have even more shoulder flexibility than other swimmers.

Breaststrokers do not need as high a level of shoulder girdle flexibility as swimmers in the other strokes, but a mobile shoulder girdle will assist them to relax in the recovery phase of the stroke cycle. They do, however, need high levels of flexibility in thigh extension and abduction. Finally, all swimmers need above-average flexibility in the thigh flexors and extensors, with high levels of plantar- and dorsiflexion.

Figure 13.3 Backstroke swimmers need high levels of trunk hypertension.

Backstrokers can be too flexible in the shoulder and elbow joints, and this can lead to a poor arm position at the catch, thereby increasing horizontal frontal resistance. In this situation swimmers must not only modify their technique, but also reduce flexibility levels using strength training exercises. Backstroke swimmers, however, do need high levels of trunk hyperextension, which enables them to have their body enter the "hole" that has already been opened by the hands and head during the start of the race (figure 13.3).

Rowers must have extensive mobility in the thigh, leg and foot flexors and extensors. To have very flexible shoulders is not essential for rowers, but they should be reasonably mobile in the shoulder girdle and trunk in order to relax this portion of the body as much as possible during each stroke. Canoe and kayak paddlers need high levels of flexibility in all the joints, but especially in the shoulder girdle. Specific exercises for aquatic athletes include the following.

- **Shoulder girdle:** For freestylers, butterflyers, backstrokers, water polo players and kayak paddlers, this region must be well stretched. Rowers can also carry out the same exercises but do not need as many of them. The following exercises should be performed to achieve a high level of mobility in this region of the body:
    - Triceps stretch (exercise 55)
    - Backward stretch (56)
    - Circular shoulder stretch (58)
    - Arm extensor stretch (82)
    - Arm adductor stretch (83)
- **Trunk:** Swimmers, water polo players, canoeists and rowers can benefit from flexibility exercises that increase trunk flexion-extension, lateral flexion and rotation. The following exercises should be performed for this purpose:
    - abdominal and hip stretch (60)
    - upper back stretch (62)
    - lateral trunk stretch (63)
    - lower back and hip stretch (64)
    - rotational trunk stretch (65)
    - trunk extensor stretch (86)

- **Pelvic girdle and legs:** Swimmers, water polo players and rowers (and canoeists to a lesser extent) need the thigh, leg and foot flexors and extensors to be well stretched. The following exercises should be performed to achieve this:
  - groin stretch 2 (67)
  - hamstring stretch 1 (71)
  - hamstring stretch 2 (72)
  - quadriceps-ankle stretch (76)
  - calf stretch (79)
  - ankle stretch (81)
  - thigh extensor stretch 1 (88)
  - thigh flexor stretch 2 (91)

SPEED AND AGILITY

A considerable body of knowledge already exists on the application of movement speed training to swimming; this has been partially based on the Dintiman (1979) sprint running program. First, specific strength and power exercises, combined with speed-resisted training, are essential. Swimming with hand paddles and pull buoys is almost universally carried out, while some coaches use the speed trainer. Speed-assisted swimming can then be done in the latter part of the preseason and during the midseason. This can take the form of using modified flippers and in some cases, where available, a sprint towing device.

In rowing and kayaking, a similar program is necessary. Strength and power programs should underpin a speed-resisted program in which heavy boats or kayaks with slightly larger oar or paddle blades are used. The speed-assisted program must be carried out in a light racing boat using slightly smaller oar or paddle blades.

Water polo players need total-body speed as well as specific limb speed. The first is obtained by methods identical to those used by sprint swimmers, while the second can be developed by specific strength and power, speed-assisted and speed-resisted training and technique coaching.

GYMNASTICS, DIVING AND POWER SPORTS

Strength and power are dominant capabilities in gymnastics as well as in diving and the power sports, but there are differences in the types of strength that these sports demand. Gymnasts must have high levels of muscular power and strength in both the upper and the lower body. Divers need to have a high degree of explosive power in the lower limbs in particular and should also emphasize forearm strength. The sport of weightlifting quintessentially requires great strength and power. Among all sports, gymnastics and diving demand the highest level of flexibility.

Limitations—Self-Selection Parameters

Morphological limitations are most prevalent among elite competitors in gymnastics, diving and power sports. Gymnasts and divers are typically the smallest and lightest of all sportspersons (Norton and Olds 1996), with a high ratio of sitting height to stature caused by shorter than average lower limb lengths. Referring to the physiques of World Championship diving competitors, Mazza and colleagues (1994) suggested that this would assist platform divers in their execution of somersault actions during flight due to a reduced moment of inertia around a transverse axis through the centre of mass. Similar relationships between size and the mechanics of rotational performance among female gymnasts were explained by Ackland, Elliott, and Richards (2003).

Though the size of power athletes (weightlifters and powerlifters) varies according to their weight category, elite performers possess unique proportions that give a competitive advantage. As outlined in chapter 6, these athletes have a high ratio of sitting height to stature caused by shorter than average upper and lower limb lengths, as well as low crural and brachial indexes (i.e., a short distal segment).

Modifiable Parameters

The following text describes at length the modifiable parameters to consider in gymnastics, diving and power sports.

BODY COMPOSITION

Gymnasts are among the leanest of athletes, since any excess body fat must be transported and accelerated throughout the competition routines. Unfortunately, many coaches are obsessive about this characteristic and often induce performance decrements, or worse still, eating disorders among their athletes as a consequence of their overzealous attempts to reach arbitrary levels for adiposity. Body composition needs to be monitored and modified carefully, especially among adolescent female gymnasts; this should involve the counsel of a specialist sport dietician (see chapter 5). This advice is equally valid for divers, although it appears that slightly higher levels of adiposity are tolerated among elite performers in diving compared to gymnasts.

Weightlifters and powerlifters who do not participate in open competition must make a weight category, and so this parameter becomes a critical focus during preparation for competition. These athletes endeavour to maximize lean tissue and minimize fat mass within the total body mass constraint for their particular event. Careful guidance is needed in the lead-up to competition so that training is not adversely affected by the implementation of strategies to modify body weight. Conversely, provided that the body mass does not impede good technique, open-class lifters are not restricted in their tissue composition. Apart from sumo wrestlers, the super-heavyweight lifters are the heaviest of all elite sport competitors (Norton and Olds 1996).

STRENGTH AND POWER

Gymnastics Gymnastics is a sport that demands a high degree of muscular power and strength throughout the entire body. Due to the wide variety of movements performed in gymnastics events, the majority of the strength and power exercises described in this chapter are suitable for these athletes. Those that are most suited include:

- ballistic squat (exercise 47)
- ballistic split squat (46)
- ballistic bench press (42)
- ballistic standing push press (48)
- plyometric push-up (53)
- depth jump (51)
- bounding (49)
- high pull (15)
- wrist curl (41)
- hammer curl (13)
- reverse curl (23)
- barbell curl (4)
- dip (10)
- thigh extension (36)
- thigh abduction (34)
- thigh adduction (35)
- abdominal rotation (1)
- lying leg raise (20)
- wide-grip chin (40)
- deadlift (9)
- inclined bench press (16)
- ankle curl (2)
- standing calf raise (30)

In fact, the importance of strength and power to gymnastics requires that an upper body–lower body split routine be adopted whereby four resistance training sessions are performed per week during the preseason. An example of a preseason resistance training routine for gymnastics is outlined in table 13.5.

Diving The sport of diving requires explosive power in the lower limbs to achieve vertical height. Better vertical jump capacity enables the diver to achieve a longer time in the air, and therefore a greater number of manoeuvres can be performed. Consequently, exercises such as ballistic squat (one- and two-leg takeoffs, exercise 47), ballistic split squat (46), depth jump (51) and squat (29) are recommended. Divers also require forearm strength so that the hands

Table 13.5 Preseason Resistance Training Routine for Gymnasts

Monday	Tuesday	Thursday	Saturday
Ballistic bench press	Ballistic squat	Ballistic standing push press	Ballistic split squat
Ballistic standing push press	Bounding	Ballistic bench press	Depth jump
Dip	High pulls	Inclined bench press	Deadlift
Wide-grip chin	Thigh extension	Dip	Wide-grip chin
Plyometric push-up	Thigh abduction	Barbell curl	Thigh adduction
Barbell curl	Ankle curl	Hammer curl	Thigh extension
Hammer curl	Calf raise	Wrist curl	Calf raise
Wrist curl	Abdominal rotation	Lying leg raise	Abdominal rotation

can maintain a firm position on entry to the water. Exercises such as wrist curl (41), reverse curl (23) and hammer curl (13) are therefore important. Additionally, platform divers should perform standing push press (31), stiff-legged deadlift (32) and lying leg raise (20) so that they can easily achieve and maintain inverted positions on the platform.

Weightlifting The sport of weightlifting has an obvious requirement for strength and power. Exercises comprise the actual competitive lifts, that is, the snatch and clean and jerk, and smaller components of these lifts including high pull (exercise 15), standing push press (31) and squat (29). To be more specific to the competitive lifts, the squat exercise is often performed with the bar held on the shoulders in front of the neck (i.e., front squat) and through a greater depth than a normal squat. Additional exercises that will assist the weightlifter include the following:

- ballistic squat (47)
- ballistic standing push press (48)
- deadlift (9)
- trunk extension (3)
- reverse curl (23)
- wrist curl (41)
- bent-over rowing (6)
- inclined sit-up (17)
- shrug (27)

Figure 13.4 Unless the body is very flexible, it is not possible to attain the aesthetic positions needed to gain high marks in gymnastics.

FLEXIBILITY

Among athletes in any of the sports, gymnasts and divers need the highest level of overall flexibility. It is not possible to attain the aesthetic positions needed to gain high marks in the various manoeuvres unless the body is extremely flexible (figure 13.4). For this reason a wide variety of flexibility exercises need to be carried out and considerable time spent in mobility training. Furthermore, modern gymnastics and diving have become very ballistic, and stretched muscles can produce the contractile force needed for high-level performance.

Gymnasts and divers will probably need additional specialized exercises to those listed here in order to reach the levels they need in these sports. They should also carry out passive stretching exercises with a skilled partner or with their coach.

- **Neck, shoulder girdle and forearms:** It is important for gymnasts and divers to keep the neck supple and the shoulder girdle, hands and forearms very flexible. The following exercises should be performed for this purpose:

- neck stretch (exercise 54)
- backward stretch (56)
- overhead stretch (57)
- circular shoulder stretch (58)
- arm extensor stretch (82)
- arm adductor stretch (83)

- **Trunk:** In gymnastics, high levels of trunk flexion-extension, lateral flexion and rotation are essential. The following exercises should be performed to achieve these characteristics:
  - abdominal and hip stretch (60)
  - back roll stretch (61)
  - upper back stretch (62)
  - lower back and hip stretch (64)
  - rotational trunk stretch (65)
  - trunk lateral flexor stretch (84)
  - split pike stretch (85)
  - trunk extensor stretch (86)
- **Pelvic girdle and legs:** All gymnasts and divers must have very high levels of flexibility in the thigh, leg and foot flexors and extensors. Extreme thigh flexion is essential if divers are to reach the classic tuck or pike positions required of them (figure 13.5). The following exercises should be performed to achieve this:
  - groin stretch 1 (66)
  - groin stretch 2 (67)
  - side split stretch (69)
  - hamstring stretch 1 (71)
  - split stretch (70)
  - hamstring stretch 3 (73)
  - groin-trunk stretch (74)
  - quadriceps stretch 1 (75)
  - quadriceps-ankle stretch (76)
  - quadriceps-groin stretch (77)
  - Achilles tendon–ankle stretch (80)
  - ankle stretch (81)
  - split pike stretch (85)
  - trunk extensor stretch (86)
  - thigh adductor stretch (87)
  - thigh extensor stretch 1 (88)
  - thigh extensor stretch 2 (89)
  - thigh flexor stretch 2 (91)

Traditionally, the training routines of weightlifters have not included much in the way of flexibility exercises. However, several authors have recently demonstrated the value of flexibility with experienced powerlifters who improved their performances after a flexibility training program. Whilst it is acknowledged that stretching prior to competition may have a negative effect on strength, flexibility training at other times can be beneficial. Clearly, though, these athletes do not need the flexibility levels required of gymnasts and divers, so weightlifters and powerlifters are advised to carry out the routines designed for contact field sport players that are discussed in a later section of this chapter. Weightlifters require very sport-specific flexibility to allow them to assume the body positions involved in the snatch and clean and jerk. For example, the snatch requires very good shoulder and trunk flexibility, as well as the ability to assume a full squat position with the heels flat on the floor while the barbell is held overhead. These specific abilities are usually addressed in the specific training of the weightlifter.

Figure 13.5 The tuck and pike positions can only be attained if the diver has a high level of thigh flexion flexibility.

SPEED AND AGILITY

Total-body speed is not an essential ingredient in gymnastics or weightlifting except for the run-up to the vault. However, some speed training can be performed if the coach feels that the gymnast would benefit. Training for specific limb speed is of great value to gymnasts who lack "snap" in their movements, but the majority of their time should be spent on technique to develop the flow that is essential in their sport.

TRACK, FIELD AND CYCLING SPORTS

Power is crucial for athletes in track, field and cycling. Runners must have high power and strength-endurance, though the relative levels of these capacities depend on the duration of the event. Jumpers need to have high levels of lower body power in particular; throwers, especially high levels of strength and power; and cyclists, great power and strength-endurance. Flexibility is essential to running and hurdling, though more so in hurdling, as well as to jumping and throwing. Different types of speed programs are appropriate for these groups of athletes.

Limitations—Self-Selection Parameters

SPRINT RUNNING

While sprint runners are of average stature with an average crural index, they generally possess a low relative lower limb length (chapter 6), although these observations appear to differ somewhat for European and African athletes. Since these are power events, sprinters must naturally be endowed with a high percentage of fast-twitch (FT) muscle fibres (chapter 8), especially in the lower limbs.

DISTANCE RUNNING

Distance runners, especially those competing in the 10,000 m and marathon events, are among the shortest and lightest of all elite athletes. These characteristics naturally affect the runners' economy of motion. As mentioned in chapter 3, the majority of elite marathon runners in recent years are from East African and Japanese origins, with progressively fewer Caucasians having the right size to compete successfully in these events. Apart from the requirement of a large proportion of slow-twitch (ST) fibres in the lower limb musculature, distance runners have relatively short lower limbs and a low crural index (see chapter 6).

FIELD EVENTS (JUMPS)

Jumpers are among the tallest of elite athletes, with a high relative lower limb length and a high crural index. These proportionality characteristics provide a mechanically advantageous lever system for jumping. These events require rapid power generation and so these athletes possess a high proportion of FT fibres.

FIELD EVENTS (THROWS)

Shot, discus, hammer and javelin competitors are generally very tall and must possess a high brachial index (chapter 6). With the exception of the javelin athletes, throwers are also among the heaviest of all elite athletes (Norton and Olds 1996). They also possess a high proportion of FT muscle fibres.

CYCLING

Though of average weight, sprint cyclists are taller than average for elite performers. As mentioned in chapter 3, increased body weight may be viewed as either advantageous (downhill) or disadvantageous (hill climbing) depending on the nature of the cycling event. As with the running athletes, the most beneficial proportion of FT and ST muscle fibres is determined largely by the event distance and whether the requirement is for speed and power or endurance and economy of motion.

Modifiable Parameters

The following text describes at length the modifiable parameters to consider in track, field and cycling sports.

BODY COMPOSITION

Due to the volume of endurance training undertaken by distance runners, excessive body fat is rarely a problem. However, such training can be quite catabolic for muscle tissue, so some hypertrophy training throughout the preseason and in-season will help to maintain muscle mass. Due to the acceleration and maximum speed requirements of sprinting and horizontal jumping, minimizing body fat is important. The power requirements of these events will benefit from high levels of muscle hypertrophy—thus the very muscular physiques especially for 100 m sprint runners. For high jumpers and pole-vaulters, total body mass must be kept low as these athletes must work against gravity to maximize jump height. Clearly, fat mass is only extra baggage and so should be minimized; but in general, increased muscle mass brings increased strength and power, which will be of benefit. Throwers have to project a fixed mass (shot, javelin, discus, hammer) and there are no weight divisions, so maximizing muscle mass and thus strength and power is critical. Similar to distance runners, cyclists generally do not have high body fat due to the volumes of training they undertake, but excess fat is certainly not of any benefit and should be monitored.

STRENGTH AND POWER

Sprint Running Running requires a high degree of power and strength-endurance; however, the relative amount of each of these capacities is dependent upon the duration of the event. The exercises most beneficial to runners include the following:

- squat (exercise 29)
- ballistic squat (one- and two-legged, 47)
- ballistic split squat (46)
- standing calf raise (30)
- bounding (49)
- thigh extension (36)
- thigh flexion (37)
- running dumbbell (25)
- lying leg raise (20)
- stiff-legged deadlift (32)
- seated row (26)
- ballistic bench press (42)
- ankle curl (2)
- leg curl (18)
- leg press (19)
- thigh abduction (34)
- thigh adduction (35)
- inclined sit-up (17)
- trunk extension (3)
- close-grip pulldown (8)

An example of a resistance training routine used by a runner is outlined in table 13.6.

Field Events (Jumps) Jumping events such as long, high and triple jump require a great deal of lower body power. Consequently the exercises most suited to jumping events include:

- squat (exercise 29)
- ballistic squat (particularly single-legged jump, 47)
- ballistic split squat (46)
- ballistic leg extension (high jump only, 44)
- depth jump (51)
- bounding (49)
- thigh flexion (37)
- lying leg raise (20)
- thigh extension (36)
- standing calf raise (30)
- ankle curl (2)
- stiff-legged deadlift (32)
- leg press (19)

Table 13.6 In-Season Resistance Training Routine for Runners

Monday	Wednesday	Friday
Ballistic squat	Ballistic split squat	Ballistic squat
Thigh extension	Leg curl	Thigh extension
Bounding	Lying leg raise	Bounding
Thigh abduction	Leg press	Thigh adduction
Ballistic bench press	Running dumbbell	Ballistic bench press
Standing leg raise	Calf raise	Standing leg raise
Calf raise	Stiff-legged deadlift	Stiff-legged deadlift
Running dumbbell	Ankle curl	Inclined sit-up
Back extension	Seated rowing	Close-grip pulldown

Performance in triple and particularly long jump is dependent on achieving a fast run-up speed; consequently the exercises outlined for runners are also applicable to these athletes. Furthermore, long and triple jumpers must absorb very high forces through their lower bodies and therefore should develop a particularly strong musculoskeletal system to avoid injury. These athletes must also develop the capacity to tolerate high loads so that such loads do not adversely affect their competitive performance. Therefore, long and triple jumpers should periodically expose themselves to relatively high eccentric loading during training by performing high-intensity plyometric work. Also, performing exercises such as the ballistic squat (47) or ballistic split squat (46), but with reduced eccentric braking as tolerated, will improve this characteristic. Depth jumps (51) from relatively high drop heights (i.e., up to 1 m) may be undertaken by more advanced athletes. Due to the extreme forces experienced, high eccentric load training should be performed only for a few sets every couple of weeks at a maximum.

Field Events (Throws) The throwing events require a great deal of strength and power. In fact, the large requirement for strength and power in these events often necessitates four or five resistance training sessions per week using an upper body–lower body split routine. The exercises most applicable to the throwing events include:

- ballistic inclined bench press (one- and two-handed, shot put only, exercise 43)
- ballistic inclined bench press (discus only, performed with a wide grip, 43)
- ballistic lying triceps extension (45) and overhead throw (javelin and shot put only, 52)
- ballistic standing push press (48)
- ballistic squat (47)
- ballistic split squat (46)
- depth jump (jumping backward for shot, 51)
- bounding (49)
- high pull (15)
- inclined bench press (16)
- pullover (javelin only, 22)
- squat (29)
- standing calf raise (30)
- bench pull (6)
- abdominal rotation (1)
- wrist curl (41)
- hammer curl (13)
- thigh adduction (35)
- thigh abduction (34)
- internal-external rotator raise (11)

Cycling Cycling requires a high degree of power and strength-endurance. The relative degree of each physical capacity is dependent upon the duration of the event. Cyclists benefit from exercises that promote explosive lower body power. In addition, upper body strength is required to provide a stable platform from which lower body power can be exerted effectively. The exercises most beneficial to cyclists include:

- ballistic squat (concentric phase only, exercise 47)
- depth jump (51)
- thigh extension (36)
- thigh flexion (37)
- squat (29)
- leg press (19)
- lying leg raise (20)
- standing calf raise (30)
- stiff-legged deadlift (32)
- wrist curl (41)
- hammer curl (13)
- seated rowing (26)
- wide-grip chin (40)
- barbell curl (4)
- trunk extension (3)

FLEXIBILITY

For more than 50 years, it has been traditional for coaches in track and field to give their athletes intensive stretching exercises. Until 10 years ago, most of the exercises were ballistic, but a gradual change has been made toward the static stretching technique in the majority of countries. Because track and field incorporates a large number of individual events, it is not possible to cover each one in detail; however, we do deal with basic stretching exercises for each sport group.

Running, Hurdling and Cycling Good technique in running, whether in sprints or middle distance events, is very dependent on high levels of flexibility. The same can be said for hurdling, but the requirement is even higher because the trail leg must have very good range of motion in order for the hurdler to stride over the hurdle rather than jump over it. Furthermore, because sprinting and hurdling are very ballistic, many athletes are prone to soft tissue damage. It is important for the major muscle groups in the leg to be well stretched and in balance, from both an explosive power and a flexibility viewpoint.

Cycling, on the other hand, does not have a history of flexibility training compared with running and hurdling, though flexibility training could improve the muscular efficiency of cyclists.

- **Shoulder girdle**: In modern sprint running and hurdling, the arms contribute greatly to the propulsive power of the movement. In middle distance running there is some assistance from the arms, but they are also used for balance. Many elite coaches believe that high levels of mobility in the shoulder girdle and trunk assist the athlete to relax during the event and therefore prescribe stretching exercises for both running and hurdling. This is also true for cycling, and therefore cyclists should also carry out stretching. The following exercises should be performed to achieve this mobility:
    - backward stretch (exercise 56)
    - overhead stretch (57)
    - arm extensor stretch (82)
    - arm adductor stretch (83)
- **Trunk:** The trunk of a runner, hurdler or cyclist must also be flexible if the athlete is to attain high levels of relaxation while competing. The following exercises should be performed to achieve this:
    - upper back stretch (62)
    - lower back and hip stretch (64)
    - trunk lateral flexor stretch (84)
    - trunk extensor stretch (86)
- **Pelvic girdle and legs:** Runners, hurdlers and cyclists must also have very high levels of thigh, leg and foot flexion and extension. The following exercises should be performed to achieve these levels:

- hamstring stretch 1 (71)
- hamstring stretch 3 (73)
- quadriceps stretch 1 (75)
- quadriceps-groin stretch (77)
- quadriceps stretch 2 (78)
- calf stretch (79)
- Achilles tendon–ankle stretch (80)
- trunk extensor stretch (86)
- thigh extensor stretch 1 (88)

The following exercises are for hurdlers only:

- groin stretch 1 (66)
- side split stretch (69)
- split stretch (70)
- groin-trunk stretch (74)
- split pike stretch (85)
- thigh adductor stretch (87)

Field Sports (Jumps) In the jumping events, the athlete needs a very high level of flexibility in all regions of the body in order to perform well.

- **Neck and shoulder girdle:** The following exercises should be performed to attain a high level of flexibility in this region:
  - neck stretch (exercise 54)
  - triceps stretch (55)
  - backward stretch (56)
  - overhead stretch (57)
  - arm adductor stretch (83)
- **Trunk:** The following exercises will assist the jumper to attain high levels of flexibility in trunk flexion-extension, lateral flexion and trunk rotation:
  - abdominal and hip stretch (60)
  - upper back stretch (62)
  - lower back and hip stretch (64)
  - rotational trunk stretch (65)
  - trunk lateral flexor stretch (84)
  - trunk extensor stretch (86)
- **Pelvic girdle and legs:** High levels of flexibility are essential in the thigh, leg and foot flexors and extensors for the jumps. The following exercises should be performed to achieve this:
  - groin stretch 3 (68)
  - hamstring stretch 1 (71)
  - hamstring stretch 3 (73)
  - quadriceps stretch 1 (75)
  - quadriceps-ankle stretch (76)
  - quadriceps-groin stretch (77)
  - calf stretch (79)
  - ankle stretch (81)
  - thigh extensor stretch 1 (88)
  - thigh extensor stretch 2 (89)
  - thigh flexor stretch 1 (90)
  - thigh flexor stretch 2 (91)

Field Sports (Throws) As is the case for the jumps, field athletes need high levels of flexibility in all regions of the body.

• Neck, shoulder girdle and forearms: An extensive range of movement and elastic energy is needed for any good throw, and therefore the neck, forearms and the shoulder girdle in particular need to be well stretched. The following exercises should be performed to achieve this:

- neck stretch (exercise 54)
- triceps stretch (55)
- backward stretch (56)
- circular shoulder stretch (58)
- hand-forearm stretch (59)
- arm extensor stretch (82)
- arm adductor stretch (83)

• Trunk: Because the trunk is an essential part of any ballistic throwing movement, it must be extremely flexible. The following exercises should be performed to achieve trunk flexibility:

- abdominal and hip stretch (60)
- upper back stretch (62)
- lower back and hip stretch (64)
- rotational trunk stretch (65)
- trunk lateral flexor stretch (84)
- trunk extensor stretch (86)

• Pelvic girdle and legs: High levels of flexibility in this region are important for throwers because they are relying on leg power in the initial part of the throw. To achieve this flexibility, the thigh, leg and foot flexors and extensors must be well stretched. The following exercises should be performed to increase mobility levels in this region of the body:

- groin stretch 1 (66)
- groin stretch 3 (68)
- hamstring stretch 1 (71)
- hamstring stretch 3 (73)
- quadriceps-groin stretch (77)
- Achilles tendon–ankle stretch (80)
- ankle stretch (81)
- thigh adductor stretch (87)
- thigh extensor stretch 1 (88)
- thigh flexor stretch 1 (90)

SPEED AND AGILITY

During the past decade, track sprinters, particularly those in the United States, have been using several of the techniques common in strength and power training programs, as well as speed-resisted and speed-assisted training as outlined in chapter 10. As this is the case, it is not necessary to outline the running speed training program here.

Middle and long distance runners may be able to benefit from a total-body speed program, particularly if they have little "kick" or natural speed for tactical running. In today's highly competitive environment, it is up to the coach to decide if athletes need to improve this capacity in order to perform better.

Jumpers and throwers need total-body speed as well as specific limb speed. Many track and field coaches are now giving their athletes training for both facets as part of their normal program. Total-body speed is now a standard program, but the specific limb speed may need to be formulated by the coach. This is not difficult to do, because there are now a large number of speed-resisted and speed-assisted activities in the literature for jump and field game athletes. The development of the optimal technique is also up to the coach, who must isolate the critical factors retarding performance and improve on these.

MOBILE FIELD SPORTS: FIELD HOCKEY, SOCCER, LACROSSE

Limitations—Self-Selection Parameters

As with the racquet sports, there are no morphology-related limiting factors that prevent players from reaching elite levels in these the mobile field sports. Tall, linear physiques provide some advantage for soccer players in certain playing positions, especially in goal-scoring opportunities from a corner kick; but the smaller, more agile player may be successful in other roles.

Modifiable Parameters

The following text describes at length the modifiable parameters to consider in mobile field sports.

BODY COMPOSITION

Given the aerobic requirements of these sports, combined with the need for speed and agility, mobile field sport players must be lean, with minimal excess body fat. However, unless the player has an obvious problem in this regard, no universal dietary intervention is warranted for these sports.

STRENGTH AND POWER

An integral factor in all mobile field sports is the ability to run fast. Thus the exercises previously outlined for runners are also applicable to these athletes. In addition, these athletes should perform the following exercises:

- ballistic leg extension (soccer only, exercise 44)
- ballistic lying triceps extension (45) and overhead throw (soccer and lacrosse only, 52)
- abdominal rotation (1)
- wrist curl (41) and hammer curl (hockey, lacrosse and soccer goalkeeper only, 13)
- head curl (soccer only, 14)
- ballistic bench press (42)
- upright row (39).

An example of a preseason resistance training routine for a soccer player is outlined in table 13.7.

Table 13.7 Preseason Resistance Training Routine for Soccer Players

Monday	Wednesday	Friday
Ballistic squat	Ballistic leg extension	Ballistic split squat
Overhead throw	Leg curl	Ballistic leg extension
Bounding	Lying leg raise	Thigh extension
Thigh adduction	Leg press	Thigh adduction
Ballistic bench press	Ballistic lying triceps extension	Bounding
Thigh extension	Calf raise	Standing leg raise
Stiff-legged deadlift	Stiff-legged deadlift	Calf raise
Running dumbbell	Ankle curl	Abdominal rotation
Head curl	Thigh abduction	Shrug

FLEXIBILITY

In mobile field sports, body contact occurs, but it is not as severe as in the contact field sports in which whole-body tackling, to impede the opponent's progress, is an important feature of the game. Athletes in the mobile field sports need an above-average level of flexibility, but do not need to be excessively flexible in any specific region of the body. Like other agility athletes, all players in this group should understand that high levels of flexibility will enable them to accelerate their stick or leg through a greater range of movement, as well as to store elastic energy in their propulsive muscles. These two features will enable them to apply more force to the ball as they hit, throw or kick it.

- **Neck, shoulder girdle and forearms:** Movements of flexion and rotation of the neck, as well as flexion-extension of the arm and flexion of the hand, are important. The following exercises will assist the athlete to improve performance of these movements:
    - neck stretch (exercise 54)
    - overhead stretch (57)
    - circular shoulder stretch (58)
    - hand-forearm stretch (59)
    - arm extensor stretch (82)
- **Trunk:** Because the trunk is constantly moving during a game, flexion-extension, lateral flexion and rotation are important movements carried out by all players. The following exercises should be performed to reach the level of flexibility needed by this group:
    - abdominal and hip stretch (60)
    - back roll stretch (61)
    - upper back stretch (62)
    - lateral trunk stretch (63)
    - lower back and hip stretch (64)
    - trunk extensor stretch (86)
- **Pelvic girdle and legs:** All agility athletes must have thigh, leg and foot flexion-extension in the field sports. To achieve this, the following exercises should be carried out:
    - groin stretch 1 (66)
    - groin stretch 3 (68)
    - hamstring stretch 1 (71)
    - hamstring stretch 2 (72)
    - quadriceps stretch 1 (75)
    - quadriceps stretch 2 (78)
    - calf stretch (79)
    - ankle stretch (81)
    - thigh extensor stretch 1 (88)
    - thigh flexor stretch 1 (90)

SPEED AND AGILITY

Sports such as field hockey, soccer and lacrosse rely on both running speed and specific limb speed. Those coaches who have persevered with a total-body speed program have been agreeably surprised with the results. Such speed enables a player to burst with greater speed and power, giving a dynamic dimension to the performance. For this reason more coaches are concentrating on this particular aspect of training, especially in the early part of the season. As the season progresses, they concentrate on specific limb speeds using simulated techniques from the sport.

CONTACT FIELD SPORTS: RUGBY CODES, AUSTRALIAN FOOTBALL, AMERICAN FOOTBALL

Limitations—Self-Selection Parameters

The contact field sports are a group that is very much affected by body size. As noted in chapter 3, the body mass of National Football League offensive and defensive tackles and quarterbacks increased at a rate of 4.6 kg (10 lb) per decade from 1920 to 1999, and the proportion of males in the normal population with the necessary height and weight for success in this sport is minute. American football linemen, ruckmen in Australian football and second row forwards in rugby are among the tallest of athletes, while American football and rugby players in general are much heavier than most athletes as well as the normal population. Despite these obvious size advantages for specialist position players, most contact field sports comprise a variety of playing roles and positions. This allows for other players of lesser size to participate in these sports at an elite level.

Modifiable Parameters

The following text describes at length the modifiable parameters to consider in contact field sports.

BODY COMPOSITION

The mix of aerobic and anaerobic requirements of the sports within this group varies considerably. Australian football has probably the highest endurance demand, particularly for the midfield players, and this necessitates a leaner body composition than would be expected for rugby forwards and American football linemen. While all players in this sport group need a mesomorphic body shape to cope with the mechanical demands of physical contact, those whose roles involve constant tackling will often be advantaged if they carry extra adipose tissue to increase their inertia. This applies more to American football, in which the distance covered per game by players is far less than for rugby or Australian football. Strategies to increase or decrease fat and muscle are discussed in chapter 5.

STRENGTH AND POWER

The contact field sports require a high degree of muscular size, strength and power. In each of these sports, some players additionally require a relatively high degree of strength-endurance. The high level of strength and power required in these sports generally necessitates that four resistance training sessions be performed per week during the preseason using an upper body–lower body split routine. The exercises most suited to contact field sports include the following:

- ballistic squat (exercise 47)
- ballistic split squat (46)
- ballistic bench press (42)
- ballistic standing push press (48)
- ballistic lying triceps extension (American football only, 45)
- ballistic leg extension (44)
- bounding (49)
- bench press (5)
- standing push press (31)
- barbell curl (4)
- hammer curl (13)
- wrist curl (41)
- bench pull (6)
- high pull (15)
- deadlift (9)
- shrug (27)
- lying leg raise (20)
- abdominal rotation (1)
- squat (29)
- thigh adduction (35)
- thigh abduction (34)
- stiff-legged deadlift (32)
- thigh extension (36)
- standing calf raise (30)
- shrug (27)
- head curl (14)

Table 13.8 Preseason Resistance Training Routine for Contact Field Sport Athletes (Rugby Codes, Australian Football, American Football)

Monday	Tuesday	Thursday	Friday
Ballistic squat	Ballistic bench press	Ballistic split squat	Ballistic inclined bench press
Ballistic leg extension	Ballistic standing push press	Ballistic leg extension	Bench press
High pull	Abdominal rotation	Bounding	Lying leg raise
Thigh extension	Head curl	Thigh extension	Standing push press
Calf raise	Inclined bench press	Calf raise	Lying triceps extension
Leg press	Close-grip bench press	Deadlift	Barbell curl
Bench pull	Barbell curl	Thigh adduction	Hammer curl
Stiff-legged deadlift	Hammer curl	Shrug	Wrist curl
Thigh abduction	Wrist curl		

As with the other routines previously outlined, the specific training program will depend upon the position played within the sport, the individual characteristics of the athlete and the particular phase of the competitive year. With these limitations in mind, readers can refer to the general example of a preseason resistance training routine for contact field sports that is outlined in table 13.8.

FLEXIBILITY

Players in contact field sports need strong joint capsules and musculature around the shoulder, knee and ankle joints so that these joints are not dislocated or easily injured. In these regions they should have normal levels of flexibility; however, the adductors of the thigh, hamstrings, quadriceps and calf groups should be well stretched and in balance so as to avoid soft tissue injuries. The contact sport athlete must also be careful not to overstretch the knee or ankle joints with additional activities such as intensive freestyle kicking, which can cause cruciate ligament and knee instability in these joints.

- **Shoulder girdle and trunk:** The following exercises should be carried out for these regions, though athletes should not strive for high levels of flexibility:
  - backward stretch (exercise 56)
  - overhead stretch (57)
  - abdominal and hip stretch (60)
  - back roll stretch (61)
  - upper back stretch (62)
  - lower back and hip stretch (64)
- **Pelvic girdle and legs:** The following exercises will enable athletes to stretch the adductors of the thigh, hamstrings, quadriceps and the calf groups:
  - groin stretch 2 (67)
  - groin stretch 3 (68)
  - hamstring stretch 1 (71)
  - hamstring stretch 3 (73)
  - quadriceps stretch 1 (75)
  - quadriceps-groin stretch (77)
  - quadriceps stretch 2 (78)
  - calf stretch (79)
  - thigh extensor stretch 1 (88)
  - thigh flexor stretch 1 (90)

SPEED AND AGILITY

Contact field sports, led by American football, have seen a heavy concentration on the development of running speed. As coaches from other contact football codes have visited North America, they have brought back many of the body movement speed techniques that are used there. However, they have not developed the total-body speed programs as well as the coaches in American football and could benefit from a more systematic application of this regimen. All players in contact football need speed training, which should assist them with their acceleration rate, stride length and the rate of their stride for open- and closed-field running. However, they also need specific limb speed training using simulated techniques from the sport.

SET FIELD SPORTS: GOLF, BASEBALL, SOFTBALL AND CRICKET

Limitations—Self-Selection Parameters

Successful participation in this group of sports, which includes golf, baseball, softball and cricket, appears unencumbered by body size and shape, although for specialist roles like fast bowling in cricket, a tall linear physique can be useful. As reported by Norton and Olds (1996), average heights and weights for elite performers in these sports do not differ greatly from those of the normal or source population.

Modifiable Parameters

The following text describes at length the modifiable parameters to consider in set field sports.

BODY COMPOSITION

In general, these sports do not require high levels of endurance capacity or agility and so appear quite tolerant of wide variations in body composition. Though a lean body has obvious health benefits, some of the greatest champions in these sports have clearly "carried a few extra pounds" throughout their careers!

STRENGTH AND POWER

Athletes involved in set field sports have not traditionally participated in resistance training to a large extent. These sports have a particularly high skill base, and thus the physical capacities of strength, power and endurance are seen to be of low priority. Nevertheless, the running and throwing ability of athletes in baseball and cricket can be enhanced by resistance training, as can hitting power in all sports. Further resistance training can also be used to remove muscular imbalances in these sports and thus help reduce the incidence of injury.

Baseball and Cricket Resistance training exercises most suited to baseball and cricket include:

- ballistic squat (exercise 47)
- ballistic split squat (46)
- bounding (49)
- ballistic bench press (42)
- ballistic lying triceps extension (45)
- thigh extension (36)
- wrist curl (41)
- internal-external rotator raise (11)
- bench pull (6)
- stiff-legged deadlift (32)
- abdominal rotation (1)
- upright row (39)
- pullover (22)
- standing calf raise (30)
- side bend (28)
- ankle curl (fast bowlers especially, 2)
- thigh adduction (35)
- thigh abduction (34)

Table 13.9 Resistance Training Routine for Cricket and Baseball Players

Monday	Thursday
Ballistic squat	Ballistic split squat
Ballistic bench press	Bounding
Ballistic lying triceps extension	Overhead throw
Thigh extension	Thigh extension
Abdominal rotation	Abdominal rotation
Wrist curl	Stiff-legged deadlift
Bench pull	Wrist curl
Internal-external rotator raise	Calf raise
Pullover	Ankle curl
Thigh adduction	Thigh abduction

An example of a resistance training routine for cricket and baseball players is outlined in table 13.9.

Golf Resistance training exercises most suited to golfers include the following:

- wrist curl (exercise 41)
- bent-over row (6)
- ballistic bench press (42)
- dip (10)
- front pulldown (12)
- pullover (22)
- wide-grip chin (40)
- chest throw (50)
- hammer curl (13)
- abdominal rotation (1)
- trunk extension (3)
- stiff-legged deadlift (32)
- upright row (39)
- side bend (28)
- lying triceps extension (21)
- reverse curl (23)
- seated row (26)
- bench press (5)
- internal-external rotator raise (11)

FLEXIBILITY

In set field sports, high levels of mobility are of great value, first because the storage of elastic energy in the well-stretched muscle will enable the player to hit, throw or bowl with a powerful action. Good flexibility also allows the player to deliver the club, bat or ball through a greater range of movement. The combination of these two phenomena enables the individual to perform the skills of these games with considerable explosive power.

- **Shoulder girdle, trunk and forearms:** All the games in this group require very high levels of arm flexion-extension, trunk flexion-extension, lateral flexion and trunk rotation as well as extension of the hands. The following exercises should be performed to achieve these capabilities:

  - backward stretch (exercise 56)
  - overhead stretch (57)
  - circular shoulder stretch (58)
  - hand-forearm stretch (59)
  - upper back stretch (62)
  - lateral trunk stretch (63)
  - lower back and hip stretch (64)
  - rotational trunk stretch (65)
  - arm extensor stretch (82)
  - trunk extensor stretch (86)

- **Pelvic girdle and legs:** The thighs and legs are used in the performance of almost all skills in the sports in this group; high levels of flexibility are also necessary in thigh, leg and foot flexion-extension. The following exercises should be performed:
    - groin stretch 3 (68)
    - hamstring stretch 1 (71)
    - hamstring stretch 2 (72)
    - quadriceps stretch 1 (75)
    - quadriceps-groin stretch (77)
    - quadriceps stretch 2 (78)
    - calf stretch (79)
    - Achilles tendon–ankle stretch (80)
    - ankle stretch (81)
    - thigh extensor stretch 1 (88)
    - thigh flexor stretch 1 (90)

SPEED AND AGILITY

Games like baseball and cricket demand high levels of running speed, but not for long distances, so rapid acceleration from a set position should receive the most emphasis in this program. Players must also be aware that specific limb speed is important in hitting, throwing and bowling; this will increase mainly through technique training, although strength, power, flexibility, speed-assisted and speed-resisted activities all play a part.

COURT SPORTS: BASKETBALL, NETBALL, VOLLEYBALL

Limitations—Self-Selection Parameters

Specific body size and proportionality characteristics for success in court sports are very prevalent among this group of elite competitors. Basketball, netball and volleyball players in the key shooting and defending positions are typically among the tallest and heaviest of all sportspersons (Norton and Olds 1996), with a low ratio of sitting height to stature caused by longer than average lower limb lengths. Referring to the physiques of World Championship female basketball players, Ackland, Schreiner, and Kerr (1997) reported clear differences between specialist position players, with scores increasing significantly for height, weight and segment breadth and girth variables between specialist guards, forwards and centres, respectively.

As outlined in chapters 3 and 6, these athletes rely on jumping prowess, and like specialist jumping athletes, must have a linear physique and a high relative lower limb length, as well as a high crural index.

Modifiable Parameters

The following text describes at length the modifiable parameters to consider in court sports.

BODY COMPOSITION

These players must be relatively lean, given the aerobic requirements of these sports, combined with the need for speed and agility around the court and jumping performance. This is especially so for the ball carriers in netball and basketball. However, other key position players such as the basketball centres are advantaged with an increased height and mass, since their role requires more physical contact during blocking and rebounding tasks. Unless the player has an obvious problem in regard to his or her body fat, no universal dietary intervention is warranted for these sports. Due to the high level of contact in the modern game of basketball, very thin athletes with low muscle mass would benefit from extra concentration on hypertrophic resistance training during the off-season and preseason.

STRENGTH AND POWER

These sports require a high degree of muscular power, particularly in the lower body. Resistance training exercises most suited to these sports include:

- squat (exercise 29)
- ballistic squat (47)
- ballistic split squat (46)
- depth jump (51)
- bounding (49)
- ballistic bench press (42)
- ballistic lying triceps extension (45)
- overhead throw (52)
- chest throw (50) and bench press (basketball and netball only, 5)
- ankle curl (2)
- standing calf raise (30)
- thigh adduction (35)
- thigh abduction (34)
- stiff-legged deadlift (32)
- thigh extension (36)
- leg press (19)
- standing push press (basketball and netball only, 31)
- wrist curl (41)
- abdominal rotation (1)
- hammer curl (13)
- lying leg raise (20)
- close-grip pulldown (basketball only, 8)

A preseason resistance training routine for basketball players is outlined in table 13.10.

FLEXIBILITY

Court sports are highly ballistic, and thus good range of motion and the ability to store and reuse elastic energy are very important. As well, one needs to be concerned with the prevention of injuries to athletes in this group, and long stretched muscles will not be injured as readily as those that are bunched and tight.

- **Shoulder girdle:** Arm flexion-extension mobility is an important capacity for athletes in this group. The following exercises should be carried out to achieve this quality:
    - overhead stretch (exercise 57)
    - circular shoulder stretch (58)
    - arm extensor stretch (82)
    - arm adductor stretch (83)

Table 13.10 Preseason Resistance Training Routine for Basketball Players

Monday	Wednesday	Friday
Ballistic squat	Ballistic split squat	Ballistic squat
Ballistic bench press	Overhead throw	Ballistic inclined bench press
Bounding	Depth jumps	Thigh extension
Ballistic lying triceps extension	Bench press	Ballistic lying triceps extension
Close-grip pulldown	Thigh extension	Standing push press
Stiff-legged deadlift	Leg press	Ankle curl
Abdominal rotation	Reverse curl	Close-grip pulldown
Wrist curl	Calf raise	Lying leg raise
Thigh abduction	Thigh adduction	Calf raise

- **Trunk:** The court sport player when leaping often twists while in the air and as a result needs not only a high degree of trunk flexion-extension, but also lateral flexion and trunk rotation mobility. The following exercises should be performed to increase flexibility in these movements:
    - abdominal and hip stretch (60)
    - upper back stretch (62)
    - lower back and hip stretch (64)
    - rotational trunk stretch (65)
    - trunk lateral flexor stretch (84)
    - trunk extensor stretch (86)
- **Pelvic girdle and legs:** All agility athletes must have the thigh, leg and foot flexors and extensors well stretched. The following exercises should be performed to this end:
    - groin stretch 2 (67)
    - hamstring stretch 1 (71)
    - hamstring stretch 2 (72)
    - quadriceps stretch 1 (75)
    - quadriceps-ankle stretch (76)
    - quadriceps stretch 2 (78)
    - calf stretch (79)
    - Achilles tendon–ankle stretch (80)
    - thigh extensor stretch 1 (88)
    - thigh flexor stretch 1 (90)

SPEED AND AGILITY

Basketball, netball and volleyball have similar demands to the racquet sports, where total-body speed and specific limb speed are needed. The reader should refer to the section on racquet sports for more information.

MARTIAL ARTS

Limitations—Self-Selection Parameters

Martial arts competitions are generally organized by weight classes or categories. Thus absolute body mass is, by definition, a selection parameter for competition within a specific class, although some changes to this parameter can be effected through diet and training as discussed next. Furthermore, the sports within this category have varying requirements for competition, thereby determining which body sizes, shapes and proportionalities are favoured. For example, a linear physique has advantages for taekwondo (assisting with fast, powerful kicking and striking techniques), whereas a shorter and broader body type is more suited to wrestling and judo competition.

Modifiable Parameters

The following text describes at length the modifiable parameters to consider in the martial arts.

BODY COMPOSITION

Success in these sports requires a combination of speed, power and agility. Thus martial artists, except those competing in the open weight categories, must be lean with minimal excess body fat. Unless the player has too little lean mass or too much adipose tissue, no universal dietary intervention is warranted for this sport. However, individuals who seek to compete within a certain weight category and need to either increase or reduce the current mass should use the specific techniques discussed in chapter 5.

STRENGTH AND POWER

Wrestling and Judo Wrestling and judo require a great deal of muscular strength and power throughout the entire body. Exercises most suited to these sports include the following:

- wrist curl (exercise 41)
- reverse curl (23)
- hammer curl (13)
- seated rowing (26)
- barbell curl (4)
- shrug (27)
- high pull (15)
- deadlift (9)
- bench press (5)
- close-grip pull-down (7)
- ballistic standing push press (48)
- dip (10)
- head curl (14)
- reverse curl (23)
- ballistic bench press (42)
- squat (29)
- ballistic squat (47)
- stiff-legged deadlift (32)
- lying leg raise (20)
- leg press (19)
- abdominal rotation (1)
- side bend (28)
- standing calf raise (30)
- bench pull (6)

A resistance training routine for these athletes is outlined in table 13.11.

Punching and Kicking Sports Sports involving kicking and punching require a high degree of speed and power. Exercises most suited to these sports include the following:

- ballistic bench press (concentric phase only, exercise 42)
- ballistic standing push press (concentric phase only, 48)
- ballistic squat (47)
- ballistic split squat (46)
- ballistic leg extension (kicking sports only, 44)
- squat (29)
- plyometric push-up (53)
- depth jump (51)
- bounding (49)
- chest throw (50)
- abdominal rotation (1)
- inclined sit-up (17)
- lying leg raise (20)
- thigh flexion (kicking sports only, 37)
- high pull (15)
- seated rowing (26)
- shrug (27)
- head curl (14)
- thigh adduction (35)
- thigh abduction (34)
- stiff-legged deadlift (32)
- bench press (5)
- standing calf raise (30)

FLEXIBILITY

Combatants in many of the grappling martial arts in the past have not been very flexible; however, the past decade has seen a very definite change to this training policy. There is now an understanding of the concept of the storage of elastic energy, as well as a realization that serious injuries can be sustained in the grappling sports if the individual does not have a flexibility level well above average, especially in the upper body. Furthermore, a supple wrestler or judoka is sometimes able to escape from a hold by slipping out of it. It is important to understand, however, that high levels of strength and explosive power must accompany this mobility. Karate and taekwondo athletes have traditionally spent considerable time in developing hip and knee flexibility to increase height, speed and power of kicking.

Table 13.11 Resistance Training Routine for Wrestlers and Judo Participants

Monday	Tuesday	Thursday	Saturday
Ballistic standing push press	High pull	Ballistic bench press	Ballistic squat
Abdominal rotation	Ballistic split squat	Close-grip pulldown	Seated rowing
Bench press	Seated rowing	Standing push press	Abdominal rotation
Dip	Shrug	Reverse curl	Deadlift
Barbell curl	Standing calf raise	Hammer curl	Thigh extension
Hammer curl	Leg press	Barbell curl	Head curl
Wrist curl	Stiff-legged deadlift	Wrist curl	Lying leg raise
	Side bend	Bench pull	Side bend

- **Neck, shoulder girdle and trunk:** All martial arts combatants require high levels of arm flexion-extension, trunk flexion-extension, lateral flexion and rotation. The following exercises should be performed to increase flexibility in these movements:
    - neck stretch (exercise 54)
    - triceps stretch (55)
    - backward stretch (56)
    - overhead stretch (57)
    - abdominal and hip stretch (60)
    - back roll stretch (61)
    - upper back stretch (62)
    - lateral trunk stretch (63)
    - lower back and hip stretch (64)
    - rotational trunk stretch (65)
- **Pelvic girdle and legs:** Even though the legs play only a supporting role in the grappling martial arts, participants should have high levels of thigh and leg flexion and extension mobility. As mentioned previously, kicking requires very high hip joint flexibility, and this needs particular emphasis for karate and taekwondo. The following exercises should be performed to achieve this:
    - groin stretch 2 (67)
    - hamstring stretch 1 (71)
    - hamstring stretch 2 (72)
    - quadriceps stretch 1 (75)
    - quadriceps-ankle stretch (76)
    - quadriceps stretch 2 (78)
    - calf stretch (79)
    - thigh extensor stretch 1 (88)
    - thigh flexor stretch 1 (90)

The following exercises are for participants in the kicking sports only: thigh extensor stretch 2 (89) and thigh flexor stretch 2 (91).

SPEED AND AGILITY

Only specific limb speed is needed in wrestling, judo and boxing. These athletes should mainly use technique training to achieve such speed; however, strength, explosive power and flexibility training, as well as speed-resisted and speed-assisted training, all play an important part. In boxing, slightly heavier gloves and a punching bag can be useful in the early season; a speed ball and lighter gloves in the speed-assisted phase of training should be used during the midseason.

Summary

Resistance training for strength and power development, flexibility, speed and agility has resulted in vast improvements in performance in a variety of sports over the past 20 years. This chapter has outlined the use of generic programs from a performance enhancement perspective. In developing effective training routines, one needs to have a thorough understanding of the training principles of overload, specificity, variation and recovery. An understanding of the advantages and limitations of current training methods and equipment is also important. While the use of strength training especially has been fundamental to performance improvement in many sports over the past 20 years, the steady progression from pure strength toward explosive power exercises will bring about far greater performance improvements in the near future.

part III

Biomechanics

Assessment and Modification of Sport Techniques

fourteen

Analysis of Sport Performance

Bruce C. Elliott, PhD; and Duane Knudson, PhD

A critical skill for coaches is the ability to analyse sport technique. Technique analysis falls on a continuum between subjective (qualitative) analyses and more objective (quantitative) analyses. Most coaching situations rely on the use of subjective or qualitative analysis or both (figure 14.1). Qualitative analysis in our context has been defined as the systematic observation and introspective judgment of the quality of human movement for the purpose of providing the most appropriate intervention to improve performance (Knudson and Morrison 2002). The main advantage of qualitative analysis is the rich body of coaching knowledge that can be quickly accessed and used to provide immediate feedback or instruction to the athlete. This knowledge must be used carefully because coaches, who try to see everything, often end up not observing key elements of performance. A structured qualitative analysis procedure, as described in this chapter, is therefore essential if a coach is to be of optimal benefit to the athlete.

Coaches also have access to technologies that allow for quantitative analysis. Quantitative analysis involves the measurement of performance variables using technology. A stopwatch or videotape replay can be used to quantify timing variables of human performance. A baseball coach might have access to a radar gun that measures the speed of the ball in flight and a camera linked to a computer for the calculation of selected kinematic aspects of pitching. Kinematic variables such as joint angle, rotational speed of an implement (e.g., tennis racquet) or segment end point (e.g., ankle in kicking skills or club head in golf) and displacement of a segment (e.g., movement of the head during a swing in baseball) precisely describe the motion used by an athlete. These are examples of quantitative analysis of sport performance. This chapter presents typical models of technique analysis and illustrates how these can assist the coach in helping athletes improve performance and reduce the risk of injury.

Figure 14.1 Most analyses of sport techniques by coaches are qualitative: Several performances are observed, and the coach evaluates strengths and weaknesses, diagnoses likely causes of poor performance and selects intervention strategies to help the athlete improve.

Models of Technique Analysis

Coaches analyze sport technique based on either qualitative or quantitative (i.e., numerical measurements) judgments.

The traditional approach to technique analysis in coaching was qualitative, and involved visual observation followed by the provision of a correction. However, this error detection and correction approach is an outdated and suboptimal approach to technique analysis. Professional coaches should be expected to provide technique analysis rather than a string of correction clichés.

The complexity of sport techniques, normal perceptual limitations, individual differences and research shows that a more expansive view of technique analysis is required if performance is to be optimised. Good technique analysis must even use higher-order thinking skills than the term "analysis" implies (breaking something down into parts). Highly effective technique analysis must evaluate the movement to identify both strengths and weaknesses and then diagnose the performance in order to prescribe intervention (Knudson and Morrison 2002). This is why some scholars have proposed that qualitative analysis of technique be called *clinical diagnosis* (Hoffman 1983). This chapter will continue to use the currently accepted term *analysis,* but readers are reminded that this kind of analysis involves much more than the term implies.

A variety of models of technique analysis have been proposed, and these are extensively reviewed by Knudson and Morrison (2002). With either kind of analysis (qualitative or quantitative), there are essentially two approaches to understanding the technique components contributing to performance. One is biomechanical, focusing on biomechanical principles (e.g., Hudson 1995; Knudson 2003) or variables using a deterministic model (Elliott and Alderson 2003; Hay 1983). This has also been called the mechanical method (Hay and Reid 1988). Figure 14.2 illustrates a deterministic model of the mechanical variables that contribute to a high jump. A track coach could use these biomechanical variables and principles in qualitative analysis to judge the quality of an athlete's long jump and prescribe technique improvements or practice. Some biomechanics researchers have used these models as a basis for planning quantitative research and statistical analyses of the technique factors that affect sport skill performance (Lees 1999). More about this model of analysis is presented in the section on quantitative analysis.

The other approach to analysis focuses on key technique points or "critical features" (Barrett 1983; Knudson and Elliott 2003). Critical features have been defined in many ways, but one can say that they are the technique points in a movement that are necessary for optimal performance and are the least modifiable for ensuring successful performance of a movement. A coach using this approach in qualitative analysis would observe a skill, focusing on a small number (three to six) of critical features. This approach has also been called a sequential model because the critical features are often organized

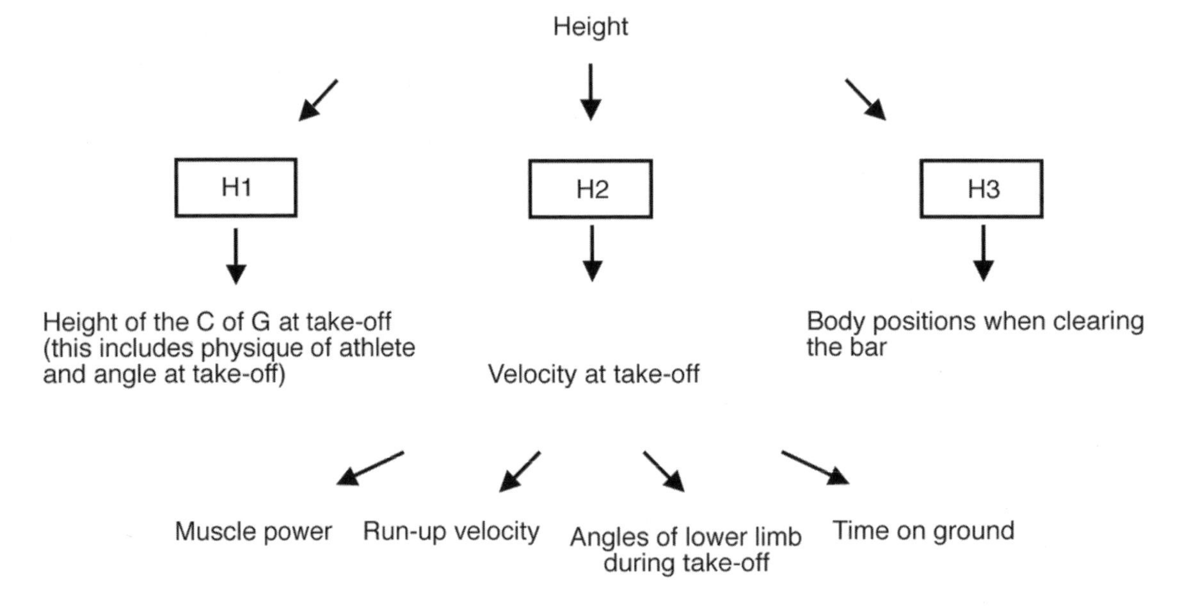

Figure 14.2 A deterministic model of the mechanical variables that contribute to high jump performance.

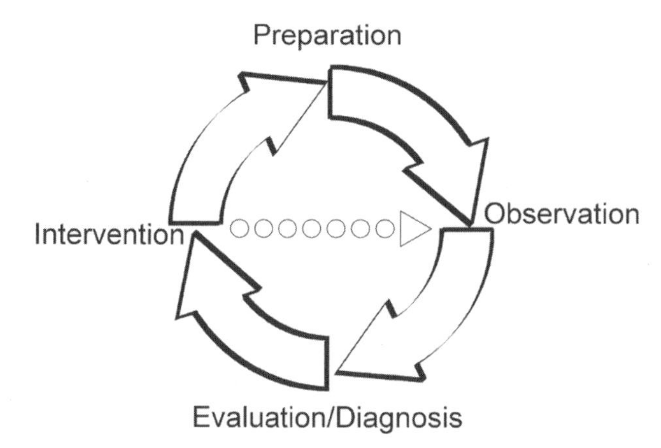

Figure 14.3 The Knudson and Morrison (2002) model of qualitative analysis of human movement emphasizes four important tasks.

Reprinted, by permission, from D. Knudson and C. Morrison, 2002, *Qualitative analysis of human movement*, 2nd ed. (Champaign, IL: Human Kinetics), 9.

in the temporal sequence (preparation, action, follow-through) of most movements (Hay and Reid 1988). Most athletic coaches use this approach to qualitative analysis because it is natural to make a comparison between an observed movement and a visual or mental image of a desirable or model technique.

The two most recent and comprehensive models of qualitative analysis recommend the use of this comparative approach to movement analysis (Knudson and Morrison 2002; McPherson 1990). These two models strive to summarize a complex process in four basic tasks or steps. The Knudson and Morrison (2002) model of qualitative analysis is illustrated in figure 14.3. We use this model as the basis for the next section on qualitative or subjective analysis of sport techniques.

Qualitative or Subjective Analysis

Sport coaches applying biomechanics in qualitative analysis must integrate this understanding of movement with other kinds of knowledge (Knudson and Morrison 2002). This integration of the various sport sciences, knowledge about the sport and knowledge of the performer is critical to the effectiveness of qualitative analysis. It is similar to what the field of motor behaviour calls an ecological perspective, which integrates knowledge of the performer, task and environment. While biomechanics provides the primary source of evidence for proper technique, this knowledge must be integrated with evidence from other sport sciences and professional experience. It would be counterproductive, for example, for a coach to focus a young player's attention on a minor weakness (based only on biomechanics) in batting technique if that weakness was symptomatic of an obvious fear of being hit by the ball (see chapter 1).

Another innovative feature of this broader vision of qualitative analysis is the emphasis on critical evaluation of the strengths and weaknesses of the movement, followed by a diagnosis of the given performance. Early models of qualitative analysis emphasized only observation and error or fault identification, followed by corrections. Remember that this low-level analysis should be avoided in favour of a model that mimics the professional diagnosis used by physicians in the clinical diagnosis of the source of pain. A coach who determines the strengths and weaknesses of performance and then diagnoses the likely causes of poor performance will provide better suggestions and greater improvement than a poor coach with a correction complex. With this holistic and interdisciplinary perspective in mind, let's see how biomechanical concepts are integrated with other sources of knowledge in the four tasks of qualitative analysis.

Preparation

The first task of qualitative analysis is preparation. In preparation, the coach gathers knowledge about the performer, sport and techniques to be analysed. Knowledge is more than just information. Knowledge is the thoughtful combination of logic and experiential and scientific observations into a theoretical and coherent whole. True professions like medicine have formalized how evidence is weighed to produce an informed treatment of patients.

In coaching, knowledge about the performer or client might take many forms, from memories of previous sessions to fitness test data to written records. Knowledge about a selected sport and the techniques used in that sport also comes from a variety of sources. First, coaches have professional experience that provides some information about the sport and what seems to work in helping athletes learn the activities and skills that contribute to successful performance in that sport. Because the responses of individuals or teams are anecdotal, coaches should check their professional opinions against the sport science research literature. Ideally coaches should read and integrate the latest research in sport science, but limitations in time and access to these journals usually require that coaches rely on professional coaching resources for summaries of the latest research findings. Professional coaching conferences and magazines like *Sports Coach, Technique* and *Track and Field Coaches Review,* to name a few examples of important periodicals, strive to integrate experiential knowledge with sport science research. The International Society of Biomechanics in Sports is dedicated to the transmission of sport science knowledge to coaches.

By constantly updating their knowledge of sport science research and integrating it with their professional experience, coaches develop a deep and accessible body of knowledge that they must use in the analysis process. This knowledge of mechanical principles or critical features gathered in preparation is important in the next task of qualitative analysis, observation of the movement. As Edgar Dale (1984, p. 58) noted, "we can only see in a picture what our experience permits us to see".

Qualitative analysis of sport technique by coaches should emulate professions like medicine where decisions are based on the integration of evidence about the movement, athlete and situation.

Observation

The second task of qualitative analysis is the observation of movement. Observation includes a systematic observational strategy that utilizes all the relevant senses, not just vision. A gymnastics coach, for example, can use the temporal information from the sounds of impact with the mat or the kinaesthetic information obtained in his or her spotting of the athlete. The purpose of observation is to gather all the relevant information there is about performance so that the coach can base subsequent evaluation and diagnosis on the best possible evidence.

A systematic observational strategy helps ensure that the tremendous perceptual demands of observing human movement are minimized. This is important because the complexity and speed of human movement make it very difficult to accurately perceive many aspects of movement (Knudson and Kluka 1997). About 50% of untrained observers can accurately estimate the minimum knee joint angle at reversal in a vertical jump (Knudson 1999; Knudson and Morrison 2000), while the vast majority of novice observers accurately estimate speed of joint movement in slow- to moderate-speed arm motions (Bernhardt, Bate, and Matyas 1998; Morrison et al. 2005).

There are four common ways to organize a systematic observational strategy:

- One way is to focus attention sequentially by the phases (preparatory, action, follow-through) of the movement.
- Another strategy is to observe with emphasis on the balance of the mover, so that observation proceeds from the base of support and origins of the movement.
- The third strategy is to observe based on the importance of the critical features to success in the movement.
- The fourth strategy is less structured and moves from general to specific. This observational strategy is based on Gestalt psychology, according to which an overall impression may provide a holistic perspective and provide cues for more specific observational foci in subsequent observation.

Whatever the observational strategy or approach to qualitative analysis employed, the observer essentially makes a visual or mental comparison of the observed movement technique and the desirable technique. How the desirable technique is defined naturally depends on the approach (mechanical or critical features) and is discussed in greater detail in the following section on evaluation and diagnosis. For these comparisons to be accurate, the coach needs to control several aspects of observation. Four important aspects of a systematic observational strategy often missed by coaches are

- the situation,
- vantage points,
- number of observations and
- improving observational power with videotape.

Since the movement technique employed and observed will be highly related to the context, the coach should ensure that the movement situation closely matches the conditions of interest. The techniques that athletes use are strongly affected by environmental, psychological or competitive conditions. Coaches need to be sure that the speed and other conditions under which observations of technique take place relate to those found in the match or competition environment. Coaches should also plan to observe multiple repetitions (five to eight) from several vantage points. This ensures that they can evaluate the consistency of performance and see movement from the perspective of various anatomical planes of motion. An effective tool for extending observational power is videotape imaging and replay.

The qualitative use of videotape replay for teaching motor skills goes back almost half a century. The greatest advantage of videotape replay to qualitative analysis is the greater temporal resolution of video compared to vision. Regular video allows body positions to be frozen in time and broken down (25/50 frames/fields per second [PAL format] or 30/60 frames/fields per second [NTSC format]). Most television or video pictures or frames are interlaced, meaning that they are composed of two half-spatial resolution fields. High-speed (sampling rates greater than 60 pictures per second) video cameras are also available. While progressive scan cameras are changing this recording format, the general principles related to qualitative analysis are held constant.

Video data can be collected from perspectives that show anatomical planes of motion (sagittal [side],

Figure 14.4 The siliconcoach Sport program allows the replay of split-screen images.

Figure 14.5 The siliconcoach Sport program allows an overlay of images compared with a model performer.

transverse [above], frontal [front or behind]) or from a combination of movement-relevant camera views. Digital video signals can now be directly imported to computers so that the video may be replayed using a variety of software packages designed for qualitative analysis of video (figures 14.4 and 14.5). Some examples of these programs are Dartfish, NEAT, siliconcoach, Simi MOStill and Quintic, with features that enable coaches to do the following:

- Present multiple videos (event synchronized) showing athlete versus model comparisons and currently up to four videos from different phases of development (figures 14.4 and 14.5)

- Bring the athlete's attention to specific aspects of performance by drawing on a frame (see chapter 15 for examples of this) or by tracing a movement over a number of frames
- Display visual reference objects, points and lines
- Calculate 2-D planar data

While these programs make some quantitative calculations, coaches need to remember that these 2-D images accurately show movements only in a plane parallel to the plane of the video image. Many of the angle and speed calculations made from these 2-D images will be inaccurate because of distortion of the 3-D nature of the motion or other scaling or optical errors. Chapter 15 on motion analysis highlights some of these issues in biomechanical research.

Evaluation and Diagnosis

In the third task of qualitative analysis there are two difficult objectives—evaluating the strengths and weakness of performance and diagnosing the causes of poor performance. Both of these objectives are necessary to focus coaches' attention on the one or two factors that are most likely to improve the player's performance. These objectives are critical for setting up the last task of qualitative analysis, providing intervention to improve performance.

Evaluation of the movement technique noted in observation involves the identification of strengths (agreements between desired and observed technique) and weaknesses (differences between desired and observed technique). This expansion of focus from "errors" in technique to all relevant technique factors allows coaches to note relationships between critical features and note improvements, setting them up to provide feedback that reinforces and improves player motivation. This larger picture of evaluation also takes into account the range of acceptability or correctness for each critical feature. Perfect technique cannot be uniquely defined, so it is useful for coaches to use an evaluation strategy that differentiates techniques but allows for some variation.

We recommend that the technique points in a movement be evaluated to three levels at a minimum. A three-level ordinal scale (figure 14.6) allows the coach to rate critical features with good reliability while also relating to an intervention that is likely to be effective. A player with excessive range of motion can be cued to decrease the intensity or extent of certain movements.

The other approach is for the coach to consider the critical feature on a linear scale so that he or she can mentally mark a point on the continuum of performance that seems to reflect the player's performance in observation. This visual analog scale (figure 14.6) allows for tracking of more subtle (much more than three in the ordinal scale) changes in performance. Coaches should select an evaluation approach that they are comfortable with and that balances the accuracy and consistency they feel they can achieve. There is no point in coaches' using an analog scale in their mind if they believe that the desirable technique for the critical feature of interest is not easily

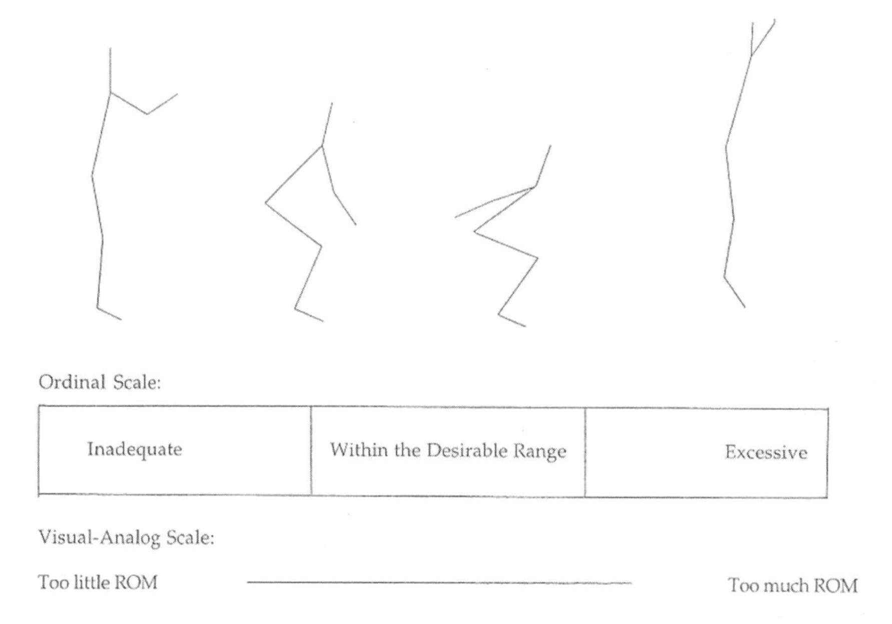

Figure 14.6 Range of motion in a vertical jump can be evaluated using an ordinal (categorical) or a visual analog scale.

Reprinted, by permission, from D. Knudson, 2000, "What can professionals qualitatively analyze?" *Journal of Physical Education, Recreation, and Dance* 71(2): 19-23.

defined, so that they can essentially detect only success or a surplus or deficit of that critical feature.

Once the strengths and weaknesses of performance have been identified, this information must be diagnosed. Diagnosis focuses on prioritising the possible interventions to identify causes of poor performance and, not unimportantly, symptomatic technique issues. Good coaches focus on the single most important technique issue rather than overburdening athletes with feedback on many technique points. Too much feedback from a coach creates "paralysis by analysis" (figure 14.7).

Unfortunately, there is no one accepted rationale for prioritising possible interventions to improve performance. Knudson and Morrison (2002) have noted that at least six different rationales have been used to diagnose movement technique.

- One rationale is to try to relate technique actions to previous actions (Hay and Reid 1988).
- Another is to focus on the technique issue that is believed likely to result in maximal improvement.
- A rationale that tends to help with motivation is to focus intervention based on difficulty. By focusing on the easiest change first, the athlete may perceive a greater sense of progress or improvement.
- Diagnosis can also be based on prioritising intervention in sequence. It is possible that improvements in preparatory motions in some skills might have carryover improvements in the later stages of the skill.
- For skills in which balance is critical, many coaches prioritise intervention from the base of support.
- The last rationale used in technique diagnosis relates to prioritising critical features over other, minor technique issues. This would mean addressing the cause of a flaw in performance rather than attempting to correct an effect that may be the direct result of the cause.

Skilled coaches avoid "paralysis by analysis" in their athletes by focusing their intervention on one technique point at a time.

In the absence of scientific data on which diagnostic rationale is best, coaches much select a rationale or a combination of rationales to use in qualitative analysis. For example, a golf coach might believe that the precision of the sport requires excellent balance and strive to diagnose swings based on two rationales, emphasizing the base of support and trying to relate weaknesses to previous or other weaknesses.

Intervention

The fourth task of qualitative analysis is the implementation of the intervention selected in evaluation and diagnosis to help the athlete improve. Traditionally, this has been in the form of verbal feedback about some correction in technique. Technique corrections, however, are just one method that good coaches can use to provide intervention. Table 14.1 lists only some of the many forms of intervention a coach can employ to help athletes improve their technique. The best coaches have a variety of intervention strategies to select from and use

Figure 14.7 Diagnosis to focus on one key intervention is important to avoid "paralysis by analysis".

Reprinted, by permission, from D. Knudson and C. Morrison, 2002, *Qualitative analysis of human movement*, 2nd ed. (Champaign, IL: Human Kinetics), 12.

Table 14.1 Intervention Strategies Often Used in Sport

Strategy	Advantages
Feedback	Focuses attention of the athlete on a desirable change
Visual modelling	Good in early stages of learning, conveys lots of information
Modifying practice	Indirectly elicits changes and challenges athlete
Manual guidance	Good for difficult positions or motions
Conditioning	Changes fitness variables influencing technique
Attentional cueing	Provides technique structure for an athlete
Overcompensation	Helps with difficult changes in technique

Adapted from D. Knudson and C. Morrison, 2002, *Qualitative analysis of human movement,* 2nd ed. (Champaign, IL: Human Kinetics).

the strategy that best fits the athlete and the situation. Chapter 16 addresses in more detail how an athlete may benefit from different learning situations to optimise performance.

Quantitative Analysis

Quantitative analysis of sport technique involves the measurement of biomechanical variables that precisely describe performance. At this level of technique analysis, coaches usually need to consult with a biomechanist (preferably a person with a PhD in sport biomechanics) to provide the expertise in data collection and analysis. Coaches' decisions on technique modification will be improved only if the measurements are accurate. This cooperation between coaches and biomechanists examining specific technique issues is the approach used by most Olympic training centres throughout the world. Coaches in different sports may request various types of information as suggested by the following examples.

The swimming coach may wish to know

- the muscle groups that need to be specifically trained for a faster recovery in the butterfly stroke,
- the angles of the hand and forearm as they enter the water in the freestyle stroke and
- the body position as it enters the water during a racing dive.

The long jump coach may wish to know

- the velocity of the body at takeoff,
- the angle of the front leg at takeoff and
- the levels of vertical, forward-and-back and side-to-side ground reaction forces at takeoff.

The volleyball coach may be interested in

- the height of the body at ball impact in a spike,
- the role of trunk rotation in velocity generation and
- the position of the hand at ball impact in the spike.

The tennis coach may be interested in

- the angles between the upper arm and trunk at impact in the serve,
- the amount of shoulder internal rotation in a forehand drive and
- the typical strength ratio of shoulder internal and external rotators in uninjured tennis players.

The measurement of biomechanical phenomena for quantitative analysis of technique generally falls into four categories: imaging, dynamometry, electromyography and simulation. These measurement techniques can provide information on kinematic (motion description) and kinetic (causes of motion) variables. The following sections briefly summarize these techniques as they apply to sport. Chapters 15 and 16 provide more information on and examples of the use of quantitative analysis in sport.

Image Analysis

Image analysis techniques in sport are characterized by use of video or opto-reflective (optical without a visual image—these produce an anatomical model using reflective markers) cameras to record a movement. The sequence of multiple images recorded over time provides position data that can be processed by computer software programs to calculate a variety of biomechanical variables (see chapter 15 for a full discussion of this topic). While the captured images on a video or anatomical model can be qualitatively analysed by the coach and scientist, the major advantage of this approach is in the quantification of kinematic and kinetic variables from the movement. Further biomechanical data such as net forces and torques at the hip or shoulder and the flow of power from the trunk to the upper limb, or up the lower limb, can also be obtained.

> There are four different kinds of quantitative biomechanical data relevant to sport technique: imaging, dynamometry, electromyography and simulation.

Another advantage of quantitative analysis using image analysis is that the data gathered may be compared with data from research related to performance or injury risk. Data on the actual movement technique measured for an athlete can be compared with normative data from similar and elite athletes, in much the same way clinicians compare data from children with cerebral palsy with normative values collected from children of similar age before deciding on the appropriate surgical procedures. The motions or forces and torques measured may also be compared with known tissue strengths to document the safety factor relative to injury-producing values.

While opto-reflective systems tend to be used exclusively in the laboratory environment, it is important for sport scientists and coaches to understand their role in providing biomechanical data for athlete development. The most accurate of all visual analysis systems (Elliott, Alderson, and Denver 2007; Richards 1999), opto-reflective systems rely on the tracking of markers through space (see figure 14.8), linked to appropriate computer software for the quantification of movement.

Figure 14.8 Opto-reflective markers on a subject ready for testing.

Analysis using these systems, discussed more fully in chapter 15, relies on a testing situation in the laboratory that is as close to the "game environment" (validity) as possible. The following are examples of anatomically based 3-D data at the knee:

- Flexion-extension, varus-valgus, internal-external angles
- Flexion-extension, varus-valgus, internal-external moments or torques

Such data provide the coach with valuable information on performance or training methods linked to the reduction of knee injury (see chapter 16). In running, patellofemoral pain has been associated more with movements in the frontal (valgus-varus: side to side) and transverse planes (internal-external rotation) than in the sagittal plane (flexion-extension), so accurately measured anatomical 3-D data are essential for understanding and reducing the risk of this common running injury (Alderson, Elliott, and Lloyd 2008).

While 3-D video systems are commonly used in the laboratory, their strength is in field-based applications. High-speed digital cameras may now be used to collect 3-D data in both match and simulated match environments. Fleisig and colleagues used this approach at the Sydney 2000 Olympics to record the technique and shoulder loading in the service action for elite tennis players competing at the highest level (Elliott et al. 2003; Fleisig et al. 2003). They also used on-court filming to identify the importance of internal rotation at the shoulder joint in the power tennis serve (Elliott, Marshall, and Noffal 1995).

Movements linked to injury in sport have also been identified using video-based prospective studies. Research using both 2-D and 3-D setups, in a combined laboratory- and field-based approach, has identified the aetiology of back injuries to fast bowlers in cricket and shown that these injuries may be reduced with technique modification (Elliott and Khangure 2002; Foster et al. 1989). The credibility of more highly controlled laboratory studies is increased when similar results can be obtained in the field.

Dynamometry

Dynamometric techniques provide direct measurement of kinetic variables (force, torque) using a variety of technologies. The most common dynamometer used in sport biomechanics is the force platform. A force platform mounted on the ground enables measurement of the three perpendicular components of the ground reaction forces (vertical [Fz], forward-backward [Fy] and side to side [Fx]; figure 14.9) and the moments of force about these axes, although few biomechanical studies have shown meaningful uses for these moments. Force platforms are valuable research tools that come in a variety of sizes and can be mounted and used in a variety of ways. Force platforms have been mounted in treadmills, stairs, exercise machines and vertically to measure the forces involved in scrumming in rugby or in swimming turns.

Many dynamometers, such as the force platform or the isokinetic dynamometer, also measure kinematic variables. For example, the force platform measures the location of the point of resultant force applied to the platform. This centre of pressure as given by two coordinates (CPx, CPy) is generally reported relative to the centre of the force plate. In the following discussion we present illustrations of the use of these measures to help teachers and coaches understand the forces associated with a number of sporting activities.

External ground force data are often used to aid in understanding of performance or as an estimate of approximate levels of internal load placed on the body. Figure 14.9 illustrates the ground reaction forces associated with a long jump. From these data it is evident that the following strategies may be adopted to help this jumper improve performance:

- The horizontal braking force should be reduced and the propulsive force increased at foot strike. A possible means of doing this would involve training the jumper to position the landing foot closer to the body at board contact and having the jumper increase the strength of the lower limb musculature.
- The magnitude of the side-to-side force should be minimized, as this does not aid jumping performance but causes lateral movement at takeoff.
- The large peak vertical impact force of approximately 6.5 BW on the board shows that long jumping places a high load on the lower extremity. Coaches of long jumpers must therefore carefully monitor the number of repetitions performed so as to reduce the likelihood of overuse injuries (see chapter 16 for a more detailed discussion of overuse injuries).

Fast bowling in cricket has also been shown to place high loads on the lower limb, with the bowler experiencing a series of "collisions" or minor impacts with the ground during the run-up, followed by a large impact when landing on the front foot during the delivery stride. A force platform can be used to identify the maximum

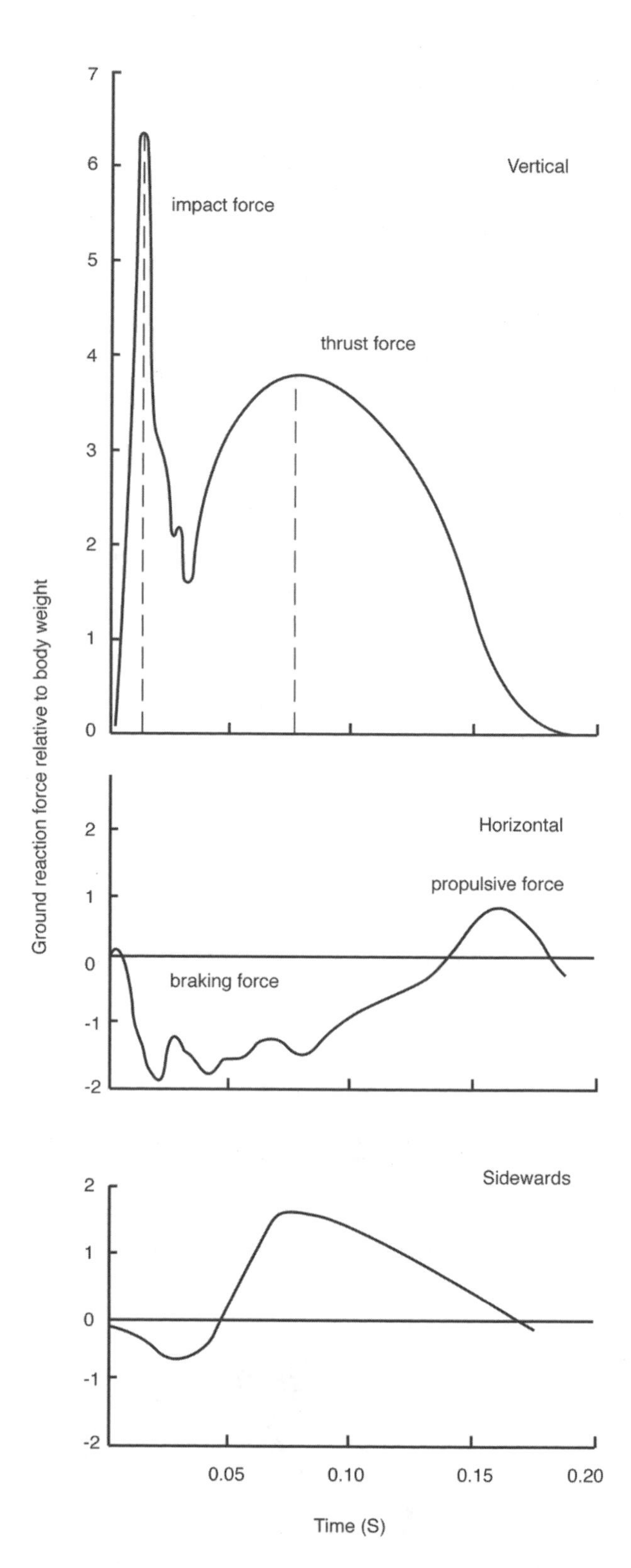

Figure 14.9 Typical orthogonal (3-D) ground reaction forces measured by a force platform during the takeoff of a long jump.

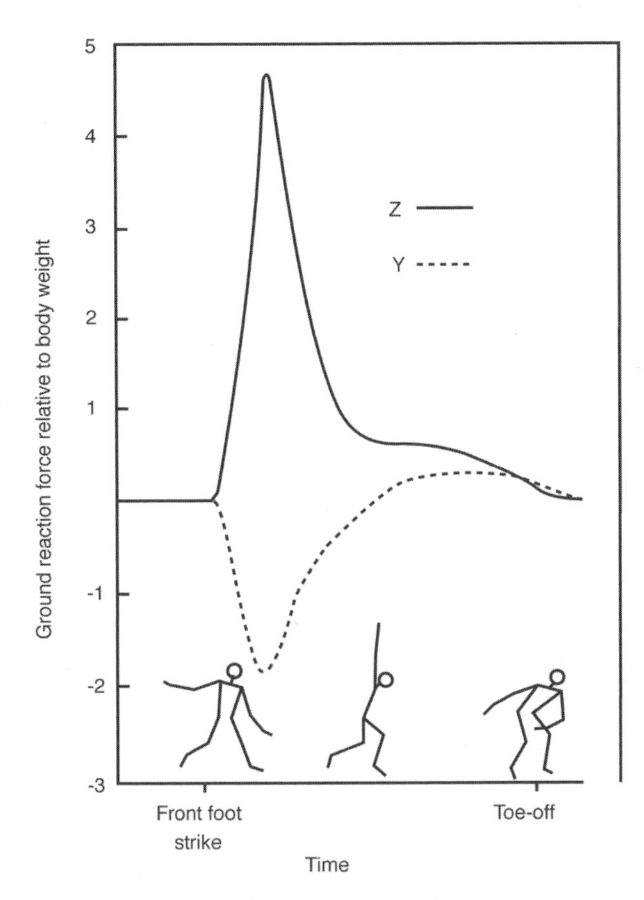

Figure 14.10 Typical vertical (Fz) and forward-backward (Fy) ground reaction forces on the front foot during cricket bowling.

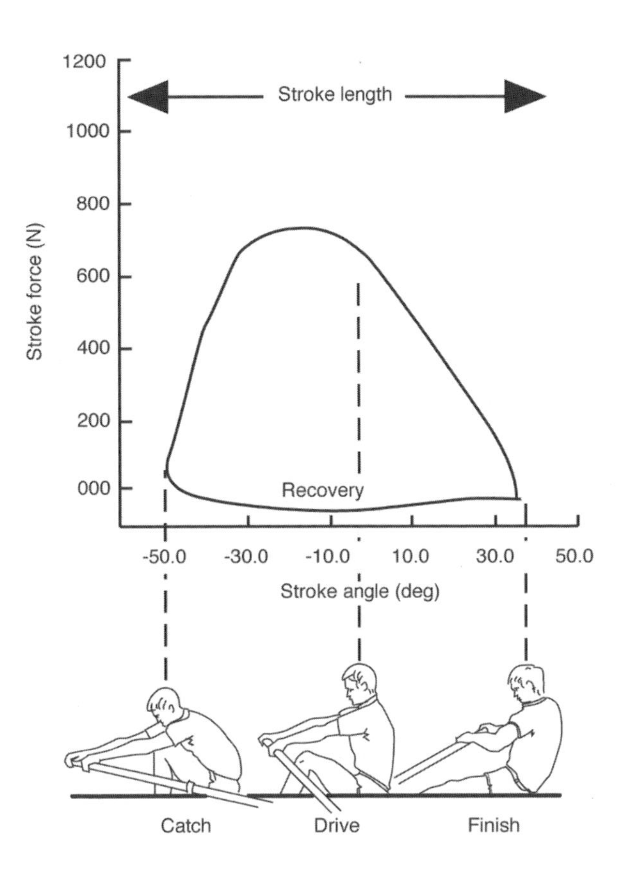

Figure 14.11 The force profile from a rowing stroke.

Adapted from R. Smith and W. Spinks, 1989, Matching technology to coaching needs: On water rowing analysis. In *Proceedings of the VIIth International Symposium of the Society of Biomechanics in Sports,* edited by W. Morrison (Melbourne, Australia: Footscray Institute of Technology Press), 277-287.

vertical and forward-back ground reaction forces at front foot impact during the delivery stride (figure 14.10). As the front foot makes impact with the platform, a peak vertical force of about five times the bowler's body weight and a peak backward force of twice the body weight are typically recorded. While research has not directly linked these forces to the development of back injuries in fast bowlers, considerable force must be transmitted through bones, cartilage and muscles to the various joints in the body, especially in the region of the lumbar spine. To avoid overuse injuries, bowlers should reduce impact forces through flexion (bending) of the front knee joint and possibly by inserting "force-absorbing" orthotic insoles in their shoes.

In rowing, the forces applied by the athlete to the oar during the stroke are of major interest to the coach, and small transducers (force-measuring devices) have been attached to the oar or to the oarlock to measure these forces. The force applied to rotate the oar is shown in figure 14.11, which illustrates how this force varies over the complete stroke. The relationship between the force applied to the oar and the boat speed (figure 14.12) provides the coach with valuable information on the mechanical effectiveness of the stroke as well as the boat's "flow" through the water. Smith and colleagues have used such approaches to improve rowing performance while at the same time reducing the risk of injury (Smith and Loschner 2002; Smith 2004). Impulse curves (such as that shown in figure 14.11) have also been used extensively in coaching with reference to matching of bow and stroke side rowers for optimal boat performance.

In relatively stationary activities such as archery and the shooting sports, body sway is an important factor to be minimized. The movement of the centre of pressure has been analysed because it represents the dynamics (forces) that control the motion of the centre of gravity of the body (sway). Figure 14.13 represents the total area over which the centre of pressure of an archer moved

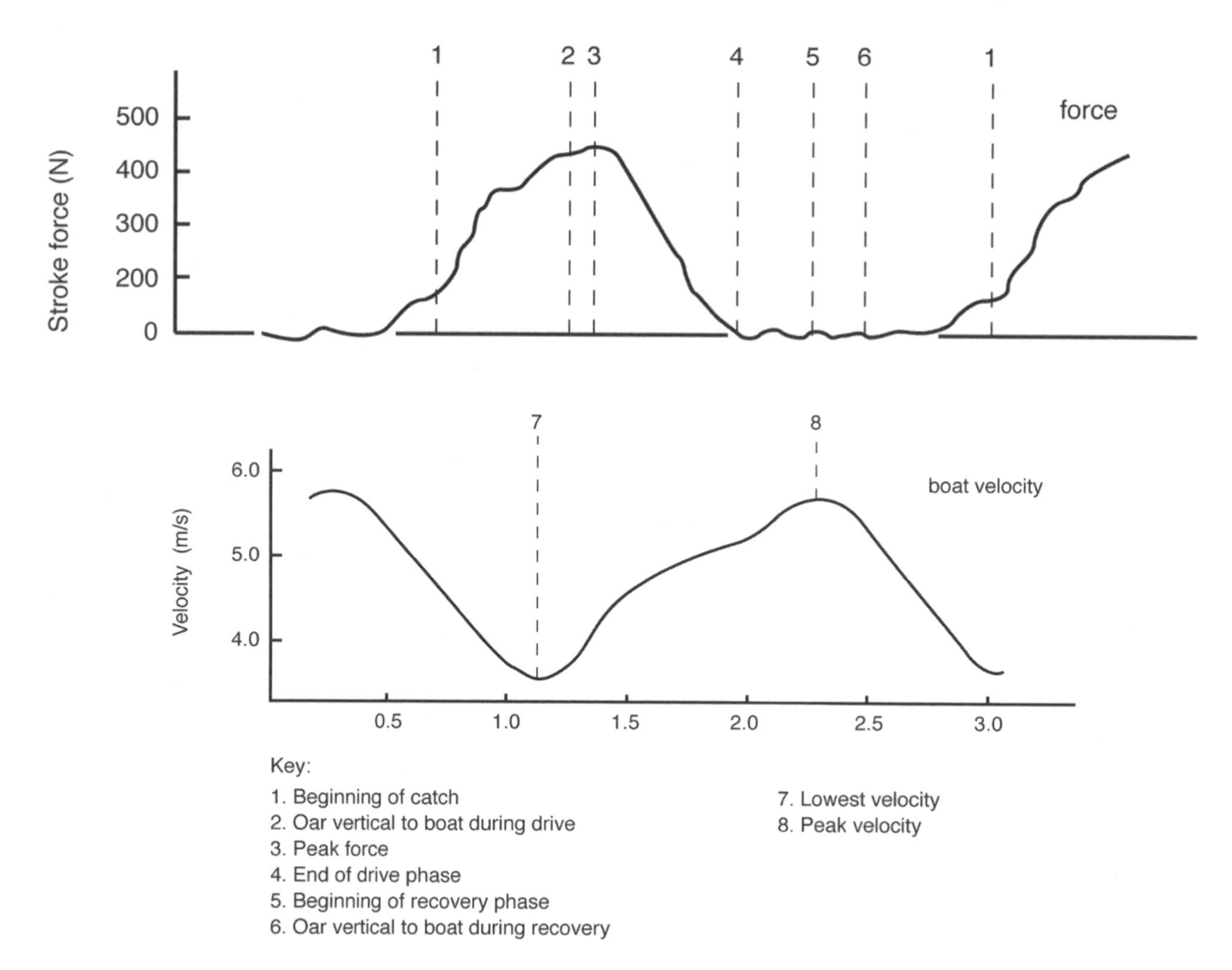

Figure 14.12 The relationship between force and boat speed during a rowing stroke.

Adapted from F. Angst, H. Gerber, and E. Stussi, 1985, *Physical and biomechanical foundations of the rowing motion*. Paper presented at the Olympic Solidarity Seminar. 10-11, Canberra, Australia.

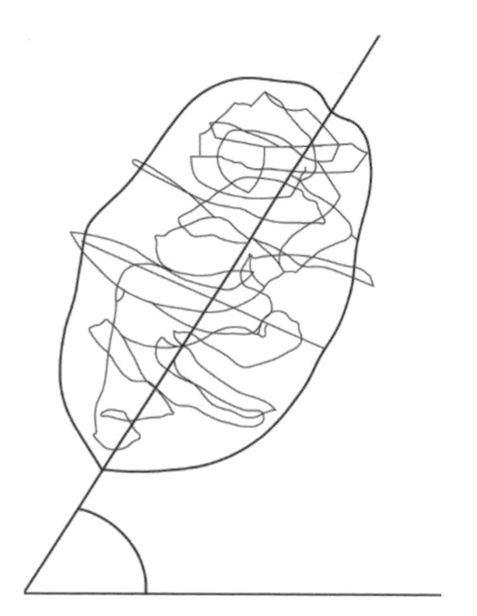

Figure 14.13 Diagram of the area of the centre of pressure motion and angle of stance of an archer.

during the aiming period and may be used to represent the control of sway of the archer. Additionally, the overall direction of the sway may be used to approximate the angle of stance with reference to the target.

The centre of pressure movement in the stance phase of running has also been used as a quantitative index of the type of foot strike pattern, although recent research has focused more on the distribution of pressure under the foot. The location of this centre of pressure referenced to the foot has been used in the following ways:

- To compare the typical patterns for midfoot and rearfoot runners from foot strike to toe-off
- To provide the position for inclusion of the ground reaction forces in the free body diagram of the foot in inverse dynamics—these data, when combined with kinematic information from the activity in question, permit the calculation of forces and torques about the ankle, knee and hip joints during gait and sporting manoeuvres

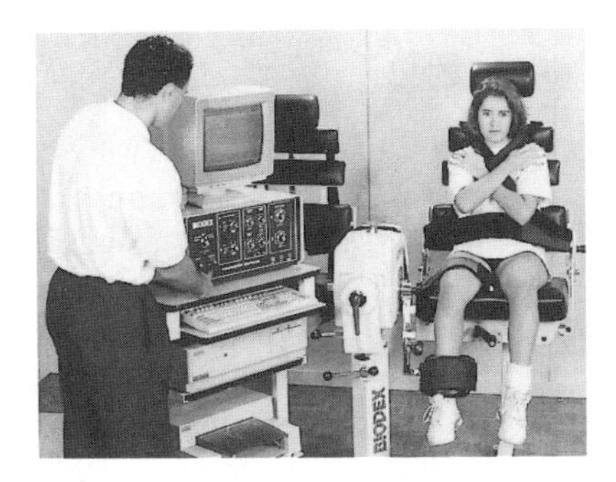

Figure 14.14 Typical isokinetic dynamometer.

Isokinetic dynamometers developed for use in exercise research and rehabilitation are also now commonly used in biomechanical analyses to assist sport performance. These devices provide force and torque (moment) data throughout the range of motion for muscle groups for most joints in the body (figure 14.14). Values may be measured in concentric (muscle shortening), isometric (motionless) or eccentric (muscle lengthening) conditions at a variety of speeds. These machines have been around for several decades, so there are extensive normative data on athletes and people of all ages. For example, isokinetic testing at several joints has been used to profile tennis players in an attempt to improve performance and reduce the risk of injury (Ellenbecker 1992; Ellenbecker and Roetert 2003).

However, isokinetic dynamometers are relatively expensive, so coaches of all but professional teams usually work with therapists, sport institutes or university faculties possessing this equipment. These measures of torque, work and other variables are useful for documenting the strength of muscle groups and strength balance (agonist-antagonist, sides of the body). It is essential that strength ratios, speeds and muscle actions between muscle groups be logically compared. For instance, in tennis, the peak torque and total torque produced during concentric internal rotation at the shoulder during the forward swing of the serve should be compared with the eccentric "strength" of the external rotators during the follow-through, whose job it is to stop this rotation. Only then will one be able to predict if the eccentric contraction of the external rotators is sufficient to stop the rotation created internally.

Electromyography

Muscle stimulation by the central nervous system results in electrical potentials in the muscle fibres; and the detection, amplification and recording of these signals is a technique known as electromyography (EMG). The EMG signal, after suitable processing, provides an indication of the timing and level of activation of a particular muscle or muscle group.

Comparisons of levels of EMG activity between muscles or different subjects cannot be directly made because of the many variables that affect the EMG signal. One technique that assists in allowing some comparison is to normalize (express as a percentage of maximum) EMG to a standard contraction for each individual. Even this form of processing has limitations because it is difficult to measure a true maximum EMG, and studies often report peak dynamic EMG values recorded during a sporting activity of over 150% to 200%!

While coaches may never use EMG with their own athletes, awareness of data from controlled studies will certainly assist in the planning of training programs. Figure 14.15 depicts the surface EMG activity of selected muscles of the lower extremity during stationary cycling. The rectus femoris (RF), a major extensor of the knee joint (and hip flexor), shows two bursts of activity. One begins just after the top of the pedal stroke and ends slightly before the leg is fully extended at the bottom of the stroke cycle. A second burst is seen starting at about the bottom of the stroke and finishing about halfway through the recovery (upward) movement. The knee flexors, represented by the biceps femoris muscle (BF), show

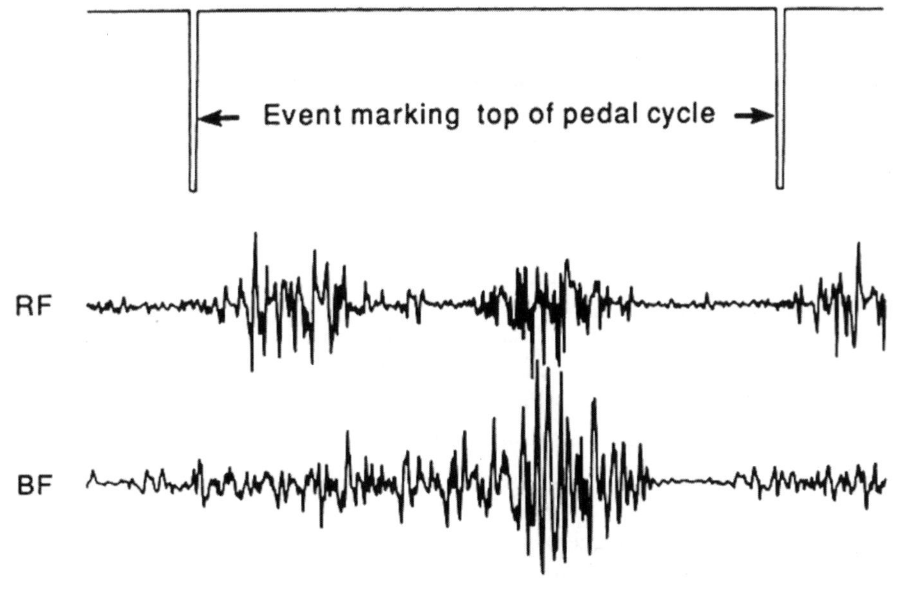

Figure 14.15 Raw electromyogram (EMG) activity of selected lower extremity muscles during stationary cycling: RF, rectus femoris; BF, biceps femoris.

Adapted from J. Bloomfield, P. Fricker, and K. Fitch. 1992. *Textbook of science and medicine in sport* (Melbourne, Australia: Blackwell Scientific Publications), 50-61.

slight to moderate activity during the propulsive (downward) stroke and a major burst at the start of the recovery movement. Remember that even more relevant data on cycling may be gathered if EMG and force applied to the pedals (collected from 3-D load cells) are simultaneously recorded.

Studies of the muscle activity that occurs during swimming, as reported by Piette and Clarys (1979), illustrate how the muscles identified as playing a role in the freestyle stroke by virtue of high levels of activation must then be the ones that are trained in a muscle strength-endurance program for freestyle swimming.

The best sport technique is difficult to determine and requires the integration of all the four kinds of quantitative biomechanical analysis. Coaches should integrate their quantitative analyses with the results of biomechanical research in making technique decisions with athletes.

Predictive Modelling

A predictive quantitative approach uses a mathematical model of the body and computer software to first run a simulation (use of a validated computer model) and then an optimisation (use of computer simulation to optimise a facet of performance). This gives the researcher the power to predict the results of changes in technique. Computer simulation and optimisation of complex biomechanical models are beginning to play an important role in the development of sporting excellence. The general aim of this approach, when applied to athletic performance, is to predict changes that would occur as a result of alterations to selected aspects of technique. The coach is then able to obtain answers to various questions, for example, What would happen in a given skill if a critical feature was modified? The following are studies that illustrate the use of this approach:

- Optimum release characteristics of the new javelin (post-1986) were investigated by Hubbard and Alaways (1987), who reported that release conditions were velocity dependent. They found that success with the new javelin depended to a large extent on strength and power and that finesse and skill were of less importance.
- In gymnastics, Yeadon (1988) simulated twisting somersaults to determine the contribution of asymmetrical arm, chest and hip actions to aerial movement. He demonstrated that twists are most frequently produced by asymmetrical arm actions and that even where some twist was evident at takeoff, the major contributions were still made by aerial techniques.
- Koh and colleagues (2003) were able to predict how an elite gymnast's Yurchenko layout vault could be improved using optimisation. The predicted optimal vault displayed greater postflight height and angular momentum compared with the gymnast's best trial performance.
- Bobbert and van Soest (1994) simulated vertical jumps and showed that merely increasing quadriceps strength will actually decrease jump height with use of the same coordination, but that jump height will increase if jumpers alter their coordination. This study supports the belief that skill practice must be maintained during heavy conditioning periods to fine-tune coordination. Subsequent simulations showed how well-timed countermovements in jumping change the ways in which the major muscle groups contribute to performance relative to static jumps (Bobbert et al. 1996; Bobbert and Casius 2005).
- Some prediction research uses EMG as input to complex biomechanical models. Lloyd and colleagues (see figure 14.16; Buchanan et al. 2004; Lloyd, Buchanan, and Besier 2005) developed an EMG-driven model linked with force and kinematic measures collected using an opto-reflective Vicon system to examine knee joint loading in different movement situations. This model can be used to predict loads during cutting and stopping manoeuvres and to assess training regimens, such as proprioceptive training, in an endeavour to reduce anterior cruciate ligament injuries in sport.

Coaches might wonder, given the fast, high-memory computers currently available, why the major technique issues in sport have not been resolved. The answer is that biomechanical modelling and simulation are very difficult tasks and that many scientists who develop these models use them to answer theoretical or clinically based questions. Preparation of elite athletes is changing: Whereas in the past, the coaches or athletes themselves often led innovation in changes in technique, usually through a trial-and-error approach, biomechanists can now help coaches predict which further modifications may be desirable.

It is important that results from all kinds of quantitative analysis methods be integrated to inform the coach as to the best techniques to use for sport skills. Coaches can help with this process by advising scientists about technique issues they want to be resolved. The high cost (time and money) of biomechanical research means that these resources must be used wisely if there is going to be an adequate amount of research relevant to coaching. Readers interested in more information on technical issues in making quantitative measurements in biomechanics are directed to the work of Robertson and colleagues (2004).

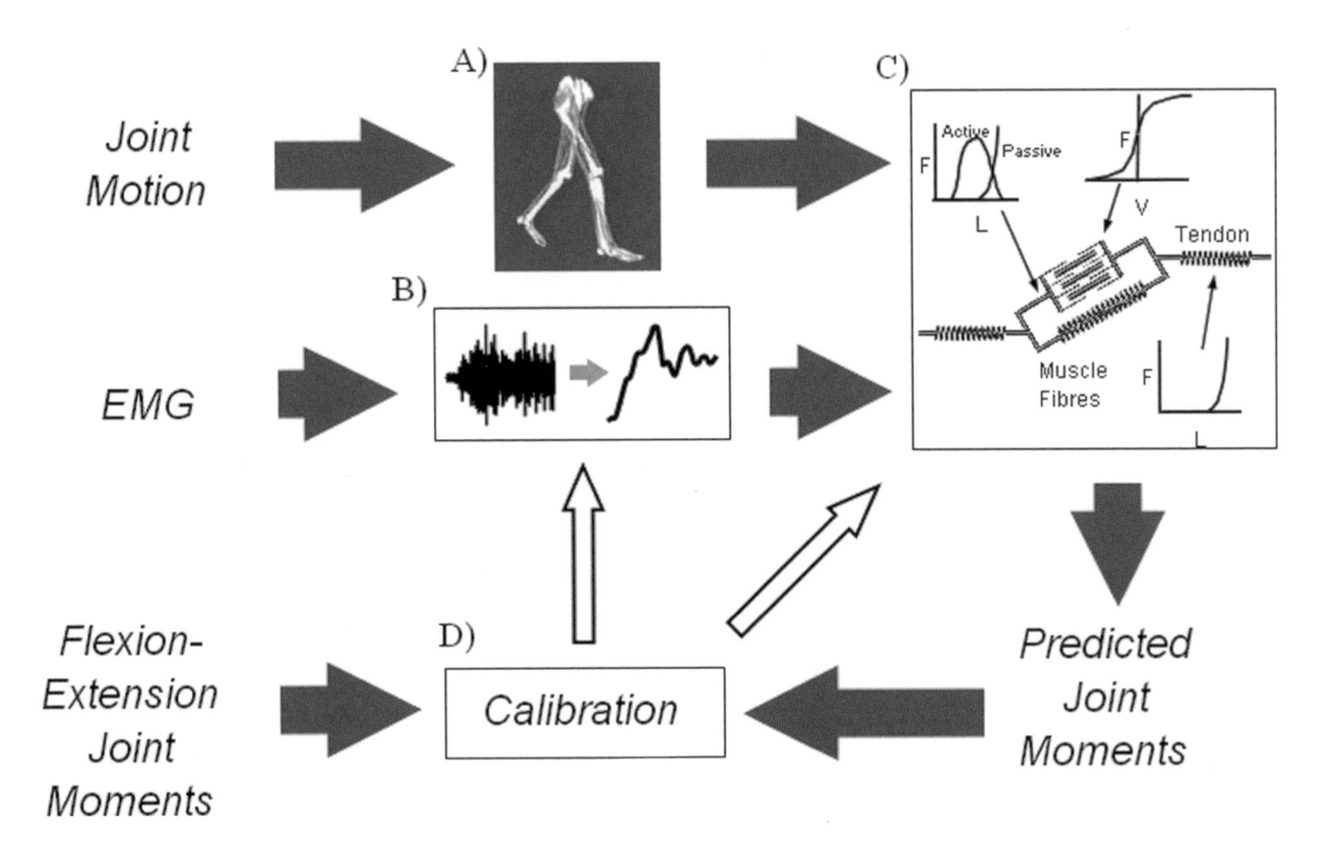

Figure 14.16 Model for the calculation of knee loading.

Reprinted, by permission, from D. Lloyd, T. Buchanan, and T. Besier, 2005, "Neuromuscular biomechanical modeling to understand knee ligament loading," *Medicine and Science in Sports and Exercise* 37: 1939-1947.

Summary

The ability to analyse sport technique using qualitative or quantitative methods is an essential skill for any coach. The best coach strives to integrate experiential and scientific knowledge in selecting the best techniques and practices for athletes. Researchers also need the help of coaches to inform and guide biomechanical research on important sport technique issues. Those coaches who use the methods described in this chapter can better assist their athletes to refine technique and improve performance and often reduce the likelihood of injury.

fifteen

Image Analysis in Sport Performance

Jacqueline A. Alderson, PhD; and Bruce C. Elliott, PhD

The influence of technology, which is playing an ever-increasing role in today's society, has certainly pervaded the sporting domain. Technique modifications to improve performance and reduce the risk of injury, based on objective data collected with the use of hi-tech equipment, are becoming ever more commonplace.

The sport biomechanist then faces a quandary in determining what image-based quantitative analysis system should be used. First, does one select 2-D (planar) or 3-D video-based systems, in which images of the performance are digitised and relevant data extracted, or select marker identification systems (passive or active) to track the body during the motion in question?

It is logical to select an analysis system according to the following considerations:

- What provides the best answer to the coaching or teaching question of interest?
- What error in measurement is acceptable with reference to the application of results from an analysis?
- What testing protocol and data format are most acceptable to the coach or teacher and the athlete?
- What system is available to the coach or manager or the athlete?

It is apparent from chapter 14 that quantitative data collected with the use of any image-based system will be relevant in the sporting setting only if

- the data collected truly represent the performance under "match conditions" and
- the coach or teacher is able to communicate these data to the athlete in a meaningful way.

This chapter aims to build on the approaches presented in chapter 14 and to discuss in more detail the varying roles that image-based techniques play in sporting movements. First, a historical overview will provide readers with an appreciation of changes in image-based procedures during recent times. A more comprehensive discussion of this topic can be found in Elliott and Alderson 2006.

Background

Sport biomechanics in the 1970s and 1980s was dominated by image-based analyses using high-speed motor-driven photography (100-500 Hz). Manually digitised marker or joint centre displacement was recorded over time using commercially available digitisers. These digitisers were linked to customized software developed "in-house" at most universities or clinical research facilities to calculate 2-D or 3-D kinematics and, to a lesser extent, kinetic data. The 1990s and the new century ushered in a sudden explosion in video technology, which has been paralleled by the advent of commercially available 2-D (e.g., Dartfish, siliconCoach, SIMI and Quintic) and 3-D (e.g., ViconPeak, Ariel and SIMI) motion analysis systems. Three-dimensional opto-reflective systems, which for the last few decades have been the gold standard image-recording system utilized in clinical gait laboratories around the world, have only recently been employed for data collection of sporting manoeuvres. Historically, these systems were inappropriate for sporting investigations due to computer and camera limitations restricting the number of markers that could be used and the speed at which movements could be captured (e.g., Vicon, Motion Analysis and Elite).

When Would You Choose 2-D Analysis?

Two-dimensional analysis is typically chosen when the motion to be captured is primarily planar (see detailed discussion of planes of motion later in the chapter). For this reason it is generally difficult to record continuous motion in sports where the skill being performed moves through multiple planes of motion (e.g., throwing). For this reason it is often better to aim to capture discrete points in time (e.g., angle or position at impact, release, foot strike, takeoff). However, for activities such as running or kicking, the coach may be interested in continuous changes to the knee flexion-extension angle. These may be recorded in 2-D, as this angle remains primarily planar (sagittal) throughout the movement cycle. In contrast, during upper body activities such as throwing, continuous angle changes about the elbow or shoulder cannot be recorded in 2-D because of the rotational (i.e., nonplanar) characteristics of the movement. However, one can strategically place cameras to record angles such as elbow flexion-extension at discrete time points during the throwing cycle, though it is necessary to ensure that the camera position is perpendicular (90°) to the angle being measured. For example, to measure elbow flexion-extension in throwing as accurately as possible in 2-D, the camera should be positioned as follows:

- At completion of backswing—camera is in the sagittal plane
- At midthrow—camera is in the frontal plane
- At ball release—camera is approximately at the intersection (45°) of the sagittal and frontal plane if it is to remain perpendicular to the elbow flexion-extension angle

When Would You Choose 3-D Analysis?

Whenever accurate measurement of long-axis rotations of limb segments (forearm, upper arm, lower leg and thigh) is required, or when the motion is clearly in a number of planes, it is necessary to utilize 3-D analysis. While it is possible to infer 3-D rotational segment positions from 2-D (e.g., measurement of maximum shoulder external rotation in throwing movements using the 2-D segmental position of the forearm), continuous assessment of shoulder external-internal rotation requires 3-D analysis. With regard to sport, it is relatively easy to generalize that the great majority of assessments should be 3-D if accuracy is important. In practice, limited access to 3-D systems and the high cost are prohibitive. Further, an understanding of the limitations and errors inherent in 2-D and 3-D data capture and analysis processes (systems and models) is essential if the application of biomechanical principles to sporting movements is to be successful.

Errors in Motion Analysis

Both 2-D and 3-D analyses are subject to error. Most frequently, errors in 2-D analysis occur when images from 3-D movements are recorded and processed in 2-D. An understanding of perspective error and refractive error can help to increase the accuracy of 2-D representations of 3-D motion. Errors in 3-D analysis have been shown mainly through comparisons of opto-reflective and video-based systems and also occur in the application of marker sets and corresponding models.

Two-Dimensional Analysis

The most common error observed in 2-D analyses occurs when images from 3-D movements, typically found in sport, are recorded and quantitatively processed in 2-D. Examples of 2-D nonplanar error in interpretation of a 3-D elbow angle (measured using the shoulder elbow and wrist joint centres) are shown in figure 15.2. In figure 15.1 (sagittal view), the elbow appears flexed, whereas in the frontal plane view the upper limb appears straight. Clearly a 3-D analysis is needed to accurately measure elbow range of movement, particularly if the upper arm is rotating about the shoulder during the motion being analysed.

As already discussed, it is essential that video footage collected for 2-D analysis accurately represent the 3-D motion being performed. The following concepts are critical for understanding 2-D limitations of motion analysis.

Figure 15.1 Two-dimensional planar images of a cricketer from *(a)* sagittal and *(b)* frontal views.

• *Perspective error* occurs when parts of the body lie outside the principal photographic plane. Although one cannot eliminate this problem completely, one can reduce it by increasing lens-to-object distance (i.e., placing the camera as far back as possible and then zooming to fill the camera's field of view) and filming perpendicular (90°) to the angle of interest.

• *Refractive error* occurs when measurements are made at the water-air interface (water-glass-air). This creates a problem when one is measuring positions involving a body part that is partially submerged in water. Regardless of the type of recording system, which may use a waterproof camera housing or underwater viewing window, a water-air interface will always exist. Researchers attempt to overcome this problem by positioning calibration structures both above and below the water line. We do not discuss this problem further, as its complexity is beyond the scope of this book.

Three-Dimensional Analysis

Research on analysis errors attributed to the 3-D system employed has primarily focused on comparing opto-reflective and video-based systems. Ehara and colleagues (1995) showed that the percentage error in predicting a known length of 90 cm increased from ~0.3% to 0.6% in passive reflective systems compared with video-based systems. Similarly, Richards (1999), in predicting a 50 cm distance, reported that the root mean square error increased from approximately 0.1 to 0.2 cm in opto-reflective compared with video-based systems. Recent research by Elliott, Alderson, and Denver (2006) showed that video systems produced larger errors (~0.5°) in measuring a known flexion-extension elbow angle compared with an opto-reflective system (~2.5°) when similar mathematical approaches were used to calculate elbow angle.

While the error contribution of the analysis system is an important consideration in determining the selection of an analysis procedure, one must also consider errors in the application of marker sets and corresponding models used in data reconstruction.

Historically, the prevalence of field testing using video-based systems, combined with the time burden of digitising images from multiple cameras (multiple views over a number of trials), has resulted in the adoption of biomechanical modelling procedures based on vector algebra principles. Joint centres digitised from at least two camera views resulted in a 3-D joint angle formed between two vectors. For instance, the line drawn from the wrist joint centre (WJC) to the elbow joint centre (EJC) representing the forearm formed the first vector, while the upper arm was similarly described as the vector from the EJC to the shoulder joint centre (SJC) (figure 15.2*a*). In real terms, the resulting 3-D vector joint angle describes a nonanatomically based angle. Initial efforts to separate out the component parts of joint motion (e.g., flexion-extension) involved vector angles projected onto a 2-D plane (figure 15.2*b*).

The accuracy of such an approach relies heavily on the premise that the motion remains relatively planar during execution. Interestingly, research by Elliott and colleagues (2006) showed that 3-D flexion-extension elbow angles reconstructed using an opto-reflective system

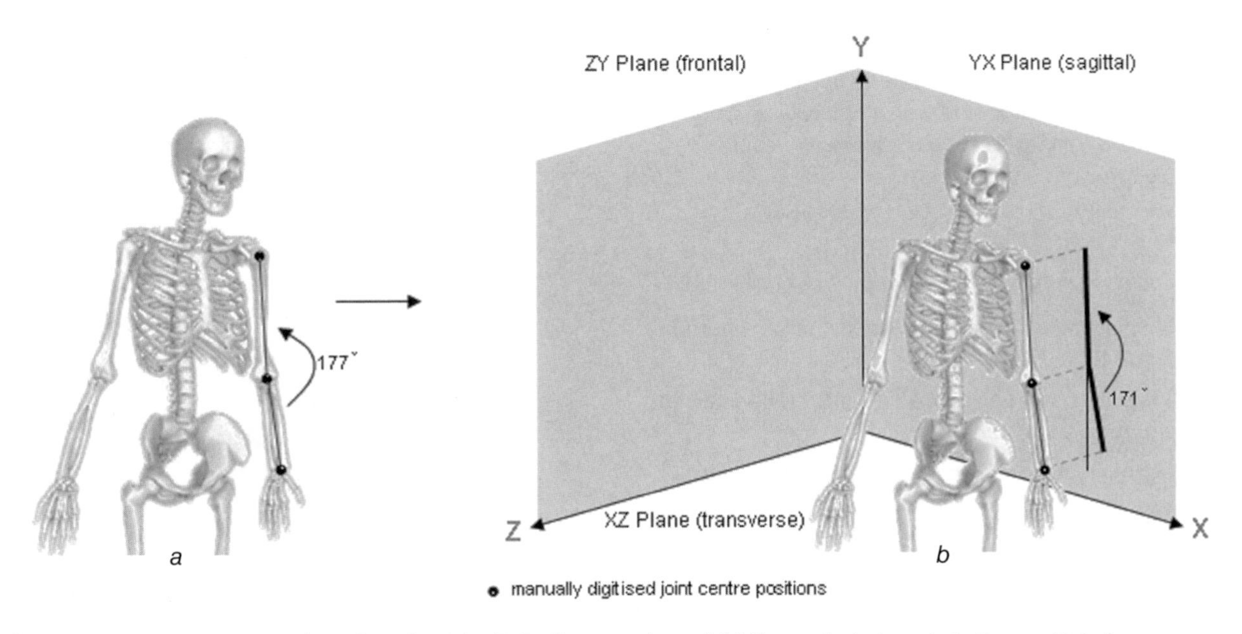

Figure 15.2 Vector approach to describe *(a)* a 3-D elbow angle and *(b)* the projected angle to the sagittal plane.

compared with a 2-D video-based projected angle were similar if the motion remained planar. However, when external-internal rotation about the shoulder joint was incorporated into the movement, errors in the 2-D projected angle from the known angle increased dramatically. In contrast, the error in the 3-D reconstruction did not increase significantly.

For sporting actions occuring at high speed, critical events (e.g., impact) may not be captured using standard video. Careful thought must be given to which equipment would be most beneficial, considering the event being recorded.

Planar Video Analysis

As discussed in chapter 14, video images recorded in a planar format are commonly used in qualitative analysis. The 2-D commercial software systems mentioned in the previous chapter may be used to obtain quantitative data from video.

Technical Requirements and Calibration

Digital video cameras should be selected that have the capacity to manually adjust shutter speeds, white balance and focus settings. This will allow fast movements, such as the leg in a kick or the hand in a karate strike, to be captured clearly, without blurring. It is important to remember that depending on geographical region, cameras record at only 50/60 fields per second (PAL versus NTSC format). However, high-speed video cameras such as Phantom, Redlake and NAC, to mention only a few, will permit an increased number of images to be recorded each second and will allow for greater opportunity to capture key features of interest.

A scaling factor must be placed in the line of motion to enable data to be converted from image to actual size. This known distance should *not* be placed in front of or behind the "line of travel or movement", as this will introduce error into the analysis.

Data

Two-dimensional data typically include the following:

- Displacements—for example, the stride width in a golf drive
- Joint angles—for example, the knee joint angle in a gymnastics manoeuvre
- Segment angles—for example, the forward trunk lean in a sprint start relative to an absolute vertical or horizontal
- Speed—for example, hand speed of a quarterback throw in American football
- Derived measures—for example, joint or segmental rotational (angular) speed, which may be calculated in Microsoft Excel

It is also possible to draw or graph these measures over the top of the image to assist understanding of performance. The image in figure 15.3 is an example of how one may draw on a screen to assist coaches' understanding of foot motion in two types of running shoes. The simple addition of a grid screen behind the golfer (which is possible in all the 2-D video systems discussed in chapter 14) may greatly enhance a player's appreciation of head movement throughout the stroke. Furthermore, split-screen or overlay images may be used to compare performers at different levels or to compare technique over time (see figures 14.4 and 14.5).

Where applicable, all of these data may be obtained in each plane of motion of the body.

Sagittal Plane

The sagittal plane is the view commonly used for qualitative analysis. With the camera positioned perpendicular to the plane of movement, images are recorded from a "side-on position". In baseball pitching, for instance, front knee angle, stride length and angle of the trunk to the ground are examples of measures that can be taken from the sagittal plane view. As a rule of thumb for straight-line motion, flexion-extension angles are generally recorded in the sagittal plane (e.g., knee flexion in running).

Frontal Plane

Cameras positioned either in front of or behind the line of movement may also collect data that is of value in coaching and teaching. In the example of the baseball pitcher, the position of the front foot with reference to the back foot (stride width) and the trunk lean to the side are examples of motion in this plane. As a rule of thumb for straight-line motion, abduction-adduction angles are generally recorded in the frontal plane (e.g., hip abduction in running).

Figure 15.3 Two-dimensional rearfoot angle of a runner wearing two types of running shoes.

Transverse Plane

The transverse plane is the view from above (or from underneath). Cameras placed on large tripods or positioned overhead on a gantry may also be used to record the rotation of the shoulders (trunk) in the baseball pitching action. Elliott and Khangure (2002) used this approach to measure the rotation of the shoulder alignment in cricket fast bowling from both research and coaching perspectives. As a rule of thumb for straight-line motion, trunk, shoulder and hip rotations are generally recorded in the transverse plane (e.g., pelvis rotation in running).

> 2-D motion analysis is most applicable when the variables of interest in a selected movement are in one plane of motion.

Three-Dimensional Motion Analysis

Three-dimensional reconstruction requires at least two cameras; researchers using opto-reflective systems usually position 6 to 12 cameras around the laboratory. Most often 16 to 20 known points are used to solve the direct linear transformation equations employed for reconstruction of 3-D displacement data. All commercially available motion analysis systems come complete with proprietary calibration equipment and algorithms to solve these equations.

Technical Requirements and Calibration

At least two cameras are needed for 3-D reconstruction, although three or four are preferable. These are obviously positioned in a manner that best permits the landmarks to be tracked at all times. The high-speed video cameras used in 3-D reconstruction are the same as those used for 2-D analysis. Figure 15.4 shows the device used to calibrate the 3-D area involved in the movement to be analysed.

Figure 15.4 Three-dimensional calibration frame.

Those working in laboratories that use opto-reflective systems typically position 6 to 12 cameras around the laboratory in a manner that best tracks the markers used in the analysis. The MX40 camera used by ViconPeak can record 10-bit grayscale using 2352 × 1728 pixels, with capture speeds of up to 1000 frames/s. Figure 15.5*a* shows a subject ready for testing using an opto-reflective approach, while figure 15.5*b* displays a 3-D reconstruction of the markers. This approach has been used to calculate 3-D elbow angles in cricket fast bowling. Figure 15.6 shows a typical image linked to a graph that would enable a coach to consider the bowling action with elbow angle over the section of the delivery he or she is interested in.

A minimum of six noncollinear points are needed to solve the direct linear transformation equations commonly used in sport biomechanics to reconstruct 3-D displacement data. However, typically 16 to 20 known

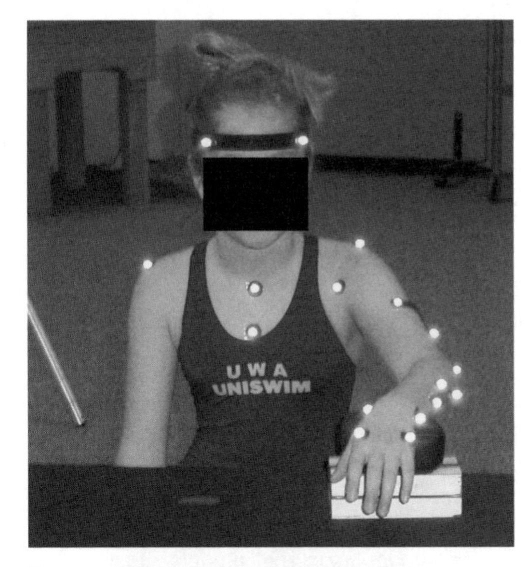

a

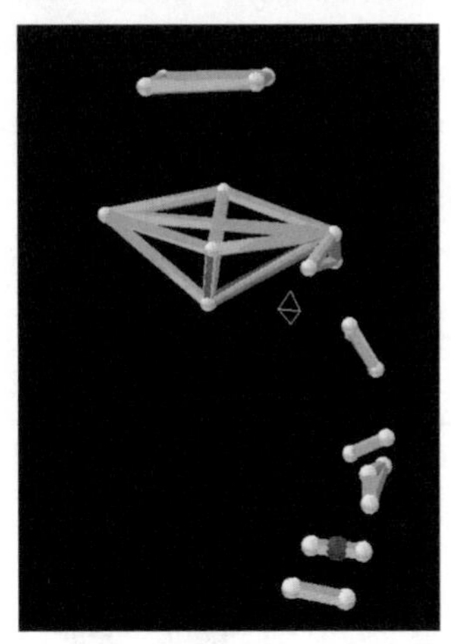

b

Figure 15.5 *(a)* Subject with opto-reflective markers attached and *(b)* resulting 3-D reconstruction of markers.

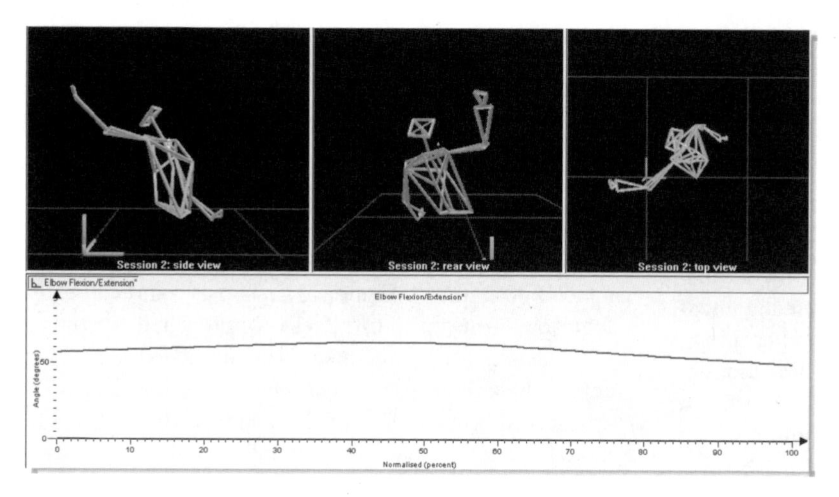

Figure 15.6 Report prepared in Vicon Polygon showing time-synchronized images from all three planes with 3-D elbow flexion-extension angle graph.

points are used depending on the intricacy and range of motion of the activity. These known points must cover the movement space to allow collection of accurate data (Challis and Kerwin 1992; Wood and Marshall 1986). "Calibration structures" either are commercially available from the manufacturer or may be custom built. The latter is often necessary for nonstandard analyses such as kayak sprinting analysis (Ong et al. 2006); in that case the structure needed to be large (6 m × 1.2 m × 2.0 m [6.7 × 1.3 × 2 yd]), capable of being floated on water and finally anchored to the bottom of a river.

Data

The data include all 3-D displacement and angle data previously discussed in relation to 2-D. Long-axis rotations (e.g., upper arm and thigh internal-external rotation, forearm pronation-supination) may be recorded with accuracy if more than three markers are placed on relevant segments. The interesting challenge with the creation of "situation-specific" models is that nonattached segments may be compared. For instance, a separation angle showing the difference in rotational angle of the shoulder alignment compared with the hips is generally of interest to coaches or teachers of hitting and throwing types of skills.

Three-dimensional kinetic data (moments and forces) are calculated using one of two inverse dynamic approaches, as follows:

- *Closed approach:* In this situation, force recorded generally from a force platform in the floor is combined with kinematic data of the lower limb recorded using an opto-reflective or video-based system. In this way, provided that the application point of the force to the foot is known, kinetics can be calculated about the ankle, knee and hip joints using Newtonian mechanics. In cycling, moments and forces about the joints of the lower limb can be calculated if the pedal is instrumented in 3-D to provide the application of force to the foot.
- *Open approach:* In this situation, the forces at a freely moving end point (e.g., the hand or a free-swinging leg) are not known. The mass and moment of inertia of the segments of the upper limb, together with their kinematics, may then be used to calculate the moments and forces at the wrist, elbow and shoulder. This open approach is more prone to error when compared with the closed approach where the input force is directly recorded, as one must rely on acceleration data to estimate force in open approach calculations. If a person has an implement in the hand, this must be introduced to the system as part of the hand (e.g., a ball) or as a segment with its own characteristics (e.g., a javelin).

These kinetic data may also be measured in 2-D if the motion analysed is planar.

Data Filtering

While all data include system and marker digitising error (due to reconstruction or manual digitising), it is beyond the scope of the chapter to cover this topic. Winter (1990) and Challis (1999) present succinct reviews of this area. However, it is important to state that this procedure is essential in the accurate calculation of derivative data. Another important point is that in analysing movements that include an impact (e.g., football kick or tennis forehand), one must also consider how to treat the impact in order to obtain valid data. One method that we have found to be acceptable for a punt kick and a tennis serve is to extrapolate data from one frame prior to five frames after impact (~0.024 s) using a spline routine. In this way, the influence of the impact is removed and the general shape of movement is retained.

Simple smoothing routines may also be incorporated into spreadsheet packages such as Excel, in which 2-D or 3-D data can be filtered and graphed.

Summary

Two-dimensional video analysis linked to a computer with the appropriate software will always be the system of choice for most coaches. However, coaches should now be in a better position to understand the limitations of 2-D analysis while also appreciating how 3-D systems may be used to improve their coaching knowledge. The choice of system will always be guided by the question that needs to be addressed.

sixteen

Application of Biomechanics in the Improvement of Sport Performance

Bruce C. Elliott, PhD; and Damian Farrow, PhD

Coaches who understand the mechanical base of a skill, who can analyse movement and who are also able to communicate with their athletes provide the best opportunity for optimal development with minimum risk of injury. It is generally accepted by coaches that successful high-level performance from a technique perspective is dependent upon mastery of a number of critical factors. Chapter 14 outlined analysis procedures promoted by sport scientists and used by coaches and teachers to observe motion and apply the most appropriate intervention. The concept of a *range of acceptability* for each of these key mechanical variables, identified in the analysis procedure, provides teachers and coaches with a clear start point in the modification of technique.

For example, in the tennis service action, descriptive research has identified the following as characteristics that are common to high-performance players.

- Approximately 120° ± 20° of leg flexion (included angle) during the preparatory movements (that is, at least 60° of flexion)
- An upper arm to trunk (abduction) angle of 100° ± 10° at impact
- An impact position in line with the front heel ±20 cm (±8 in.) for a power serve

This chapter discusses how technique is modified using such data as a basis for change. A performance environment in which the potential for injury is reduced is essential at all levels of sporting development, and this topic is also presented later in the chapter.

Instructional Approaches in Technique Modification

Coaching a new technique or modifying an existing technique is one of a coach's primary responsibilities, and feedback and the use of practice are the two major tools at the disposal of coaches to achieve this aim. This section provides some general guidelines regarding the content and frequency of feedback and the influence of practice organization on skill learning.

Feedback

Athletes crave feedback from their coach when learning or refining a new skill, and certainly research has demonstrated that feedback enhances learning. However, while many believe that too much is never enough, there are some guidelines that should be followed to promote effective learning, which in some instances requires coaches to simply say nothing.

- Feedback generally occurs as a natural consequence of completing a movement. For example, a golfer can see the path of the ball. However, in many instances, more detailed information such as the length of a shot or, importantly, the club position at the top of the backswing and at impact is not immediately available. Hence, coaches must consider what elements of the skill they are going to provide feedback about. Remember, the learner should be able to act on the feedback provided.

- Providing feedback after every practice trial has a beneficial short-term influence on performance but is detrimental to longer-term learning. Learners tend to become reliant on receiving feedback from their coach or some external source such as a video rather than relying on their own intrinsically produced feedback. This obviously becomes problematic in the performance environment, where players must problem solve for themselves.

- The frequency of feedback should decrease as a player's skill level improves. To counteract the problems with feedback frequency, a number of strategies can be employed. *Summary feedback,* as the name implies, requires the coach to provide feedback after a predetermined number of practice attempts. This approach, which has the advantage that the performer is not overloaded with information, ensures that the feedback is based on a range of practice attempts rather than reactions to the completion of every individual practice trial. *Bandwidth feedback* requires the coach to provide feedback only when the movement or skill component of interest falls outside of some predetermined bandwidth. The advantage of this approach is that as the athlete's skill level improves, feedback is faded naturally. Then, as learning progresses, the coach can set a narrower bandwidth if required.

- As skill level improves, the precision of the feedback provided can be increased. Even though skilled performers are usually able to act upon their own intrinsic feedback, it is still sometimes necessary for the coach to provide precise information to further refine a technique.

For more detailed reviews of these concepts, see Magill 2001 or Swinnen 1996.

There are many information sources that can be used as feedback. While traditionally, verbal instructions have been the predominant feedback choice of coaches, technological innovations have offered many alternatives. The use of video feedback to show performers their movement patterns has been prevalent since video cameras became readily accessible. In more recent times, software programs have allowed coaches to display or overlay two video clips so that a direct comparison can be drawn between the movement patterns of an expert performer and someone lesser skilled (see chapters 14 and 15). While this is a valuable tool in allowing the learner to distinguish between the two actions, it is rare that learners can or should directly model their technique on someone else's (Bartlett 1999).

In the sport environment, the desired skill outcome is usually the coordinated interaction between a number of body segments. Feedback information derived from movement kinematics such as displacement and velocity profiles has been trialled, with results generally supporting the use of such information over and above other feedback methods (Swinnen et al. 1993). For example, a recent development is the use of video goggles that allow a rower to view, in real time, an overlay of stroke force and boat velocity. While technologically exciting, such tools are not immune to the general feedback guidelines outlined previously. Importantly, the skill level of the performer should dictate whether more complex kinematic information of this kind should be presented. Likewise, there is mixed evidence supporting the use of concurrent feedback, as it has the potential to draw the performer's attention away from the critical task characteristics (Lintern 1991).

Remember to provide positive biomechanical feedback as part of this process. Athletes need to hear that they have responded in an appropriate manner through modification of their technique.

Practice Considerations

One of the primary objectives of a coach is to maximize the available practice time. It is common to organize practice of more than one skill in a given training session. At the conclusion of the session the coaches also want to feel confident that the practice schedule generated as much skill improvement as possible. Sport scientists have investigated this issue for some time, specifically examining the effects of a theory known as *contextual interference.* Contextual interference refers to the interference that results from performing various skills within a practice session. In simple terms it can be equated to the amount of mental effort a learner is required to use when practicing a sport skill. Generally, the greater the mental effort a learner uses, the better the learning of the practiced skill (Battig 1966).

Two methods of practice have been compared to examine the impact of contextual interference. High contextual interference usually involves random practice or a practice approach in which two or more skills or skill variations are practiced alternately in a session. For example, football training might involve performing a kick and then a handball, another kick and so on. Neither the kick nor the handball is practiced repeatedly by itself. Alternatively, low contextual interference may involve a blocked practice schedule in which one skill is practiced continuously before another skill is practiced. This is commonly seen at a golf driving range, where a player may hit a bucket of balls at a specific target using only one club. Research has shown that blocked practice leads to better performance of the skills in the short term than random practice (figure 16.1). This would seem logical due to the ability of players to get into the "groove" on a given skill during the session. However, when the skills are examined over the longer term with the aim of determining whether the training performance

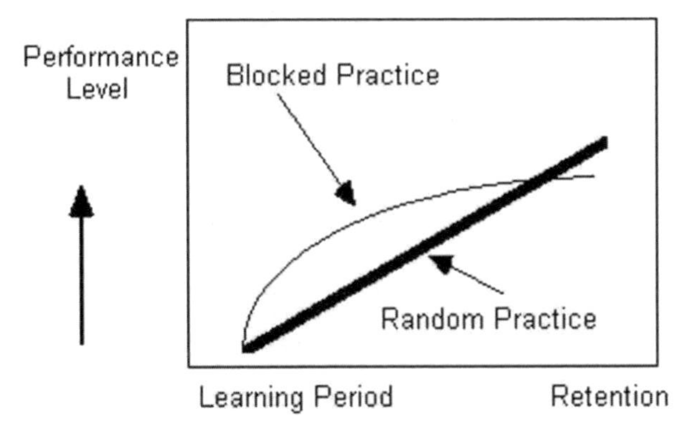

Figure 16.1 The contextual interference effect.

is permanent, it is observed that random practice produces improved retention or learning of the skill (Goode and Magill 1986; Hall, Domingues, and Cavazos 1994; Magill and Hall 1990).

While the majority of research supports the contention that random practice generates more learning than blocked practice, when the characteristics of a learner are considered, opinions are divergent. Some researchers argue that beginners, who have no experience or little skill in the tasks to be practiced, benefit more from blocked practice than random practice. Beginners need the opportunity to get an idea of the movement and establish a basic movement pattern (Gentile 1972) before engaging in random practice. Blocked practice provides this opportunity as the learner can reinforce a desirable outcome or correct an error from the previous practice attempt without the interference of having to change to a totally different skill. However, when the related issue of movement variability is considered, the evidence is not so clear.

Traditionally, coaches like to minimize the amount of movement variability displayed by their players; hence they prefer a blocked practice approach. However, research suggests that one should encourage movement variability when people are learning a sport technique (Handford et al. 1997). Technique variability can be seen as a logical means by which the body organizes itself to deal with complex environments where functional movement solutions are a necessity. Indeed, some have argued that the unorthodox yet extremely functional batting technique of Sir Donald Bradman (a famous cricket batsman) was the result of a more variable or random practice environment (Glazier et al. 2005). Bradman's often-cited practice of repeatedly hitting a golf ball with a cricket stump against a corrugated iron water tank is an excellent example of a variable practice activity. Hence it has been argued that coaches should consider a more aggressive approach to skill development and, instead of progressing systematically from a blocked to random practice approach, start with random practice approaches like that seen when children engage in backyard games (Williams and Hodges 2005). In such circumstances movement variability is usually a positive by-product.

> Feedback and the use of practice are the two major tools at the disposal of coaches to influence the development of technical skill. Coach-driven feedback needs to be faded as learning progresses so that athletes do not become dependent on its availability. Practice should encourage movement variability, particularly in early learning.

Armed with these procedures, the coach or teacher may then use a broad biomechanical approach to the way they modify technique. We next discuss this approach as used by athletes involved in high-speed activities.

Preparation of Athletes Involved in High-Speed Activities

An important characteristic in modern athletic performance is the development of explosive movements (see chapter 9 for the physiological development of this physical attribute). Athletes are challenged to improve various aspects of their skills so that they are able to perform movements such as the following:

- Hit or kick a ball with a higher velocity while still maintaining an acceptable level of control—this often occurs in such sports as tennis, golf, squash, baseball, soccer and rugby
- Generate higher limb velocities, which will in turn produce higher release velocities in throwing or in discus and javelin events
- Rotate segments of the body more quickly to apply more force to an implement, either with a bat, club, oar or pedal or in a medium such as water in which the body needs to be propelled

An outcome of many of these performance changes is that athletes in time-stressed sports such as tennis, soccer, baseball and cricket need to become biomechanists in their own right in order to "buy" themselves time to anticipate the likely outcome of the high-velocity movements just described. For example, it has been estimated that in a soccer penalty kick situation, the average time from ball contact to the time the ball crosses the goal line is approximately 0.6 s. In comparison, the movement time of goalkeepers, measured from their first movement until any part of their body crosses the flight path of the ball, is between 0.5 and 0.7 s. Like professional tennis players trying to return an opponent's first serve or a cricket batsman facing a fast bowler, soccer goalkeepers need to initiate their response before their opponent contacts the ball (Abernethy and Russell 1983). In this section we discuss the mechanical information sources available to a performer and, importantly, how coaches can use this information to improve a player's anticipatory capabilities.

Invariant Movement Kinematics: A Key Anticipatory Information Source

One source of information that elite players in time-stressed sports use to anticipate an opponent's intentions is that arising from the opponent's movement patterns. The biomechanical properties of most high-velocity sport skills ensure that performers must adhere to a relatively predetermined sequence of movements if they are to produce a biomechanically efficient action. For example, hitting a powerful tennis serve whilst maintaining an acceptable level of accuracy requires a player to achieve the coordination of the legs-hips-trunk-arm (kinetic chain) needed to effectively produce a high service speed.

This constraint on a server guarantees that at some point before racquet-ball contact, an opponent will receive invariant movement pattern information predictive of the forthcoming service direction and spin. Of equal importance is that in many instances, novice players do not appear to be attuned to the same information sources as experts (Abernethy et al. 2001).

A variety of "open" sport skills have been examined to determine the information usage patterns of expert and novice athletes. The importance placed on the specific biomechanical features of a particular movement pattern has meant that experimentation has typically sought methods to isolate the predictive or anticipatory role of the various skill components within a motor action. A prominent experimental methodology, *the temporal occlusion approach,* relies on the use of sport-specific dynamic visual images filmed from the perspective of a player during competition (e.g., a tennis server filmed from the perspective of the receiver). These images are then selectively edited to provide differing amounts of advance and ball flight information (figure 16.2). Within this approach, participants are typically required to predict the direction or depth (or both) of the opponent's shot from available information. A significant change in the prediction accuracy from one occlusion window to the next is assumed to be indicative of information pickup from within the additional viewing period.

For example, Abernethy and Russell (1987), investigating the anticipatory skill of badminton players, found

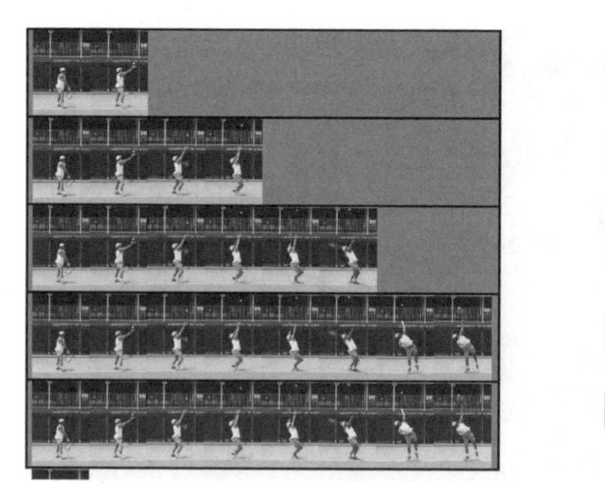

Figure 16.2 Progressive occlusion sequence for the tennis serve.

that the motion of the racquet and of the arm holding the racquet comprised information used only by expert performers; novices were reliant solely on the motion of the racquet as a source of anticipatory information. The use of the progressive temporal occlusion paradigm has repeatedly shown a visual-perceptual advantage in anticipation for expert players in comparison to lesser-skilled counterparts, across a variety of sports (for a review of this topic see Abernethy, Wann, and Parks 1998). While obviously the information sources differ between sports, the pickup of predictive movement pattern information from earlier time-windows within a perceptual display by expert relative to lesser-skilled performers remains a robust finding.

The temporal occlusion paradigm, reliant on "normal" video-based footage, provides an excellent means of identifying the key time-windows in which information pickup is likely to occur. However, it does not necessarily provide information to substantiate that a direct relationship exists between anticipatory skill and the kinematics of the action being viewed. Intuitively, the significant information pickup observed within earlier time periods for experts points to the use of more proximal information sources, thereby supporting the notion of proximal-distal kinematic information extraction. That is, larger proximal (core) muscle groups are recruited first, followed by recruitment of the smaller, faster-acting, peripherally located muscle groups. However, one cannot demonstrate conclusive evidence of a direct relationship between movement pattern kinematics and perceptual expertise when relying simply on normal video footage using the traditional temporal occlusion approach (Abernethy et al. 2001).

The use of a point-light display offers an alternative approach to providing more direct evidence concerning the link between perceptual expertise and the kinematics of the movement being viewed. The point-light display technique, most famously utilized by Johansson (1973, 1975), involves presentation of the motion of the joint centres of a performer with the normal display features (i.e., pictorial cues for form, size and orientation) removed (figure 16.3). Each joint centre is represented by a single point light that is disconnected and when static is completely ambiguous. However, Johansson demonstrated that only brief visual exposure to the motion of these joint centres was necessary for the type of motion (i.e., gait) to be identified. Abernethy and colleagues (2001) applied a similar methodology to the sport of squash in an attempt to isolate the key information (biomechanical) sources used by expert and novice players to predict the direction and depth of a variety of squash strokes. Findings revealed that the same information pickup and expert–novice differences occurred within the degraded point-light displays as in the normal video display, thereby supporting the existence of a direct link between perceptual expertise and the kinematics of the action being viewed.

Temporal occlusion is a common experimental approach used to examine the perceptual skills of athletes in time-stressed situations. The perception of invariant movement kinematics is a key source of anticipatory information in reactive sports. Anticipatory skill can be developed through the use of video-based training approaches that temporally occlude the action.

Anticipatory Skill Using Video

A logical extension to the research just described, aimed at identifying key anticipatory cues used by experts to help them return high-velocity movements, is to examine whether these perceptual skills can be developed in lesser-skilled players. One of the most promising methods for developing this anticipation is via vision-based perceptual training. Such training requires players to learn the essential link between early cues (e.g., arm motion) and eventual event outcome (e.g., tennis service direction). In a growing number of instances such approaches have successfully improved the perceptual speed or accuracy (or both) of sport performers (e.g., Abernethy, Wood, and Parks 1999; Farrow et al. 1998; Williams, Ward, and Chapman 2003).

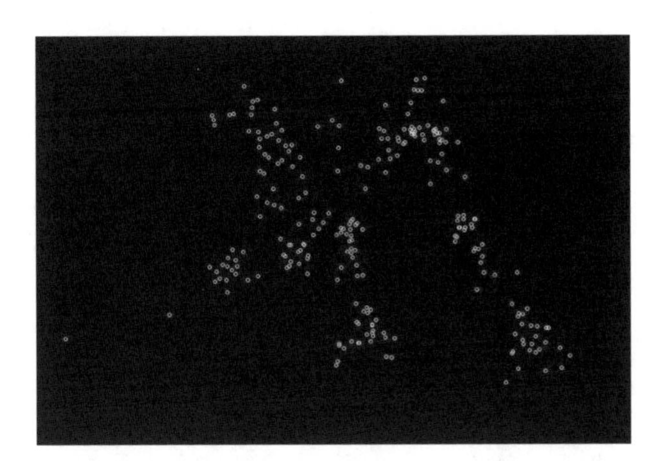

Figure 16.3 An example of the use of the point-light technique. While it is difficult to resolve the image in this static form, when presented dynamically it becomes easy to identify as a hockey drag flick.

Similar to the temporal occlusion methodology, vision-based training generally involves the presentation of a player performing a particular action from the opponent's usual viewing perspective; this presentation is then edited at a point just before the occurrence of a particular cue. Participants have to respond by predicting the outcome of the full play sequence. For example, players watch a tennis server from the receiver's perspective; and as the ball toss reaches its highest point, the vision is occluded or paused so that no more cues are provided. The participants then report their estimate of the direction and spin of the serve. Feedback is then given on that predicted outcome either verbally, or more typically, by having players view the postocclusion action.

Training activities have often involved a combination of temporally occluded video footage of particular strokes (i.e., tennis serve and field hockey flick) and specific instruction (guidance) concerning the relationship between the various perceptual information sources and subsequent ball flight direction (e.g., an instruction such as "a ball toss placed over the server's head indicates a topspin service is likely"). Interestingly, more recent work has examined whether players can *implicitly* (unconsciously, without direct instruction) pick up the key anticipatory information available (Farrow and Abernethy 2002). Irrespective of the instructional approach used, improvements in perceptual skill have been demonstrated across a variety of high-velocity movements within both racquet and field sports (see Williams and Ward 2003 for a comprehensive review of this topic).

Biomechanical Considerations in Reducing Sporting Injury Rates

Unfortunately, it may be that large-impact injuries are all but inevitable. Nevertheless, a few studies have shown that proprioceptive neuromuscular facilitation training can reduce the incidence of certain types of high-impact injuries. Overuse injuries, on the other hand, can be prevented or their incidence decreased with the use of a careful approach to training and performance. These injuries generally result from one of two types of loads: those associated with a high number of repetitions of an activity (e.g., in swimming, fast bowling in cricket, netball, basketball) and those associated with a smaller number of repetitions of an activity that requires high levels of skill or technique and power (e.g., gymnastics, jumping, javelin throwing). In both cases, general guidelines will help coaches keep athletes free of injury to the extent possible.

Impact Injuries

It is extremely difficult to protect the body from large-impact injuries that are likely to overload human tissue. However, a number of papers have shown that it is possible to significantly reduce the number of anterior cruciate ligament injuries in soccer and handball through proprioceptive neuromuscular training (Caraffa et al. 1996; Myklebust et al. 2003). This training supposedly prepares the knee such that cocontracture occurs before and during impact with the ground, thus protecting the ligaments from injury.

Similarly, Besier and colleagues (2001) reported that unanticipated cutting manoeuvres "loaded" the knee more than preplanned movements. That is, movements such as a side step performed without adequate planning may increase the risk of noncontact knee ligament injury due primarily to increased varus-valgus and internal-external rotation moments at the knee. Again, proprioceptive training was an effective means of preparing the knee for these unanticipated loads (Cochrane et al. 2003). Specific forms of impact injuries, such as the debilitating knee injury, may therefore be reduced through a carefully planned training program (see also chapter 12).

Overuse Injuries

Greater participation by children and adolescents in highly organized sport has resulted in an increased number of injuries, particularly of the overuse type. These are generally due to loading of the musculoskeletal system whereby a number of repetitive forces, each lower than the acute injury threshold for any tissue, produce a combined fatigue effect in that tissue over time. An important consideration in the cause of sporting injuries is the load on the body during training or performance; a diagram of the principal factors influencing load in sporting movements is shown in figure 16.4. Key contributing factors to overuse injuries include "poor technique", an

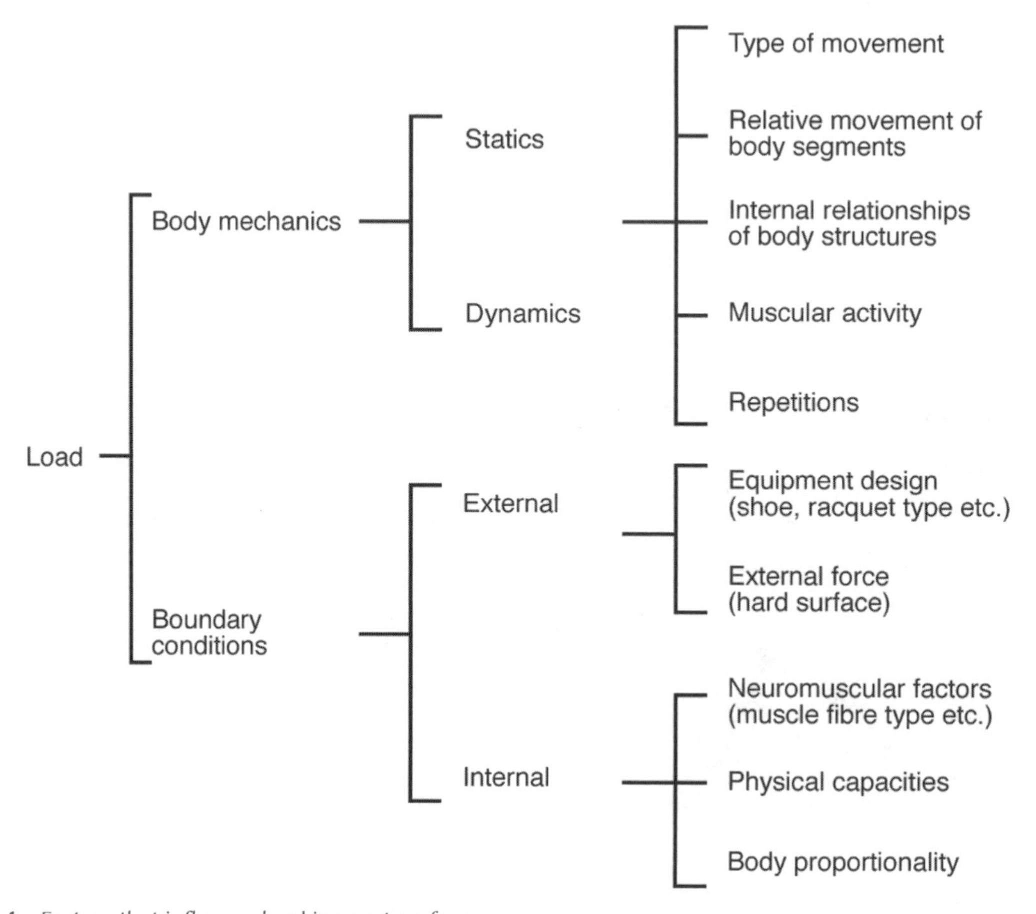

Figure 16.4 Factors that influence load in sport performance.

Adapted from B. Nigg et al., 1984, Load, sports shoes and playing surfaces. In *Sport shoes and playing surfaces*, edited by E. Frederick (Champaign, IL: Human Kinetics Publishers).

excessive number of attempts at an activity, musculoskeletal immaturity, a body build that may predispose an athlete to injury in a particular sport or activity or some combination of these factors.

Sports success often requires high levels of mechanical loading to perform optimally. Coaches and teachers must then "train to play" and manage these high loads so that optimal performance can be achieved in a relatively injury-free environment.

The potential of a specific activity or technique to cause an overuse injury is best determined by epidemiological and prospective studies. In epidemiological studies, injury statistics are carefully assessed following a change in technique, equipment, playing surface or the rules of play. In prospective studies, factors such as the mechanical characteristics of a given technique or the physical capacities of athletes (or both) are measured prior to the commencement of the season, and any resulting injuries are then related back to these characteristics. In general terms, studies investigating sporting injury may have four phases (van Mechelen, Hlobil, and Kemper 1992):

1. Determining the extent and severity of the injury, or the *epidemiology,* to ascertain whether or not further work is warranted
2. Developing research questions that will enable determination of the cause, or the *aetiology,* of the injury, normally through prospective studies
3. Developing an educational, community-based program to inform the population about how to avoid the injury—*education*
4. *Reassessing* the incidence or severity of the injury across a population to determine if the educational program has worked

Finch (2006) proposed a six-step process (Translating Research into Injury Prevention Practice, or TRIPP) that followed a similar structure to that above. The principal difference was that more emphasis was placed on the evaluation of the intervention strategies.

As such studies are difficult to administer with high-performance athletes and are also costly to run, biomechanists often use the forces acting on the body during a particular performance as an indicator of the potential for overuse injury. It must be understood, however, that external forces (which are usually measured using a force platform; see chapter 14) of a given magnitude and direction may produce different internal forces, at least in part because of the morphology of the athlete. That is, the athlete with a flat foot arch would generally experience greater forces in the Achilles tendon than would an athlete who has a high foot arch with equivalent external force acting on the ball of the foot. The higher arch is thus able to absorb the external force more effectively than a low arch.

The loads experienced in sport are generally of two types. The first is associated with a high number of repetitions of an activity, as in endurance training or fast bowling; in these activities, forces from the ground (ground reaction forces, GRF) and associated turning forces (torques) must be absorbed by footwear and the body at each foot strike. Similarly, in swimming, the high number of upper limb rotations places stress on the shoulder region. The second type of load is associated with a smaller number of repeated performances requiring high levels of skill or technique and power, as in gymnastics or in jumping; in these activities, much higher forces are usually generated than in the activities in the other category.

High-Repetition, Moderate-Force Activities

The type of overuse injury to muscle, tendon, bursa or bone associated with high-repetition, moderate-force activities is generally caused either by frictional forces (as in swimming) or by repetitive loading of a high number of exposures of the body to a force. Examples of these repetitive loading activities include fast bowling in cricket; running, both as an activity in itself or as part of a training program for another sport; netball and basketball, in which a player is required to land on a hard surface after jumping and catching a pass; and pitching or throwing for speed in baseball, in which large internal muscle forces are needed to produce very fast movements of body segments.

- **Swimming:** The potential for overuse injuries in swimming, particularly in the shoulder region, is obvious, as swimmers may rotate each arm 2500 times while completing 5000 m of freestyle. Tenosynovitis of the long head of the biceps brachii tendon or of the supraspinatus muscle attachment, as well as inflammation of the subacromial bursa ("swimmer's shoulder"), can result from repeated upper limb movements, particularly during freestyle and butterfly swimming (Fricker 1992).

- **Fast bowling in cricket:** Fast bowling is an impact activity in which the bowler experiences a series of "collisions" or minor impacts with the grass surface in the run-up. This is followed by two larger impacts during landing on the rear foot and then on the front foot on a very hard turf or concrete surface during the delivery stride. Peak vertical GRF of approximately five times the bowler's body weight (BW) have been recorded when the front foot is planted. These impact forces are transmitted through bones, cartilages, tendons and muscles to the joints of the body (Nigg et al. 1984). Stress fractures in the spine (primarily to the L4-L5 region) are the most serious injury a fast bowler can sustain, although injuries to the body musculature and joints, particularly of the lower limb and back, are also commonplace (Foster et al. 1989).

- **Walking, jogging, running:** While all of these movement patterns may stress the cardiorespiratory system, differences are apparent in force data recorded during the stance phase. Walking (≈1.4 BW) typically produces lower peak vertical GRF than jogging (≈2.2 BW) and running (≈2.4 BW). Unfortunately, the internal forces associated with these activities can easily be a multiple of the external GRF; therefore, depending on the direction of the acting forces and the geometry of the locomotor system, the stress pattern on the body can be greater than the GRF would signify (Nigg 1986). However, shock-absorbing insoles were found not to be effective in reducing the incidence of lower limb stress fractures (Gardner et al. 1988), and no significant differences in short-term running injuries were recorded between high-, medium- and low-impact force peaks (Bahlsen 1989). The level of forces for these activities is one reason walking is a favoured activity of persons who are elderly.
- **Netball and basketball:** These ball games require players to continually accelerate and decelerate and to jump and land in an attempt to receive a pass or gather a rebound. It is not surprising that injuries occur, particularly in the lower limb. Steele and Lafortune (1989) reported that in netball, peak vertical GRF of 5.2 BW (heel landing) and 5.7 BW (forefoot landing) were recorded for high-performance players landing on one foot after catching a ball propelled to a point 20 cm (8 in.) above the head. In basketball, peak vertical GRF of 2.3 to 7.1 BW have been recorded on landing following a rebound (Valiant and Cavanagh 1985). The majority of injuries in these sports are the result of landing in an inappropriate manner; thus many may be classified as either an impact injury, if contact occurs with another player, or an overuse injury.
- **Throwing:** Studies by Adams (1968) and by Torg, Pollack, and Sweterlitsch (1972) on young pitchers have clearly shown that overuse injuries are very apparent in baseball pitching, the most common being "elbow" soreness and separation of the medial epicondylar epiphysis. If a high number of repetitions are combined with poor technique, an overuse injury is almost inevitable (English et al. 1984). The underlying mechanism for injury primarily to the elbow region appears to be related to the large forces associated with the forward-swing phase of the upper limb during the pitching action, in conjunction with the position of the forearm with respect to the upper arm during this phase. Very high moments are needed to cause the rapid extension of the forearm at the elbow joint, a characteristic of pitching and high-velocity throwing (Gainor et al. 1980).

General guidelines for the prevention of overuse injuries in high-repetition, moderate-force activities are as follows:

- Have players warm up prior to and cool down at the completion of practice and competition.
- Include specific flexibility and muscle endurance training as part of the general program.
- Emphasize good technique.
- Ensure that increases in training load or training location (hill running; change from aerobic to anaerobic conditioning) are introduced gradually.
- Vary training to place the load on different areas of the body.
- Have players practice landing skills while catching a ball (if appropriate to the sport) to develop *kinaesthetic sense* with relation to where the body is in space prior to and upon landing.
- Carefully monitor the number of repetitions (bowling, throwing) both in practice and during competition.
- Make sure that athletes use footwear appropriate to the playing or training surface.

Moderate-Repetition, High-Force Activities

In overuse injuries from moderate-repetition, high-force activities, the high forces associated with each performance limit the number of repetitions that can be performed if injury, particularly to the musculoskeletal system, is to be avoided. With these activities, coaches must develop highly specific lead-up drills that enable athletes to practice selected aspects of the final performance without exposing themselves to the potential for injury. Safety precautions, including the use of specialized equipment associated with each of these activities, must be strictly adhered to.

- **Gymnastics:** The magnitudes of takeoff and landing forces clearly show that extremely high forces are associated with gymnastics activities. Peak single-limb vertical GRF at takeoff for a running forward somersault of 13.6 BW (Miller and Nissinen 1987) and between 8.8 and 14.4 BW on landing after a double back somersault (Panzer et al. 1988) have been recorded. Bruggemann (1987) recorded peak vertical takeoff forces of between 3.4 and 5.6 BW for a back somersault following a round-off, which translated into much larger internal forces in the Achilles tendon. While it has been shown that asymmetrical landings produce higher forces than symmetrical ones, there were only minor reductions in GRF in landings with a flexed leg compared to a competition landing with the legs fully extended (Panzer et al. 1988). Therefore, if injuries are to be avoided, particularly those that may disturb normal growth, it is imperative that correct equipment, such as an appropriate floor and good matting, be used for teaching gymnastics.
- **Jumping activities:** Ramey (1970), in a study of the long jump takeoff, recorded peak vertical GRF of almost

7 BW for a jump of only 4.2 m (4.6 yd). These GRF increased to levels between 7 and 12 BW in the takeoff phases for the triple jump (Ramey and Williams 1985). In a high jump of approximately 2.2 m (2.4 yd), peak vertical impact forces of between 8.4 and 8.9 BW and peak horizontal retarding impact forces of 5.6 to 6.5 BW (Deporte and van Gheluwe 1989) show the very high loads associated with this activity. It is therefore not surprising that athletes involved in jumping activities or physical education classes emphasizing jumping must be very careful to avoid overuse or even misuse injuries.

> Most sporting injuries, irrespective of whether they are impact or overuse in nature, have a mechanical cause. The coach or teacher in attempting to achieve optimal performance must always be aware of the loading associated with performance.

- **Javelin throwing:** A mean peak vertical GRF at front foot impact of 9 BW for javelin throwing (Deporte and van Gheluwe 1989) showed that this activity must not be repeated excessively if injury to the lower limb or back is to be avoided. The high-velocity requirement of the upper limb segments prior to javelin release also means that high internal forces are acting and that repeated maximum efforts should be avoided.

General guidelines for reducing injuries in moderate-repetition, high-force activities are as follows:

- Ensure that athletes carry out a thorough warm-up and cool-down, including both flexibility and moderate power activities, prior to and after performance.
- Stress that good technique must be used at all levels of performance.
- Carefully monitor the number of repetitions both at practice and during competition.
- Physically prepare athletes for the selected activity.
- Vary training so that emphasis is placed on different skills and activities throughout each session.
- Have athletes perform in shoes and on surfaces that assist in the reduction of the impact forces at ground contact.

Prevention is the key to the alleviation of the majority of overuse injuries, because in many instances they can be avoided or reduced if a sensible approach to training and performance is adopted. When symptoms of overuse injuries such as continual pain are reported, teachers and coaches must evaluate the athlete's total program. The technique being used must be analysed, the physical preparation evaluated and the intensity of training or play reconsidered. An effective coach or teacher will of course have considered all these aspects of the program prior to the start of each season.

It may also be prudent to advise a young athlete to undergo a specific medical examination prior to making a total commitment to any sport that requires a vast increase in training. Furthermore, athletes with segment malalignments and inappropriate body dimensions (discussed in chapter 6) may then follow a program designed to prepare the body for the forces expected in their sport. Today, a high degree of perfection in performance is needed for competition at the international level, and therefore long hours of practice are a normal phenomenon of modern training. It is hoped that a greater understanding of the forces that the body must absorb during various activities will enable coaches to allow all athletes to reach their full potential in an environment as free of injury as possible.

Summary

The role of biomechanics when linked with motor learning in improving sport performance should not be underestimated. A knowledge of biomechanics is not only essential for technique development and modification in order to achieve excellence, but also important in helping athletes to avoid injuries.

seventeen

Mechanics in Sport

Specific Applications

Bruce C. Elliott, PhD; Timothy R. Ackland, PhD;
and Jacqueline A. Alderson, PhD

Chapter 14 provides the reader with a framework that may be used in the analysis of sporting movements, and chapter 16 addresses how biomechanical principles may be presented to athletes to improve their performance. This chapter considers the most important biomechanical principles (these form the basis of the mechanical analysis models in chapter 14) pertaining to a number of sports within the nine generic categories discussed in this book (table 17.1). General biomechanical concepts are presented and linked to specific sports from the table. It is not our intention to reflect what would typically be found in a biomechanics text, but to further stimulate a student's (or athlete's) appreciation of biomechanics from a primarily qualitative perspective. Remember, it is generally the application of these principles that is modified during the intervention phase of motion analysis as discussed in chapters 14 and 16. More detailed presentations of biomechanics in sport can be found in texts such as *Biomechanical Basis of Human Movement* (Hamill and Knutzen 2003) and *Fundamentals of Biomechanics* (Knudson 2003).

Interventions related to the biomechanical principles presented, within an endeavour to improve performance, may take a number of different approaches. Generally in sporting movements these may be classified as follows:

- **Technical:** This is the approach commonly used by coaches and teachers, in which selected aspects of movement are modified from a technical perspective.

Table 17.1 Generic Grouping of Sports With Respect to Biomechanical Principles

Generic sport groups	Subsections	Specific sports discussed
Racquet		Tennis, squash, badminton
Aquatic		Swimming, water polo, synchronized swimming
Gymnastics, power	Gymnastics	Artistic gymnastics, diving
	Power sports	Weightlifting, powerlifting
	Rowing and paddling	Rowing, canoeing, kayaking
Track, field, cycling	Track	Running
	Field	Jumping and throwing
	Cycling	Track cycling
Mobile field		Hockey, soccer
Set field		Baseball, softball, curling, tenpin bowling
Court		Basketball, volleyball, netball
Contact field		American and Australian football, rugby, ice hockey
Martial arts		Karate, taekwondo, judo, kobudo, boxing

- **Physical:** The relevant physical capacity (e.g., strength, power, flexibility) is often required before a skill can be performed correctly. Improvement in performance will not occur until this capacity is developed to the appropriate level.
- **Tactical:** Drills with a specific tactical requirement are often presented to assist athletes in modifying their technique with a view to improve performance (see chapter 1).

The reader is challenged to think about how these approaches may be used to modify the mechanical variables discussed in the examples provided in this chapter.

Balance

In sporting movements an athlete needs to have enduring balance *(static stability)* when the skill to be performed (e.g., shooting) requires a high level of stability. The archer practices coordinating the release of an arrow to the time the body is extremely stable, whereas the squash player moving quickly about the court, or the swimmer at the start of a race, needs to be in a position that is balanced only momentarily *(dynamic stability)*. That is, athletes who exhibit dynamic or momentary stability are on the verge of movement.

The position of the centre of gravity with reference to the base of support (height, anterior-posterior location and side-to-side location) will primarily determine the type of stability achieved. In general terms, an athlete lowers the centre of gravity, which is centrally based to achieve enduring stability but is positioned closer to the edge of the base of support to achieve momentary stability. The feet are normally positioned approximately shoulder-width apart as a start point in most sporting endeavours.

Centre of Gravity

An athlete's body weight is the product of the body's mass and the acceleration due to gravity. Centre of gravity is the balancing point of the body, the place that represents the weight centre of the athlete. It is the point where the sum of all the moments (or torques, as discussed later) equals zero.

Gymnastics

In many tumbling and vaulting manoeuvres, the centre of gravity is used as the point about which the body rotates (figure 17.1). Once a gymnast leaves the ground, the centre of gravity follows a parabolic path, and all rotations and twists occur about this point as the axis of rotation.

Figure 17.1 A gymnast rotating about her centre of gravity.

Martial Arts

In judo, players lower their centre of gravity, making them more stable and therefore more difficult to rotate. Furthermore, by adopting a wide stance during grappling, the judo player increases his or her static stability. Similarly, by spreading the feet long and wide in forward stance, the martial artist increases the base of support and improves stability. The defensive "cat stance" adopted in many forms of martial arts is generally quite narrow for open-hand sparring (for dynamic stability), but needs to be longer and wider (increasing the base of support) when the player is using a heavy weapon such as the bo in kobudo.

> Thinking about how balancing, both static and dynamic, develops from childhood to adulthood is important in understanding how balance encourages athletic performance.

Tennis, Squash and Badminton

Court players move the centre of gravity forward by placing their weight on the balls of the feet. This brings the line of gravity close to the edge of the support base so as to enable quick movement in response to the requirements of the game.

Centre of Buoyancy

The supporting force of a body in water is termed buoyancy. This force acts upward at a point termed the *centre of buoyancy.* The centre of buoyancy is located superior to the centre of gravity (i.e., higher in the trunk) because of the larger volume of the upper compared with the lower body, as well as the entrapment of air within the lungs and trachea of the upper trunk.

Swimming

The centre of buoyancy is higher for males compared with females because of the greater muscle mass in the lower limbs of men. This causes the legs to sink more than for females, which has the potential to increase the drag force in males during swimming.

Synchronized Swimming

The many postures adopted in synchronized swimming require the positions of the centres of gravity and buoyancy to be constantly manipulated for optimal efficiency and the performance of skills. While the centre of buoyancy remains relatively fixed, strategic positioning of the arms and legs can significantly alter the location of the centre of gravity and hence the body's orientation.

Newton's Laws of Motion

Isaac Newton's laws of motion deal with the relationship between force and motion. These laws are known as the laws of inertia, acceleration, and action-reaction. Each is relevant to many actions that take place in sport, as the following examples suggest.

First Law

Newton's first law states, *A body continues in its state of rest or uniform motion in a straight line, unless acted upon by an external force.* This law refers to creating motion through application of an internal (e.g., muscle) or external force (e.g., golf club hitting a ball) or modifying previously created motion (e.g., gravity).

Running

In a track start, the athlete waits in the blocks *(momentary stability)* and applies forces primarily from the muscles about the hip, knee and ankle joints to drive the body from its state of rest. Gravity also rotates the body forward as the supporting hands are removed from the ground (figure 17.2).

Figure 17.2 With the hands lifted from the ground, the centre of gravity is outside the base of support, thus facilitating forward rotation.

Golf

In golf, the ball will remain on the tee (unless blown off by the wind) until impact is made by an external force applied by the club.

Ice Hockey

In ice hockey, the velocity applied to the puck is not as greatly reduced during its travel across the ice as would be the case when a ball rolls over the ground, as the friction between the surfaces is not as high.

Second Law

Newton's second law states, *A change of motion is proportional to the applied force and takes place in the line of action of the force.*

This law has two broad applications in regard to sport. One primarily concerns performance (the ability to produce acceleration; *force = mass × acceleration*), and the other concerns force dissipation *(force = change in momentum/time,* where *momentum = mass × velocity).*

Baseball

If the mass of a bat is constant, then the acceleration of the bat (and hence its speed at impact) is related to the force applied to the bat. Consider the force application needed to perform a bunt compared with that required to rotate the bat very quickly and hit a home run (figure 17.3).

Figure 17.3 Increasing the speed of the bat is directly related to the force applied to the bat, and thus affects the end result (consider a bunt as opposed to a home run hit).

Jumping

In the high jump, the momentum of the athlete as she or he falls from the peak of the jump must be dissipated over as long a time as possible at landing to avoid injury (figure 17.4). A jumper landing on concrete (where momentum is stopped very quickly) must absorb a very high peak force over a minimal time; hence the likelihood of injury is high. Obviously, one prefers to land on a large "crash mat" (permitting a longer time over which momentum is slowed), thus reducing the possibility of injury. Remember, the impulse *(force × time)* is the same for both situations, whereas the rate of application and size of the peak force are the factors more related to injury.

Gymnastics and Judo

During the performance of advanced routines on the various apparatus, gymnasts risk falling from a great height, especially from the rings, high bar or uneven bars. Not only do the competition venues have padded floors and crash mats to help absorb the forces of impact, but gymnasts are trained to actively contact the ground after a fall to increase the impact surface area and reduce the peak impact forces. Judo players are similarly skilled in landing from a throw or takedown without suffering serious injury.

Figure 17.4 The "crash mat" allows the momentum to slow, thus diminishing the likelihood of injury to the athlete.

Third Law

Newton's third law states, *To every action there is an equal and opposite reaction.*

This is a very important concept that has relevance to almost all sporting actions. That is, the action of all dynamic movements of the body, such as throwing, hitting and kicking, causes a counteraction of the body as a whole. This "reaction" force needs to be countered or stabilized (e.g., via contact friction with the ground) in order for an effective dynamic action to result.

Tennis

In the serve in tennis, the player applies a "leg drive" to the court to create an equal and opposite force that causes the body to leave the ground for impact (added height is a great advantage in the service action). Similarly, a right-handed player running wide to return a forehand applies force to the court with the right leg to stabilize the body for effective stroke production and to "push" back into the court to prepare for the next stroke.

Softball

In softball, the ball applies a force to the bat, and an equal and opposite force is applied to the ball (some energy from the collision is lost during this transfer). Since the mass and moment of inertia of the bat predominate over the ball, the latter leaves the collision with great velocity. The bat, however, often continues to move forward due to the large momentum that was developed during the forward-swing phase, but its rotational velocity will decrease as a result of the collision with the ball.

Karate and Taekwondo

Powerful punching and striking techniques in karate and taekwondo require, among other elements, the development of high linear velocity of the impact point (knuckles, hand or fingers). Both these martial arts emphasize a strong countermovement of the nonstriking limb to enhance the punch or strike (figure 17.5). This countermovement serves to stabilize the body, which has a tendency to rotate in a direction opposing that of the punching or striking limb.

Linking of Linear and Angular Motion

Appreciating the relationship between linear and angular motion requires an understanding of both types of motion. Distance (scalar) and displacement (vector) are technically different measures; however, they are often used interchangeably in coaching. For instance, if an athlete runs one circumference of a 400 m track, returning to the start point, the distance covered is 400 m whereas the displacement is 0 m.

Figure 17.5 The countermovement of pulling the nonstriking limb back in martial arts serves to stabilize the body and enhances the striking action.

While speed (no indication of direction) is the rate of change in distance covered *(d/t)*, it may be considered at one instant (e.g., between two frames from a video) or as an average value over a given distance (e.g., 100 m travelled in 10 s = 10 m/s). Remember that both these variables have angular equivalents:

- Distance (m): angular distance (measured in radians or degrees)
- Velocity (m/s): angular velocity (measured in radians/s or degrees/s)

If one wishes to link angular motion and linear motion, one needs to multiply the *length of the lever* × *angular velocity of the segment* (ω). In tenpin bowling, one would then add the horizontal velocity of the shoulder (V_{sh}) to the length of the arm (L_{arm}) multiplied by the rate of rotation of the arm (ω_{arm}) to calculate the velocity of the ball at release, as follows:

$$V_{ball} = V_{sh} + (L_{arm} \times \omega_{arm})$$

However, for this calculation, the angular motion must be measured in *radians,* not degrees (simply divide the distance rotated in degrees by 57.3° to convert to radians).

Rowing

In rowing, while mean velocity over a race is the measure of success, the instantaneous velocity during the various phases of a stroke will give a coach the best indication of where to improve technique (see figure 14.12). As with many sports, such as breaststroke swimming, the most efficient stroke is one in which the level of velocity fluctuation is reduced.

Boxing, Karate, Taekwondo

To create a good striking force, the competitor in boxing, karate and taekwondo must rotate the forearm through an angle of ~90° in a minimal period to attain a high angular velocity for impact; this also provides a high linear velocity of the point of impact. As discussed in chapter 6, the linear velocity of the end point is determined by how quickly the lever rotates as well as by the length of that lever (body segment).

Cricket Bowling

The linear velocity of release is a combination of the velocity of the shoulder plus the length of the arm multiplied by the rate of rotation of the arm. This is why fast bowlers run into the delivery relatively quickly, are generally tall (with long arms) and rotate the delivery arm quickly to achieve a high release velocity.

When thinking about the linking of linear and angular motion, start the process at the shoulder for throwing and hitting skills and at the hip for lower limb based movements.

Projectile Motion

In projectile motion, a body leaves the ground as a result of jumping, or an implement leaves the ground by being thrown into the air. The primary forces that must be taken into account are gravity (which operates toward the centre of the earth at 9.8 m/s) and air resistance (which slows down progression; see later section on fluid dynamics). The following are other factors that affect the trajectory and distance covered:

- Projection angle
- Release or takeoff velocity
- Height of release or takeoff

Golf

In golf, the goal of the shot will determine the club to be selected. Long irons, for example, have a long shaft that serves to increase club head velocity at impact and a lower angle of the face for a flatter trajectory (e.g., compare a 3 iron with a 9 iron). Furthermore, when driving into a strong headwind, the golfer tees the ball lower in order to achieve a reduced trajectory to minimize the effect of the wind and maximize ground roll.

Jumping

In the long jump, the athlete takes off at ~20°, as this angle provides an appropriate trajectory while still allowing the jumper to maintain an optimal horizontal velocity to achieve the greatest horizontal distance (figure 17.6). The theoretical "optimal" takeoff angle of ~45° for projectile motion applies only when the takeoff and the landing heights are the same, which is not the case for long jumpers. Furthermore, in order to increase the takeoff angle, long jumpers would have to sacrifice too much horizontal velocity, thereby further reducing the final jump distance.

Basketball

In basketball, the player performs a jump shot to increase the height of release in order to avoid blocking of the shot by a defender. When at maximum height (vertical velocity close to zero), the ball is released at a relatively constant angle (~55°), with varying velocity depending on the distance from the basket.

Momentum

Momentum is the quantity of motion that a body possesses. The two types are linear momentum (the quantity of linear motion) and angular momentum (the quantity of angular momentum or rotation). An obvious example of linear momentum in sport is running; and various motions in diving and throwing are clear examples of angular momentum.

Linear Momentum

Linear momentum is a combination of mass and velocity. From Newton's second law ($F = m \times a$), one can derive the following:

$$F = m\,(V_f - V_i)\,/\,t$$

where F = force, m = mass, V_f = final velocity, V_i = initial velocity and t = time. From this equation the term impulse (Ft) is derived, which equates to the change in final and initial momentum values.

Figure 17.6 Taking off at the correct angle allows the long jumper to maintain an optimal horizontal velocity.

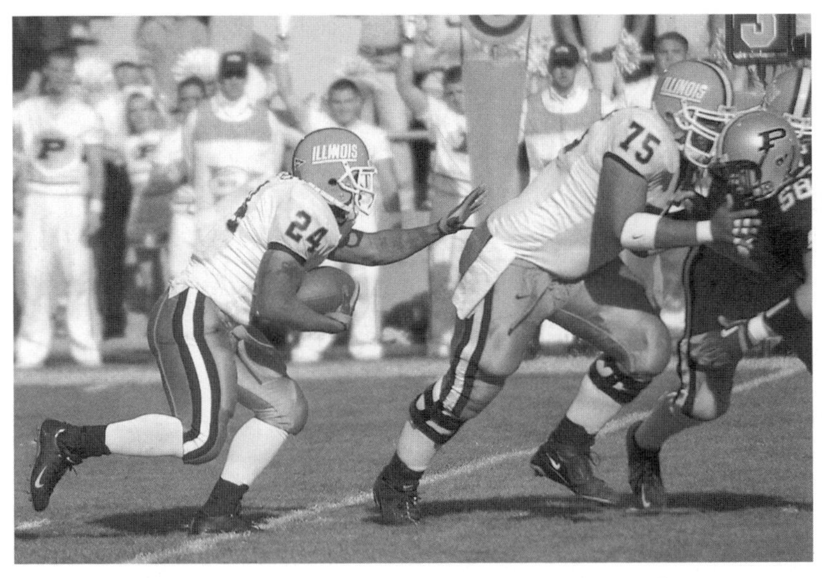

Figure 17..7 An athlete of substantial mass is naturally hard to bring to a stop, given that he is able to develop high velocity.

American Football

Running backs, who are large (with substantial mass) and are also able to develop high velocity, can produce a high level of momentum (figure 17.7). These athletes are naturally difficult to stop during a collision situation.

Weightlifting

In performing a snatch, a lifter initially develops vertical linear momentum in the bar, using the large muscles of the lower limb, before finally using the upper limb and trunk to complete the lift. The clean and jerk lift similarly requires the development of vertical linear momentum in the clean phase in order for the lifter to raise the bar to the chest.

Angular Momentum

Angular momentum may simply be calculated as

$$\text{Angular momentum} = I \times \omega$$

where I = moment of inertia (the angular equivalent of mass) and ω = angular velocity.

Moment of inertia is determined by where the mass is distributed and is an indication of how difficult an implement or body part is to rotate. For calculation of the moment of inertia, the distance of mass particles from the axis of rotation is squared (I = sum of all masses × distance^2). Consider swinging a hammer from the metal end (where most of the mass is located close to the axis of rotation) compared with swinging it from the handle (most of the mass is farther from the axis of rotation); in the latter case it is more difficult to rotate.

While angular momentum may be treated simply as a product of moment of inertia and angular velocity, in sporting activities the role of the angular momentum of different segments compared with the body centre must be acknowledged. Think of how a runner balances the trunk by having the two arms move in different directions. More information on this topic is available in the work of Hamill and Knutzen (2003) and other biomechanics texts.

Diving

The start point for various dives is, at least in part, dictated by the moment of inertia involved in the dive (tuck, pike or layout). That is, it is easier to complete a double somersault in the tuck compared with the layout position.

Tennis

The tennis racquet selected by a child should have a short handle to decrease the moment of inertia (swing weight) and thus make the implement easier to manoeuvre (figure 17.8). For a given angular momentum, a child will be able to rotate this shorter racquet with greater velocity.

Figure 17.8 Having a child use a racquet with a short handle decreases the swing weight so that it is easier for the child to manipulate the racquet.

Throwing Actions

In throwing, the upper limb is initially flexed so that its moment of inertia is low, thus making it easier to rotate. The elbow is then extended to increase the length of the lever near release to combine the high rotational velocity with a long lever for optimal release speed.

Soccer

The soccer player flexes at the knee joint to reduce the lower limb system moment of inertia and thus make it easier to rotate this system forward for impact. As with throwing, the knee is then extended to increase the length of the lever (and to add another segment to the motion).

Moment of Force or Torque

How do you rotate an implement or a body segment? To accomplish this, you need to apply a force away from the centre of gravity of the object or body segment. Moment of force or torque is therefore the product of *force* and the *perpendicular distance* from the line of action of the force to the axis of rotation. In coaching and teaching we must recognize that we can modify this internally (i.e., modify how a muscle force is applied to a bony segment such as the leg) or externally when a force is applied to an object such as an oar in rowing.

Cycling

The ability to rotate the crank arm is obviously critical to success in cycling. Torque (developed primarily about the hip and knee joints) exerted on the crank is greatest when the crank is positioned midway between top and bottom dead centre (i.e., at 90°; figure 17.9). At other crank positions, the applied force is not directed at right angles (largest perpendicular distance), so the resulting torque is reduced.

Figure 17.9 When the crank is positioned midway, the applied force is directed at right angles, thus increasing torque for greater cycling success.

Weightlifting

In lifts in which the bar moves in front of the body, it must be kept relatively close to the body to reduce the perpendicular distance to the weight. Lesser moments are thus needed to lift the bar. Compare performing a biceps curl with the elbow flexed (bar close to the body) and with the elbow fully extended (bar a greater distance from the body).

Rowing

In pulling the oar, the rower applies a force to the handle such that the system rotates about the blade, which is locked in the water (though there may be slight slippage in the water). In this movement the force is applied to the boat at the oar gate.

Energy

Three forms of energy are of interest to coaches and teachers: potential, kinetic and rotational. The first two of these are more generally discussed and therefore form the basis of this section.

Potential Energy

Potential energy is energy that relates to an object's position and equates to *mass* × *gravity* × *position.*

Archery

When the archer pulls the string on a bow, potential energy is stored that is then used to drive the arrow forward (figure 17.10). Depending on the material characteristics of the bow, more energy is stored when the strings are pulled back farther, in turn imparting a greater velocity to the arrow at release.

Trampolining

In performing manoeuvres on a trampoline, an athlete in the air possesses potential energy at the height of the bounce that decreases as the distance to the bed decreases. At impact with the bed, energy is again potential (or strain) as the bed is displaced and the springs store energy.

Diving

The more height a diver achieves in the hurdle step prior to impact with the board, the greater the potential energy the athlete possesses. While the magnitude of the displacement of the board following impact gives an indication of the potential energy now stored in the board, the diver may alter the flexibility of the board to modify the stored energy.

Figure 17.10 Upon releasing the arrow, the previously stored energy (stored when the athlete pulled the strings back) is then used to propel the arrow forward.

Kinetic Energy

Kinetic energy is energy of motion and is equal to $0.5 \times mass \times velocity^2$.

Baseball

A player running from first to second base must develop as much kinetic energy as possible and then quickly reduce this energy on approaching the base (slide), such that he or she maintains contact with the bag.

Tumbling

In trampolining, diving or gymnastics, an athlete at the top of the flight pattern has maximum potential energy (minor levels of kinetic energy if he or she is moving forward). This decreases such that halfway to the bed (or water or floor), the body has both potential and kinetic energy and on impact it has only kinetic energy (no potential energy).

Elastic Energy—a Special Form of Potential Energy

The way in which elastic energy is stored in muscles and tendons is similar to the way energy is stored in an elastic band when it is stretched. This creates a prestretched condition, thereby placing muscles in a better position to contract forcibly than if they were in a rested state. On movement reversal during the forward swing, the muscles that have been stretched as a result of an eccentric contraction and the tendons recoil to their original shape, and a portion of the stored energy is used to assist the movement (Walshe, Wilson, and Ettema 1998). The key to the recovery of this energy is the timing within the stretch-shorten movement. As a general rule, 50% of the stored energy is "lost" after a 1 s pause, so the pause between the backswing and forward-swing phases of a movement should be minimal (Wilson, Elliott, and Wood 1991).

Figue 17.11 In the backswing of a pitch, the shoulder internal rotation muscles are put "on stretch".

Hitting and Throwing

Coaches of athletes who perform hitting and throwing should ensure that the internal rotator muscles at the shoulder are eccentrically contracted during the backswing to store energy (figure 17.11). Then, during internal rotation, this energy at least partially assists in creating angular velocity during the forward-swing phase of the movement.

Running

During running, the calf musculature stores elastic energy as the body moves over the foot during stance. When the foot is flat on the ground and the body continues moving forward, the calf musculature is eccentrically stretched and stores elastic energy (is also placed in a better position to contract concentrically). This stored energy then contributes to the concentric contraction of the calf musculature to assist in rapid plantarflexion of the foot at toe-off.

Coefficient of Restitution

In many sporting activities involving collisions, the coefficient of restitution (e) can be used to estimate the energy loss of one object with reference to another. This dimensionless number varies from near 1.0 (for a very elastic collision, as when a superball bounces on the ground) to 0.0 (for a perfectly inelastic collision, as when a car collides with a concrete wall). The most common use of coefficient of restitution is in defining the elastic properties of balls used in sport—a tennis ball, for instance. Think of the varying bounce properties of different squash balls (different colours of dots). If one drops a basketball on the ground, it returns to a lesser vertical height after each bounce. This relationship can be simply calculated from the equation

$$Vup = Vdown \times e$$

where *Vup* = velocity after the bounce, *Vdown* = velocity just prior to impact and *e* = coefficient of restitution.

It is obvious that the velocity after the bounce will be less than that before the bounce if *e* is less than 1.0, as is generally the case. The interaction between a bat and a ball will show the same effect.

Baseball

The different properties of aluminium and wooden baseball bats influence the rebound velocity of the ball (figure 17.12). The pitcher is certainly in more danger of being hit with a ball struck by an aluminium bat because of the higher coefficient of restitution of aluminium compared with wood.

Football

The value of the coefficient of restitution is altered by the state of a ball (dry vs. wet; fully inflated vs. deflated; hot vs. cold), and this will influence the distance it is kicked. As a squash ball warms, it will bounce better, showing that temperature affects the coefficient of restitution.

Figure 17.12 A batter near impact.

Tennis

The fact that the coefficient of restitution changes during play is the reason tennis balls are replaced at regular times throughout tournament matches. As a ball experiences wear, the coefficient of restitution changes, thus altering the nature of play.

Coordination and Sequencing of Body Segments

Coordination of various body segments enables athletes to control an implement or a body part and at the same time achieve high speed, which is important in velocity sports such as squash and boxing. The body segments that participate in the motion typically move one after the other in a proximal to distal fashion. Accuracy tends to relate inversely to the number of segments involved in a movement; thus when accuracy is more important than velocity, the body segments should move more or less as one unit.

Velocity Sports

In sports that require high speed yet good control of an implement, a number of body segments must be coordinated in such a way that this high speed is achieved at impact. The sequencing of segments participating in high-speed striking skills is generally proximal to distal. This coordination occurs such that the movement of one segment begins as the speed of the previously moving segment has reached its maximum, in a "staircase effect". The sequential rotation of the body segments is usually measured in striking or throwing sports by the speed of the segment end points (the wrist joint is the end of the forearm segment and so on). A movement pattern that exhibits a coordinated action of lower limb, trunk and upper limb segments is usually the sign of a mature action.

Squash

The forward-swing phase of the squash forehand drive consists of the following movements:

- A forward step and trunk rotation
- Continuous trunk rotation as the upper arm moves forward
- Extension and pronation of the forearm at the elbow and upper arm inward rotation
- Minor hand flexion at the wrist to complete the forward swing .

Taekwondo

Developing maximum velocity of the heel (side kick) or the ball of the foot (turning or roundhouse kicks) requires skilled sequencing of body segment actions beginning with a strong pivot of the hips. The motion of the thigh segment is added with the leg still flexed at the knee. As the thigh reaches maximum angular velocity, it is rapidly decelerated with a strong eccentric contraction of the hamstrings group, such that inertia causes the lower leg (and foot) to accelerate toward the target. Body weight must continue to shift toward the target to add to the momentum of the striking point.

Boxing

In throwing a punch, the boxer coordinates forward movement of the trunk with rotations of the upper limb to achieve optimal velocity of the glove (i.e., optimal linear momentum) at impact. Given time, the legs would also play a role.

Accuracy Results

In movements requiring accuracy, the number of segments involved in the movement is reduced (decreasing the number of degrees of freedom), and these can then be coordinated more as a unit to produce controlled endpoint velocity of the distal segment.

Netball

In netball, as players are permitted to shoot only inside a perimeter, the body segments should "move as one" in order for players to attain optimal accuracy since power is not generally a prerequisite for shooting. However, the recreational player, when shooting from or near the perimeter, is often forced to used a "velocity-based approach", resulting in a loss of accuracy.

Basketball

In the basketball jump shot, ball velocity is generated primarily by two segments (elbow extension and wrist flexion). The accuracy of shooting is decreased if a player uses the forward and upward movement of the body to generate ball velocity rather than shooting from the peak of a vertical jump, where the shooting shoulder velocity is close to zero (figure 17.13).

Figure 17.13 When the athlete releases the ball at the peak of a vertical jump, she increases shooting accuracy.

Friction

Friction is a force that acts parallel to the interface between two surfaces that are in contact, during motion or impending motion. This friction force is proportional to the *normal force* between the surfaces (i.e., the force pushing the bodies together):

$$F_f = \mu \times N$$

where F_f = the force needed to create motion, μ = the coefficient of friction (this coefficient will change depending on movement circumstances—it may be static [no movement], sliding or rolling [kinetic friction]) and N = force perpendicular to the surface.

Tennis

Players require shoes with soles of different patterns and materials to provide the ideal frictional level between the shoe and courts with varying surfaces (e.g., grass, clay, board and hard court). While professional players have the luxury of wearing shoes with characteristics that suit the playing surface, this is often a problem for the recreational player, who may own only one pair.

Football

In football, cleats of different configurations and lengths are used to increase the friction between the player and the ground and therefore permit quick changes of direction. We have all seen the example of players slipping because they have chosen cleats that are too short and do not grip the surface.

Curling

In curling, the path of the stone across the ice is modified by a player who manipulates the coefficient of friction between the ice and the stone by sweeping a curling broom to alter the path of the stone.

Centripetal Force

Athletes who wish to increase their velocity when running in a straight line must apply more force to the ground to accelerate, in accordance with Newton's second law ($F = m \times a$). Similarly, when athletes run a curve, they must apply force to increase their velocity; however, they must also apply another force that will enable them to change direction (i.e., run the curve). This acceleration toward the centre of the circle of movement is termed *centripetal acceleration* (radial acceleration), and the radial force toward the centre is termed the *centripetal force.* The magnitude of this "centre-seeking" force is

$$F_c = m \times v^2 / r$$

where F_c = centripetal force *(equal and opposite to centrifugal force)*, m = mass of the system, v = velocity and r = radius of the curve or track.

Cycling and Car Racing

In track cycling, the bank of the track enables the cyclist to remain more upright (figure 17.14), whereas in running the athlete typically leans into the curve to create this centripetal force. In NASCAR racing, higher velocities may be achieved if the track is banked.

Figure 17.14 The bank of a cycling track allows the cyclist to remain more upright; thus higher velocity may be achieved.

Hammer Throwing

The hammer thrower leans backward during rotations prior to the throw to enable the hammer to be rotated at a high angular velocity. A greater lean angle is required to accommodate a high rotational velocity of the hammer.

Fluid Dynamics

In many sporting endeavours, motion is affected by the fluid in which the activity is performed. The density and viscosity of the fluid are characteristics central to this scenario. Density is the mass per unit volume, with denser fluids having a greater effect on performance (e.g., compare humid and normal air). Viscosity is a measure of the fluid's resistance to flow (e.g., compare water and air). The force that opposes motion in a fluid is termed *drag,* which operates in the same direction as the flow of the fluid past the object and in the opposite direction to the object's motion. The force at right angles to the flow of the fluid is termed *lift.*

Water

In water, drag is generally classified as one of three forms:

- **Surface:** This is the drag created by the surface of the object and is the reason swimmers shave their bodies and wear full body suits.
- **Wave:** This is the drag created by the motion of the body or craft at the surface in response to excessive pitch, roll and yaw motions.
- **Form:** This is related to the frontal surface area of the object moving through the water.

Swimming

On turning from the wall, a swimmer pushes off 0.2 to 0.4 m beneath the surface and maintains a streamlined position to reduce drag. While maintaining this streamlined position, the swimmer then begins to kick and angles gradually to the surface such that stroking can resume when velocity has decreased to near "race pace".

Kayaking and Canoeing

Paddlers use appropriate techniques to reduce excessive yaw, pitch and roll in an endeavour to lower drag. Paddling technique is such that it optimises lift and drag forces to maximize propulsion.

Air

The speed of air flow over an object also affects *pressure drag*. In air, the level of drag changes with the velocity of the athlete (and of course the direction of the wind). Generally drag increases to the square for any change in velocity. That is, if a runner doubles his or her velocity (sprint vs. jog), the level of drag will quadruple (i.e., $2^2 = 4$). The effect of the speed of movement influences the *boundary layer* of air about an object (e.g., baseball, cricket ball or runner). Unfortunately, the effects of laminar and turbulent flow about an object on drag are beyond the scope of this section, but this topic is well covered in the texts recommended earlier in this chapter.

Cycling

Cyclists conserve energy (~30%) by riding in a line *(slip streaming)*, with the wheels relatively close to those of other cyclists, to reduce drag. It is for this reason that this practice is banned in selected triathlon cycling events, as triathlon is deemed an individual sport. Cyclists also use equipment (bike and riding gear) that has been specifically developed (e.g., in wind tunnels) to reduce drag.

Running

Sprinters, who once wore loose-fitting clothes, now wear molded garments to ensure that drag forces are minimized (even the head may now be covered). The difference between winning and coming second is now very small, so any advantage from reduced air resistance is considered important.

Discus Throwing

The discus is thrown at a specific angle to optimise lift and at the same time to reduce drag. If throwers were concerned only with reduced drag, they would position the discus in line with the oncoming air flow (0° angle).

Summary

While it is broadly acknowledged that sporting injuries have a mechanical cause (Whiting and Zernicke, 1998), it should be now evident to the reader that sport performance and mechanics are also closely linked. The teacher or coach who understands the importance of mechanics to performance will provide athletes with the best opportunity to reach their potential in an injury-free environment.

part IV

Practical Example

eighteen

A Practical Example

Pole Vault

Timothy R. Ackland, PhD; Andrew Lyttle, PhD; and Bruce C. Elliott, PhD

The ability to assess an athlete's capabilities—to determine his or her strengths and weaknesses—is integral to successful coaching. The model presented in chapter 1 (figure 1.1) provides a framework whereby the coach or sport scientist can optimise performance by targeting an individual's specific physical and technical components in relation to the sport.

This chapter focuses on the progression of an elite athlete over the course of five years as specific interventions were trialled in an attempt to improve pole-vaulting performances. In this sport the athlete must not only master the difficult technique of vaulting, but also have the ability to generate a high running speed during the approach and the strength to transfer the high forces at the "pole plant" to strain energy in the pole. Therefore, limitations in any of these facets can become a major hindrance to successful vaulting performance.

The subject (PB) (figure 18.1) trained with the Western Australian Institute of Sport (WAIS) and during the years under discussion emerged as one of the world's top pole-vaulters. He is one of the few competitors to have reached the exclusive "6 m Club" by clearing a height of 6.00 m, a feat he accomplished in 2005. He was ranked number two in the world for the 2006 season. Table 18.1 outlines the progression of "season best jumps" after PB took up the sport in 1994. This table demonstrates considerable improvement in his best vaulting performances after 2000 (apart from some problems in 2003), but does not show the improvement he made in consistency, which is the hallmark of the top competitor. As of 2006, not only was PB achieving consistently high performances, but also the starting height for competition had increased over the previous two years.

Figure 18.1 Elite pole-vaulter PB during competition.

Table 18.1 Pole-Vault Performance Record for PB

Year	Season best height (m)	Key international results
1994	3.90	
1995	5.25	
1996	5.35	1st: World Junior Championships
1997	5.51	
1998	5.60	2nd: Commonwealth Games
1999	5.50	
2000	5.60	
2001	5.71	2nd: East Asian Games
2002	5.75	2nd: Commonwealth Games
2003	5.55	
2004	5.77	
2005	6.00	Australian all-comers record
2006	5.92	

Assessment of Physical Capacity and Intervention Program

Having entered the sport of pole-vaulting from a background in gymnastics, PB was a relatively short, compact athlete with a height of 184.0 cm (72.4 in.). Since this sport requires the athlete's body to be raised a significant distance from the ground, body mass (and especially fat mass) must be minimized without compromising strength and power. Figure 18.2 displays body mass and sum of skinfold data for PB for the five years ending in 2006. Small fluctuations in body mass over this period (range: 83-88 kg [183-194 lb]) coincide with periods of increased adiposity, especially from late 2002 through 2003.

It is not surprising that poor performances were recorded during the 2003 season given the 30% rise in adiposity at that time. At this point, PB was placed on a rigid eating plan with the goal to improve his diet in order to lose body fat but maintain sufficient energy for training. So, despite concurrent improvements in strength (as discussed later), the athlete was able to reduce body mass and effect a significant decrease in body fat during 2004 and 2005.

This athlete was always relatively strong in the upper body as a result of his gymnastics background. However, a specific intervention program with the goal to improve strength and power was implemented in 2004. The data in figure 18.3 show the positive outcome of this intervention, with improvements in standard 1-repetition maximum (1RM) strength tests of bench press, chins and snatch over the last two seasons.

Biomechanical Assessment and Technique Modification

Biomechanical assessment of PB's vaulting technique centred on approach velocity; stride consistency and visual targeting; the timing of each phase of the vault with respect to the total skill; and the energy components of performance, such as horizontal kinetic energy and potential energy. These elements of the skill were then specifically addressed through technique modification and training.

Approach Velocity

Although fast running speed is essential, the important factor for vaulting is termed approach velocity (the ability

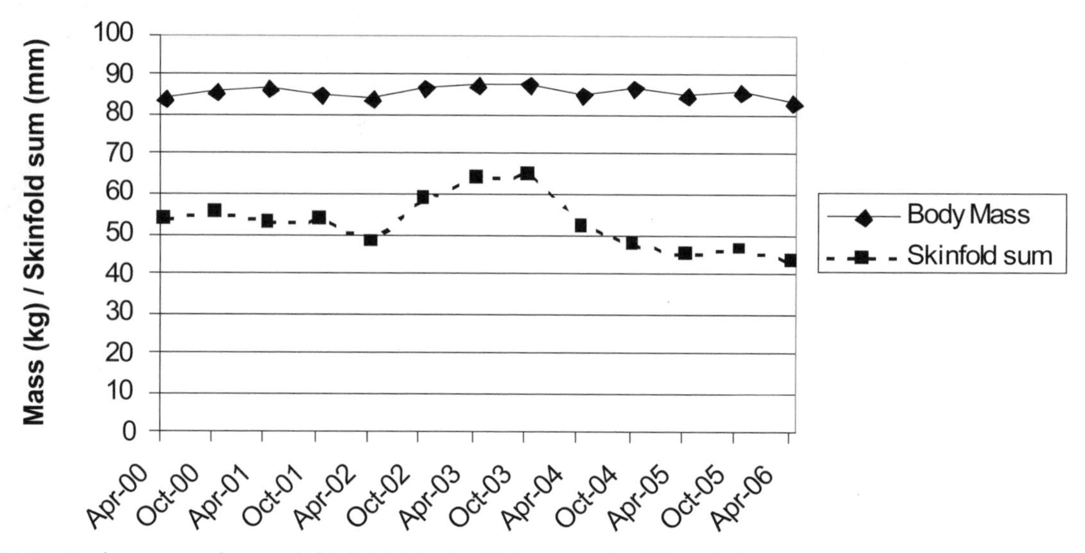

Figure 18.2 Body mass and sum of skinfold data for PB from April 2000 to 2006.

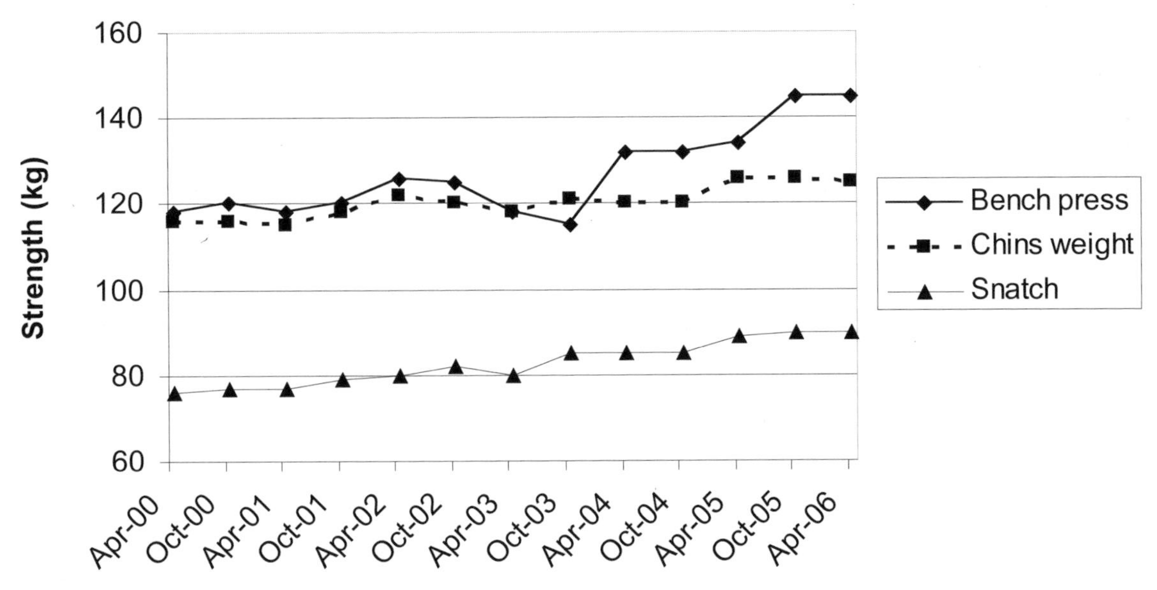

Figure 18.3 One-repetition maximum strength data for PB from April 2000 to 2006.

to run at high speeds carrying a pole in the lead-up to pole plant). This is a discriminating variable between vaulters of varying levels (although not necessarily between elite competitors, who achieve consistently high scores on this measure). Higher approach velocities translate to greater *horizontal kinetic energy* (i.e., due to the momentum of the athlete) that can be converted into *strain energy* at pole plant, which then is returned to the vaulter as *vertical kinetic energy* plus *potential energy*. Approach velocity is monitored continually in competition and during training sessions.

The data in figure 18.4 show run-up velocity from the start of the run-up (approximately 45 m [49 yd] from the base of the supports; far right of the display) to just after pole plant (at approximately 5 m [5.5 yd] from the base of the supports). Although timing gates can be used for this process, the smoothed data in this figure were derived from a laser gun set up behind the athlete, which allows determination of the speed profile rather than just average speeds between sections. Though this trial was not particularly fast, the athlete demonstrated a stable acceleration over the first 10 m (11 yd) and an ability to reach or maintain peak velocity toward pole plant. As a consequence of his specific speed training after early 2004, PB improved the approach velocity to around 9.55 m/s while maintaining control of the pole plant (see figure 18.5).

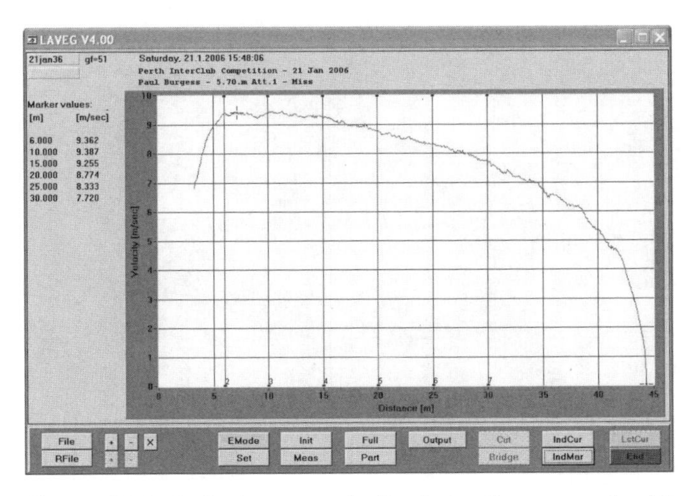

Figure 18.4 Instantaneous velocity during the run-up for PB during competition in 2006.

Stride Consistency and Visual Targeting

In addition to good running speed, achieving a high horizontal velocity at pole plant is dependent on the athlete's ability to maintain a consistent running rhythm by hitting the control markers and the takeoff point during the approach. Using a 2D video analysis system, the sport biomechanists at WAIS were able to monitor this aspect of PB's technique. The data in figure 18.6*a* show the variability in stride length during the final eight foot falls prior to takeoff over four consecutive competition attempts (at 5.6 and 5.8 m). The approach path and consistency of foot strikes over these same four attempts are displayed in figure 18.6*b*. This athlete was able to arrive at a consistent takeoff point during successive attempts—a mark of the elite-level vaulter.

Temporal Analysis

Given the complexities of the vaulting action, it is often useful to break down the skill into its components or

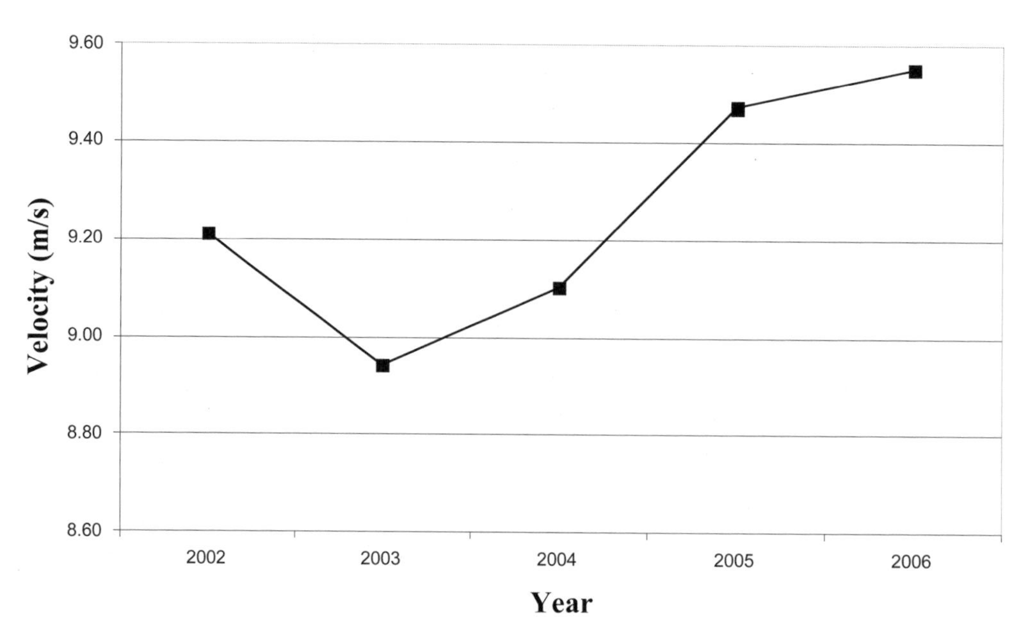

Figure 18.5 Progression of maximum approach velocity from 2002 to 2006.

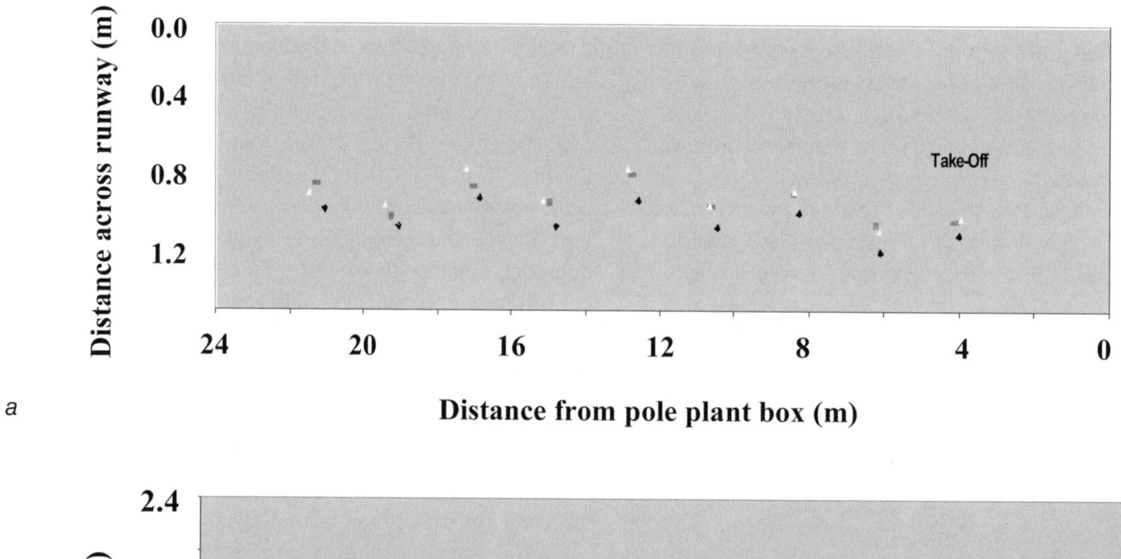

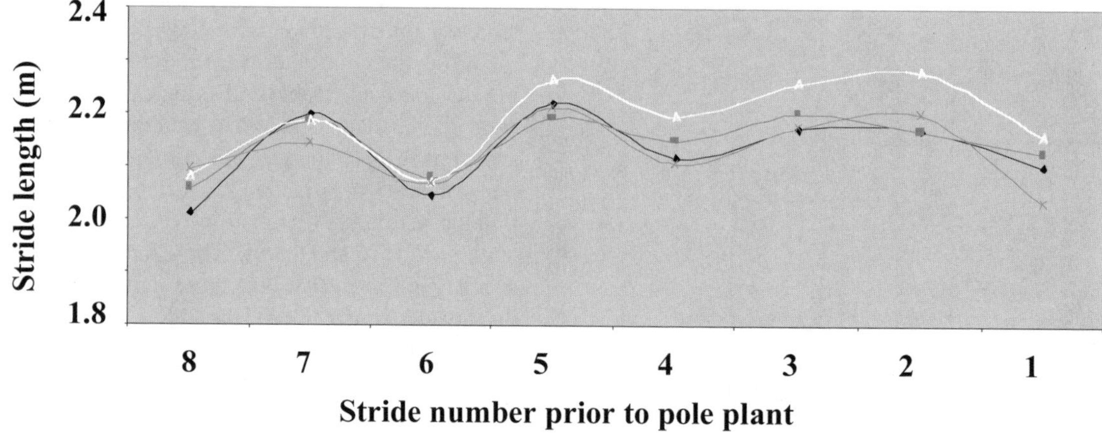

Figure 18.6 *(a)* Stride length consistency and *(b)* successful visual targeting of control markers and takeoff point during four successive competition attempts.

phases. Temporal analysis is used to assess the relative contribution (in regard to time) of each phase with respect to the total skill. The graph in figure 18.7 shows the breakdown of PB's vault over five consecutive trials with attempts at 5.3 and 5.5 m. Of these trials, only the second attempt at the 5.3 m height was successful, and this was characterized by a longer final section—"maximum hip height" (indicating that PB's hip was increasing in height for a longer period of time after release of the pole).

Energy Analysis

The ability to first store (by bending) and then successfully utilize the strain energy within the pole is an essential aspect of elite pole-vault performance. The data in figure 18.8 were derived from 2D kinematics during the 2003 season when PB's performances were down. This graph of the energy transfer shows the horizontal kinetic energy, potential energy and total energy components during an attempt at a 5.55 m vault. These data show how the potential and horizontal kinetic energy interact from a point just prior to takeoff in order to produce PB's total energy. The horizontal kinetic energy trace drops sharply from the initial high energy levels, due to the vaulter's speed during the last stride before takeoff (at time = 0.12 s), to low levels as the forward speed slows during the pole bend phase (maximum pole bend at time = 0.72 s). Conversely, the potential energy rapidly increases as the pole begins to extend and the vaulter gains height (note that this graph does not include the vertical kinetic energy component).

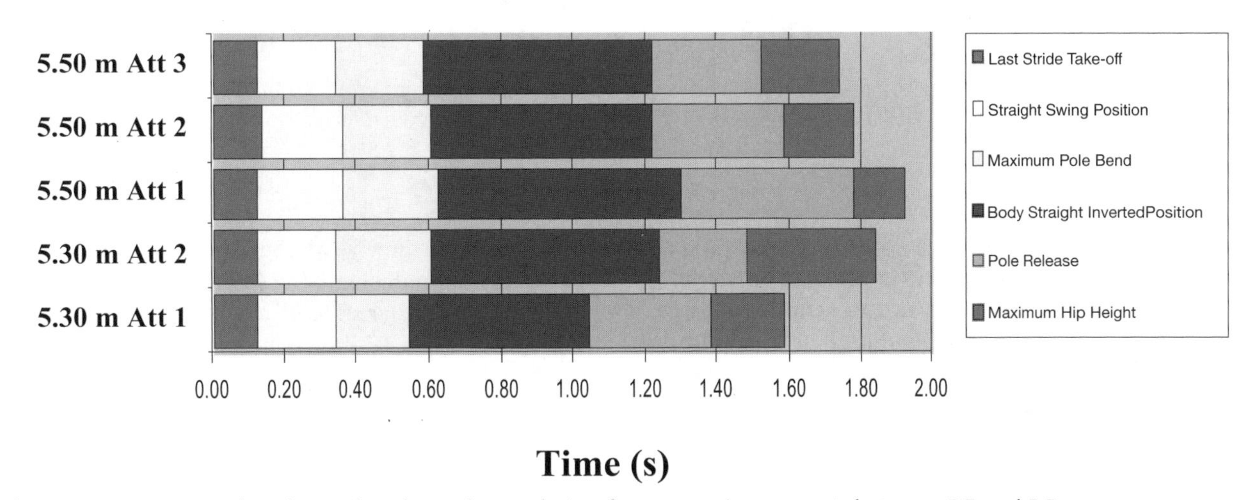

Figure 18.7 Temporal analysis of vaulting phases during five successive attempts between 5.3 and 5.5 m.

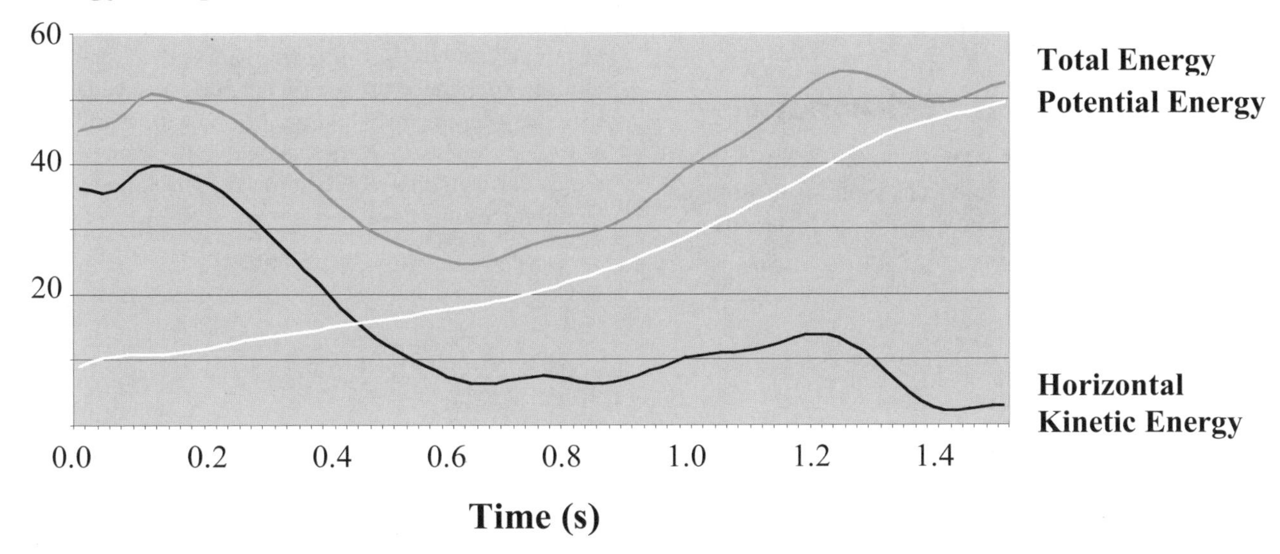

Figure 18.8 Pole-vault energy analysis for PB during a 5.55 m attempt in competition.

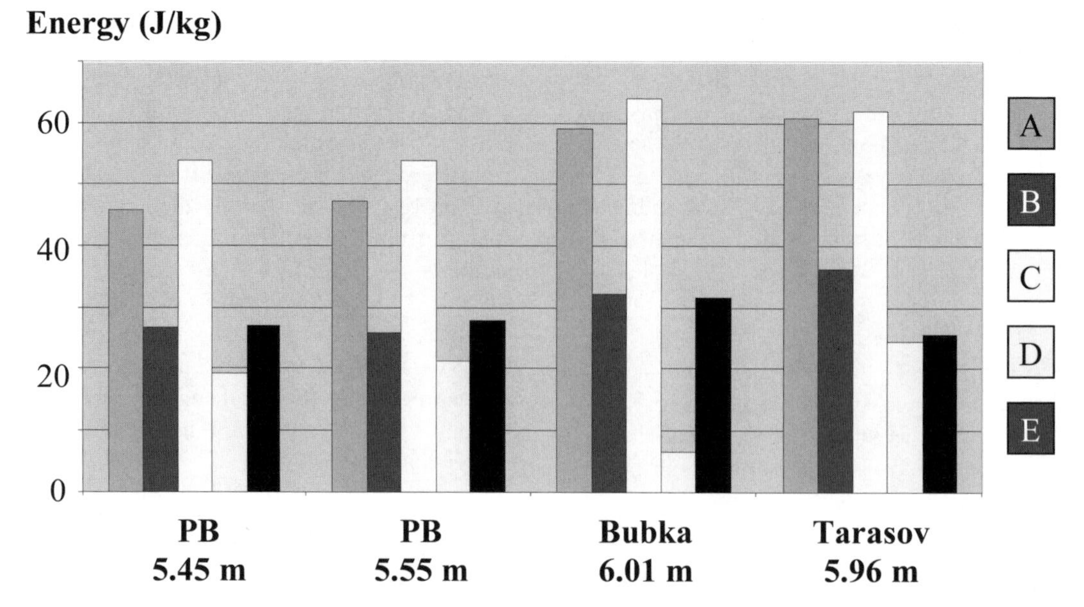

Figure 18.9 Energy analysis comparisons. A = Initial energy; B = energy at maximum pole bend; C = energy at pole release; D = energy decrease (depends on the stiffness of the pole); and E = energy increase (shows the athlete's work done on the pole).

Energy analysis comparisons with higher-ranked performers (Bubka and Tarasov, neither of whom is currently competing) then helped the coach to identify areas of weakness in PB's technique (figure 18.9). It is clear from this graph that the major discrepancy resulted from the much lower initial energy values, which were caused by the slower approach velocity that PB was able to achieve. Similar *energy increase* values were found between PB and Bubka, indicating that PB was producing world-class efforts in "active work" on the pole during pole extension (thereby adding extra energy into the system). Hence, an increase in approach velocity and ability to translate that into strain energy of the pole were determined to be areas of focus for PB's subsequent training.

Influence on Performance

Clearly this athlete benefited from the very focused physical and mechanical assessment and profiling undertaken by his coach in association with the sport scientists at WAIS. Over the three years ending in 2006 in particular, PB worked on reducing body fat and concurrently improving his muscular strength and power. The latter had a clear effect on his ability to increase approach velocity and to achieve significant pole bend to maximize the strain energy stored in the pole.

Improvements in other aspects of technique also complemented these changes. PB achieved uniformity in the run-up, so that the control and takeoff markers were "hit" with minimum variation, which in turn led to consistency in performance. The combined effect of these improvements in physique and technique was a substantial gain in vaulting height—not just in the season's best vaults, but also in vaulting consistency. During the last three years of the time period under discussion, this athlete experienced the significant performance benefits that can be derived when quality coaching and sport science expertise are combined within the framework of the assessment and modification model outlined at the beginning of this book.

references

Chapter 1

Bloomfield, J., and B. Blanksby. 1973. Anatomical features involved in developing the ideal swimming stroke. *International Swimmer* 8: 8-9.

Pollock, M., A. Jackson, and R. Pate. 1980. Discriminant analysis of physiological differences between good and elite distance runners. *Research Quarterly for Exercise and Sport* 51: 521-532.

Chapter 2

Ackland, T. 2001. Physical profiling and talent identification for sporting success. *Proceedings of the 2001 KNUPE International Symposium* 20: 57-63.

Ackland, T.R. 2006. Built for success: Homogeneity in elite athlete morphology. In *Kinanthropometry IX,* edited by M. Marfell-Jones, A. Stewart, and T. Olds. Potchefstroom: ISAK.

Ackland, T., and J. Bloomfield. 1996. Stability of human proportions through adolescent growth. *Australian Journal of Science and Medicine in Sport* 28: 57-60.

Alabin, V., G. Nischt, and W. Jefimov. 1980. Talent selection. *Modern Athlete and Coach* 18: 36-37.

Baker, J., and S. Horton. 2004. A review of primary and secondary influences on sport expertise. *High Ability Studies* 15(2): 211-228.

Battaerd, C. 2004. *Feasibility of a System that Supports Talent Transfer.* Canberra: Australian Sports Commission.

Bayer, L., and N. Bayley. 1959. *Growth Diagnosis.* Chicago: University of Chicago Press.

Beunen, G.P., R.M. Malina, J. Lefevre, A.L. Claessens, R. Renson, and J. Simons. 1997. Prediction of adult stature and noninvasive assessment of biological maturation. *Medicine and Science in Sports and Exercise* 29(2): 225-230.

Bloomfield, J. 1992. Talent identification and profiling. In *Textbook of Science and Medicine in Sport,* edited by J. Bloomfield, P.A. Fricker, and K. Fitch. Melbourne: Blackwell Scientific.

Cockerill, I.M. 2005. They think it's all over, but it may not be! *British Journal of Sports Medicine* 39: 880-883.

Cratty, B.J. 1960. A comparison of fathers and sons in physical ability. *Research Quarterly* 31: 12-15.

Ericsson, K.A., R.Th. Krampe, and C. Tesch-Romer. 1993. The role of deliberate practice in the acquisition of expert performance. *Psychological Review* 100(3): 363-406.

Gayagay, G., B. Yu, B. Hambly, T. Boston, A. Hahn, D.S. Celermajer, and R.J. Trent. 1998. Elite endurance athletes and the ACE I allele—the role of genes in athletic performance. *Human Genetics* 103(1): 48-50.

Greulich, W., and S. Pyle. 1959. *Radiographic Atlas of Skeletal Development of the Hand and Wrist.* Palo Alto, CA: Stanford University Press.

Gulbin, J. 2001. From novice to national champion. *Sports Coach* 24(1): 24-26.

Gulbin, J. 2004. Paradigm shifts in talent identification. *Proceedings of the 2004 Pre-Olympic Congress, Thessaloniki, Greece,* 1: L.105.

Hahn, A. 1990. Identification and selection of talent in Australian rowing. *Excel* 6: 5-11.

Halson, S., D.T. Martin, A.S. Gardner, K. Fallon, and J. Gulbin. 2006. Persistent fatigue in a female sprint cyclist after a talent–transfer initiative. *International Journal of Sports Physiology and Performance* 1: 65-69.

Hardeman, E.C. 2001. Genetic manipulation of muscle function and its application to performance enhancement. In: *2001: A Sports Medicine Odyssey, Proceedings of the Australian Conference of Science and Medicine in Sport,* edited by T. Ackland and C. Goodman. Canberra: Sports Medicine Australia.

Hausswirth, C., J.M. Vallier, D. Lehenaff, J. Brisswalter, D. Smith, G. Millet, and P. Dreano. 2001. Effect of two drafting modalities in cycling and running performance. *Medicine and Science in Sports and Exercise* 33(3): 485-492.

Helsen, W.F., J. Van Winckel, and A.M. Williams. 2005. The relative age effect in youth soccer across Europe. *Journal of Sports Sciences* 23(6): 629-636.

Henderson, J., J.M. Withford-Cave, D.L. Duffy, S.J. Cole, N.A. Sawyer, J.P. Gulbin, A. Hahn, R. Trent, and B. Yu. 2005. The EPAS1 gene influences the aerobic-anaerobic contribution in elite endurance athletes. *Human Genetics* 118(3/4): 416-423.

Hogan, K., and K. Norton. 2000. The "price" of Olympic gold. *Journal of Science and Medicine in Sport* 3(2): 203-218.

Hohmann, A. 2004. Scientific aspects of talent development. In *Talent Identification, Selection and Development: Problems and Perspectives,* edited by H. Ziemainz, A. Rutten, and U. Roeger. Butzbach: Afra-Verlag.

Klissouras, V. 2001. The nature and nurture of human performance. *European Journal of Sport Sciences* 1(2): 1-10.

Komadel, L. 1988. The identification of performance potential. In *The Olympic Book of Sports Medicine I,* edited by A. Dirix, H. Knuttgen, and K. Tittel. Oxford: Blackwell Scientific.

Lowery, G. 1978. *Growth and Development of Children* (7th ed.). Chicago: Year Book Medical.

Malina, R., and C. Bouchard. 1991. *Growth, Motivation and Physical Activity.* Champaign, IL: Human Kinetics Books.

Oldenziel, K.E., F. Gagne, and J.P. Gulbin. 2004. Factors affecting the rate of athlete development from novice to senior elite: How applicable is the 10-year rule? *Proceedings of the 2004 Pre-Olympic Congress, Thessaloniki, Greece,* 1: O.027.

Rankinen, T., M.S. Bray, J.M. Hagberg, L. Perusse, S.M. Roth, B. Wolfarth, and C. Bouchard. 2006. The human gene map for performance and health-related fitness phenotypes: The

2005 update. *Medicine and Science in Sports and Exercise* 38(11): 1863-1888.

Rowland, T. 1998. Predicting athletic brilliancy, or the futility of training 'til the Salchow's come home. *Pediatric Exercise Science* 10: 197-201.

Sherar, L.B., R.L. Mirwald, A.D. Baxter-Jones, and M. Thomis. 2005. Prediction curves of adult height using maturity-based cumulative height velocity curves. *Journal of Pediatrics* 147(4): 508-514.

Tanner, J. 1989. *Foetus into Man.* Ware, UK: Castlemead.

Williams, A., and T. Reilly. 2000. Talent identification and development in soccer. *Journal of Sports Sciences* 18: 657-667.

Yang, N., D.G. MacArthur, J.P. Gulbin, A.G. Hahn, A.H. Beggs, S. Easteal, and K. North. 2003. ACTN3 genotype is associated with human elite athletic performance. *American Journal of Human Genetics* 73(3): 627-631.

Ziemainz, H., and J. Gulbin. 2002. Talent selection, identification, and development exemplified in the Australian Talent Search program. *New Studies in Athletics* 3/4: 27-32.

Chapter 3

Himes, J.H. 1979. Secular changes in body proportions and composition. In *Secular Trends in Human Growth, Maturation and Development,* edited by A.F. Roche. Chicago: Society for Research in Child Development.

International Society for the Advancement of Kinanthropometry. 2001. *International Standards for Anthropometric Assessment.* Adelaide: University of South Australia.

Meredith, H.V. 1976. Findings from Asia, Australia, Europe and North America on secular change in mean height of children, youths and young adults. *American Journal of Physical Anthropology* 44: 315-326.

Norton, K.I., N.P. Craig, and T.S. Olds. 1999. The evolution of Australian football. *Journal of Science and Medicine in Sport* 2(4): 389-404.

Norton, K.I., and T.S. Olds. 2000. The evolution of the size and shape of athletes: Causes and consequences. In *Kinanthropometry VI: Proceedings of the Sixth Conference of the International Society for the Advancement of Kinanthropometry,* edited by K. Norton, T. Olds, and J. Dollman. Adelaide: ISAK.

Norton, K.I., T.S. Olds, S.C. Olive, and N.P. Craig. 1996. Anthropometry and sports performance. In *Anthropometrica,* edited by K.I. Norton and T.S. Olds. Sydney: UNSW Press.

Olds, T.S. 2001. The evolution of physique in male Rugby Union players in the twentieth century. *Journal of Sports Sciences* 19: 253-262.

Olds, T.S., and S.C. Olive. 1999. Methodological considerations in the determination of projected frontal area in cyclists. *Journal of Sports Sciences* 17: 335-345.

Chapter 4

Ackland, T., and J. Bloomfield. 1995. Functional anatomy. In *Textbook of Science and Medicine in Sport* (2nd ed.), edited by J. Bloomfield, P. Fricker, and K. Fitch. Melbourne: Blackwell Scientific.

Ackland, T., D. Kerr, P. Hume, K. Norton, B. Ridge, S. Clark, E. Broad, and W. Ross. 2001. Anthropometric normative data for Olympic rowers and paddlers. In *2001: A Sports Medicine Odyssey: Proceedings of the Sports Medicine Australia Annual Conference,* edited by T.R. Ackland and C. Goodman. Canberra: Sports Medicine Australia.

Borms, J., W.D. Ross, W. Duquet, and J.E.L. Carter. 1986. Somatotypes of world class body builders. In *Perspectives in Kinanthropometry,* edited by J.A.P. Day. Champaign, IL: Human Kinetics.

Carlson, B.R., J.E.L. Carter, P. Patterson, K. Petti, S.M. Orfanos, and G.J. Noffal. 1994. Physique and motor performance characteristics of US national rugby players. *Journal of Sports Sciences* 12: 403-412.

Carter, J.E.L. 1984. Somatotypes of Olympic athletes from 1948 to 1976. In *Physical Structure of Olympic Athletes Part II,* edited by J.E.L. Carter. Basel: Karger.

Carter, J.E.L. 1985. Morphological factors limiting human performance. In *Limits of Human Performance, The American Academy of Physical Education Papers, No. 18,* edited by H.M. Eckert and D.H. Clarke. Champaign, IL: Human Kinetics.

Carter, J.E.L. 1996. Somatotyping. In *Anthropometrica,* edited by K. Norton and T. Olds. Sydney: UNSW Press.

Carter, J.E.L. 2003. Anthropometry of team sports. In *Kinanthropometry VIII,* edited by T. Reilly and M. Marfell-Jones. London: Routledge.

Carter, J.E.L., T.R. Ackland, D.A. Kerr, and A.B. Stapff. 2005. Somatotype and size of elite female basketball players. *Journal of Sports Sciences* 23(10): 1057-1063.

Carter, J.E.L., S.P. Aubry, and D.A. Sleet. 1982. Somatotypes of Montreal Olympic athletes. In *Physical Structure of Olympic Athletes Part I,* edited by J.E.L. Carter. Basel: Karger.

Carter, J.E.L., and B.H. Heath. 1990. *Somatotyping—Development and Applications.* Cambridge: Cambridge University Press.

Carter, J.E.L., and M. Marfell-Jones. 1994. Somatotypes. In *Kinanthropometry in Aquatic Sports,* edited by J.E.L. Carter and T. Ackland. Champaign, IL: Human Kinetics.

Carter, J.E.L., L.A. Powell-Santi, and C. Rodriquez Alonzo. 1994. Physique and performance of USA volleyball players. In *Access to Active Living, Proceedings of the 10th Commonwealth and International Scientific Congress,* edited by F.I. Bell and G.H. Van Gyn. Victoria, Canada: University of Victoria.

Carter, J.E.L., E.G. Rienzi, P.S.C. Gomes, and A.D. Martin. 1998. Somatotipo y tamaño corporal. In *Futbolista Sudamericano de Elite: Morphfologia, Analisis del Juego y Performance,* edited by E. Rienzi. Rosario: Biosystem Servicio Educativo.

Claessens, A.L., G. Beunen, J. Lefevre, G. Martens, and R. Wellens. 1986. Body structure, somatotype, and motor fitness of top-class Belgian judoists and karateka: A comparative study. In *Kinanthropometry III,* edited by T. Reilly, J. Watkins, and J. Borms. London: Spon.

Claessens, A.L., J. Bourgois, J. Lefevre, B. Van Renterghem, R. Philippaerts, R. Loos, M. Janssens, M. Thomis, and J. Vrijens. 2001. Body composition and somatotype characteristics of elite male junior rowers in relation to competition level, rowing style and boat type. *Journal of Sports Sciences* 19(8): 611.

Claessens, A.L., J. Bourgois, K. Pintens, J. Lefevre, B. Van Renterghem, R. Philippaerts, R. Loos, M. Janssens, M. Thomis, and J. Vrijens. 2002. Body composition and somatotype characteristics of elite female junior rowers in relation to competition level, rowing style, and boat type. *Humanbiologiae Budapestinensis* 27: 159-165.

Claessens, A.L., F.M. Veer, V. Stijnen, J. Lefevre, H. Maes, G. Steens, and G. Beunen. 1991. Anthropometric characteristics of outstanding male and female gymnasts. *Journal of Sports Sciences* 9: 53-74.

Cyrino, E.S., N. Maestá, D.A. dos Reis, N. Nardo Jr., M.Y.G. Morelli, J.M. Santarém, and R.C. Burini. 2002. Perfil antropométrico de culturistas Brasilieras de elite. *Revista Paulista Educacao Fisica, São Paulo* 16(1): 27-34.

De Ridder, J.H. 1992. A morphological profile of junior and senior Cravenweek rugby players. PhD thesis, Potchefstroom University for CHE, Potchefstroom, South Africa.

De Ridder, J.H., L.O. Amusa, K.D. Monyeki, A.L. Toriola, and J.E.L. Carter. 2001. Kinanthropometry in African sports: Body composition and somatotypes of world class female African middle-, long distance and marathon runners. *African Journal for Physical, Health Education, Recreation and Dance* 7(1): 1-13.

De Ridder, J.H., D. Monyeki, L. Amusa, A. Toriola, M. Wekesa, and L. Carter. 2000. Kinanthropometry in African sports: Body composition and somatotypes of world class male African middle-distance, long distance and marathon runners. In *Kinanthropometry VI: Proceedings of the Sixth Conference of the International Society for the Advancement of Kinanthropometry,* edited by K. Norton, T. Olds, and J. Dollman. Adelaide: ISAK.

De Ridder, J.H., K.D. Monyeki, L.O. Amusa, A.L. Toriola, M. Wekesa, and J.E.L. Carter. 2000. Kinanthropometry in African sports: Somatotypes of female African athletes. *African Journal for Physical, Health Education, Recreation and Dance* 6(1): 1-15.

De Ridder, J.H., and J. Peens. 2000. Morphological prediction functions for South African club championship cricket players. *African Journal for Physical, Health Education, Recreation and Dance* 6(1): 65-74.

De Rose, E.H., S.M. Crawford, D.A. Kerr, R. Ward, and W.D. Ross. 1989. Physique characteristics of Pan American Games lightweight rowers. *International Journal of Sports Medicine* 4(10): 292-297.

Duquet, W., and J.E.L. Carter. 2008. Somatotyping. In *Kinanthropometry and Exercise Physiology Laboratory Manual: Tests, Procedures and Data* (3rd ed.), edited by R. Eston and T. Reilly. London: Routledge.

Gavilan, C.A., and P.J. Godoy. 1994. Las categorías de peso en el Judo y el levantamiento olímpico de pesas. Su efecto en el somatotipo y la composición corporal. [The weight categories of judo and olympic weightlifting. Its effect on the somatotype and the body composition.] *Archivos de la Sociedad Chilena de Medicina del Deporte* 39: 102-106.

Goulding, M. 2002. *Somatotype—Calculation and Analysis.* Mitchell Park, South Australia: Sweat Technologies.

Heath, B.H., and J.E.L. Carter. 1967. A modified somatotype method. *American Journal of Physical Anthropology* 27: 57-74.

Hebbelinck, M., L. Carter, and A. De Garay. 1975. Body build and somatotype of Olympic swimmers, divers and water polo players. In *Swimming II,* edited by L. Lewillie and J. Clarys. Baltimore: University Park Press.

Hebbelinck, M., W.D. Ross, J.E.L. Carter, and J. Borms. 1980. Anthropometric characteristics of female Olympic rowers. *Canadian Journal of Applied Sport Sciences* 5: 255-262.

International Society for Advancement of Kinanthropometry. 2001. *International Standards for Anthropometric Assessment.* Potchefstroom: ISAK.

Kang, S.J. 2001. Somatotype characteristics of national level combat sport players. *Proceedings of 2001 KNUPE International Symposium.* Seoul: Korean National University of Physical Education.

Kieffer, S., L. Carter, M. Held-Sturman, P. Patterson, and R. Carlson. 2000. Physique characteristics of USA national and university level rugby players. In *Kinanthropometry VI: Proceedings of the Sixth Conference of the International Society for the Advancement of Kinanthropometry,* edited by K. Norton, T. Olds, and J. Dollman. Adelaide: ISAK.

Landers, G.J., B.A. Blanksby, T.R. Ackland, and D. Smith. 1999. Kinanthropometric differences between world championship senior and junior elite triathletes. *Proceedings of the Gatorade International Triathlon Science II Conference.* Noosa, Queensland: Gatorade.

Leake, C.N., and J.E.L. Carter. 1991. Comparison of body composition and somatotype of trained female triathletes. *Journal of Sports Sciences* 9: 125-135.

Malina, R., and C. Bouchard. 1991. *Growth, Maturation, and Physical Activity.* Champaign, IL: Human Kinetics.

Mészáros, J., and J. Mohácsi. 1982. The somatotype of Hungarian male and female class I paddlers and rowers. *Anthropologiai Kozlemenyek* 26: 175-179.

Norton, K., and T. Olds. 1996. *Anthropometrica.* Sydney: UNSW Press.

Parnell, R.W. 1958. *Behaviour and Physique.* London: Edward Arnold.

Pyke, F., and G. Watson. 1978. *Focus on Running.* London: Pelham Books.

Rienzi, E., M. Pérez, M. Stefani, C. Maiuri, and G. Rodríguez. 2003. Anthropometric characteristics of Rugby Sevens players: An evaluation during two consecutive international events. In *Kinanthropometry VIII,* edited by T. Reilly and M. Marfell-Jones. London: Routledge.

Ross, W.D., S.R. Brown, J.W. Yu, and R.A. Faulkner. 1977. Somatotypes of Canadian figure skaters. *Journal of Sports Medicine and Physical Fitness* 17: 195-205.

Semple, S., N. Neveling, and R. Roussouw. 2003. Physical and performance-related profile of elite male South African distance runners. *African Journal for Physical, Health Education, Recreation and Dance* 9(2): 238-249.

Sheldon, W.H., C.W. Dupertuis, and E. McDermott. 1954. *Atlas of Men.* New York: Harper and Brothers.

Sheldon, W.H., S.S. Stevens, and W.B. Tucker. 1940. *The Varieties of Human Physique.* New York: Harper and Brothers.

Štěpnička, J. 1974. Typology of sportsmen. *Acta Universitatis Carolinae Gymnica* 1: 67-90.

Štěpnička, J. 1986. Somatotype in relation to physical performance, sports and body posture. In *Kinanthropometry III,* edited by T. Reilly, J. Watkins, and J. Borms. London: Spon.

Tanner, J.M. 1964. *The Physique of the Olympic Athlete.* London: Allen and Unwin.

Travill, A.L., J.E.L. Carter, and K.P. Dolan. 1994. Anthropometric characteristics of elite male triathletes. In *Access to Active Living: Proceedings of the 10th Commonwealth and International Scientific Congress,* edited by F.I. Bell and G.H. Van Gyn. Victoria, Canada: University of Victoria.

Tucker, W.B., and W.A. Lessa. 1940. Man: A constitutional investigation. *Quarterly Review of Biology* 15: 411-455.

Underhay, C., J.H. De Ridder, L.O. Amusa, A.L. Toriola, A.P. Agbonjinmi, and J.O. Adeogun. 2005. Physique characteristics of world-class African long distance runners. *African Journal for Physical, Health Education, Recreation and Dance* 11(1): 6-16.

White, J., G. Quinn, M. Al-Dawalibi, and J. Mulhall. 1982. Seasonal changes in cyclists' performance. Part 1, the British Olympic road race squad. *British Journal of Sports Medicine* 16: 4-12.

Wilders, C.J., and J.H. De Ridder. 2001. Somatotype differences in playing positions among South African senior provincial rugby players. *African Journal for Physical, Health Education, Recreation and Dance* 7(1): 51-60.

Wildschutt, P.J., A.L. Travill, L. Leach, and L. Burrell. 2002. Anthropometric and physiological characteristics of South African triathletes. *African Journal for Physical, Health Education, Recreation and Dance* 8(2): 297-308.

Withers, R., N. Craig, and K. Norton. 1986. Somatotypes of South Australian male athletes. *Human Biology* 58: 337-356.

Withers, R., N. Wittingham, K. Norton, and M. Dutton. 1987. Somatotypes of South Australian female games players. *Human Biology* 59: 575-589.

Chapter 5

Ackland, T.R., B.A. Blanksby, and J. Bloomfield. 1994. Physical growth and motor performance of adolescent males. In *Athletics, Growth and Development in Children: The University of Western Australia Study,* edited by B.A. Blanksby, J. Bloomfield, T.R. Ackland, B.C. Elliott, and A.R. Morton. Chur, Switzerland: Harwood Academic Press.

Ackland, T.R., B.A. Blanksby, G. Landers, and D. Smith. 1998. Anthropometric profiles of elite triathletes. *Journal of Science and Medicine in Sport* 1(1): 51-56.

Ackland, T., B. Elliott, and J. Richards. 2003. Growth in body size affects rotational performance in women's gymnastics. *Sports Biomechanics* 2: 163-176.

Ackland, T.R., A.B. Schreiner, and D.A. Kerr. 1997. Absolute size and proportionality characteristics of world championship female basketball players. *Journal of Sports Sciences* 15(5): 485-490.

Adams, J., M. Mottola, K.M. Bagnall, and K.D. McFadden. 1982. Total body fat content in a group of professional football players. *Canadian Journal of Applied Sport Sciences* 7: 36-40.

Bailey, D.A., H.A. McKay, R.L. Mirwald, P.R. Crocker, and R.A. Faulkner. 1999. A six-year longitudinal study of the relationship of physical activity to bone mineral accrual in growing children: The University of Saskatchewan bone mineral accrual study. *Journal of Bone and Mineral Research* 14: 1672-1679.

Ballard, T.P., L. Fafara, and M.D. Vukovich. 2004. Comparison of Bod Pod and DXA in female collegiate athletes. *Medicine and Science in Sports and Exercise* 36: 731-735.

Bass, S., G. Pearce, M. Bradney, E. Hendrich, P.D. Delmas, A. Harding, and E. Seeman. 1998. Exercise before puberty may confer residual benefits in bone density in adulthood: Studies in active prepubertal and retired female gymnasts. *Journal of Bone and Mineral Research* 13: 500-507.

Baumgartner, R.N. 1996. Electrical impedance and total body electrical conductivity. In *Human Body Composition,* edited by A.R. Roche, S.B. Heymsfield, and T.G. Lohman. Champaign, IL: Human Kinetics.

Bedell, G.N., R. Marshall, A.B. Dubois, and J.H. Harris. 1956. Measurement of the volume of gas in the gastrointestinal tract; values in normal subjects and ambulatory patients. *Journal of Clinical Investigation* 35: 336-345.

Blumberg, J. 1994. Nutrient requirements of the healthy elderly—should there be specific RDAs? *Nutrition Review* 52: S15-18.

Borst, S.E. 2004. Interventions for sarcopenia and muscle weakness in older people. *Age and Ageing* 33: 548-555.

Bouchard, C. 1997. Genetic determinants of regional fat distribution. *Human Reproduction* 12 Suppl 1: 1-5.

Bouchard, C., and L. Perusse. 1993. Genetics of obesity. *Annual Review of Nutrition* 13: 337-354.

Bouchard, C., L. Perusse, C. Leblanc, A. Tremblay, and G. Theriault. 1988. Inheritance of the amount and distribution of human body fat. *International Journal of Obesity* 12: 205-215.

Brozek, J. 1960. The measurement of body composition. In *A Handbook of Anthropometry,* edited by M.F.A. Montagu. Springfield, IL: Charles C Thomas.

Brozek, J., F. Grande, J.T. Anderson, and A. Keys. 1963. Densitometric analysis of body composition: revision of some quantitative assumptions. *Annals of the New York Academy of Sciences* 110: 113-140.

Cameron, J.R., and J. Sorenson. 1963. Measurement of bone mineral in vivo: An improved method. *Science* 142: 230-232.

Carmeli, E., R. Coleman, and A.Z. Reznick. 2002. The biochemistry of aging muscle. *Experimental Gerontology* 37: 477-489.

Carter, J.E.L., and T.R. Ackland. 1994. *Kinanthropometry in Aquatic Sports.* Champaign, IL: Human Kinetics.

Clark, R.R., C. Bartok, J.C. Sullivan, and D.A. Schoeller. 2004. Minimum weight prediction methods cross-validated by the four-component model. *Medicine and Science in Sports and Exercise* 36: 639-647.

Clarys, J.P., A.D. Martin, and D.T. Drinkwater. 1984. Gross tissue weights in the human body by cadaver dissection. *Human Biology* 56: 459-473.

Collins, M.A., M.L. Millard-Stafford, P.B. Sparling, T.K. Snow, L.B. Rosskopf, S.A. Webb, and J. Omer. 1999. Evaluation of the BOD POD for assessing body fat in collegiate football players. *Medicine and Science in Sports and Exercise* 31: 1350-1356.

De Lorenzo, A., I. Bertini, L. Iacopino, E. Pagliato, C. Testolin, and G. Testolin. 2000. Body composition measurement in highly trained male athletes. A comparison of three methods. *Journal of Sports Medicine and Physical Fitness* 40: 178-183.

Dempster, P., and S. Aitkens. 1995. A new air displacement method for the determination of human body composition. *Medicine and Science in Sports and Exercise* 27: 1692-1697.

DuBois, D., and E.F. DuBois. 1916. Clinical calorimetry: A formula to estimate the approximate surface area if stature and weight are known. *Archives of Internal Medicine* 17: 863-871.

Durnin, J.V., and J. Womersley. 1974. Body fat assessed from total body density and its estimation from skinfold thickness: Measurements on 481 men and women aged from 16 to 72 years. *British Journal of Nutrition* 32: 77-97.

Epstein, S. 1988. Serum and urinary markers of bone remodeling: Assessment of bone turnover. *Endocrine Review* 9: 437-449.

Evans, E.M., B.M. Prior, S.A. Arngrimsson, C.M. Modlesky, and K.J. Cureton. 2001. Relation of bone mineral density and content to mineral content and density of the fat-free mass. *Journal of Applied Physiology* 91: 2166-2172.

Fiatarone, M.A., E.F. O'Neill, N.D. Ryan, K.M. Clements, G.R. Solares, M.E. Nelson, S.B. Roberts, J.J. Kehayias, L.A. Lipsitz, and W.J. Evans. 1994. Exercise training and nutritional supplementation for physical frailty in very elderly people. *New England Journal of Medicine* 330: 1769-1775.

Fuller, N.J., M.A. Laskey, and M. Elia. 1992. Assessment of the composition of major body regions by dual-energy X-ray absorptiometry (DEXA), with special reference to limb muscle mass. *Clinical Physiology* 12: 253-266.

Hansen, N.J., T.G. Lohman, S.B. Going, M.C. Hall, R.W. Pamenter, L.A. Bare, T.W. Boyden, and L.B. Houtkooper. 1993. Prediction of body composition in premenopausal females from dual-energy X-ray absorptiometry. *Journal of Applied Physiology* 75: 1637-1641.

He, Q., M. Horlick, J. Thornton, J. Wang, R.N. Pierson Jr., S. Heshka, and D. Gallagher. 2004. Sex-specific fat distribution is not linear across pubertal groups in a multiethnic study. *Obesity Research* 12: 725-733.

Holliday, M. 1978. Body composition and energy needs during growth. In *Human Growth,* Vol. 2, edited by F. Falkner and J. Tanner. London: Bailliere Tindall.

Iuliano-Burns, S., L. Saxon, G. Naughton, K. Gibbons, and S.L. Bass. 2003. Regional specificity of exercise and calcium during skeletal growth in girls: A randomized controlled trial. *Journal of Bone and Mineral Research* 18: 156-162.

Jebb, S.A., G.R. Goldberg, and M. Elia. 1993. DXA measurements of fat and bone mineral density in relation to depth and adiposity. In *Human Body Composition,* edited by K.J. Ellis and J.D. Eastman. New York: Plenum Press.

Kerr, D., T. Ackland, B. Maslen, A. Morton, and R. Prince. 2001. Resistance training over 2 years increases bone mass in calcium-replete postmenopausal women. *Journal of Bone and Mineral Research* 16: 175-181.

Kerr, D., A. Morton, I. Dick, and R. Prince. 1996. Exercise effects on bone mass in postmenopausal women are site-specific and load-dependent. *Journal of Bone and Mineral Research* 11: 218-225.

Kohrt, W.M., S.A. Bloomfield, K.D. Little, M.E. Nelson, and V.R. Yingling. 2004. American College of Sports Medicine position stand: Physical activity and bone health. *Medicine and Science in Sports and Exercise* 36: 1985-1996.

Lemon, P.W. 2000. Beyond the zone: Protein needs of active individuals. *Journal of the American College of Nutrition* 19: 513S-521S.

Lohman, T., S. Going, R. Pamenter, M. Hall, T. Boyden, L. Houtkooper, C. Ritenbaugh, L. Bare, A. Hill, and M. Aickin. 1995. Effects of resistance training on regional and total bone mineral density in premenopausal women: A randomized prospective study. *Journal of Bone and Mineral Research* 10: 1015-1024.

MacKelvie, K.J., K.M. Khan, and H.A. McKay. 2002. Is there a critical period for bone response to weight-bearing exercise in children and adolescents? A systematic review. *British Journal of Sports Medicine* 36: 250-257.

MacKelvie, K.J., K.M. Khan, M.A. Petit, P.A. Janssen, and H.A. McKay. 2003. A school-based exercise intervention elicits substantial bone health benefits: A 2-year randomized controlled trial in girls. *Pediatrics* 112: e447.

Malina, R. 1978. Growth of muscle tissue and muscle mass. In *Human Growth,* Vol. 2, edited by F. Falkner and J. Tanner. London: Bailliere Tindall.

Malina, R.M. 1987. Bioelectric methods for estimating body composition: An overview and discussion. *Human Biology* 59: 329-235.

Marshall, W.A. 1978. Puberty. In *Human Growth,* Vol. 2, edited by F. Falkner and J. Tanner. London: Bailliere Tindall.

Martin, A.D., W.D. Ross, D.T. Drinkwater, and J.P. Clarys. 1985. Prediction of body fat by skinfold caliper: Assumptions and cadaver evidence. *International Journal of Obesity* 9 Suppl 1: 31-39.

Mazess, R.B., H.S. Barden, J.P. Bisek, and J. Hanson. 1990. Dual-energy x-ray absorptiometry for total-body and regional bone-mineral and soft-tissue composition. *American Journal of Clinical Nutrition* 51: 1106-1112.

McCrory, M.A., T.D. Gomez, E.M. Bernauer, and P.A. Mole. 1995. Evaluation of a new air displacement plethysmograph for measuring human body composition. *Medicine and Science in Sports and Exercise* 27: 1686-1691.

Menkes, A., S. Mazel, R.A. Redmond, K. Koffler, C.R. Libanati, C.M. Gundberg, T.M. Zizic, J.M. Hagberg, R.E. Pratley, and B.F. Hurley. 1993. Strength training increases regional bone mineral density and bone remodeling in middle-aged and older men. *Journal of Applied Physiology* 74: 2478-2484.

Mitchell, A.D., J.M. Conway, and W.J. Potts. 1996. Body composition analysis of pigs by dual-energy x-ray absorptiometry. *Journal of Animal Science* 74: 2663-2671.

Nelson, M.E., M.A. Fiatarone, C.M. Morganti, I. Trice, R.A. Greenberg, and W.J. Evans. 1994. Effects of high-intensity strength training on multiple risk factors for osteoporotic fractures. A randomized controlled trial. *Journal of the American Medical Association* 272: 1909-1914.

Nguyen, T.V., G.M. Howard, P.J. Kelly, and J.A. Eisman. 1998. Bone mass, lean mass, and fat mass: Same genes or same environments? *American Journal of Epidemiology* 147: 3-16.

Nindl, B.C., K.E. Friedl, L.J. Marchitelli, R.L. Shippee, C.D. Thomas, and J.F. Patton. 1996. Regional fat placement in physically fit males and changes with weight loss. *Medicine and Science in Sports and Exercise* 28: 786-793.

Nord, R., and R. Payne. 1995. Body composition by dual energy X-ray absorptiometry—a review of the technology. *Asia Pacific Journal of Clinical Nutrition* 4: 167-171.

Norris, A., T. Lundy, and N. Shock. 1963. Trends in selected indices of body composition between the ages 30 and 80 years. *Annals of the New York Academy of Science* 110: 623.

Norton, K. 1996. Anthropometric estimation of body fat. In *Anthropometrica,* edited by K. Norton and T. Olds. Sydney: UNSW Press.

Pietrobelli, A., C. Formica, Z. Wang, and S.B. Heymsfield. 1996. Dual-energy X-ray absorptiometry body composition model: Review of physical concepts. *American Journal of Physiology* 271: E941-951.

Poulter, N.R. 2001. Birthweights, maternal cardiovascular events, and Barker hypothesis. *Lancet* 357: 1990-1991.

Prince, R., A. Devine, I. Dick, A. Criddle, D. Kerr, N. Kent, R. Price, and A. Randell. 1995. The effects of calcium supplementation (milk powder or tablets) and exercise on bone density in postmenopausal women. *Journal of Bone and Mineral Research* 10: 1068-1075.

Prior, B.M., K.J. Cureton, C.M. Modlesky, E.M. Evans, M.A. Sloniger, M. Saunders, and R.D. Lewis. 1997. In vivo validation of whole body composition estimates from dual-energy X-ray absorptiometry. *Journal of Applied Physiology* 83: 623-630.

Prior, B.M., C.M. Modlesky, E.M. Evans, M.A. Sloniger, M.J. Saunders, R.D. Lewis, and K.J. Cureton. 2001. Muscularity and the density of the fat-free mass in athletes. *Journal of Applied Physiology* 90: 1523-1531.

Raisz, L.G. 1999. Physiology and pathophysiology of bone remodeling. *Clinical Chemistry* 45: 1353-1358.

Rennie, M.J., and K.D. Tipton. 2000. Protein and amino acid metabolism during and after exercise and the effects of nutrition. *Annual Review of Nutrition* 20: 457-483.

Ross, W.D., S.M. Crawford, D.A. Kerr, R. Ward, D.A. Bailey, and R.M. Mirwald. 1988. Relationship of the body mass index with skinfolds, girths, and bone breadths in Canadian men and women aged 20-70 years. *American Journal of Physical Anthropology* 77: 169-173.

Ryan, A.S., M.S. Treuth, M.A. Rubin, J.P. Miller, B.J. Nicklas, D.M. Landis, R.E. Pratley, C.R. Libanati, C.M. Gundberg, and B.F. Hurley. 1994. Effects of strength training on bone mineral density: Hormonal and bone turnover relationships. *Journal of Applied Physiology* 77: 1678-1684.

Segal, K.R. 1996. Use of bioelectrical impedance analysis measurements as an evaluation for participating in sports. *American Journal of Clinical Nutrition* 64: 469S-471S.

Segal, K.R., B. Gutin, E. Presta, J. Wang, and T.B. Van Itallie. 1985. Estimation of human body composition by electrical impedance methods: A comparative study. *Journal of Applied Physiology* 58: 1565-1571.

Siri, W.E. 1961. Body composition from fluid spaces and density: Analysis of methods. In *Techniques for Measuring Body Composition,* edited by J. Brozek and A. Henschel. Washington, DC: National Academy of Sciences.

Snow-Harter, C., M.L. Bouxsein, B.T. Lewis, D.R. Carter, and R. Marcus. 1992. Effects of resistance and endurance exercise on bone mineral status of young women: A randomized exercise intervention trial. *Journal of Bone and Mineral Research* 7: 761-769.

Stewart, A. 1999. Body composition of athletes assessed by dual x-ray absorptiometry and other methods. Unpublished PhD thesis. Edinburgh: The University of Edinburgh.

Stewart, A. 2003. Fat patterning: Indicators and implications. *Nutrition* 19: 559-560.

Stewart, A.D., and W.J. Hannan. 2000. Prediction of fat and fat-free mass in male athletes using dual X-ray absorptiometry as the reference method. *Journal of Sports Sciences* 18: 263-274.

Stewart, A.D., and J. Hannan. 2000. Sub-regional tissue morphometry in male athletes and controls using dual x-ray absorptiometry (DXA). *International Journal of Sport Nutrition and Exercise Metabolism* 10: 157-169.

Stunkard, A.J., J.R. Harris, N.L. Pedersen, and G.E. McClearn. 1990. The body-mass index of twins who have been reared apart. *New England Journal of Medicine* 322: 1483-1487.

Svendsen, O.L., J. Haarbo, C. Hassager, and C. Christiansen. 1993. Accuracy of measurements of body composition by dual-energy x-ray absorptiometry in vivo. *American Journal of Clinical Nutrition* 57: 605-608.

Tarnopolsky, M. 2004. Protein requirements for endurance athletes. *Nutrition* 20: 662-668.

Tarnopolsky, M.A., G. Parise, N.J. Yardley, C.S. Ballantyne, S. Olatinji, and S.M. Phillips. 2001. Creatine-dextrose and protein-dextrose induce similar strength gains during training. *Medicine and Science in Sports and Exercise* 33: 2044-2052.

Tipton, K.D., T.A. Elliott, M.G. Cree, S.E. Wolf, A.P. Sanford, and R.R. Wolfe. 2004. Ingestion of casein and whey proteins result in muscle anabolism after resistance exercise. *Medicine and Science in Sports and Exercise* 36: 2073-2081.

Tittel, K. 1978. Tasks and tendencies of sport anthropometry's development. In *Biomechanics of Sport and Kinanthropometry,* edited by F. Landry and W.A. Orban. Miami: Symposia Specialists.

Ward, R., W.D. Ross, A.J. Leyland, and S. Selbie. 1989. *The Advanced O-Scale Physique Assessment System.* Burnaby, Canada: Kinemetrix Inc.

Welle, S., C. Thornton, and M. Statt. 1995. Myofibrillar protein synthesis in young and old human subjects after three months of resistance training. *American Journal of Physiology* 268: E422-427.

Wessel, J., A. Ufer, W. Van Huss, and D. Cederquist. 1963. Age trends of various components of body composition and functional characteristics in women aged 20-69 years. *Annals of the New York Academy of Science* 110: 608.

Wilmore, J.H., and A.R. Behnke. 1969. An anthropometric estimation of body density and lean body weight in young men. *Journal of Applied Physiology* 27: 25-31.

Wilmore, J.H., and D.L. Costill. 1988. *Training for Sport and Activity: The Physiological Basis of the Conditioning Process.* Dubuque, IA: Brown.

Withers, R.T., N.P. Craig, P.C. Bourdon, and K.I. Norton. 1987. Relative body fat and anthropometric prediction of body density of male athletes. *European Journal of Applied Physiology and Occupational Physiology* 56: 191-200.

Withers, R.T., K.I. Norton, N.P. Craig, M.C. Hartland, and W. Venables. 1987. The relative body fat and anthropometric prediction of body density of South Australian females aged 17-35 years. *European Journal of Applied Physiology and Occupational Physiology* 56: 181-190.

Woolford, S., P. Bourdon, N. Craig, and T. Stanef. 1993. Body composition and its effects on athletic performance. *Sports Coach* 16: 24-30.

World Health Organization. 1994. *Assessment of Fracture Risk and Its Application to Screening for Post-Menopausal Women* (Technical Report Series No. 843, WHO Scientific Study Group). Geneva: World Health Organization.

Yeager, K.K., R. Agostini, A. Nattiv, and B. Drinkwater. 1993. The female athlete triad: Disordered eating, amenorrhea, osteoporosis. *Medicine and Science in Sports and Exercise* 25: 775-777.

Chapter 6

Ackland, T.R. 1999. Talent identification: What makes a champion swimmer? In *International Congress on Biomechanics in Sport.* Perth: International Society of Biomechanics in Sport.

Ackland, T.R., and J. Bloomfield. 1995. Functional anatomy. In *Textbook of Science and Medicine in Sport* (2nd ed.), edited by J. Bloomfield, P. Fricker, and K. Fitch. Melbourne: Blackwell Scientific.

Ackland, T.R., and J. Bloomfield. 1996. Stability of human proportions through adolescent growth. *Australian Journal of Science and Medicine in Sport* 28: 57-60.

Ackland, T., J. Mazza, and J. Carter. 1994. Summary and implications. In *Kinanthropometry in Aquatic Sports,* edited by J.E.L. Carter and T.R. Ackland. Champaign, IL: Human Kinetics.

Ackland, T., K. Ong, D. Kerr, and B. Ridge. 2003. Morphological characteristics of Olympic sprint canoe and kayak paddlers. *Journal of Science and Medicine in Sport* 6: 285-294.

Amar, J. 1920. *The Human Motor.* New York: Dutton.

Arnold, A. 1931. *Korperent Kicklung und Leibesubungen fur Schul und Sportarzte.* Leipzig: Johann Barth.

Behnke, A., O. Guttentag, and C. Brodsky. 1959. Quantification of body weight and configuration from anthropometric measurements. *Human Biology* 31: 213-234.

Behnke, A., and J. Wilmore. 1974. *Evaluation and Regulation of Body Build and Composition.* Englewood Cliffs, NJ: Prentice-Hall.

Bloomfield, J. 1979. Modifying human physical capacities and technique to improve performance. *Sports Coach* 3: 19-25.

Bloomfield, J., and P. Sigerseth. 1965. Anatomical and physiological differences between sprint and middle distance swimmers at the university level. *Journal of Sports Medicine and Physical Fitness* 5: 76-81.

Boardman, R. 1933. World's champions run to types. *Journal of Health and Physical Education* 4: 32.

Carter, J.E.L. 1984. Physical structure of Olympic athletes. In *Medicine and Sports Science,* Vol. 18. Basel: Karger.

Carter, J.E.L., and T.R. Ackland. 1994. *Kinanthropometry in Aquatic Sports.* Champaign, IL: Human Kinetics.

Cureton, T. 1951. *Physical Fitness of Champion Athletes.* Urbana, IL: University of Illinois Press.

de Garay, A., L. Levine, and J. Carter. 1974. *Genetic and Anthropological Studies of Olympic Athletes.* New York: Academic Press.

De Ridder, J.H., L.O. Amusa, K.D. Monyeki, A.L. Toriola, and J.E.L. Carter. 2001. Kinanthropometry in African sports: Body composition and somatotypes of world class female African middle-, long distance and marathon runners. *African Journal for Physical, Health Education, Recreation and Dance* 7(1): 1-13.

De Ridder, J.H., K.D. Monyeki, L.O. Amusa, A.L. Toriola, M. Wekesa, and J.E.L. Carter. 2000. Kinanthropometry in African sports: Body composition and somatotypes of world class male African middle-, long distance and marathon runners. In *Kinanthropometry XI,* edited by K. Norton, T. Olds, and J. Dollman. Adelaide: ISAK.

De Ridder, J.H., E.M. Smith, C.J. Wilders, and C. Underhay. 2003. Sexual dimorphism in elite middle-distance runners: 1995 All-Africa Games (Project HAAGKiP). In *Kinanthropometry XII,* edited by H. De Ridder and T. Olds. Brisbane: ISAK.

De Ridder, J.H., C. Underhay, L.O. Amusa, A.L. Toriola, A.P. Agbonjinmi, and J.O. Adeogun. 2005. Sexual dimorphism in world-class African long distance runners. *African Journal for Physical, Health Education, Recreation and Dance.* Under review.

Dintiman, G., and R. Ward. 1988. *Sport Speed.* Champaign, IL: Leisure Press.

Eiben, O. 1972. *The Physique of Women Athletes.* Budapest: Hungarian Scientific Council for Physical Education.

Garrett, J., and W. Kennedy. 1971. *A Collation of Anthropometry,* Vols. 1 and 2. Springfield, VA: National Technical Information Services.

Gomes, P., and J. Mazza. 1998. Proporcionalidad corporal de futbolistas Sudamericanos. In *Futbolistas Sudamericano de Elite: Morphfologica, Analisis del Juego y Performance,* edited by E. Rienzi, J.C. Mazza, J.E.L. Carter, and T. Reilly. Rosario: Biosystem Servicio Educativo.

Hart, C., T. Ward, and J. Mayhew. 1991. Anthropometric correlates of bench press performance following resistance training. *Sports Training, Medicine and Rehabilitation* 2: 89-95.

Hills, A. 1991. *Physical Growth and Development of Children and Adolescents.* Brisbane: Queensland University of Technology.

Kerr, D., W. Ross, K. Norton, P. Hume, M. Kagawa, and T. Ackland. 2006. Olympic lightweight and open rowers possess distinctive physical and proportionality characteristics. *Journal of Sports Sciences.* Under review.

Kohlrausch, W. 1929. Zusammenlgne von korpenform und leistung—ergebruise der anthropoinetrischen messungen an der athleten der Amsterdamer Olympiade. *Arbeitsphysiologie* 2: 129.

Kruger, A., J.H. De Ridder, H.W. Grobbelaar, and C. Underhay. 2005. A kinanthropometric profile and morphological prediction functions of elite international male javelin throwers. In *Kinanthropometry IX,* edited by M. Marfell-Jones, A. Stewart, and T. Olds. Thessaloniki: ISAK.

Malina, R., and C. Bouchard. 1991. *Growth, Maturation, and Physical Activity.* Champaign, IL: Human Kinetics.

Metheny, E. 1939. Some differences in bodily proportions between the American Negro and white male college students as related to athletic performance. *Research Quarterly* 10: 41-53.

Meyer, E., J.H. De Ridder, W.G. Schulze, and S. Ellis. 2005. The evolution of physique in male South African Springbok rugby players: 1896-2004. Unpublished manuscript.

Olds, T. Centre for Applied Anthropometry, University of South Australia. Personal communication.

Ross, W., R.M. Leahy, J.C. Mazza, and D.T. Drinkwater. 1994. Relative body size. In *Kinanthropometry in Aquatic Sports,* edited by J.E.L. Carter and T.R. Ackland. Champaign, IL: Human Kinetics.

Ross, W., and M. Marfell-Jones. 1991. Kinanthropometry. In *Physiological Testing of the High Performance Athlete* (2nd ed.), edited by J. MacDougall, H. Wenger, and H. Green. Champaign, IL: Human Kinetics.

Ross, W., and N. Wilson. 1974. A stratagem for proportional growth assessment. *Acta Paediatrica Belgica* 28: 169-182.

Sargent, D. 1887. The physical characteristics of the athlete. *Scribners II* 5: 541-561.

Tanner, J. 1964. *The Physique of the Olympic Athlete.* London: Allen and Unwin.

Tanner, J. 1989. *Foetus into Man.* Ware, UK: Castlemead.

Underhay, C., J. De Ridder, L. Amusa, A. Toriola, A. Agbonjinmi, and J. Adeogun. 2005. Physique characteristics of world class African long distance runners. *African Journal for Physical, Health Education, Recreation and Dance* 11: 6-16.

Chapter 7

Ackland, T., and J. Bloomfield. 1992. Functional anatomy. In *Textbook of Science and Medicine in Sport,* edited by J. Bloomfield, P. Fricker, and K. Fitch. Melbourne: Blackwell Scientific.

Adams, R., A. Daniel, and L. Rullman. 1975. *Games, Sports and Exercises for the Physically Handicapped.* Philadelphia: Lea & Febiger.

Bloomfield, J. 1979. Modifying human physical capacities and technique to improve performance. *Sports Coach* 3: 19-25.

Brodecker, P. 1952. *Physical Build vs Athletic Ability in American Sports.* Chicago: Athletic Ability.

Chaffin, D., and G. Andersson. 1984. *Occupational Biomechanics.* New York: Wiley.

Hills, A. 1991. *Physical Growth and Development of Children and Adolescents.* Brisbane: Queensland University of Technology.

Kendall, F.P., E.K. McCreary, P.G. Provance, M.M. Rodgers, and W.A. Romani. 2005. *Muscles, Testing and Function with Posture and Pain* (5th ed.). Baltimore: Lippincott, Williams & Wilkins.

Krogman, W. 1951. The scars of human evolution. *Scientific American* 185: 54-57.

Lorenzton, R. 1988. Causes of injuries: Intrinsic factors. In *The Olympic Book of Sports Medicine I,* edited by A. Dirix, H. Knuttgen, and K. Tittle. Oxford: Blackwell Scientific.

Lun, V., W.H. Meeuwisse, P. Stergiou, and D. Stefanyshyn. 2004. Relation between running injury and static lower limb alignment in recreational runners. *British Journal of Sports Medicine* 38: 576-580.

Napier, J. 1967. The antiquity of human walking. *Scientific American* 3: 38-48.

Rasch, P., and R. Burke. 1978. *Kinesiology and Applied Anatomy.* Philadelphia: Lea & Febiger.

Sheldon, W., S. Stevens, and W. Tucker. 1940. *The Varieties of Human Physique.* New York: Harper and Brothers.

Sinclair, D. 1973. *Human Growth after Birth.* London: Oxford University Press.

Tanner, J. 1964. *The Physique of the Olympic Athlete.* London: Allen and Unwin.

Watson, A. 1992. Children in sport. In *Textbook of Science and Medicine in Sport,* edited by J. Bloomfield, P. Fricker, and K. Fitch. Melbourne: Blackwell Scientific.

Webster, F. 1948. *The Science of Athletics.* London: Nicholas Kaye.

Chapter 8

Aagaard, P. 2003. Training-induced changes in neural function. *Exercise and Sport Sciences Reviews* 31: 61-67.

Ahtiainen, J.P., A. Pakarinen, W.J. Kraemer, and K. Häkkinen. 2003. Acute hormonal and neuromuscular responses and recovery to forced vs. maximum repetitions multiple resistance exercises. *International Journal of Sports Medicine* 24: 410-418.

American College of Sports Medicine. 2002. Position stand: Progression models in resistance training for healthy adults. *Medicine and Science in Sports and Exercise* 34: 364-380.

Anton, M.M., W.W. Spirduso, and H. Tanaka. 2004. Age-related declines in anaerobic muscular performance: Weightlifting and powerlifting. *Medicine and Science in Sports and Exercise* 36: 143-147.

Augustsson, J., A. Esko, R. Thomee, and U. Svantesson. 1998. Weight training of the thigh muscles using closed vs. open kinetic chain exercises: A comparison of performance enhancement. *Journal of Orthopedics and Sports Physical Therapy* 27: 3-8.

Augustsson, J., and R. Thomee. 2000. Ability of closed and open kinetic chain tests of muscular strength to assess functional performance. *Scandinavian Journal of Medicine and Science in Sports* 10: 164-168.

Baker, D., G. Wilson, and R. Carlyon. 1994. Periodization: The effect on strength of manipulating volume and intensity. *Journal of Strength and Conditioning Research* 8: 235-242.

Blackburn, J.R., and M.C. Morrissey. 1998. The relationship between open and closed kinetic chain strength of the lower limb and jumping performance. *Journal of Orthopedics and Sports Physical Therapy* 27: 430-435.

Blazevich, A.J., N.D. Gill, R. Bronks, and R.U. Newton. 2003. Training-specific muscle architecture adaptation after 5-wk training in athletes. *Medicine and Science in Sports and Exercise* 35: 2013-2022.

Bloomfield, J., B. Blanksby, T. Ackland, and G. Allison. 1990. The influence of strength training on overhead throwing velocity of elite water polo players. *Australian Journal of Science and Medicine in Sport* 22: 63-67.

Bloomfield, J., P.A. Fricker, and K.D. Fitch. 1992. *Textbook of Science and Medicine in Sport.* Melbourne: Blackwell Scientific.

Brandenburg, J.P., and D. Docherty. 2002. The effects of accentuated eccentric loading on strength, muscle hypertrophy, and neural adaptations in trained individuals. *Journal of Strength and Conditioning Research* 16: 25-32.

Campos, G.E.R., T.L. Luecke, H.K. Wendeln, K. Toma, F.C. Hagerman, T.F. Murray, K.E. Ragg, N.A. Ratamess, W.J. Kraemer, and R.S. Staron. 2002. Muscular adaptations in response to three different resistance-training regimens: Specificity of repetition maximum training zones. *European Journal of Applied Physiology* 88: 50-60.

Carroll, T.J., S. Riek, and R.G. Carson. 2001. Neural adaptations to resistance training: Implications for movement control. *Sports Medicine* 31: 829-840.

Drinkwater, E.J., T.W. Lawton, R.P. Lindsell, D.B. Pyne, P.H. Hunt, and M.J. McKenna. 2005. Training leading to repetition failure enhances bench press strength gains in elite junior athletes. *Journal of Strength and Conditioning Research* 19: 382-388.

Durell, D.L., T.J. Pujol, and J.T. Barnes. 2003. A survey of the scientific data and training methods utilized by collegiate strength and conditioning coaches. *Journal of Strength and Conditioning Research* 17: 368-373.

Ebben, W.P., and R.L. Jensen. 2002. Electromyographic and kinetic analysis of traditional, chain, and elastic band squats. *Journal of Strength and Conditioning Research* 16: 547-550.

Escamilla, R. 1988. The use of powerlifting aids in the squat. *Powerlifting USA* 12: 14-15.

Escamilla, R.F., G.S. Fleisig, N. Zheng, S.W. Barrentine, K.E. Wilk, and J.R. Andrews. 1998. Biomechanics of the knee during closed kinetic chain and open kinetic chain exercises. *Medicine and Science in Sports and Exercise* 30: 556-569.

Escamilla, R.F., K.P. Speer, G.S. Fleisig, S.W. Barrentine, and J.R. Andrews. 2000. Effects of throwing overweight and underweight baseballs on throwing velocity and accuracy. *Sports Medicine* 29: 259-272.

Evans, W.J. 2000. Exercise strategies should be designed to increase muscle power. *Journals of Gerontology: Biological Sciences* 55: M309-M310.

Faigenbaum, A.D., R.L. Loud, J. O'Connell, S. Glover, J. O'Connell, and W.L. Westcott. 2001. Effects of different resistance training protocols on upper-body strength and endurance development in children. *Journal of Strength and Conditioning Research* 15: 459-465.

Faulkner, J., D. Claflin, and K. McCully. 1986. Power output of fast and slow fibers from human skeletal muscles. In *Human Muscle Power,* edited by N. Jones, N. McCartney, and A. McComas. Champaign, IL: Human Kinetics.

Fleck, S.J., and W.J. Kraemer. 2004. *Designing Resistance Training Programs* (3rd ed.). Champaign, IL: Human Kinetics.

Folland, J.P., C.S. Irish, J.C. Roberts, J.E. Tarr, and D.A. Jones. 2002. Fatigue is not a necessary stimulus for strength gains during resistance training. *British Journal of Sports Medicine* 36: 370-374.

Fry, A.C. 2004. The role of resistance exercise intensity on muscle fiber adaptations. *Sports Medicine* 34: 663-679.

Fry, A.C., and W.J. Kraemer. 1991. Physical performance characteristics of American football players. *Journal of Applied Sport Science Research* 5: 126-139.

Fry, A.C., W.J. Kraemer, C.A. Weseman, B.P. Conroy, S.E. Gordon, J.R. Hoffman, and C.M. Maresh. 1991. Effects of an off-season strength and conditioning program on starters and non-starters in women's collegiate volleyball. *Journal of Applied Sport Science Research* 5: 174-181.

Fry, A.C., B.K. Schilling, R.S. Staron, F.C. Hagerman, R.S. Hikida, and J.T. Thrush. 2003. Muscle fiber characteristics and performance correlates of male Olympic-style weightlifters. *Journal of Strength and Conditioning Research* 17: 746-754.

Fry, A.C., J.M. Webber, L.W. Weiss, M.P. Harber, M. Vaczi, and N.A. Pattison. 2003. Muscle fiber characteristics of competitive powerlifters. *Journal of Strength and Conditioning Research* 17: 402-410.

Giorgi, A., G.J. Wilson, R.P. Weatherby, and A.J. Murphy. 1998. Functional isometric weight training: Its effects on the development of muscular function and the endocrine system over an 8-week training period. *Journal of Strength and Conditioning Research* 12: 18-25.

Glass, S.C., and D.R. Stanton. 2004. Self-selected resistance training intensity in novice weightlifters. *Journal of Strength and Conditioning Research* 18: 324-327.

Glowacki, S.P., S.E. Martin, A. Maurer, W. Back, J.S. Green, and S.F. Crouse. 2004. Effects of resistance, endurance, and concurrent exercise on training outcomes in men. *Medicine and Science in Sports and Exercise* 36: 2119-2127.

Goertzen, M., K. Schoppe, G. Lange, and K.P. Schulitz. 1989. Injuries and damage caused by excess stress in body building and power lifting. *Sportverletzung Sportschaden* 3: 32.

Griffin, J. 1987. Differences in elbow flexion torque measured concentrically, eccentrically and isometrically. *Physical Therapy* 67: 1205-1209.

Haff, G.G., A. Whitley, L.B. McCoy, H.S. O'Bryant, J.L. Kilgore, E.N. Haff, K. Pierce, and M.H. Stone. 2003. Effects of different set configurations on barbell velocity and displacement during a clean pull. *Journal of Strength and Conditioning Research* 17: 95-103.

Häkkinen, K., M. Alén, and P. Komi. 1984. Neuromuscular, anaerobic and aerobic performance characteristics of elite power athletes. *European Journal of Applied Physiology* 53: 97-105.

Häkkinen, K., M. Alén, W.J. Kraemer, E. Gorostiaga, M. Izquierdo, H. Rusko, J. Mikkola, A. Häkkinen, H. Valkeinen, E. Kaarakainen, S. Romu, V. Erola, J. Ahtiainen, and L. Paavolainen. 2003. Neuromuscular adaptations during concurrent strength and endurance training versus strength training. *European Journal of Applied Physiology* 89: 42-52.

Häkkinen, K., P. Komi, and M. Alén. 1985. Effect of explosive type strength training on isometric force- and relaxation-time, electromyographic and muscle fiber characteristics of leg extensor muscles. *Acta Physiologica Scandinavica* 125: 587-600.

Häkkinen, K., P. Komi, M. Alén, and H. Kauhanen. 1987. EMG, muscle fiber and force production characteristics during a one year training period in elite weight-lifters. *European Journal of Applied Physiology* 56: 419-427.

Häkkinen, K., W.J. Kraemer, A. Pakarinen, T. Triplett-McBride, J.M. McBride, A. Häkkinen, M. Alén, M.R. McGuigan, R. Bronks, and R.U. Newton. 2002. Effects of heavy resistance/power training on maximal strength, muscle morphology, and hormonal response patterns in 60-75-year-old men and women. *Canadian Journal of Applied Physiology* 27: 213-231.

Hansen, S., T. Kvorning, M. Kjaer, and G. Sjogaard. 2001. The effect of short-term strength training on human skeletal muscle: The importance of physiologically elevated hormone levels. *Scandinavian Journal of Medicine and Science in Sports* 11: 347-354.

Hart, C., T. Ward, and J. Mayhew. 1991. Anthropometric correlates of bench press performance following resistance training. *Sports Training, Medicine and Rehabilitation* 2: 89-95.

Henneman, E., H. Clamann, J. Gillies, and R. Skinner. 1974. Rank order of motorneurons within a pool: Law of combination. *Journal of Neurophysiology* 37: 1338-1349.

Hoeger, W.W., S.L. Barette, D.F. Hale, and D.R. Hopkins. 1987. Relationship between repetitions and selected percentages of one repetition maximum. *Journal of Applied Sport Science Research* 1: 11-13.

Hoffman, J.R., and J. Kang. 2003. Strength changes during an in-season resistance-training program for football. *Journal of Strength and Conditioning Research* 17: 109-114.

Hoffman, J.R., M. Wendell, J. Cooper, and J. Kang. 2003. Comparison between linear and nonlinear in-season training programs in freshman football players. *Journal of Strength and Conditioning Research* 17: 561-565.

Ikai, M., and A. Steinhaus. 1961. Some factors modifying the expression of human strength. *Journal of Applied Physiology* 16: 157-163.

Jackson, A., T. Jackson, J. Hnatek, and J. West. 1985. Strength development: Using functional isometrics in an isotonic strength training program. *Research Quarterly for Exercise and Sport* 56: 234-237.

Jones, K., P. Bishop, G. Hunter, and G. Fleissig. 2001. The effects of varying resistance-training loads on intermediate- and high-velocity-specific adaptations. *Journal of Strength and Conditioning Research* 15: 349-356.

Kemmler, W.K., D. Lauber, K. Engelke, and J. Weineck. 2004. Effects of single- vs. multiple-set resistance training on maximum strength and body composition in trained postmenopausal women. *Journal of Strength and Conditioning Research* 18: 689-694.

Keogh, J.W.L., G.J. Wilson, and R.P. Weatherby. 1999. A cross-sectional comparison of different resistance training techniques in the bench press. *Journal of Strength and Conditioning Research* 13: 247-258.

Knapik, J.J., C.T. Bauman, D.H. Jones, J.M. Harris, and L. Vaughan. 1991. Preseason strength and flexibility imbalances associated with athletic injuries in female collegiate athletes. *American Journal of Sports Medicine* 19: 76-81.

Kraemer, W.J., and L.A. Gotshalk. 2000. Physiology of American football. In *Exercise and Sport Science,* edited by W.E. Garrett and D.T. Kirkendall. Philadelphia: Lippincott, Williams & Wilkins.

Kraemer, W.J., B.C. Nindl, N.A. Ratamess, L.A. Gotshalk, J.S. Volek, S.J. Fleck, R.U. Newton, and K. Häkkinen. 2004. Changes in muscle hypertrophy in women with periodized resistance training. *Medicine and Science in Sports and Exercise* 36: 697-708.

Kraemer, W.J., J.F. Patton, S.E. Gordon, E.A. Harman, M.R. Deschenes, K. Reynolds, R.U. Newton, N.T. Triplett, and J.E. Dziados. 1995. Compatibility of high-intensity strength and endurance training on hormonal and skeletal muscle adaptations. *Journal of Applied Physiology* 78: 976-989.

Kulig, K., J.G. Andrews, and J.G. Hay. 1984. Human strength curves. *Exercise and Sport Sciences Reviews* 12: 417-466.

Lander, J.E., R.L. Simonton, and J.K.F. Giacobbe. 1990. The effectiveness of weight-belts during the squat exercise. *Medicine and Science in Sports and Exercise* 22: 117-126.

Lawton, T., J. Cronin, E. Drinkwater, R. Lindsell, and D. Pyne. 2004. The effect of continuous repetition training and intra-set rest training on bench press strength and power. *Journal of Sports Medicine and Physical Fitness* 44: 361-367.

Leveritt, M., and P.J. Abernethy. 1999. Acute effects of high-intensity endurance exercise on subsequent resistance activity. *Journal of Strength and Conditioning Research* 13: 47-51.

Mazzetti, S.A., W.J. Kraemer, J.S. Volek, N.D. Duncan, N.A. Ratamess, A.L. Gomez, R.U. Newton, K. Häkkinen, and S.J. Fleck. 2000. The influence of direct supervision of resistance training on strength performance. *Medicine and Science in Sports and Exercise* 32: 1175-1184.

McBride, J.M., N.T. Triplett-McBride, A. Davie, and R.U. Newton. 1999. A comparison of strength and power characteristics between power lifters, Olympic lifters, and sprinters. *Journal of Strength and Conditioning Research* 13: 58-66.

McBride, J.M., N.T. Triplett-McBride, A. Davie, and R.U. Newton. 2002. The effect of heavy- vs. light-load jump squats on the development of strength, power, and speed. *Journal of Strength and Conditioning Research* 16: 75-82.

McCarthy, J.P., M.A. Pozniak, and J.C. Agre. 2002. Neuromuscular adaptations to concurrent strength and endurance training. *Medicine and Science in Sports and Exercise* 34: 511-519.

McDonagh, M., and C. Davies. 1984. Adaptive response of mammalian skeletal muscle to exercise with high loads. *European Journal of Applied Physiology* 52: 139-155.

Minetti, A.E., and L.P. Ardigo. 2002. Halteres used in ancient Olympic long jump. *Nature* 420: 141-142.

Miyashita, M., and H. Kanehisa. 1979. Dynamic peak torque related to age, sex and performance. *Research Quarterly* 50: 249-255.

Mookerjee, S., and N.A. Ratamess. 1999. Comparison of strength differences and joint action durations between full and partial range-of-motion bench press exercise. *Journal of Strength and Conditioning Research* 13: 76-81.

Munn, J., R.D. Herbert, and S.C. Gandevia. 2004. Contralateral effects of unilateral resistance training: A meta-analysis. *Journal of Applied Physiology* 96: 1861-1866.

National Strength and Conditioning Association. 2000. *Essentials of Strength Training and Conditioning* (2nd ed.), edited by T.R. Baechle and R.W. Earle. Champaign, IL: Human Kinetics.

Newton, R.U., K. Häkkinen, A. Häkkinen, M. McCormick, J.S. Volek, and W.J. Kraemer. 2002. Mixed-methods resistance training increases power and strength of young and older men. *Medicine and Science in Sports and Exercise* 34: 1367-1375.

Newton, R.U., W.J. Kraemer, and K. Häkkinen. 1998. Effects of ballistic training on preseason preparation of elite volleyball players. *Medicine and Science in Sports and Exercise* 31: 323-330.

O'Shea, K.L., and J.P. O'Shea. 1989. Functional isometric weight training: Its effects on static and dynamic strength. *Journal of Applied Sport Science Research* 3: 30-33.

Peterson, M.D., M.R. Rhea, and B.A. Alvar. 2004. Maximizing strength development in athletes: A meta-analysis to determine the dose-response relationship. *Journal of Strength and Conditioning Research* 18: 377-382.

Phillips, S.M. 2000. Short-term resistance training: When do repeated bouts of resistance exercise become training? *Canadian Journal of Applied Physiology* 25: 185-193.

Poliquin, C. 2001. *Modern Trends in Strength Training.* Vol. 1: *Reps and Sets* (2nd ed.). www.CharlesPoliquin.net.

Raske, A., and R. Norlin. 2002. Injury incidence and prevalence among elite weight and power lifters. *American Journal of Sports Medicine* 30: 248-256.

Ratamess, N.A., W.J. Kraemer, J.S. Volek, M.R. Rubin, A.L. Gomez, D.N. French, M.J. Sharman, M.R. McGuigan, T. Scheett, K. Häkkinen, R.U. Newton, and F. Diouguardi. 2003. The effects of amino acid supplementation on muscular performance during resistance training overreaching: Evidence of an effective overreaching protocol. *Journal of Strength and Conditioning Research* 17: 250-258.

Rhea, M.R., S.D. Ball, W.T. Phillips, and L.N. Burkett. 2002. A comparison of linear and daily undulating periodized programs with equated volume and intensity for strength. *Journal of Strength and Conditioning Research* 16: 250-255.

Rhea, M.R., W.T. Phillips, L.N. Burkett, W.J. Stone, S.D. Ball, B.A. Alvar, and A.B. Thomas. 2003. A comparison of linear and daily undulating periodized programs with equated volume and intensity for local muscular endurance. *Journal of Strength and Conditioning Research* 17: 82-87.

Rooney, K.J., R.D. Herbert, and R.J. Balnave. 1994. Fatigue contributes to the strength training stimulus. *Medicine and Science in Sports and Exercise* 26: 1160-1164.

Rutherford, O., C. Greig, A. Sargent, and D. Jones. 1986. Strength training and power output: Transference effects in the human quadriceps muscle. *Journal of Sports Sciences* 4: 101-107.

Rutherford, O.M., and D.A. Jones. 1986. The role of learning and coordination in strength training. *European Journal of Applied Physiology* 55: 100-105.

Sale, D. 1991. Testing strength and power. In *Physiological Testing of the High Performance Athlete* (2nd ed.), edited by J. MacDougall, H. Wenger, and H. Green. Champaign, IL: Human Kinetics.

Sale, D.G., I. Jacobs, J.D. MacDougall, and S. Garner. 1990. Comparison of two regimens of concurrent strength and endurance training. *Medicine and Science in Sports and Exercise* 22: 348-356.

Schmidtbleicher, D. 1988. Muscular mechanics and neuromuscular control. *Swimming Science V International Series Sport Science.* Champaign, IL: Human Kinetics.

Selye, H. 1956. *The Stress of Life.* New York: McGraw-Hill.

Sforzo, G.A., and P.R. Touey. 1996. Manipulating exercise order affects muscular performance during a resistance exercise training session. *Strength and Conditioning Research* 10: 21-24.

Sharp, R., J. Troup, and D. Costill. 1982. Relationship between power and sprint freestyle swimming. *Medicine and Science in Sports and Exercise* 14: 53-56.

Simao, R., P.T.V. Farinatti, M.D. Polito, A.S. Maior, and S.J. Fleck. 2005. Influence of exercise order on the number of repetitions performed and perceived exertion during resistance exercises. *Journal of Strength and Conditioning Research* 19: 152-156.

Simmons, L. 1996. What if? *MILO* 4: 25-29.

Smerdu, V., I. Karsch-Mizrachi, M. Campione, L. Leinwand, and S. Schiaffino. 1994. Type IIx myosin heavy chain transcripts are expressed in type IIb fibers of human skeletal muscle. *American Journal of Physiology: Cell Physiology* 267: C1723-C1728.

Staron, R.S., D.L. Karapondo, W.J. Kraemer, A.C. Fry, S.E. Gordon, J.E. Falkel, F.C. Hagerman, and R.S. Hikida. 1994. Skeletal muscle adaptations during early phase of heavy-resistance training in men and women. *Journal of Applied Physiology* 76: 1247-1255.

Stone, M., G. Moir, M. Glaister, and R. Sanders. 2002. How much strength is necessary? *Physical Therapy in Sport* 3: 88-96.

Stone, M.H., K. Sanborn, H.S. O'Bryant, M. Hartman, M.E. Stone, C. Proulx, B. Ward, and J. Hruby. 2003. Maximum strength-power-performance relationships in collegiate throwers. *Journal of Strength and Conditioning Research* 17: 739-745.

Stone, M.H., W.A. Sands, K.C. Pierce, J. Carlock, M. Cardinale, and R.U. Newton. 2005. Relationship of maximum strength to weightlifting performance. *Medicine and Science in Sports and Exercise* 37: 1037-1043.

Sweet, W. 1987. *Sport and Recreation in Ancient Greece.* Oxford: Oxford University Press.

Tyler, T.F., S.J. Nicholas, R.J. Campbell, and M.P. McHugh. 2001. The association of hip strength and flexibility with the incidence of adductor muscle strains in professional ice hockey players. *American Journal of Sports Medicine* 29: 124-128.

Vanderburgh, P.M., and C. Dooman. 2000. Considering body mass differences, who are the world's strongest women? *Medicine and Science in Sports and Exercise* 32: 197-201.

Viitasalo, J.T., S. Saukkonen, and P.V. Komi. 1980. Reproducibility of measurements of selected neuromuscular performance variables in man. *Electromyography and Clinical Neurophysiology* 20: 487-501.

Wilmore, J. 1974. Alterations in strength, body composition and anthropometric measurements consequent to a 10-week weight training program. *Medicine and Science in Sport* 6: 133-139.

Wilmore, J.H., and D.L. Costill. 2004. Aging in sport and exercise. In *Physiology of Sport and Exercise* (3rd ed.). Champaign, IL: Human Kinetics.

Wilson, G., B. Elliott, and G. Wood. 1992. Stretch-shorten cycle performance enhancement through flexibility training. *Medicine and Science in Sports and Exercise* 24: 116-123.

Yang, N., D.G. MacArthur, J.P. Gulbin, A.G. Hahn, A.H. Beggs, S. Easteal, and K. North. 2003. ACTN3 genotype is associated with human elite athletic performance. *American Journal of Human Genetics* 73: 627-631.

Zatsiorsky, V. 1995. *Science and Practice of Strength Training.* Champaign, IL: Human Kinetics.

Chapter 9

Aagaard, P., E.B. Simonsen, S.P. Magnusson, B. Larsson, and P. Dyhre-Poulsen. 1998. A new concept for isokinetic hamstring: Quadriceps muscle strength ratio. *American Journal of Sports Medicine* 26(2): 231-237.

Aagaard, P., E.B. Simonsen, M. Trolle, J. Bangsbo, and K. Klausen. 1995. Isokinetic hamstring/quadriceps strength ratio: Influence from joint angular velocity, gravity correction and contraction mode. *Acta Physiologica Scandinavica* 154(4): 421-427.

Adams, K., J.P. O'Shea, K.L. O'Shea, and M. Climstein. 1992. The effect of six weeks of squat, plyometric and squat-plyometric training on power production. *Journal of Applied Sport Science Research* 6(1): 36-41.

Atha, J. 1981. Strengthening muscle. In *Exercise and Sport Sciences Reviews,* edited by D.I. Miller. Philadelphia: Franklin Institute Press.

Baker, D. 2001. Acute and long-term power responses to power training: Observations on the training of an elite power athlete. *Strength and Conditioning Journal* 23(1): 47-56.

Baker, D., S. Nance, and M. Moore. 2001. The load that maximizes the average mechanical power output during explosive bench press throws in highly trained athletes. *Journal of Strength and Conditioning Research* 15(1): 20-24.

Baker, D., S. Nance, and M. Moore. 2001. The load that maximizes the average mechanical power output during jump squats in power-trained athletes. *Journal of Strength and Conditioning Research* 15(1): 92-97.

Bauer, T., R.E. Thayer, and G. Baras. 1990. Comparison of training modalities for power development in the lower extremity. *Journal of Applied Sport Science Research* 4(4): 115-121.

Behm, D.G., and D.G. Sale. 1993. Intended rather than actual movement velocity determines velocity-specific training response. *Journal of Applied Physiology* 74(1): 359-368.

Berger, R.A. 1962. Optimum repetitions for the development of strength. *Research Quarterly* 33(3): 334-338.

Berger, R.A. 1963. Effects of dynamic and static training on vertical jumping ability. *Research Quarterly* 34(4): 420-424.

Blakey, J.B., and D. Southard. 1987. The combined effects of weight training and plyometrics on dynamic leg strength and leg power. *Journal of Applied Sport Science Research* 1(1): 14-16.

Blazevich, A.J., N.D. Gill, R. Bronks, and R.U. Newton. 2003. Training-specific muscle architecture adaptation after 5-wk training in athletes. *Medicine and Science in Sports and Exercise* 35(12): 2013-2022.

Bobbert, M.F., K.G. Gerritsen, M.C. Litjens, and A.J. van Soest. 1996. Why is countermovement jump height greater than squat jump height? *Medicine and Science in Sports and Exercise* 28(11): 1402-1412.

Bobbert, M.F., and A.J. van Soest. 1994. Effects of muscle strengthening on vertical jump height: A simulation study. *Medicine and Science in Sports and Exercise* 26(8): 1012-1020.

Bompa, T.O., and J. Fox. 1990. *Theory and Methodology of Training: The Key to Athletic Performance* (2nd ed.). Dubuque, IA: Kendall/Hunt.

Bosco, C., and P.V. Komi. 1979. Potentiation of the mechanical behavior of the human skeletal muscle through prestretching. *Acta Physiologica Scandinavica* 106(4): 467-472.

Bosco, C., J.T. Viitasalo, P.V. Komi, and P. Luhtanen. 1982. Combined effect of elastic energy and myoelectrical potentiation during stretch-shortening cycle exercise. *Acta Physiologica Scandinavica* 114(4): 557-565.

Chu, D.A. 1992. *Jumping into Plyometrics*. Champaign, IL: Leisure Press.

Clutch, D., M. Wilton, C. McGown, and G.R. Bryce. 1983. Effect of depth jumps and weight training on leg strength and vertical jump. *Research Quarterly for Exercise and Sport* 54(1): 5-10.

Delbridge, A., and J.R.L. Bernard. 1988. *The Macquarie Concise Dictionary*. Sydney: Macquarie Library Pty Ltd.

Di Brezzo, R., I.L. Fort, and R. Diana. 1988. The effects of a modified plyometric program on junior high female basketball players. *Journal of Applied Research in Coaching and Athletics* 3(3): 172-181.

Duchateau, J., and K. Hainaut. 1984. Isometric or dynamic training: Differential effects on mechanical properties of a human muscle. *Journal of Applied Physiology: Respiratory, Environmental and Exercise Physiology* 56(2): 296-301.

Duke, S., and D. BenEliyahu. 1992. Plyometrics: Optimizing athletic performance through the development of power as assessed by vertical leap ability: An observational study. *Chiropractic Sports Medicine* 6(1): 10-15.

Elliott, B.C., G.J. Wilson, and G.K. Kerr. 1989. A biomechanical analysis of the sticking region in the bench press. *Medicine and Science in Sports and Exercise* 21(4): 450-462.

Ettema, G.J., A.J. van Soest, and P.A. Huijing. 1990. The role of series elastic structures in prestretch-induced work enhancement during isotonic and isokinetic contractions. *Journal of Experimental Biology* 154: 121-136.

Faulkner, J.A., D.R. Claflin, and K.K. McCully. 1986. Power output of fast and slow fibers from human skeletal muscles. In *Human Muscle Power*, edited by N.L. Jones, N. McCartney, and A.J. McComas. Champaign, IL: Human Kinetics.

Garhammer, J. 1993. A review of power output studies of Olympic and powerlifting: Methodology, performance prediction, and evaluation tests. *Journal of Strength and Conditioning Research* 7(2): 76-89.

Gollhofer, A., and H. Kyröläinen. 1991. Neuromuscular control of the human leg extensor muscles in jump exercises under various stretch-load conditions. *International Journal of Sports Medicine* 12(1): 34-40.

Gollnick, P.D., and A.W. Bayly. 1986. Biochemical training adaptation and maximal power. In *Human Muscle Power*, edited by N.L. Jones, N. McCartney, and A.J. McComas. Champaign, IL: Human Kinetics.

Grimby, L., and J. Hannerz. 1977. Firing rate and recruitment order of toe extensor motor units in different modes of voluntary contraction. *Journal of Physiology* 264: 865-879.

Häkkinen, K. 1989. Neuromuscular and hormonal adaptations during strength and power training. A review. *Journal of Sports Medicine and Physical Fitness* 29(1): 9-26.

Häkkinen, K., and P.V. Komi. 1985. Changes in electrical and mechanical behavior of leg extensor muscles during heavy resistance strength training. *Scandinavian Journal of Sports Sciences* 7(2): 55-64.

Häkkinen, K., and P.V. Komi. 1985. Effect of explosive type strength training on electromyographic and force production characteristics of leg extensor muscles during concentric and various stretch-shortening cycle exercises. *Scandinavian Journal of Sports Sciences* 7(2): 65-76.

Häkkinen, K., P.V. Komi, and P.A. Tesch. 1981. Effect of combined concentric and eccentric strength training and detraining on force-time, muscle fiber and metabolic characteristics of leg extensor muscles. *Scandinavian Journal of Sports Sciences* 3(2): 50-58.

Hannerz, J. 1974. Discharge properties of motor units in relation to recruitment order in voluntary contraction. *Acta Physiologica Scandinavica* 91: 374-385.

Hasson, C.J., E.L. Dugan, T.L.A. Doyle, B. Humphries, and R.U. Newton. 2004. Neuromechanical strategies employed to increase jump height during the initiation of the squat jump. *Journal of Electromyography and Kinesiology* 14: 515-521.

Hatfield, F.C. 1989. *Power: A Scientific Approach*. Chicago: Contemporary Books.

Holtz, J., J. Divine, and C. McFarland. 1988. Vertical jump improvement following preseason plyometric training. *Journal of Applied Sport Science Research* 2(3): 59.

Humphries, B.J., R.U. Newton, and G.J. Wilson. 1995. The effect of a braking device in reducing the ground impact forces inherent in plyometric training. *International Journal of Sports Medicine* 16(2): 129-133.

Kanehisa, H., and M. Miyashita. 1983. Specificity of velocity in strength training. *European Journal of Applied Physiology and Occupational Physiology* 52(1): 104-106.

Kaneko, M., T. Fuchimoto, H. Toji, and K. Suei. 1983. Training effect of different loads on the force-velocity relationship and mechanical power output in human muscle. *Scandinavian Journal of Sports Sciences* 5(2): 50-55.

Knuttgen, H.G., and W.J. Kraemer. 1987. Terminology and measurement in exercise performance. *Journal of Applied Sport Science Research* 1(1): 1-10.

Komi, P.V. 1986. The stretch-shortening cycle and human power output. In *Human Muscle Power*, edited by N.L. Jones, N. McCartney, and A.J. McComas. Champaign, IL: Human Kinetics.

Komi, P.V., and K. Häkkinen. 1988. Strength and power. In *The Olympic Book of Sports Medicine*, edited by A. Dirix, H.G. Knuttgen, and K. Tittel. Boston: Blackwell Scientific.

Komi, P.V., H. Suominen, E. Heikkinen, J. Karlsson, and P. Tesch. 1982. Effects of heavy resistance and explosive-type strength training methods on mechanical, functional, and metabolic aspects of performance. In *Exercise and Sport Biology*, edited by P.V. Komi, R.C. Nelson, and C.A. Morehouse. Champaign, IL: Human Kinetics.

Kraemer, W.J. 1992. Involvement of eccentric muscle action may optimize adaptations to resistance training. In *Sports Science Exchange.* Chicago: Gatorade Sports Science Institute.

Lesmes, G.R., D.L. Costill, E.F. Coyle, and W.J. Fink. 1978. Muscle strength and power changes during maximal isokinetic training. *Medicine and Science in Sports* 10(4): 266-269.

MacDougall, J.D. 1986. Morphological changes in human skeletal muscle following strength training and immobilization. In *Human Muscle Power,* edited by N.L. Jones, N. McCartney, and A.J. McComas. Champaign, IL: Human Kinetics.

Mateyev, L. 1972. *Periodisierang des sportlichen training. [Periodizing Sport Training.]* Berlin: Berles and Wernitz.

Mayhew, J.L., J.L. Prinster, J.S. Ware, D.L. Zimmer, J.R. Arabas, and M.G. Bemben. 1995. Muscular endurance repetitions to predict bench press strength in men of different training levels. *Journal of Sports Medicine and Physical Fitness* 35(2): 108-113.

McBride, J.M., T. Triplett-McBride, A. Davie, and R.U. Newton. 1999. A comparison of strength and power characteristics between power lifters, Olympic lifters, and sprinters. *Journal of Strength and Conditioning Research* 13(1): 58-66.

McBride, J.M., N.T. Triplett-McBride, A. Davie, and R.U. Newton. 2001. The effect of heavy versus light load jump squats on the development of strength, power and speed. *Journal of Strength and Conditioning Research* 16(1): 75-82.

Medvedyev, A. 1988. Several basics on the methodics of training weightlifters. *Soviet Sports Review* 22: 203-206.

Moritani, T., M. Muro, K. Ishida, and S. Taguchi. 1987. Electrophysiological analyses of the effects of muscle power training. *Research Journal of Physical Education in Japan* 1(1): 23-32.

Murphy, A.J., and G.J. Wilson. 1996. The assessment of human dynamic muscular function: A comparison of isoinertial and isokinetic tests. *Journal of Sports Medicine and Physical Fitness* 36(3): 169-177.

Murphy, A.J., and G.J. Wilson. 1996. Poor correlations between isometric tests and dynamic performance: Relationship to muscle activation. *European Journal of Applied Physiology and Occupational Physiology* 73(3-4): 353-357.

Murphy, A.J., and G.J. Wilson. 1997. The ability of tests of muscular function to reflect training-induced changes in performance. *Journal of Sports Sciences* 15(2): 191-200.

Newton, R.U. and E. Dugan. 2002. Application of strength diagnosis. *Strength and Conditioning Journal* 24(5): 50-59.

Newton, R.U., and W.J. Kraemer. 1994. Developing explosive muscular power: implications for a mixed methods training strategy. *Strength and Conditioning Journal* 16(5): 20-31.

Newton, R.U., W.J. Kraemer, and K. Häkkinen. 1999. Effects of ballistic training on preseason preparation of elite volleyball players. *Medicine and Science in Sports and Exercise* 31(2): 323-330.

Newton, R.U., W.J. Kraemer, K. Häkkinen, B.J. Humphries, and A.J. Murphy. 1996. Kinematics, kinetics, and muscle activation during explosive upper body movements. *Journal of Applied Biomechanics* 12: 31-43.

Newton, R.U., and G.J. Wilson. 1993. Reducing the risk of injury during plyometric training: The effect of dampeners. *Sports Medicine, Training and Rehabilitation* 4: 159-165.

O'Shea, K.L., and J.P. O'Shea. 1989. Functional isometric weight training: Its effects on dynamic and static strength. *Journal of Applied Sport Science Research* 3(2): 30-33.

Patel, T.J., and R.L. Lieber. 1997. Force transmission in skeletal muscle: From actomyosin to external tendons. *Exercise and Sport Sciences Reviews* 25: 321-363.

Sale, D.G. 1992. Neural adaptation to strength training. In *Strength and Power in Sport,* edited by P.V. Komi. Boston: Blackwell Scientific.

Schmidtbleicher, D. 1992. Training for power events. In *Strength and Power in Sport,* edited by P.V. Komi. Boston: Blackwell Scientific.

Schmidtbleicher, D., A. Gollhofer, and U. Frick. 1988. Effects of a stretch-shortening type training on the performance capability and innervation characteristics of leg extensor muscles. In *Biomechanics XI-A,* edited by G. de Groot, A. Hollander, P. Huijing, and G. van Ingen Schenau. Amsterdam: Free University Press.

Williams, D.R. 1991. The effect of weight training on performance in selected motor activities for prepubescent males. *Journal of Applied Sport Science Research* 5(3): 170.

Wilson, G.J., A.D. Lyttle, K.J. Ostrowski, and A.J. Murphy. 1995. Assessing dynamic performance: A comparison of rate of force development tests. *Journal of Strength and Conditioning Research* 9(3): 176-181.

Wilson, G.J., A.J. Murphy, and A.D. Walshe. 1997. Performance benefits from weight and plyometric training: Effects of initial strength level. *Coaching and Sport Science Journal* 2(1): 3-8.

Wilson, G.J., R.U. Newton, A.J. Murphy, and B.J. Humphries. 1993. The optimal training load for the development of dynamic athletic performance. *Medicine and Science in Sports and Exercise* 25(11): 1279-1286.

Winter, D.A. 1990. *Biomechanics and Motor Control of Human Movement.* New York: Wiley.

Young, W. 1993. Training for speed/strength: Heavy vs. light loads. *National Strength and Conditioning Association Journal* 15(5): 34-42.

Young, W.B. 1995. Laboratory strength assessment of athletes. *New Studies in Athletics* 10(1): 89-96.

Young, W.B., and G.E. Bilby. 1993. The effect of voluntary effort to influence speed of contraction on strength, muscular power and hypertrophy development. *Journal of Strength and Conditioning Research* 7(3): 172-178.

Young, W.B., J.F. Pryor, and G.J. Wilson. 1995. Effect of instructions on characteristics of countermovement and drop jump performance. *Journal of Strength and Conditioning Research* 9(4): 232-236.

Zatsiorsky, V.M. 1995. *Science and Practice of Strength Training.* Champaign, IL: Human Kinetics.

Chapter 10

Aagaard, P., J.L. Andersen, P. Dyhre-Poulsen, A.M. Leffers, A. Wagner, S.P. Magnusson, J. Halkjaer-Kristensen, and E.B. Simonsen. 2001. A mechanism for increased contractile strength of human pennate muscle in response to strength training: Changes in muscle architecture. *Journal of Physiology* 534: 613-623.

Aagaard, P., E.B. Simonsen, J.L. Andersen, S.P. Magnusson, J. Halkjaer-Kristensen, and P. Dyre-Poulsen. 2000. Neural inhibition during maximal eccentric and concentric quadriceps contraction: Effects of resistance training. *Journal of Applied Physiology* 89(6): 2249-2257.

Abe, T., K. Kumagai, and W.F. Brechue. 2000. Fascicle length of leg muscles is greater in sprinters than distance runners. *Medicine and Science in Sports and Exercise* 32(6): 1125-1129.

Abernethy, P.J., J. Jurimae, P.A. Logan, A.W. Taylor, and R.E. Thayer. 1994. Acute and chronic response of skeletal muscle to resistance exercise. *Sports Medicine* 17(1): 22-38.

Alway, S.E., W.H. Grumbt, J. Stray-Gundersen, and W.J. Gonyea. 1992. Effects of resistance training on elbow flexors of highly competitive bodybuilders. *Journal of Applied Physiology* 72(4): 1512-1521.

Andersen, L.L., J.L. Andersen, S.P. Magnusson, C. Suetta, J.L. Madsen, L.R. Christensen, and P. Aagaard. 2005. Changes in the human muscle force-velocity relationship in response to resistance training and subsequent detraining. *Journal of Applied Physiology* 99: 87-94.

Aura, O., and P. Komi. 1987. Coupling time in stretch-shortening cycle: Influence on mechanical efficiency and elastic characteristics of leg extensor muscles. In *Biomechanics X*. Champaign, IL: Human Kinetics.

Baker, D., and S. Nance. 1999. The relation between running speed and measures of strength and power in professional rugby league players. *Journal of Strength and Conditioning Research* 13(3): 230-235.

Bamman, M.M., J.R. Shipp, J. Jiang, B.A. Gower, G.R. Hunter, A. Goodman, C.L. McLafferty Jr., and R.J. Urban. 2001. Mechanical load increases muscle IGF-1 and androgen receptor mRNA concentrations in humans. *American Journal of Physiology* 280: E383-390.

Blazevich, A.J., N. Gill, R. Bronks, and R.U. Newton. 2003. Training-specific muscle architecture adaptation after 5-wk training in athletes. *Medicine and Science in Sports and Exercise* 35(12): 2013-2022.

Blazevich, A.J., D. Cannavan, D.R. Coleman, and S. Horne. 2007. Influence of concentric and eccentric resistance training on architectural adaptation in human quadriceps muscles. *Journal of Applied Physiology* 103: 1565-1575.

Brocherie, F., N. Babault, G. Cometti, N. Maffiuletti, and J.C. Chatard. 2005. Electromyostimulation training effects on the physical performance of ice hockey players. *Medicine and Science in Sports and Exercise* 37(3): 455-460.

Brooks, B.P., D.E. Merry, H.L. Paulson, A.P. Lieberman, D.L. Kolson, and K.H. Fischbeck. 1998. A cell culture model for androgen effects in motor neurons. *Journal of Neurochemistry* 70: 1054-1060.

Caiozzo, V.J., R.E. Herrick, and K.M. Baldwin. 1992. Response of slow and fast muscle to hypothyroidism: Maximal shortening velocity and myosin isoforms. *American Journal of Physiology* 263: C86-C94.

Canepari, M., V. Cappelli, M.A. Pellegrino, M.C. Zanadi, and C. Reggiani. 1998. Thyroid hormone regulation of MHC isoform composition and myofibrillar ATPase activity in rat skeletal muscles. *Archives of Physiological Biochemistry* 106(4): 308-315.

Cardone, A., F. Angelini, T. Esposito, R. Comitato, and B. Varriale. 2000. The expression of androgen receptor messenger RNA is regulated by triiodothyronine in lizard testis. *Journal of Steroid Biochemistry and Molecular Biology* 72: 133-141.

Cronin, J.B., and K.T. Hansen. 2005. Strength and power predictors of sports speed. *Journal of Strength and Conditioning Research* 19(2): 349-357.

Delecluse, C., H. Van Coppenolle, E. Willems, M. Van Leemputte, R. Diels, and M. Goris. 1995. Influence of high resistance and high-velocity training on sprint performance. *Medicine and Science in Sports and Exercise* 27(8): 1203-1209.

Dowson, M.N., M.E. Nevill, H.K. Lakomy, A.M. Nevill, and R.J. Hazeldine. 1998. Modelling the relationship between isokinetic muscle strength and sprint running performance. *Journal of Sports Sciences* 16(3): 257-265.

Dudley, G.A., P.A. Tesch, B.J. Miller, and P. Buchanan. 1991. Importance of eccentric actions in performance adaptations to resistance training. *Aviation and Space Environmental Medicine* 62(6): 543-550.

Esbjörnsson, M., I. Holm, C. Sylvén, and E. Jansson. 1996. Different responses of skeletal muscle following sprint training in men and women. *European Journal of Applied Physiology* 74: 375-383.

Ferro, A., A. Rivera, I. Pagola, M. Ferreruela, and V. Rocandio. 2001. Biomechanical analysis of the 7th world championships in athletics Seville 1999. *New Studies in Athletics* 16(1/2): 25-60.

Fry, A.C. 2004. The role of resistance exercise intensity on muscle fibre adaptations. *Sports Medicine* 34(10): 663-679.

Gondin, J., M. Guette, Y. Ballay, and A. Martin. 2005. Electromyostimulation training effects on neural drive and muscle architecture. *Medicine and Science in Sports and Exercise* 37(8): 1291-1299.

Häkkinen, K., M. Alén, and P.V. Komi. 1985. Changes in isometric force- and relaxation-time, electromyographic and muscle fibre characteristics of human skeletal muscle during strength training and detraining. *Acta Physiologica Scandinavica* 125: 573-585.

Häkkinen, K., P.V. Komi, and M. Alén. 1985. Effect of explosive type strength training on isometric force- and relaxation-time, electromyographic and muscle fibre characteristics of leg extensor muscles. *Acta Physiologica Scandinavica* 125: 587-600.

Häkkinen, K., P.V. Komi, M. Alén, and H. Kauhanen. 1987. EMG, muscle fibre and force production characteristics during a 1 year training period in elite weight-lifters. *European Journal of Applied Physiology* 56: 419-426.

Hay, J. 1993. *The Biomechanics of Sports Techniques.* Englewood Cliffs, NJ: Prentice-Hall.

Herzog, W. 2005. Force enhancement following stretch of activated muscle: Critical review and proposal for mechanisms. *Medical and Biological Engineering and Computing* 43(2): 173-180.

Jacobs, I., M. Esbjörnsson, C. Sylvén, I. Holm, and E. Jansson. 1987. Sprint training effects on muscle myoglobin, enzymes, fiber types, and blood lactate. *Medicine and Science in Sports and Exercise* 19(4): 368-374.

Kadi, F., P. Bonnerud, A. Eriksson, and L.E. Thornell. 2000. The expression of androgen receptors in human neck and limb muscles: Effects of training and self-administration of androgenic-anabolic steroids. *Histochemistry and Cell Biology* 113: 25-29.

Kawakami, Y., T. Abe, S-Y. Kuno, and T. Fukunaga. 1995. Training-induced changes in muscle architecture and specific tension. *European Journal of Applied Physiology* 72: 37-43.

Klinge, K., S. Magnusson, E. Simonsen, P. Aagaard, K. Klausen, and M. Kjaer. 1997. The effect of strength and flexibility training on skeletal muscle electromyographic activity, stiffness, and viscoelastic stress relaxation response. *American Journal of Sports Medicine* 25(5): 710-716.

Komi, P. 1986. Training of muscle strength and power: Interaction of neuromotoric, hypertrophic, and mechanical factors. *International Journal of Sports Medicine* 7: 10-15.

Kraemer, W.J., A.C. Fry, B.J. Warren, M.H. Stone, S.J. Fleck, J.T. Kearney, B.P. Conroy, C.M. Maresh, C.A. Weseman, N.T. Triplett, and S.E. Gordon. 1992. Acute hormonal responses in elite junior weightlifters. *International Journal of Sports Medicine* 13: 103-109.

Kraemer, W.J., K. Häkkinen, R.U. Newton, B.C. Nindl, J.S. Volek, M. McCormick, L.A. Gotshalk, S.E. Gordon, S.J. Fleck, W.W. Campbell, M. Putukian, and W.J. Evans. 1999. Effects of heavy-resistance training on hormonal response patterns in younger vs. older men. *Journal of Applied Physiology* 87(3): 982-992.

Kraemer, W.J., and N.A. Ratamess. 2005. Hormonal responses and adaptations to resistance training. *Sports Medicine* 35(4): 339-361.

Kubo, K., H. Akima, J. Ushiyama, I. Tabata, H. Fukuoka, H. Kanehisa, and T. Fukunaga. 2004. Effects of 20 days of bed rest on the viscoelastic properties of tendon structures in lower limb muscles. *British Journal of Sports Medicine* 38: 324-330.

Kubo, K., H. Kanehisa, and T. Fukunaga. 2002. Effect of stretching training on the viscoelastic properties of human tendon structures in vivo. *Journal of Applied Physiology* 92(2): 595-601.

Kubo, K., H. Kanehisa, Y. Kawakami, and T. Fukunaga. 2001. Influence of static stretching on viscoelastic properties of human tendon structure in vivo. *Journal of Applied Physiology* 90: 520-527.

Kubo, K., H. Kanehisa, Y. Kawakami, and T. Fukunaga. 2001. Influences of repetitive muscle contractions with different modes on tendon elasticity in vivo. *Journal of Applied Physiology* 91: 277-282.

Kumagai, K., A. Takashi, W.F. Brechue, T. Ryushi, S. Takano, and M. Mizuno. 2000. Sprint performance is related to muscle fascicle length in male 100-m sprinters. *Journal of Applied Physiology* 88: 811-816.

Le Roith, D., C. Bondy, S. Yakar, J-L. Liu, and A. Butler. 2001. The somatomedin hypothesis: 2001. *Endocrine Reviews* 22(1): 53-74.

Lyons, G.E., A.M. Kelly, and N.A. Rubinstein. 1986. Testosterone-induced changes in contractile protein isoforms in the sexually dimorphic temporalis muscle of the guinea pig. *Journal of Biological Chemistry* 261(28): 13278-13284.

Maffiuletti, N.A., G. Cometti, I.G. Amiridis, A. Martin, M. Pousson, and J.C. Chatard. 2000. The effects of electromyostimulation training and basketball practice on muscle strength and jump ability. *International Journal of Sports Medicine* 21(6): 437-443.

Maffiuletti, N.A., M. Pensini, and A. Martin. 2002. Activation of human plantar flexor muscles increases after electromyostimulation training. *Journal of Applied Physiology* 92: 1383-1392.

Malatesta, D., F. Cattaneo, S. Dugnani, and N.A. Maffiuletti. 2003. Effects of electromyostimulation training and volleyball practice on jumping ability. *Journal of Strength and Conditioning Research* 17(3): 573-579.

Marron, T.U., V. Guerini, P. Rusmini, D. Sau, T.A.L. Brevini, L. Martini, and A. Poletti. 2005. Androgen-induced neurite outgrowth is mediated by neuritin in motor neurones. *Journal of Neurochemistry* 92(1): 10-20.

McKenna, M.J., T.A. Schmidt, M. Hargreaves, L. Cameron, S.L. Skinner, and K. Kjeldsen. 1993. Sprint training increases human skeletal muscle Na^+-K^+-ATPase concentration and improves K^+ regulation. *Journal of Applied Physiology* 75(1): 173-180.

Mero, A., and P.V. Komi. 1986. Force-, EMG-, and elasticity-velocity relationships at submaximal, maximal and supramaximal running speeds in sprinters. *European Journal of Applied Physiology* 55: 553-561.

Mero, A., P.V. Komi, and R.J. Gregor. 1992. Biomechanics of sprint running. *Sports Medicine* 13(6): 376-392.

Nagaya, N., and A.A. Herrera. 1995. Effects of testosterone on synaptic efficacy at neuromuscular junctions in asexually dimorphic muscle of male frogs. *Journal of Physiology* 483: 141-153.

Nardone, A., C. Romano, and M. Schieppati. 1989. Selective recruitment of high-threshold human motor units during voluntary isotonic lengthening of active muscles. *Journal of Physiology* 409: 451-471.

Nardone, A., and M. Schieppati. 1987. Shift of activity from slow to fast muscle during voluntary lengthening contractions of the triceps surae muscle in humans. *Journal of Physiology* 395: 363-381.

Nesser, T.W., R.W. Latin, K. Berg, and E. Prentice. 1996. Physiological determinants of 40-meter sprint performance in young male athletes. *Journal of Strength and Conditioning Research* 10(4): 263-267.

Newsholme, E., E. Blomstarnd, and B. Ekblom. 1992. Physical and mental fatigue: Metabolic mechanisms and the importance of plasma amino acids. *British Medical Bulletin* 43(3): 447-495.

Paddon-Jones, D., M. Leveritt, A. Lonergan, and P. Abernethy. 2001. Adaptation to chronic eccentric exercise in humans: The influence of contraction velocity. *European Journal of Applied Physiology* 85: 466-471.

Pakarinen, A., K. Häkkinen, and M. Alén. 1991. Serum thyroid hormones, thyrotropin, and thyroxine binding globulin in elite athletes during very intense strength training of one week. *Journal of Sports Medicine and Physical Fitness* 31: 142-146.

Parry, T.E., P. Henson, and J. Cooper. 2003. Lateral foot placement analysis of the sprint start. *New Studies in Athletics* 18(1): 13-22.

Passelergue, P., A. Robert, and G. Lac. 1995. Salivary cortisol and testosterone variations during an official and a simulated weight-lifting competition. *International Journal of Sports Medicine* 16(5): 298-303.

Sale, D.G., J.D. MacDougall, A.R.M. Upton, and A.J. McComas. 1983. Effect of strength training upon motoneuron excitability in man. *Medicine and Science in Sports and Exercise* 15(1): 57-62.

Saplinskas, J.S., M.A. Chobotas, and I.I. Yashchaninas. 1980. The time of completed motor acts and impulse activity of single motor units according to the training level and sport specialization of tested persons. *Electromyography and Clinical Neurophysiology* 20: 529-539.

Schmidt, R.A. 1987. *Motor Control and Learning: A Behavioral Emphasis* (2nd ed.). Champaign, IL: Human Kinetics.

Semmler, J.G. 2002. Motor unit synchronization and neuromuscular performance. *Exercise and Sport Sciences Reviews* 30(1): 8-14.

Seyennes, O.R., M. de Boer, and M.Y. Narici. 2007. Early skeletal muscle hypertrophy and architectural changes in response to high-intensity resistance training. *Journal of Applied Physiology* 102: 368-373.

Tesch, P.A., P.V. Komi, and K. Häkkinen. 1987. Enzymatic adaptations consequent to long-term strength training. *International Journal of Sports Medicine* Suppl 1: 66-69.

Upton, A.R.M., and P.F. Radford. 1975. Motoneurone excitability in elite sprinters. In *Biomechanics 5A*, edited by P.V. Komi. Baltimore: University Park Press.

Weston, A.R., K.H. Myburgh, F.H. Lindsay, S.C. Dennis, T.D. Noakes, and J.A. Hawley. 1997. Skeletal muscle buffering capacity and endurance performance after high-intensity interval training by well-trained cyclists. *European Journal of Applied Physiology and Occupational Physiology* 75(1): 7-13.

Widrick, J.J., S.W. Trappe, C.A. Blaser, D.L. Costill, and R.H. Fitts. 1996. Isometric force and maximal shortening velocity of single muscle fibers from elite master runners. *American Journal of Physiology* 271: C666-C675.

Wilson, G.J., B.C. Elliott, and G.A. Wood. 1992. Stretch shorten cycle performance enhancement through flexibility training. *Medicine and Science in Sports and Exercise* 24(1): 116-123.

Wood, G.A., K.P. Singer, and G.C. Cresswell. 1986. Electromechanical adaptations to muscular strength training. Paper presented at the North American Congress of Biomechanics, Montreal.

Young, W., D. Benton, G. Duthie, and J. Pryor. 2001. Resistance training for short sprints and maximum-speed sprints. *Strength and Conditioning Journal* 23(2): 7-13.

Young, W., B. McLean, and J. Ardagna. 1995. Relationship between strength qualities and sprinting performance. *Journal of Sports Medicine and Physical Fitness* 35(1): 13-19.

Zehr, P.E., and D.G. Sale. 1994. Ballistic movement: Muscle activation and neuromuscular adaptation. *Canadian Journal of Applied Physiology* 19(4): 363-378.

Chapter 11

Avela, J., H. Kyrolainen, and P.V. Komi. 1999. Altered reflex sensitivity after repeated and prolonged passive muscle stretching. *Journal of Applied Physiology* 86(4): 1283-1291.

Bandy, W., and J. Irion. 1994. The effect of time on static stretch on the flexibility of the hamstring muscles. *Physical Therapy* 74(9): 845-850.

Bandy, W., J. Irion, and M. Briggler. 1997. The effect of time and frequency of static stretching on flexibility of the hamstring muscles. *Physical Therapy* 77(10): 1090-1096.

Bandy, W., J. Irion, and E.A. Briggler. 1998. The effect of static stretch and dynamic range of motion training on the flexibility of the hamstring muscles. *Journal of Orthopedics and Sports Physical Therapy* 27(4): 295-300.

Behm, D.G., D.C. Button, and J.C. Butt. 2001. Factors affecting force loss with prolonged stretching. *Canadian Journal of Applied Physiology* 26(3): 261-272.

Bohannon, R., R. Gajdosik, and B. LeVeau. 1985. Contribution of pelvic and lower limb motion to increases in the angle of passive straight leg raising. *Physical Therapy* 65(4): 474-476.

Burke, D.G., C.J. Culligan, and L.E. Holt. 2000. The theoretical basis of proprioceptive neuromuscular facilitation. *Journal of Strength and Conditioning Research* 14(4): 496-500.

Church, J.B., M.S. Wiggins, E.M. Moode, and R. Crist. 2001. Effects of warm-up and flexibility treatments on vertical jump performance. *Journal of Strength and Conditioning Research* 15(3): 332-336.

Corbin, C., and L. Noble. 1980. Flexibility: A major component of physical fitness. *Journal of Physical Education and Recreation* 51(23-24): 57-60.

Cornbleet, S., and N. Woosley. 1996. Assessment of hamstring muscle length in school aged children using the sit and reach test and the inclinometer measure of hip joint angle. *Physical Therapy* 76(8): 850-855.

Cornelius, W.L., R.L. Jensen, and M.E. Odell. 1995. Effects of PNF stretching phases on acute arterial blood pressure. *Canadian Journal of Applied Physiology* 20(2): 222-229.

Cornwell, A., A.G. Nelson, and B. Sidaway. 2002. Acute effects of stretching on the neuromechanical properties of the triceps surae muscle complex. *European Journal of Applied Physiology* 86(5): 428-434.

Cornwell, A. and A.G. Nelson. 1997. The acute effects of passive stretching on active musculotendinous stiffness. *Medicine and Science in Sports and Exercise* 29(Suppl): 281.

Craib, M.W., V.A. Mitchell, K.B. Fields, T.R. Cooper, R. Hopewell, and D.W. Morgan. 1996. The association between flexibility and running economy in sub-elite male distance runners. *Medicine and Science in Sports and Exercise* 28(6): 737-743.

DeVries, H. 1986. *Physiology of Exercise—for Physical Education and Athletics.* Dubuque, IA: Brown.

Dubravcic-Simunjak, S., M. Pecina, H. Kuipers, J. Moran, and M. Haspl. 2003. The incidence of injuries in elite junior figure skaters. *American Journal of Sports Medicine* 31(4): 511-517.

Enoka, R.M., R.S. Hutton, and E. Eldred. 1980. Changes in excitability of tendon tap and Hoffmann reflexes following voluntary contractions. *Electromyography and Clinical Neurophysiology* 48(6): 664-672.

Feland, J., J. Myer, and R. Merrill. 2001. Acute changes in hamstring flexibility: PNF versus static stretch in senior athletes. *Physical Therapy in Sport* 2: 186-193.

Feldman, D., I. Shrier, M. Rossignol, and L. Abenhaim. 1999. Adolescent growth is not associated with changes in flexibility. *Clinical Journal of Sports Medicine* 9: 24-29.

Fowles, J.R., D.G. Sale, and J.D. MacDougall. 2000. Reduced strength after passive stretch of the human plantarflexors. *Journal of Applied Physiology* 89(3): 1179-1188.

Funk, D., A. Swank, K. Adams, and D. Treolo. 2001. Efficacy of moist heat pack application over static stretching on hamstring flexibility. *Journal of Strength and Conditioning Research* 15(1): 123-126.

Gajdosik, R. 1991. Effects of static stretching on the maximal length and resistance to passive stretch of short hamstrings muscles. *Journal of Orthopedics and Sports Physical Therapy* 14(6): 250-255.

Gajdosik, R. 2001. Passive extensibility of skeletal muscle: Review of the literature with clinical implications. *Clinical Biomechanics* 16(2): 87-101.

Gajdosik, R.L., C.A. Giuliani, and R.W. Bohannon. 1990. Passive compliance and length of the hamstring muscles of healthy men and women. *Clinical Biomechanics* 5: 23-29.

Gajdosik, R., and G. Lusin. 1983. Hamstring muscle tightness: Reliability of an active knee extension test. *Physical Therapy* 63(7): 1085-1090.

Gillette, T., G. Holland, W. Vincent, and S. Loy. 1991. Relationship of body core temperature and warm up to knee range of motion. *Journal of Orthopedics and Sports Physical Therapy* 13(3): 126-131.

Gleim, G.W., and M.P. McHugh. 1997. Flexibility and its effects on sports injury and performance. *Sports Medicine* 24(5): 289-299.

Godges, J.J., P.G. MacRae, and K.A. Engelke. 1993. Effects of exercise on hip range of motion, trunk muscle performance, and gait economy. *Physical Therapy* 73(7): 468-477.

Gollhofer, A., A. Schopp, W. Rapp, and V. Stroinik. 1998. Changes in reflex excitability following isometric contraction in humans. *European Journal of Applied Physiology* 77: 89-97.

Gribble, P., K. Guskiewicz, W. Prentice, and E. Shields. 1999. Effects of static and hold relax stretching on hamstring range of motion using the FlexAbility LE1000. *Journal of Sport Rehabilitation* 8(3): 195-208.

Gribble, P., J. Hertel, and C. Denegar. 2005. Reliability and validity of a 2-D video digitising system during a static and a dynamic task. *Journal of Sport Rehabilitation* 14(2): 137-149.

Halbertsma, J., N. Ludwig, and M. Goeken. 1994. Stretching exercises: Effect on passive extensibility and stiffness in short hamstrings of healthy subjects. *Archives of Physical Medicine and Rehabilitation* 75 (September): 976-981.

Halbertsma, J.P.K., A.L. van Bolhuis, and L.N.H. Goeken. 1996. Sport stretching: Effects on passive muscle stiffness of short hamstrings. *Archives of Physical Medicine and Rehabilitation* 77: 688-692.

Handel, M., T. Horstmann, H.H. Dickhuth, and R.W. Gulch. 1997. Effects of contract-relax stretching training on muscle performance in athletes. *European Journal of Applied Physiology* 76(5): 400-408.

Hartig, D., and J. Henderson. 1999. Increasing hamstring flexibility decreases lower extremity overuse injuries in military basic trainees. *American Journal of Sports Medicine* 27(2): 173-176.

Harvey, L., R.D. Herbert, and J. Crosbie. 2002. Does stretching induce lasting increases in joint ROM? A systematic review. *Physiotherapy Research International* 7(1): 1-13.

Henricson, A.S., K. Fredriksson, I. Persson, R. Pereira, and N.E. Westlin. 1984. The effects of heat and stretching on the range of hip motion. *Journal of Orthopedics and Sports Physical Therapy* 6(2): 110-115.

Herbert, R.D., and M. Gabriel. 2002. Effects of stretching before and after exercising on muscle soreness and risk of injury: Systemic review. *British Journal of Sports Medicine* 325: 1-5.

High, D.M., E.T. Howley, and B.D. Franks. 1989. The effects of static stretching and warm-up on prevention of delayed-onset muscle soreness. *Research Quarterly for Exercise and Sport* 60(4): 357-361.

Hume, P.A., M. Stechman, J. Thomson, J. Cronin, C. McCullough, G. Mawston, K. O'Meara, and C. Barrett. 2004. *Rhythmic Gymnastics Developmental Assessment Programme (RG DAP) Testing Manual.* Auckland: Institute of Sport and Recreation Research New Zealand.

Hunter, J.P., and R.N. Marshall. 2002. Effects of power and flexibility training on vertical jump technique. *Medicine and Science in Sports and Exercise* 34(3): 478-486.

Johansson, P.H., L. Lindstrom, G. Sundelin, and B. Lindstrom. 1999. The effects of preexercise stretching on muscular soreness, tenderness and force loss following heavy eccentric exercise. *Scandinavian Journal of Medicine and Science in Sports* 9(4): 219-225.

Klinge, K., S.P. Magnusson, E.B. Simonsen, P. Aagaard, K. Klausen, and M. Kjaer. 1997. The effects of strength and flexibility training on skeletal muscle electromyographic activity, stiffness, and viscoelastic stress relaxation response. *American Journal of Sports Medicine* 25(5): 710-716.

Knudson, D., K. Bennett, R. Corn, D. Leick, and C. Smith. 2001. Acute effects of stretching are not evident in the kinematics of the vertical jump. *Journal of Strength and Conditioning Research* 15(1): 98-101.

Kokkonen, J., A.G. Nelson, and A. Cornwell. 1998. Acute muscle stretching inhibits maximal strength performance. *Research Quarterly for Exercise and Sport* 69(4): 411-415.

Kubo, K., H. Kanehisa, and T. Fukunaga. 2001. Is passive stiffness in human muscles related to the elasticity of tendon structures? *European Journal of Applied Physiology* 85: 226-232.

Kubo, K., H. Kanehisa, and T. Fukunaga. 2002. Effects of resistance and stretching training programmes on the viscoelastic properties of human tendon structures in vivo. *Journal of Physiology* 538(1): 219-226.

Laur, D.J., T.B. Anderson, G. Geddes, A. Crandall, and D.M. Pincivero. 2003. The effects of acute stretching on hamstring muscle fatigue and perceived exertion. *Journal of Sports Sciences* 21: 163-170.

Leard, J. 1984. Flexibility and conditioning in the young athlete. In *Pediatric and Adolescent Sports Medicine,* edited by L. Micheli (p. 198). Boston: Little, Brown.

Liebesman, J., and E. Cafarelli. 1994. Physiology of range of motion in human joints: A critical review. *Critical Reviews in Physical and Rehabilitation Medicine* 6(2): 131-160.

Lund, H., P. Vestergaard-Poulsen, I.L. Kanstrup, and P. Sejrsen. 1998. The effects of passive stretching on delayed onset muscle soreness, and other detrimental effects following eccentric exercise. *Scandinavian Journal of Medicine and Science in Sports* 8: 216-221.

Magnusson, S.P. 1998. Passive properties of human skeletal muscle during stretch manoeuvres. *Medicine and Science in Sports and Exercise* 8: 65-77.

Magnusson, S. P., P. Aagaard, B. Larsson, and M. Kjaer. 2000. Passive energy absorption by human muscle-tendon unit is unaffected by increase in intramuscular temperature. *Journal of Applied Physiology* 88: 1215-1220.

Magnusson, S.P., E.B. Simonsen, P. Aagaard, P. Dyhre-Poulsen, M.P. McHugh, and M. Kjaer. 1996. Mechanical and physical responses to stretching with and without preisometric contraction in human skeletal muscle. *Archives of Physical Medicine and Rehabilitation* 77(4): 373-378.

Magnusson, S.P., E.B. Simonsen, P. Aagaard, G.W. Gleim, M.P. McHugh, and M. Kjaer. 1995. Viscoelastic response to repeated static stretching in the human hamstring muscle. *Scandinavian Journal of Medicine and Science in Sports* 5(6): 342-347.

Magnusson, S.P., E.B. Simonsen, P. Aagaard, H. Sorensen, and M. Kjaer. 1996. A mechanism for altered flexibility in human skeletal muscle. *Journal of Physiology* 497(1): 291-298.

Magnusson, S.P., E.B. Simonson, P. Dyhre-Poulsen, P. Aagaard, T. Mohr, and M. Kjaer. 1996. Viscoelastic stress relaxation during static stretch in human skeletal muscle in the absence of EMG activity. *Medicine and Science in Sports and Exercise* 6: 323-328.

Marieb, E. 2001. *Human Anatomy and Physiology.* San Francisco: Benjamin Cummings.

McHugh, M.P., D.A.J. Connolly, R.G. Eston, I.J. Kremenic, S.J. Nicholas, and G.W. Gleim. 1999. The role of passive muscle stiffness in symptoms of exercise-induced muscle damage. *American Journal of Sports Medicine* 27(5): 594-599.

McHugh, M.P., I.J. Kremenic, M.B. Fox, and G.W. Gleim. 1998. The role of mechanical and neural restraints to joint range of motion during passive stretch. *Medicine and Science in Sports and Exercise* 30(6): 928-932.

McHugh, M.P., S.P. Magnusson, G.W. Gliem, and J.A. Nicholas. 1992. Viscoelastic stress relaxation in human skeletal muscle. *Medicine and Science in Sports and Exercise* 24(12): 1375-1382.

McNair, P. J., E. W. Dombroski, D. J. Hewson and S. N. Stanley. 2000. Stretching at the ankle joint: viscoelastic responses to holds and continuous passive motion. *Medicine and Science in Sports and Exercise* 33(3): 354-358.

McNair, P., E. Dombroski, and S.N. Stanley. 2001. Stretching at the ankle joint: Viscoelastic responses to holds and continuous passive motion. *Medicine and Science in Sports and Exercise* 33(3): 354-358.

McNair, P., and S. Stanley. 1996. Effect of passive stretching and jogging on the series muscle stiffness and range of motion of the ankle joint. *British Journal of Sports Medicine* 30: 313-318.

Micheli, L., and A. Fehlandt. 1992. Overuse injuries to tendons and apophyses in children and adolescents. *Clinics in Sports Medicine* 11(4): 713-726.

Mohr, K.J., M.M. Pink, C. Elsner, and R.S. Kvitne. 1998. Electromyographic investigation of stretching: The effect of warm-up. *Clinical Journal of Sports Medicine* 8(3): 215-220.

Moore, M., and R. Hutton. 1980. Electromyographic investigation of muscle stretching techniques. *Medicine and Science in Sports and Exercise* 12: 322-329.

Moore, M.A., and C.G. Kukulka. 1991. Depression of Hoffmann reflexes following voluntary contraction and implications for proprioceptive neuromuscular facilitation therapy. *Physical Therapy* 71(4): 321-333.

Muir, I.W., B.M. Chesworth, and A.A. Vandervoort. 1999. Effect of a static calf-stretching exercise on the resistive torque during passive ankle dorsiflexion in healthy subjects. *Journal of Orthopedics and Sports Physical Therapy* 29(7): 421-424.

Nelson, A.G., I.K. Guillory, A. Cornwell, and J. Kokkonen. 2001. Inhibition of maximal voluntary isokinetic torque production following stretching is velocity-specific. *Journal of Strength and Conditioning Research* 15(2): 241-246.

Nelson, A.G., and J. Kokkonen. 2001. Acute ballistic muscle stretching inhibits maximal strength performance. *Research Quarterly for Exercise and Sport* 72(4): 415-419.

Nelson, A.G., J. Kokkonen, C. Eldredge, A. Cornwell, and E. Glickman-Weiss. 2001. Chronic stretching and running economy. *Scandinavian Journal of Medicine and Science in Sports* 11(5): 260-265.

Osternig, L.R., R. Robertson, R. Troxel, and P. Hansen. 1987. Muscle activation during proprioceptive neuromuscular facilitation (PNF) stretching techniques. *American Journal of Physical Medicine and Rehabilitation* 66(5): 298-307.

Osternig, L., R. Robertson, R. Troxel, and P. Hansen. 1990. Differential responses to proprioceptive neuromuscular facilitation (PNF) stretch techniques. *Medicine and Science in Sports and Exercise* 22(1): 106-111.

Phillips, M. 1955. Analysis of results from the Kraus-Weber test of minimum muscular fitness in children. *Research Quarterly* 26: 314-323.

Pope, R.P., R.D. Herbert, and J.D. Kirwan. 1998. Effects of ankle dorsiflexion range and pre-exercise calf muscle stretching on injury risk in army recruits. *Australian Journal of Physiotherapy* 44: 165-177.

Pope, R.P., R.D. Herbert, J.D. Kirwan, and B.J. Graham. 2000. A randomised trial of pre-exercise stretching for prevention of lower-limb injury. *Medicine and Science in Sports and Exercise* 32: 271-277.

Prentice, W. 1983. A comparison of static stretching and PNF stretching for improving hip joint flexibility. *Athletic Training* (Spring): 56-59.

Purslow, P. 1989. Strain induced reorientation of an intramuscular connective tissue network: Implications for passive muscle elasticity. *Journal of Biomechanics* 22(1): 21-31.

Reid, D. and P. J. McNair. 2004. Passive force, angle and stiffness changes after stretching of hamstring muscles. *Medicine and Science in Sports and Exercise* 36(11): 1944-1948.

Roberts, J., and K. Wilson. 1999. Effect of stretching duration on active and passive range of motion in the lower extremity. *British Journal of Sports Medicine* 33: 259-263.

Rodenburg, J.B., D. Steenbeek, P. Schiereck, and P.R. Bar. 1994. Warm-up, stretching and massage diminish harmful effects of eccentric exercise. *International Journal of Sports Medicine* 15: 414-419.

Ross, M. 1999. Effect of lower extremity position and stretching on hamstring muscle flexibility. *Journal of Strength and Conditioning Research* 13(2): 124-129.

Sady, S.P., M. Wortman, and D. Blanke. 1982. Flexibility training: Ballistic, static or proprioceptive neuromuscular facilitation? *Archives of Physical Medicine and Rehabilitation* 63: 261-263.

Shellock, F.G., and W.E. Prentice. 1985. Warming-up and stretching for improved physical performance and prevention of sports-related injuries. *Sports Medicine* 2: 267-278.

Shepherd, R., M. Berridge, and W. Montelpare. 1990. On the generality of the "sit and reach" test: An analysis of the flexibility data for an aging population. *Research Quarterly for Exercise and Sport* 61: 326-330.

Shrier, I., D. Ehrmann-Feldman, M. Rossignol, and L. Abenhaim. 2001. Risk factors for the development of lower limb pain in adolescents. *Journal of Rheumatology* 28: 604-609.

Smith, C.A. 1994. The warm-up procedure: To stretch or not to stretch. A brief review. *Journal of Orthopedics and Sports Physical Therapy* 19(1): 12-17.

Smith, L.L., M.H. Brunetz, T.C. Chenier, M.R. McCammon, J.A. Houmard, M.E. Franklin, and R.G. Israel. 1993. The effects of static and ballistic stretching on delayed onset muscle soreness. *Research Quarterly for Exercise and Sport* 64(1): 103-107.

Soper, C., D. Reid, and P. Hume. 2004. Reliable passive ankle range of motion measures correlate to ankle motion achieved during ergometer rowing. *Physical Therapy in Sport* 5: 75-83.

Spernoga, S.G. 2001. Duration of maintained hamstring flexibility after one-time, modified hold-relax stretching protocol. *Journal of Athletic Training* 36(1): 44-48.

Sullivan, M., J. Dejulia, and T. Worrell. 1992. Effect of pelvic position and stretching method on hamstring flexibility. *Medicine and Science in Sports and Exercise* 24(12): 1383-1389.

Taylor, D.C., D.E. Brooks, and J.B. Ryan. 1997. Viscoelastic characteristics of muscle: Passive stretching versus muscular contractions. *Medicine and Science in Sports and Exercise* 29(12): 1619-1624.

Taylor, D.C., J.D. Dalton, A.V. Seaber, and W.E. Garrett. 1990. Viscoelastic properties of muscle-tendon units: The biomechanical effects of stretching. *American Journal of Sports Medicine* 18: 300-309.

Toft, E., T. Sinkjaler, S. Kalund, and G. T. Espersen. 1989. Biomechanical properties of the human ankle relation to passive stretch. *Journal of Biomechanics* 22(11-12): 1129-1132.

van Mechelen, W., H. Hlobil, H.C. Kemper, W.J. Voorn, and H.R. de Jongh. 1993. Prevention of running injuries by warm-up, cool-down, and stretching exercise. *American Journal of Sports Medicine* 21(5): 711-719.

Vujnovich, A.L., and N.J. Dawson. 1994. The effects of therapeutic muscle stretch on neural processing. *Journal of Orthopedics and Sports Physical Therapy* 20(3): 145-153.

Webright, W., B. Randolph, and D. Perrin. 1997. Comparison of nonballistic active knee extension in neural slump position and static stretch technique on hamstring flexibility. *Journal of Orthopedics and Sports Physical Therapy* 26(1): 7-13.

Weerapong, P., P. Hume, and G.S. Kolt. 2005. Stretching: Mechanisms and benefits on performance and injury prevention. *Physical Therapy Reviews* (December, 9): 189-206.

Weerapong, P., P.A. Hume, and G. Kolt. 2006. Pre-exercise activities: The effects of warm-up, stretching, and massage on performance. *Journal of Science and Medicine in Sport.* Under review.

Weldon, S.M., and R.H. Hill. 2003. The efficacy of stretching for prevention of exercise-related injury: A systematic review of the literature. *Manual Therapy* 8(3): 141-150.

White, S., and S. Sahrmann. 1994. A movement systems balance approach to the management of musculoskeletal pain. In *Physical Therapy of the Cervical and Thoracic Spine,* edited by R. Grant. Edinburgh: Churchill Livingstone.

Wiemann, K., and K. Hahn. 1997. Influences of strength, stretching and circulatory exercises on flexibility parameters of the hamstrings. *International Journal of Sports Medicine* 18(5): 340-346.

Wiktorsson-Moller, M., B. Oberg, J. Ekstrand, and J. Gillquist. 1983. Effects of warming up, massage, and stretching on range of motion and muscle strength in the lower extremity. *American Journal of Sports Medicine* 11(4): 249-252.

Williford, H., J. East, F. Smith, and L. Burry. 1986. Evaluation of warm up for improvement in flexibility. *American Journal of Sports Medicine* 14(4): 316-319.

Willy, R., B. Kyle, S. Moore, and G. Chleboun. 2001. Effect of cessation and resumption of static hamstring muscle stretching on joint range of motion. *Journal of Orthopedics and Sports Physical Therapy* 31: 138-144.

Wilson, G.J., B.C. Elliott, and G.A. Wood. 1992. Stretch shorten cycle performance enhancement through flexibility training. *Medicine and Science in Sports and Exercise* 24(1): 116-123.

Wilson, G.J., A.J. Murphy, and J.F. Pryor. 1994. Musculotendinous stiffness: Its relationship to eccentric, isometric and concentric performance. *Journal of Applied Physiology* 76(6): 2714-2719.

Witvrouw, E., L. Danneels, and P. Asselmann. 2003. Muscle flexibility as a risk factor of developing muscle injuries in professional male soccer players. *American Journal of Sports Medicine* 31(1): 41-46.

Witvrouw, E., N. Mahieu, L. Danneels, and P. McNair. 2004. Stretching and injury prevention—an obscure relationship. *Sports Medicine* 34(7): 443-449.

Worrell, T.W., T.L. Smith, and J. Winegardner. 1994. Effects of hamstring stretching on hamstring muscle performance. *Journal of Orthopedics and Sports Physical Therapy* 20(3): 154-159.

Young, W.B., and D.G. Behm. 2003. Effects of running, static stretching and practice jumps on explosive force production and jumping performance. *Journal of Sports Medicine and Physical Fitness* 43(1): 21-27.

Zito, M., D. Driver, C. Parker, and R. Bohannon. 1997. Lasting effects of one bout of two 15-second passive stretches on ankle dorsiflexion range of motion. *Journal of Orthopedics and Sports Physical Therapy* 26(4): 214-221.

Chapter 12

Bennett, D.J., M. Gorassini, and A. Prochazka. 1994. Catching a ball: Contributions of intrinsic muscle stiffness, reflexes, and higher order responses. *Canadian Journal of Physiology and Pharmacology* 72: 525-534.

Besier, T.F., D.G. Lloyd, and T.R. Ackland. 2003. Muscle activation strategies at the knee during running and cutting maneuvers. *Medicine and Science in Sports and Exercise* 35: 119-127.

Besier, T.F., D.G. Lloyd, J.L. Cochrane, and T.R. Ackland. 2001. Anticipatory effects on knee joint loading during running and cutting maneuvers. *Medicine and Science in Sports and Exercise* 33: 1176-1181.

Besier, T.F., D.G. Lloyd, J.L. Cochrane, and T.R. Ackland. 2001. External loading of the knee joint during running and cutting maneuvers. *Medicine and Science in Sports and Exercise* 33: 1168-1175.

Bloomfield, J., P.A. Fricker, and K.D. Fitch. 1992. *Textbook of Science and Medicine in Sport*. Melbourne: Blackwell Scientific Publications.

Caraffa, A., G. Cerulli, M. Projetti, G. Aisa, and A. Rizzo. 1996. Prevention of anterior cruciate ligament injuries in soccer. A prospective controlled study of proprioceptive training. *Knee Surgery, Sports Traumatology, Arthroscopy* 4: 19-21.

Dempsey, A.R., D.G. Lloyd, J.R. Steele, B.C. Elliott, and K.A. Russo. 2007. The effect of technique change on knee loads during sidestepping. *Medicine and Science in Sports and Exercise*, 39: 1765-1773.

Dintiman, G., and R. Ward. 1988. *Sport Speed.* Champaign, IL: Leisure Press.

Draper, J., and M. Lancaster. 1985. The 505: A test for agility in the horizontal plane. *Australian Journal of Science and Medicine in Sport* 17: 15-18.

Draper, J., and M. Lancaster. 2000. The 505: A test for agility in the horizontal plane. In *Physiological Tests for Elite Athletes/Australian Sports Commission,* edited by C. Gore. Champaign IL: Human Kinetics.

Draper, J., B. Minikin, and R. Telford. 1991. *Test Methods Manual,* Section 3. Canberra: Australian Sports Commission.

Gabbett, T.J. 2006. A comparison of physiological and anthropometric characteristics among playing positions in sub-elite rugby league players. *Journal of Sports Sciences* 24(12): 1273-1280.

Gabbett, T.J. 2007. Physiological and anthropometric characteristics of elite women rugby league players. *Journal of Strength Conditioning Research* 21(3):875-881.

Gabbett, T.J., J.N. Kelly, and J.M. Sheppard. 2008. Speed, change of direction speed, and reactive agility of rugby league players. *Journal of Strength and Conditioning Research* 22(1):174-181.

Getchell, B. 1985. *Physical Fitness: A Way of Life* (3rd ed.). New York: Macmillan.

Groppel, J., J. Loehr, D. Melville, and A. Quinn. 1989. *Science of Coaching Tennis.* Champaign, IL: Leisure Press.

Hasan, Z. 2005. The human motor control system's response to mechanical perturbation: Should it, can it, does it ensure stability? *Journal of Motor Behavior* 37: 484-493.

Hewett, T.E., T.N. Lindenfeld, J.V. Riccobene, and F.R. Noyes. 1999. The effect of neuromuscular training on the incidence of knee injury in female athletes. A prospective study. *American Journal of Sports Medicine* 27: 699-706.

Johansson, H. 1991. Role of knee ligaments in proprioception and regulation of muscle stiffness. *Journal of Electromyography and Kinesiology* 1: 158-179.

Lloyd, D.G. 2001. Rationale for training programs to reduce anterior cruciate ligament injuries in Australian football. *Journal of Orthopedics and Sports Physical Therapy* 31: 645-654.

Lloyd, D.G., and T.S. Buchanan. 2001. Strategies of the muscular contributions to the support of static varus and valgus loads at the human knee. *Journal of Biomechanics* 34: 1257-1267.

Lloyd, D.G., T.S. Buchanan, and T.F. Besier. 2005. Neuromuscular biomechanical modeling to understand knee ligament loading. *Medicine and Science in Sports and Exercise* 37: 1939-1947.

Markolf, K.L., D.M. Burchfield, M.M. Shapiro, M.F. Shepard, G.A. Finerman, and J.L. Slauterbeck. 1995. Combined knee loading states that generate high anterior cruciate ligament forces. *Journal of Orthopaedic Research* 13: 930-935.

Markolf, K.L., J.S. Mensch, and H.C. Amstutz. 1976. Stiffness and laxity of the knee—the contributions of the supporting structures. *Journal of Bone and Joint Surgery* 58A: 583-593.

McLean, S.G., X. Huang, A. Su, and A.J. Van Den Bogert. 2004. Sagittal plane biomechanics cannot injure the ACL during sidestep cutting. *Clinical Biomechanics* 19: 828-838.

McLean, S.G., X. Huang, and A.J. Van Den Bogert. 2005. Association between lower extremity posture at contact and peak knee valgus moment during sidestepping: Implications for ACL injury. *Clinical Biomechanics* 20: 863-870.

McLean, S.G., S.W. Lippert, and A.J. Van Den Bogert. 2004. Effect of gender and defensive opponent on the biomechanics of sidestep cutting. *Medicine and Science in Sports and Exercise* 36: 1008-1016.

Myklebust, G., L. Engebrestsen, I.H. Braekken, A. Skjolberg, O.E. Olsen, and R. Bahr. 2003. Prevention of anterior cruciate ligament injuries in female team handball players: A prospective intervention study over three seasons. *Clinical Journal of Sports Medicine* 13: 71-78.

Sheppard, J.M., W.B. Young, T.L.A. Doyle, T.A. Sheppard, and R.U. Newton. 2006. An evaluation of a new test of reactive agility and its relationship to sprint speed and change of direction speed. *Journal of Science and Medicine in Sport* 9: 342-349.

Shumway-Cook, A., and M. Woolacott. 1995. Development of posture and balance control. In *Motor Control: Theory and Practical Applications,* edited by A. Shumway-Cook and M. Woolacott. Baltimore: Williams & Wilkins.

Tittel, K. 1988. Coordination and balance. In *The Olympic Book of Sports Medicine,* edited by H. Dirix, H. Knuttgen, and K. Tittel. Oxford: Blackwell Scientific.

Winter, D.A. 1995. Human balance and posture control during standing and walking. *Gait and Posture* 3: 193-214.

Chapter 13

Ackland, T.R. 1999. Talent identification: What makes a champion swimmer? In *International Congress on Biomechanics in Sport.* Perth: International Society of Biomechanics in Sport.

Ackland, T.R., B.C. Elliott, and J.E. Richards. 2003. Growth in body size affects rotational performance in women's gymnastics. *Sports Biomechanics* 2(2): 163-176.

Ackland, T.R., A.B. Schreiner, and D.A. Kerr. 1997. Technical note: Anthropometric normative data for female international basketball players. *Australian Journal of Science and Medicine in Sport* 29(1): 24-26.

Dintiman, G. 1979. *How to Run Faster: A Do-it-yourself Book for Athletes in All Sports.* Richmond, VA: Champion Athlete.

Humphries, B.J., R.U. Newton, and G.J. Wilson. 1994. The effect of a braking device in reducing the ground impact forces inherent in plyometric training. *International Journal of Sports Medicine* 16(2): 129-133.

Mazza, J.C., T.R. Ackland, T.M. Bach, and P. Cosolito. 1994. Absolute body size. In *Kinanthropometry in Aquatic Sports,* edited by J.E.L. Carter and T.R. Ackland. Champaign, IL: Human Kinetics.

McBride, J.M., T. Triplett-McBride, A. Davie, and R.U. Newton. 2002. The effect of heavy versus light load jump squats on the development of strength, power and speed. *Journal of Strength and Conditioning Research* 16(1): 75-82.

Newton, R.U., and W.J. Kraemer. 1994. Developing explosive muscular power: Implications for a mixed methods training strategy. *Strength and Conditioning Journal* 16(5): 20-31.

Newton, R.U., W.J. Kraemer, K. Häkkinen, B.J. Humphries, and A.J. Murphy. 1996. Kinematics, kinetics and muscle activation during explosive upper body movements: Implications for power development. *Journal of Applied Biomechanics* 12: 31-43.

Newton, R.U., A.J. Murphy, B.J. Humphries, G.J. Wilson, W.J. Kraemer, and K. Häkkinen. 1997. Influence of load and stretch shortening cycle on the kinematics, kinetics and muscle activation during explosive upper body movements. *European Journal of Applied Physiology and Occupational Physiology* 75(4): 333-342.

Norton, K., and T. Olds. 1996. *Anthropometrica.* Sydney: UNSW Press.

Chapter 14

Alderson, J., B. Elliott, and D. Lloyd. 2008. Mechanical correlates of patello-femoral pain in female runners. *Medicine and Science in Sports and Exercise,* under review.

Angst, F., H. Gerber, and E. Stussi. 1985. Physical and biomechanical foundations of the rowing motion. Paper presented at the Olympic Solidarity Seminar, Canberra.

Barrett, K.R. 1983. A hypothetical model of observing as a teaching skill. *Journal of Teaching in Physical Education* 3: 22-31.

Bernhardt, J., P.J. Bate, and T.A. Matyas. 1998. Accuracy of observational kinematic assessment of upper-limb movements. *Physical Therapy* 78: 259-270.

Bloomfield, J., P. Fricker, and K. Fitch, eds. 1992. *Textbook of Science and Medicine in Sport* Melbourne: Blackwell Scientific.

Bobbert, M.F., and L.J. Casius. 2005. Is the effect of a countermovement on jump height due to active state development? *Medicine and Science in Sports and Exercise* 37: 440-446.

Bobbert, M.F., K. Gerritsen, M. Litjens, and A. van Soest. 1996. Why is countermovement jump height greater than squat jump height? *Medicine and Science in Sports and Exercise* 28: 1402-1412.

Bobbert, M.F., and A. van Soest. 1994. Effects of muscle strengthening on vertical jump height: A simulation study. *Medicine and Science in Sports and Exercise* 26: 1012-1020.

Buchanan, T., D. Lloyd, K. Manal, and T. Besier. 2004. Neuromusculoskeletal modeling: Estimation of muscle forces and joint moments and movements from measurements of neural command. *Journal of Applied Biomechanics* 20: 367-395.

Dale, E. 1984. *The Educator's Quotebook.* Bloomington, IN: Phi Delta Kappa Educational Foundation.

Ellenbecker, T. 1992. Shoulder internal and external rotation strength and range of motion of highly skilled junior tennis players. *Isokinetics and Exercise Science* 2(2): 65-72.

Ellenbecker, T., and P. Roetert. 2003. Isokinetic profile of elbow flexion and extension in elite junior tennis players. *Journal of Orthopedics and Sports Physical Therapy* 33(2): 79-84.

Elliott, B., and J. Alderson. 2003. Biomechanical performance models: The basis for stroke analysis. In *Biomechanics of Advanced Tennis Coaching,* edited by B. Elliott, M. Reid, and M. Crespo (pp. 155-176). London: International Tennis Federation.

Elliott, B., J. Alderson, and E. Denver. 2007. System and modelling errors analysis: Implications for the measurement of the elbow angle in cricket bowling. *Journal of Biomechanics* 40(12): 2679-2685.

Elliott, B., G. Fleisig, R. Nicholls, and R. Escamilla. 2003. Technique effects on upper limb loading in the tennis serve. *Journal of Science and Medicine in Sport* 6(1): 76-87.

Elliott, B., and M. Khangure. 2002. Disk degeneration and the cricket fast bowler: An intervention study. *Medicine and Science in Sports and Exercise* 34(11): 1714-1718.

Elliott, B.C., R.N. Marshall, and G.J. Noffal. 1995. Contributions of upper limb segment rotations during the power serve in tennis. *Journal of Applied Biomechanics* 11(4): 433-442.

Fleisig, G., R. Nicholls, B. Elliott, and R. Escamilla. 2003. Kinematics used by world class tennis players to produce high-velocity serves. *Sports Biomechanics* 2(1): 51-64.

Foster, D., D. John, B. Elliott, T. Ackland, and K. Fitch. 1989. Back injuries to fast bowlers in cricket: A prospective study. *British Journal of Sports Medicine* 23(3): 150-154.

Hay, J. 1983. A system for the qualitative analysis of a motor skill. In *Collected Papers on Sports Biomechanics,* edited by G.A. Wood (pp. 97-116). Perth: University of Western Australia Press.

Hay, J. 1985. *The Biomechanics of Sports Techniques* (3rd ed.). Englewood Cliffs, NJ: Prentice-Hall.

Hay, J.G., and J.G. Reid. 1988. *Anatomy, Mechanics, and Human Motion* (2nd ed.). Englewood Cliffs, NJ: Prentice-Hall.

Hoffman, S.J. 1983. Clinical diagnosis as a pedagogical skill. In *Teaching in Physical Education,* edited by T.J. Templin and J.K. Olson (pp. 35-45). Champaign, IL: Human Kinetics.

Hudson, J. 1995. Core concepts in kinesiology. *Journal of Physical Education, Recreation and Dance* 66(5): 54-55, 59-60.

Hubbard, M., and L. W. Alaways. 1987. Optimal release conditions for the new rules javelin. *International Journal of Sport Biomechanics* 3: 207-221.

Knudson, D. 1999. Validity and reliability of visual ratings of the vertical jump. *Perceptual and Motor Skills* 89: 642-648.

Knudson, D. 2000. What can professionals qualitatively analyze? *Journal of Physical Education, Recreation and Dance* 71(2): 19-23.

Knudson, D. 2003. *Fundamentals of Biomechanics.* New York: Kluwer Academic/Plenum Press.

Knudson, D., and B.C. Elliott. 2003. Analysis of advanced stroke production. In *Biomechanics of Advanced Tennis Coaching,* edited by B. Elliott, M. Reid, and M. Crespo (pp. 137-154). London: International Tennis Federation.

Knudson, D., and D. Kluka. 1997. The impact of vision and vision training on sport performance. *Journal of Physical Education, Recreation and Dance* 68(4): 17-24.

Knudson, D., and C. Morrison. 2000. Visual ratings of the vertical jump are weakly correlated with perceptual style. *Journal of Human Movement Studies* 39: 33-44.

Knudson, D., and C. Morrison. 2002. *Qualitative Analysis of Human Movement* (2nd ed.). Champaign, IL: Human Kinetics.

Koh, M., L. Jennings, B. Elliott, and D. Lloyd. 2003. A predicted optimal performance of the Yurchenko layout vault in women's artistic gymnastics. *Journal of Applied Biomechanics* 19: 187-204.

Lees, A. 1999. Biomechanical assessment of individual sports for improved performance. *Sports Medicine* 28: 299-305.

Lloyd, D., T. Buchanan, and T. Besier. 2005. Neuromuscular biomechanical modelling to understand knee ligament loading. *Medicine and Science in Sports and Exercise* 37: 1939-1947.

McPherson, M. 1990. A systematic approach to skill analysis. *Sports Science Periodical on Research and Technology in Sport* 11(1): 1-10.

Morrison, C., D. Knudson, C. Clayburn, and P. Haywood. 2005. Accuracy of visual estimates of joint angle and angular velocity using criterion movement. *Perceptual and Motor Skills* 100: 599-606.

Piette, G., and J. Clarys. 1979. Telemetric electromyography of the front crawl movement. In *Swimming III,* edited by J. Terauds and E. Bedingfield (pp. 153-159). Baltimore: University Park Press.

Richards, J. 1999. The measurement of human motion: A comparison of commercially available systems. *Human Movement Science* 18: 589-602.

Robertson, D., G. Caldwell, J. Hamill, G. Kamen, and S. Whittlesey. 2004. *Research Methods in Biomechanics.* Champaign, IL: Human Kinetics.

Smith, R. 2004. Application of mechanical modeling to sport biomechanics. In *Proceedings of the XXIInd International Symposium on Biomechanics in Sports,* edited by M. Lamontagne, G. Robertson, and H. Sveistrup (pp. 563-566). Ottawa: University of Ottawa Press.

Smith, R., and C. Loschner. 2002. Biomechanics feedback for rowing. *Journal of Sports Sciences* 20(10): 783-791.

Smith, R., and W. Spinks. 1989. Matching technology to coaching needs: On water rowing analysis. In *Proceedings of the VIIth International Symposium of the Society of Biomechanics in Sports,* edited by W. Morrison (pp. 277-287). Melbourne: Footscray Institute of Technology Press.

Winter, D. 2004. *Biomechanics and Motor Control of Human Movement* (3rd ed.). New York: Wiley.

Yeadon, M. 1988. Techniques used in twisting somersaults. In *Biomechanics XI-B,* edited by G. de Groot, P. Hollander, A. Huijing, and G. van Ingen Schenau (pp. 740-741). Amsterdam: Free University Press.

Chapter 15

Challis, J. 1999. A procedure for the automatic determination of filter cutoff frequency for the processing of biomechanical data. *Journal of Applied Biomechanics* 15(3): 303-317.

Challis, J., and D. Kerwin. 1992. Accuracy assessment and control point configuration when using the DLT for photogrammetry. *Journal of Biomechanics* 25(9): 1053-1058.

Ehara, Y., H. Fujimoto, S. Miyazaki, S. Tanaka, and S. Yamamoto. 1995. Comparison of the performance of 3D camera systems. *Gait and Posture* 3: 166-169.

Elliott, B., and J. Alderson. 2006. Laboratory vs field testing: A review of current and past practice. *Sports Biomechanics,* accepted for publication.

Elliott, B., J. Alderson, and E. Denver. 2006. System and modelling errors in motion analysis: Implications for the measurement of the elbow angle in cricket bowling. *Medicine and Science in Sports and Exercise,* under review.

Elliott, B., and M. Khangure. 2002. Disk degeneration and the cricket fast bowler: An intervention study. *Medicine and Science in Sports and Exercise* 34(11): 1714-1718.

Ong, K.B., B. Elliott, T.R. Ackland, and A. Lyttle. 2006. Paddler morphology and boat set-up in elite kayaking. *Sports Biomechanics,* accepted for publication.

Richards, J.G. 1999. The measurement of human motion: A comparison of commercially available systems. *Human Movement Science* 18: 589-602.

Winter, D. 1990. *Biomechanics and Motor Control of Human Movement* (2nd ed.). New York: Wiley.

Wood, G., and R. Marshall. 1986. The accuracy of DLT extrapolation in three-dimensional film analysis. *Journal of Biomechanics* 19: 781-785.

Chapter 16

Abernethy, B., D. Gill, S.L. Parks, and S.T. Packer. 2001. Expertise and the perception of kinematic and situational probability information. *Perception* 30: 233-252.

Abernethy, B., and D.G. Russell. 1983. Skill in tennis: Considerations for talent identification and skill development. *Australian Journal of Sports Sciences* 3(1): 3-12.

Abernethy, B., and D.G. Russell. 1987. Expert-novice differences in an applied selective attention task. *Journal of Sport Psychology* 9: 326-345.

Abernethy, B., J. Wann, and S. Parks. 1998. Training perceptual motor skills for sport. In *Training for Sport: Applying Sport Science,* edited by B. Elliott. Chichester: Wiley.

Abernethy, B., J.M. Wood, and S. Parks. 1999. Can the anticipatory skills of experts be learned by novices? *Research Quarterly for Exercise and Sport* 70: 313-318.

Adams, J. 1968. Bone injuries in very young athletes. *Clinical Orthopaedics and Related Research* 58: 129-140.

Bahlsen, A. 1989. The etiology of running injuries, a longitudinal prospective study. PhD thesis, University of Calgary, Canada.

Bartlett, R.M. 1999. *Sports Biomechanics: Reducing Injury and Improving Performance.* London: Spon.

Battig, W.F. 1966. Facilitation and interference. In *Acquisition of Skill,* edited by E.A. Bilodeau. New York: Academic Press.

Besier, T., D. Lloyd, T.R. Ackland, and J. Cochrane. 2001. Anticipatory effects on the knee joint loading during running and cutting maneuvers. *Medicine and Science in Sports and Exercise* 33(7): 1176-1181.

Bruggemann, P. 1987. Biomechanics in gymnastics. In *Current Research in Sports Biomechanics,* edited by B. van Gheluwe and J. Atha. Sydney: Karger.

Caraffa, A., G. Cerulli, M. Projetti, and A. Rizzo. 1996. Prevention of anterior cruciate ligament injuries in soccer: A prospective controlled study of proprioceptive training. *Knee Surgery, Sports Traumatology, Arthroscopy* 4(1): 19-21.

Cochrane, J., D. Lloyd, T. Besier, T. Ackland, and B. Elliott. 2003. Training to increase muscular support to reduce the risk of anterior cruciate ligament injury. *Congress Proceedings International Society of Biomechanics* , New Zealand.

Deporte, E., and B. van Gheluwe. 1989. Ground reaction forces in elite high jumping. In *Congress Proceedings, XII International Congress of Biomechanics,* edited by R. Gregor, R. Zernicke, and W. Whiting. Los Angeles: UCLA Press.

English, W., D. Young, R. Moss, and P. Raven. 1984. Chronic muscle overuse syndrome in baseball. *Physician and Sports Medicine* 12: 111-115.

Farrow, D., and B. Abernethy. 2002. Can anticipatory skills be learned through implicit video-based perceptual training? *Journal of Sports Sciences* 20: 471-485.

Farrow, D., P. Chivers, C. Hardingham, and S. Sasche. 1998. The effects of video-based perceptual training on the tennis return of serve. *International Journal of Sports Psychology* 23: 231-242.

Finch, C. 2006. A new framework for research leading to sports injury prevention. *Journal of Science and Medicine in Sport* 9: 3-9.

Foster, D., D. John, B. Elliott, T. Ackland, and K. Fitch. 1989. Back injuries to fast bowlers in cricket: A prospective study. *British Journal of Sports Medicine* 23: 150-154.

Fricker, P. 1992. Injuries to the shoulder. In *Textbook of Science and Medicine in Sport,* edited by J. Bloomfield, P. Fricker, and K. Fitch. Melbourne: Blackwell Scientific.

Gainor, B., G. Piotrowski, J. Puhl, W. Allen, and R. Hagen. 1980. The throw: Biomechanics and acute injury. *American Journal of Sports Medicine* 8: 114-118.

Gardner, L., J. Dziados, B. Jones, J. Brundage, J. Harris, R. Sullivan, and P. Gill. 1988. Prevention of lower extremity stress fractures: A controlled study of a shock absorbent insole. *American Journal of Public Health* 78: 1563-1567.

Gentile, A.M. 1972. A working model of skill acquisition with application to teaching. *Quest* 17: 3-23.

Glazier, P., K. Davids, I. Renshaw, and C. Button. 2005. Uncovering the secrets of The Don: Bradman reassessed. *Sport Health* 22(4): 16-21.

Goode, S., and R.A. Magill. 1986. Contextual interference effects in learning three badminton serves. *Research Quarterly for Exercise and Sport* 57: 308-314.

Hall, K.G., D.A. Domingues, and R. Cavazos. 1994. Contextual interference effects with skilled baseball players. *Perceptual and Motor Skills* 78: 835-841.

Handford, C., K. Davids, S. Bennett, and C. Button. 1997. Skill acquisition in sport: Some applications of an evolving practice ecology. *Journal of Sports Sciences* 15: 621-640.

Johansson, G. 1973. Visual perception of biological motion and a model for its analysis. *Perception and Psychophysics* 14: 201-211.

Johansson, G. 1975. Visual motion perception. *Scientific American* 232(6): 76-88.

Lintern, G. 1991. An informational perspective on skill transfer in human-machine systems. *Human Factors* 33: 251-266.

Magill, R.A. 2001. Augmented feedback in motor skill acquisition. In *Handbook of Sport Psychology* (2nd ed.), edited by R.N. Singer, H.A. Hausenblas, and C.M. Janelle. New York: Wiley.

Magill, R.A., and K.G. Hall. 1990. A review of the contextual interference effect in motor skill acquisition. *Human Movement Science* 9: 241-289.

Miller, D., and M. Nissinen. 1987. Critical examination of ground reaction force in the running forward somersault. *International Journal of Sports Biomechanics* 3: 189-207.

Myklebust, G., L. Engebretsen, I. Hoff-Braekken, A. Skjolberg, O. Olsen, and R. Bahr. 2003. Prevention of anterior cruciate ligament injuries in female team handball players: A prospective intervention study over three seasons. *Clinical Journal of Sports Medicine* 13: 71-78.

Nigg, B. 1986. Biomechanical aspects of running. In *Biomechanics of Running Shoes,* edited by B. Nigg. Champaign, IL: Human Kinetics.

Nigg, B., J. Denoth, B. Kerr, S. Leuthi, D. Smith, and A. Stacoff. 1984. Load, sports shoes and playing surfaces. In *Sport Shoes and Playing Surfaces,* edited by E. Frederick. Champaign, IL: Human Kinetics.

Panzer, V., G. Wood, B. Bates, and B. Mason. 1988. Lower extremity loads in landings of elite gymnasts. In *Biomechanics XI-B,* edited by G. de Groot, P. Hollander, P. Huijing, and G. van Ingen Schenau. Amsterdam: Free University Press.

Ramey, M. 1970. Force relationships of the running long jump. *Medicine and Science in Sport* 2: 146-151.

Ramey, M., and K. Williams. 1985. Ground reaction forces in the triple jump. *International Journal of Sports Biomechanics* 1: 233-239.

Steele, J., and M. Lafortune. 1989. A kinetic analysis of footfall patterns at landing in netball: A follow-up study. In *Proceedings of VII International Symposium of Biomechanics in Sports,* edited by W. Morrison. Melbourne: Footscray Institute of Technology Press.

Swinnen, S.P. 1996. Information feedback for motor skill learning: A review. In *Advances in Motor Learning and Control,* edited by H.N. Zelaznik. Champaign, IL: Human Kinetics.

Swinnen, S.P., C.B. Walter, T.D. Lee, and D.J. Serrien. 1993. Acquiring bimanual skills: Contrasting forms of information feedback for interlimb decoupling. *Journal of Experimental Psychology: Learning, Memory, and Cognition* 19: 1321-1344.

Torg, J., H. Pollack, and P. Sweterlitsch. 1972. The effects of competitive pitching on the shoulders and elbows of pre-adolescent baseball players. *Pediatrics* 49: 267-270.

Valiant, G., and P. Cavanagh. 1985. A study of landing from a jump: Implications for the design of a basketball shoe. In *Biomechanics IX-B,* edited by D. Winter, R. Norman, R. Wells, K. Hayes, and R. Patla. Champaign, IL: Human Kinetics.

van Mechelen, W., H. Hlobil, and H. Kemper. 1992. Incidence, severity, aetiology and prevention of sports injuries. *Sports Medicine* 14(2): 82-99.

Williams, A.M., and N.J. Hodges. 2005. Practice, instruction and skill acquisition in soccer: Challenging tradition. *Journal of Sports Sciences* 23(6): 637-650.

Williams, A.M., and P. Ward. 2003. Developing perceptual expertise in sport. In *Recent Advances in Research on Sport Expertise,* edited by K.A. Ericsson and J.L. Starkes. Champaign, IL: Human Kinetics.

Williams, A.M., P. Ward, and C. Chapman. 2003. Training perceptual skill in field hockey: Is there transfer from the laboratory to the field? *Research Quarterly for Exercise and Sport* 74(1): 98-103.

Chapter 17

Hamill, J., and K. Knutzen. 2003. *Biomechanical Basis of Human Movement* (2nd ed.). New York: Lippincott, Williams & Wilkins.

Knudson, D. 2003. *Fundamentals of Biomechanics.* New York: Kluwer Academic/Plenum Press.

Walshe, A., G. Wilson, and G. Ettema. 1998. Stretch-shorten cycle compared with isometric preload: Contributions to enhanced muscular performance. *Journal of Applied Physiology* 84: 97-106.

Whiting, W., and R. Zernicke. 1998. *Biomechanics of Musculoskeletal Injury.* Champaign, IL: Human Kinetics.

Wilson, G.J., B.C. Elliott, and G.A. Wood. 1991. The effect on performance of imposing a delay during a stretch-shorten cycle movement. *Medicine and Science in Sports and Exercise* 23: 364-370.

index

Note: The italicized *f* and *t* following page numbers refer to figures and tables, respectively.

C

D

T

U

V

W

about the editors

Timothy R. Ackland, PhD, is a professor of functional anatomy and biomechanics in the School of Sport Science, Exercise, and Health at the University of Western Australia. His research interests include the mechanics of human movement with themes spanning exercise rehabilitation, high-performance sport and human performance in industry. Dr. Ackland has published more than 70 peer-reviewed papers as well as three academic books and 20 book chapters. Currently, Dr. Ackland is a director of Sports Medicine Australia and is on the review boards of four international journals. He recently served as the scientific chair for the fifth IOC World Congress on Sport Sciences for the 2000 Sydney Olympics and as the 2001 conference cochair for Sports Medicine Australia in Perth. Since 1991, he has been the principal of Ackland Marshall and Associates, Ergonomics Consultants.

Bruce C. Elliott, PhD, FACHPER, FASMF, FISBS, FAAKPE, is the senior biomechanist and head of the School of Sport Science, Exercise, and Health at the University of Western Australia. He has a keen interest in performance optimisation and injury reduction in sport and has published more than 170 refereed articles, 50 refereed conference proceedings and 40 books or book chapters in this general area. He is an editorial board member of *Sports Biomechanics, Journal of Applied Biomechanics, Journal of Sports Sciences* and the *International Journal of Sport and Health Sciences.*

Elliott is a sought-after international speaker on the application of biomechanics to sport. He was the keynote speaker at the first World Congress on Racket Sports, the first World Congress of Cricket and the third World Congress of Medicine and Science in Tennis. A fellow of the International Society of Biomechanics in Sports, Sports Medicine Australia and the American Academy of Kinesiology and Physical Education, he was also the inaugural chair of the Western Australian Institute of Sport and inaugural vice-president of the Australian Association of Exercise and Sport Science. In addition, Elliot served as the scientific chair for the fifth IOC World Congress on Sport Sciences and was the organiser of the research projects at the Sydney 2000 Olympics. He was the president-elect of the International Society of Biomechanics in Sports and sits on the coaching advisory panel of Tennis Australia and the research board of Cricket Australia.

John Bloomfield, PhD, is an emeritus professor at the University of Western Australia and a former director of the Hollywood Functional Rehabilitation Clinic at Hollywood Private Hospital in Perth. He left Australia in 1960 on a Fulbright scholarship to pursue postgraduate study at the University of Oregon in the United States. After receiving his PhD in 1968, he returned to Australia to take a position at the University of Western Australia.

Throughout his notable career, Bloomfield has served as president of Sports Medicine Australia, chairman of the Australian Sports Science Council, chairman of the Australian Institute of Sport, and cochairman of the Australian Sports Commission. He has authored more than 100 scientific papers and five books in sport and sport science as well as three major government reports on the development of sport in Australia. He was a National Surf Lifesaving champion and elite-level swimming coach in Australia and the United States. Bloomfield's research interests and experiences have led him to be a highly regarded lecturer and consultant in 22 countries.

In 1979, Bloomfield was awarded Citizen of the Year in Western Australia and, in 1982, he received Member of the Order of Australia.